ADVANCES IN PRECONCENTRATION AND DEHYDRATION OF FOODS

Edited by

ARNOLD SPICER

Scientific Adviser to RHM Research Ltd.,
The Lord Rank Research Centre,
High Wycombe, Bucks., U.K.

APPLIED SCIENCE PUBLISHERS LTD
LONDON

APPLIED SCIENCE PUBLISHERS LTD
RIPPLE ROAD, BARKING, ESSEX, ENGLAND

ISBN: 0 85334 599 6

WITH 55 TABLES AND 129 ILLUSTRATIONS

Printed in Great Britain by Galliard (Printers) Ltd, Great Yarmouth

ADVANCES IN PRECONCENTRATION AND DEHYDRATION OF FOODS

A symposium organised by IUFoST
and held at Selsdon Park Hotel, Croydon, Surrey, U.K.

Foreword

The increasing world-wide demand for convenience foods has fostered an accelerating research effort to discover and develop new methods for the preconcentration and dehydration of foods. This effort is being maintained both in University Departments and other academic institutions, and in industrial research and development organisations. The field is relatively unexplored compared to other areas of science, and offers wide scope for the investigation of the fundamental mechanisms affecting the removal of water from natural products, and the application of new and improved techniques for the preconcentration and dehydration of foods.

This Symposium, under the auspices of IUFoST, was called to review and discuss advances in the field, and to promote further research and development in new techniques. Delegates included industrial and academic scientists and engineers from all parts of the world and the lecture topics selected covered a wide spectrum of current work and thinking from the fundamental theories of water removal to the design and commercial implementation of large scale processes.

The Organising Chairman for the Symposium was Professor Arnold Spicer, who for ten years was Director of Research to the Ranks Hovis McDougall Group. The RHM Group was formed in 1962, and Arnold Spicer was given the task of instituting adequate research and development facilities to service the wide requirements of RHM in the fields of agriculture, flour milling, bread and flour confectionery, and grocery products. With the support of the late Lord Rank, Professor Spicer established in The Lord Rank Research Centre, High Wycombe, a research and development facility both to meet the commercial requirements of the RHM Group, and to carry out basic research in food biochemistry, human and animal nutrition, plant physiology, and allied subjects, to provide a sound scientific basis for future food processing advances. To this end he set up firm links with Universities and Research Institutes, both in Europe and North America, to aid the understanding of fundamental science and its utilisation in the service of man.

The support of the RHM Group to the Symposium is acknowledged.

Contents

FOREWORD v

OPENING REMARKS 1
Arnold Spicer

SESSION I—FUNDAMENTALS
(*Chairman:* Dr. S. A. Goldblith)

INTRODUCTION 7
S. A. Goldblith (*Underwood-Prescott Professor of Food Science and Associate Head, Department of Nutrition and Food Science, Massachusetts Institute of Technology, Cambridge, Massachusetts, U.S.A.*)

FUNDAMENTALS OF CONCENTRATION PROCESSES 13
H. A. C. Thijssen (*Physical Technology Laboratory, Department of Chemical Engineering, Eindhoven University of Technology, Eindhoven, The Netherlands; and Director of Research, D.E.J. International Research Co. B.V., The Netherlands*)

FUNDAMENTALS OF DEHYDRATION PROCESSES 45
M. Karel (*Professor of Food Engineering, Department of Nutrition and Food Science, Massachusetts Institute of Technology, Cambridge, Massachusetts, U.S.A.*)

ESSENCE RECOVERY ON CITRUS EVAPORATORS 95
D. T. Shore (*A.P.V. Ltd., Crawley, Sussex, U.K.*)

RHEOLOGICAL ASPECTS OF FRUIT JUICE EVAPORATION 101
G. D. Saravacos (*Professor of Chemical Engineering, National Technical University, Athens, Greece*)

SESSION II—NON-MEMBRANE CONCENTRATION

(*Chairman:* Dr. A. I. Morgan, Jr.)

INTRODUCTION 111
A. I. Morgan, Jr. (*Director, Western Regional Research Laboratory, U.S. Department of Agriculture, California, U.S.A.*)

FREEZE-CONCENTRATION 115
H. A. C. Thijssen (*Physical Technology Laboratory, Department of Chemical Engineering, Eindhoven University of Technology, Eindhoven, The Netherlands; and Director of Research, D.E.J. International Research Co., B.V., The Netherlands*)

NON-MEMBRANE CONCENTRATION 151
C. H. Mannheim and Mrs. N. Passy (*Department of Food Engineering and Biotechnology, Technion, Israel Institute of Technology, Haifa, Israel*)

A NEW CONCEPT FOR THE FREEZE-CONCENTRATION PROCESS . . 195
W. E. L. Spiess, W. Wolf, W. Buttmi and G. Jung (*Bundesforschungsanstalt, Karlsruhe, W. Germany*)

SUCROSE DEHYDRATION BY THE HEAT OF CRYSTALLISATION . . 203
W. M. Nicol (*Group R and D, Tate and Lyle Ltd., Reading, Berks, U.K.*)

SESSION III—MEMBRANE CONCENTRATION

(*Chairman:* Professor D. I. C. Wang)

INTRODUCTION 209
D. I. C. Wang (*Associate Professor of Biochemical Engineering, Department of Nutrition and Food Science, Massachusetts Institute of Technology, Cambridge, Massachusetts, U.S.A.*)

TAILORED MEMBRANES 213
A. S. Michaels (*President, Alza Research, and Vice-President, Alza Corporation, Palo Alto, California, U.S.A.*)

MEMBRANE CONCENTRATION 251
R. F. Madsen (*A/S De Danske Sukkerfabrikker, Driftteknisk Laboratorium, Nakskov, Denmark*)

STUDIES OF MEMBRANE DEPOSITS IN ULTRAFILTRATION . . . 303
P. Dejmek and B. Hallström (*Chemical Centre, Department of Food Engineering, Lund University, Lund, Sweden*)

INDUSTRIAL-SCALE ULTRAFILTRATION AND REVERSE OSMOSIS PLANTS IN THE FOOD INDUSTRIES 309
B. S. Horton (*Abcor Inc., Brussels, Belgium*)

SESSION IV—SPRAY DRYING
(*Chairman:* Professor E. Seltzer)

INTRODUCTION 317
E. Seltzer (*Food Process Engineering, Department of Food Science, Rutgers University, New Brunswick, New Jersey, U.S.A.*)

EFFECTS OF THE LATEST DEVELOPMENTS ON DESIGN AND PRACTICE OF SPRAY DRYING 321
O. G. Kjaergaard (*Chemical Engineer, Niro Atomizer Ltd., Copenhagen, Denmark*)

THEORETICAL MODELLING OF THE DRYING BEHAVIOUR OF DROPLETS IN SPRAY DRYERS 349
P. J. A. M. Kerkhof and W. J. A. H. Schoeber (*Physical Technology Laboratory, Department of Chemical Engineering, Eindhoven University of Technology, Eindhoven, The Netherlands*)

WHISKY POWDERS AND THEIR APPLICATIONS TO FOODS 399
T. Yamada (*Chuo University, Tokyo, Japan*)

SPRAY DRY COATING IN FLUIDISED BEDS 401
T. Yamada (*Chuo University, Tokyo, Japan*)

GENERAL DISCUSSION 403

SESSION V—FREEZE-DRYING AND OTHER NOVEL DEHYDRATION TECHNIQUES
(*Chairman:* Dr. N. Bengtsson)

INTRODUCTION 409
N. Bengtsson (*Section Head, Swedish Institute for Food Preservation Research, Göteborg, Sweden*)

NEW DIRECTIONS IN FREEZE-DRYING 413
J. Lorentzen (*Manager, R & D, A/S Atlas, Ballerup, Denmark*)

NOVEL DEHYDRATION TECHNIQUES 435
C. J. King (*Professor and Chairman, Department of Chemical Engineering, University of California, Berkeley, California, U.S.A.*)

CYCLIC-PRESSURE FREEZE-DRYING IN PRACTICE 489
J. D. Mellor (*Senior Research Scientist, Division of Food Research, C.S.I.R.O., North Ryde, New South Wales, Australia*)

THE ECAL PROCESS—A NOVEL DEHYDRATION PROCESS 499
A. L. Moeller (*Technical Director, Ecal Nateko AB, Munka, Ljungby, Sweden*)

MICROWAVE HEATING IN VACUUM DRYING 505
N. Meisel (*General Manager, Les Micro Ondes Industrielles, Epône, France*)

GENERAL DISCUSSION 511

INDEX 513

Opening Remarks

ARNOLD SPICER

Scientific Adviser to RHM Research Ltd.,
The Lord Rank Research Centre,
High Wycombe, Bucks., U.K.

Ladies and gentlemen, it is my privilege to welcome you all, the Chairmen of the Sessions, the Speakers and the listening participants, to the Symposium on Preconcentration and Dehydration of Foods.

When, three years ago, Professor von Sydow, Secretary General of the International Union of Food Science and Technology, invited me to organise this meeting it seemed a life-time away—far too distant to worry about it. As it came closer, the problems presented themselves, and decisions had to be made. I prepared to stage a small, intimate meeting where for a few days we would be members of a family whose common bond must be the most intense and successful pursuit of the subjects under discussion.

Many of us travel across the globe in the course of a year from conference to conference. Just as journals and publications have proliferated to such an extent that we can no longer read or digest everything that is in print, so we shall have to become more and more discerning about the meetings we can attend and the benefits we are likely to gain from such gatherings. Our thirst for more and new knowledge obviously remains unquenched. How could it be otherwise when we look at the vast territory of food science and technology and its prime importance to man and beast and its dominating position in an industrial connotation? We shall witness the rise or fall of the so-called traditional industries. At times, a gloom may hang over the cotton industries, or the wool or the shipbuilding industries; but there can be nothing but a worldwide boom in the food industry.

We must take into account the phenomenal rise in the number of its potential customers and the steady rise in income levels expressing itself in changing eating habits with a preference towards manufactured food and drink. Jean-Jacques Rousseau once said: 'Everything is good when it leaves the Creator's hand and everything degenerates in the hands of man'. Relating this saying to food would mean that everything is good as nature produced it but it is debased in the hands of the processor. I don't think we need argue here about the fallacy of such a broadly based pronouncement. We know that some foods suffer cruel blows from the field, via the factory to the consumer; but some, on the other hand, are considerably enhanced in texture and flavour and in nutritional value from the treatment they receive: and all of them will and must benefit from a real advance in science and technology.

Yet science is a late comer in the food arena. Whilst engineering benefited from science as early as the eighteenth century, food remained a craft—by and large—up to the second half of the nineteenth century. The real, close integration of food science and technology probably dates only from three or four decades back. In food, the housewife, and the chef, called the tune; and still do so today. The advertisements of our industry play up to them by claiming merely that this or that product is as good as 'Mum can make it', or the chef at the Ritz, or the Imperial.

Few companies would have the courage to claim, even if they were entitled to do so, that their food item surpasses Mother's efforts; but being as good as hers, they save her drudgery and give her convenience. Dried or concentrated food, or products combining both processes, take a considerable share in the daily shopping list and we can claim that this form of food preservation goes back to prehistoric times. The drying of meat and fish had come a long way by the time civilised communities were founded. Our present-day contribution to it has been a considerable refining and sophistication of the processes involved: and what is equally, or even more important, an understanding not only of the mechanical but of the biological and biochemical aspects of the processes. This has enabled us not only to design, but also to control procedures and thus lay the foundation to a constant control.

Lack of knowledge must lead to neglect; and neglect, as Benjamin Franklin said, may breed mischief, since 'for want of a nail, the shoe was lost, and for want of a shoe the horse was lost, and for want of a horse the rider was lost'. The next few days should, in the area under discussion, help us to avoid missing the nail, and thus the shoe, and thus the horse. We have arranged the programme so that, following the morning's work, the afternoon will be free for informal meetings, or discussions, or for leisure. But the evenings will see us burning the midnight candle, but, ladies and gentlemen, remember: 'that the heights reached and kept—by great men were not attained by sudden flight—but they, whilst their companions slept—were toiling upwards in the night'.

Professors Karel and Thijssen are going to start the proceedings on Monday by restating the basic principles of heat and mass transfer. In subsequent sessions the Science and Technology of Moisture Removal by a number of different processes is going to unfold. We need to reduce the water content of food to prevent biological and chemical deterioration in storage between harvest and consumption. Without food preservation an adequate year-round supply of the required nutrients for the present and future populations of this globe could not be provided for. To retain in the vast variety of processed foods the inherent desirable qualities requires the employment of a number of processes and a considerable modification of conditions which must stay under constant review by scientists and technologists. Social and economic pressures force a continuous reappraisal of our methodologies and I sincerely hope that on Saturday, when we conclude this meeting, we shall have found it not only memorable but also worthwhile.

Session I

FUNDAMENTALS

Introduction

S. A. GOLDBLITH

Department of Nutrition and Food Science,
Massachusetts Institute of Technology,
Cambridge, Massachusetts, U.S.A.

Fundamental to the development of all process industries are the basic principles upon which the industry is founded.

When urbanisation began and the source of food supply became more remote from the consumer, man found it necessary to develop two basic operations. First transport, and second a means of processing food in such a way that it would be usable by the consumer at a distance from the supplier or grower.

With the development of technology and science transformed from a pastime to a craft, technology began as an important factor in man's life in the eighteenth and nineteenth centuries, and production by man was transformed to production by machines as a result of the Industrial Revolution. As Bernal (1971) points out, the movement of capitalism in manufacture made itself felt and thus began the transition of the traditional sciences to the chemical, thermal and electrical sciences of the eighteenth century. Thus science and its handmaiden, technology, began to pervade all human thought and endeavour and in particular the food industry.

In nineteenth-century Europe, more people entered manufacturing trades, commerce and associated activities. While none of these contributed to the food supply, their influence was felt in agriculture and food production, since the new technologies were applicable to food preservation and processing, and inexpensive transport made it feasible and profitable for the industrialised countries to obtain their food supplies elsewhere and mechanisation made more food available using less labour.

Unlike the developments of this century in electronics where scientific discovery preceded technology, early food factories were largely engineered and based on developments by craftsmen. Classical examples of this are canning, dehydration, curing, milk pasteurisation and baking.

Food process engineering began as an offshoot of chemical engineering and adopted its concept of unit operations whereby the various process patterns with a common denominator, such as heating, are put together into one unit operation. Thus the manifold and complex series of operations that make up a food factory began to be systemised and unravelled by putting them together into a manageable and understandable small series of unit operations. Each operation was based upon a fundamental

FIG. 1. (Courtesy of the National Canners Association, Washington, D.C.)

basic principle or a process pattern, such as heat transfer, product flow, etc., to achieve 'unity in diversity' and a synthesis of the entire process using modern tools such as the computer.

In order to understand a unit operation, it is first necessary to define it in terms of physical or chemical parameters to show that it depends on two basic and important laws: the conservation of mass and the conservation of energy; it is also necessary to be able to measure the parameters. Thus food engineering in this century has been transformed from a qualitative descriptive term to a series of unit operations, each dependent on quantifying the data. Kepler once said 'to measure is to know' and Lord Kelvin once stated that when you can measure a thing and express it by number, then you know something about it.

NOTE ADDITIONNELLE.

MOYEN PRATIQUE ET ECONOMIQUE DE CHAUFFER LE VIN EN FÛTS.

Beaucoup de personnes me demandent d'indiquer le moyen qui me paraîtrait le plus pratique pour l'application en grand du procédé de conservation que j'ai déduit de mes études sur les causes des maladies des vins, consistant dans une élévation préalable de la température à 50 degrés environ.

J'ai déjà dit que c'était à l'industrie et au commerce de faire cette recherche. Pour moi, si j'avais à pratiquer des essais sur une grande échelle, voici le mode de chauffage que je voudrais tenter tout d'abord : soit un générateur de vapeur, grand ou petit, suivant les besoins; que l'on visse ou que l'on adapte, par un moyen quelconque, sur le tube de sortie de la vapeur, un tube serpentin avec branche de retour pareil à celui de la figure. Il serait en cuivre, ou mieux en cuivre argenté extérieurement. Introduisez ce tube dans le tonneau, par l'ouverture de la bonde, et faites glisser le bouchon *a b* de façon à couvrir l'orifice sans le fermer hermétiquement pour que le vin de dilatation puisse s'échapper. La vapeur, en circulant dans le serpentin, échauffera le vin, et elle sortira par l'orifice *o*, d'où elle pourra se rendre dans un autre serpentin pareil, placé dans un tonneau voisin, et ainsi de suite; ou bien elle viendra échauffer l'eau d'une caisse en tôle, formant bain-marie, pour le chauffage du vin en bouteilles.

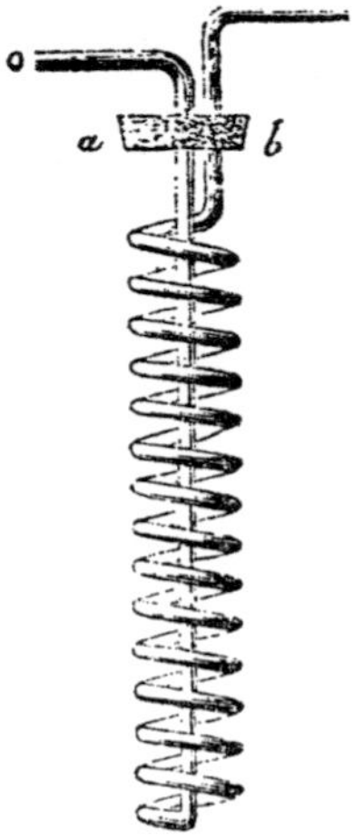

Que l'on imagine dans une filature les bassines a dévider les cocons remplacées par des tonneaux, et le tube a robinet d'admission de la vapeur communiquant avec les serpentins dont je parle, et l'on comprendra toute la facilité de l'opération du chauffage.

Sans doute il ne faut pas que la vapeur se condense directement dans le vin. Pourtant il ne faudrait pas rejeter *a priori* un tel procédé. Il est possible que pour les vins communs ce soit la plus simple et la plus économique des méthodes, car je ne pense pas que l'on puisse nuire ainsi sensiblement a la qualité de tels vins, tant il faudrait un faible poids de vapeur pour atteindre la température voulue.

FIG. 2. (From Pasteur, 1866)

In the first session of this Symposium we are going to discuss heat and mass transfer, two unit operations that go hand in hand as, for example, in freeze-drying, and we must be prepared to speak quantitatively of mass, length, time, force, temperature, and flow rates. All of these are fundamental to Kepler's and to Lord Kelvin's theses and to our discussions.

Historically, we have come a long way in the one hundred and fifty years of canning in the United States, as illustrated in Fig. 1, and since the heat exchanger designed by Louis Pasteur (Fig. 2) but Pasteur's and another French compatriot Appert's work are not old enough as examples of unit operations.

The Egyptians were constantly beset by the fear that their dead would go hungry believing that one of the dark areas of the sky was 'the field of viands'. They conceived it as having streams which, like the canals of the Nile, overflowed at the time of the annual flood. There, they believed, grew the 'grain of the dead' which was reaped, threshed and ground just like the

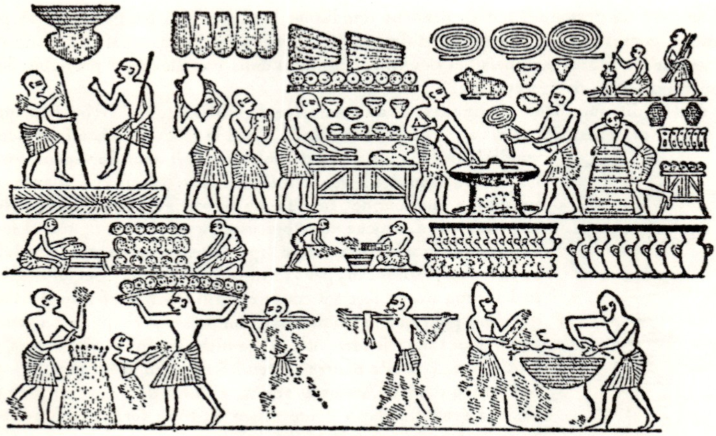

THE BAKERY OF KING RAMSES
(Egyptian tomb painting)

FIG. 3. (From Jacob, 1944.)

wheat of the living. According to Jacob (1944) the Egyptians worried about the dead not finding that field. It was therefore more prudent to provide them with victuals and allay the fear of omission, the fear of committing some error that made living eternally anxious for them. (Oh, if 'Librium' or 'Valium' had only been invented then!) But it is to this anxiety that we owe our knowledge of the minutia of their lives. Realistic scenes of labour were painted on the walls of the tombs to amuse and ward off dangers, as Fig. 3 shows in the bakery of King Ramses, indeed a series of unit operations—the first, in the upper left-hand corner shows two men with long poles trampling the dough. Centuries later, Herodotus mocked a people 'who knead dough with their feet and mud with their hands'.

Let me close by giving you one of the earliest illustrations of a series of unit operations which belie the relative newness of food engineering.

Glabau and Goldman (1937) in writing of baking in ancient Egypt described an interesting discovery of a bakery and a brewery house in the same building. In the brewery, the grain was ground and made into dough and then moulded into loaves. These loaves were dried, ground into meal which was then mixed with sprouted grain and the mashing process carried on to obtain conversion of the gelatinised starch into sugar, whereupon the mash was spread over the leaves from the sycamore tree resting on gratings, then filtered and leached. Then the wort was put into wooden casks and fermented.

Now the adjacent bakery closely resembled the brewery and there the grain was crushed in the same way and made into dough, then divided into loaves, moulded, fermented, and baked in rectangular ovens. Models of them are now at the Metropolitan Museum of Art, New York. Thus we see common unit operations in adjacent factories dealing with heat and mass transfer, crushing, fermentation, etc., all invented in past millennia.

REFERENCES

Bernal, J. D. (1971). *Science in History*, Vol. 2, *The Scientific and Industrial Revolution*, M.I.T. Press, Cambridge, Mass.

Jacob, H. E. (1944). *Six Thousand Years of Bread: Its Holy and Unholy History*, Doubleday, Doran and Co., Garden City, N.Y.

Glabau, C. A. and Goldman, P. F. (1937). 'Some physical and chemical properties of Egyptian Bread,' *Bakers Weekly*, **15**, 295.

Pasteur, L. (1866). *Etudes sur le Vin, ses Maladies: Causes qui les Provoquent, Procedes Nouveaux pour le Conserver et pour le Viellir*, Victor Masson et Fils, Paris.

Fundamentals of Concentration Processes

H. A. C. THIJSSEN

Physical Technology Laboratory,
Department of Chemical Engineering,
Eindhoven University of Technology,
Eindhoven, The Netherlands

INTRODUCTION

Water is removed from foods to provide microbiological stability, to reduce deteriorative chemical reactions and to reduce storage and transportation costs. Dehydration processes can be divided into concentration and dehydration processes. In concentration processes the water is removed from a fluid by molecular and eddy transport. The minimum water content is about 30 wt%. The process is generally steady state. This as distinct from drying where the water content is reduced to less than 10 wt% by unsteady state molecular diffusion from more or less rigid slabs, droplets or particles.

Concentration processes have to be inert with respect to the chemically unstable constituents and to be selective, all components except water being retained in the concentrate. The selectivity is of special importance for foods containing aromas. Almost all aroma components are more volatile than water.

A distinction can be made between equilibrium and non-equilibrium concentration processes. In equilibrium processes maximum separation is obtained at phase equilibrium of all components between the concentrated phase and water extracting phase. The processes include crystallisation, clathration and evaporation with, if necessary, aroma recovery by distillation. In the second class the two phases are separated by a membrane that is selectively permeable to water. Membrane processes include processes without a phase change, namely, direct osmosis, reverse osmosis, ultrafiltration and electrophoresis and one process with a phase change, namely pervaporation.

These processes except for electrophoresis will shortly be described. Electrophoresis is of little importance for the dehydration of liquid foods. In the next section (p. 16) the thermodynamics of phase equilibria relevant to concentration processes will be treated. The third section (p. 24) deals with the kinetics of the processes mentioned and with the effect of process conditions on the selectivity of the dehydration. The effect of process conditions on quality loss by deteriorative chemical reactions of unstable compounds is dealt with in short in the fourth section (p. 35). In the fifth section (p. 37) the effect of process conditions on the selectivity of the concentration process is discussed. The final section (p. 39) deals with a comparison of the performance and energy consumption of the various processes.

Evaporation processes

Almost all liquid foods are concentrated by evaporation. In order to prevent chemical deterioration, some liquid foods have to be concentrated at temperatures below 50°C and at a very short residence time and small residence time distribution of the liquid in the evaporator. Most widely used are multi-effect long tube falling film evaporators. To reduce the residence time distribution the liquid has to flow in a single pass (without recirculation) through the evaporators. Shorter residence times are obtained in wiped film evaporators. The residence time is the shortest in centrifugal film evaporators.

Volatile aromas are lost almost quantitatively with the vapour. With some liquid foods, such as milk, that may contain malodorous mercaptans the loss of volatiles results in a quality improvement. Generally, however, the loss of volatiles is accompanied by a loss in quality. Aroma losses can be partly compensated for in juice concentration by diluting the aroma-free concentrate of 60–70 wt% with fresh juice to a final concentration of about 40 wt%. Yet this cut back concentrate contains on dry solids basis, in the case of a feed concentration of 10%, only 10–12.5% of the aromas of the feed.

Much higher aroma retentions can be obtained by separating in a distillation column the volatiles from the water vapour which escapes from the evaporator and feeding this aroma concentrate back to the concentrate of non-volatiles. The methods of aroma recovery are described by Thijssen (1970). Very volatile aromas, however, are lost with the inert gases and aromas with a volatility about equal to—or less than—the volatility of water in the solution cannot be recovered from the vapour. Partly because of the incomplete aroma recovery, being generally not more than 60%, and partly because of degradative thermal reactions, the quality of thermally concentrated aromatic liquid foods, including most fruit juices and coffee and tea extracts, is always somewhat impaired. Technically, however, evaporation is the best developed concentration technique.

Pervaporation

Pervaporation is an evaporation process in which the liquid and the gas phase are separated by a membrane that is selectively permeable to water.

It has been shown in laboratory experiments that fruit juices can be concentrated by pervaporation with hardly any loss of volatiles. The best results are obtained with washed cellophane and with membranes formed *in situ* from the dissolved solids in the feed (Thijssen, 1966). The water at the vapour side of the membrane can be removed by a current of relatively dry gas or by a vacuum well below the equilibrium vapour pressure of the water in the membrane. High aroma retentions can only be obtained if the water concentration at the vapour side of the membrane is below about 10%. The heat of evaporation can be supplied to the circulating liquid or to the gas phase. Because of the low dehydration fluxes, being less than 2 kg/m^2 h, and the high energy consumption, being at least one ton of steam per ton of water removal, pervaporation will never become competitive with the other concentration processes.

Direct osmosis

In direct osmosis, too, the liquid to be concentrated is separated from the water withdrawing phase by a selectively permeable membrane. The difference is that in direct osmosis the water withdrawing phase is a liquid instead of a gas. The concentration of liquid foods by direct osmosis has been done on a laboratory scale only. A serious drawback is that the water absorbing solution has again to be concentrated in an evaporation process. An obvious advantage over evaporation and also pervaporation is that dehydration can be performed at low temperatures. A prerequisite is the availability of membranes with both a high selectivity for the liquid to be concentrated and a high permeability to water. Good results are reported for tomato juice with cellophane as a separating membrane and polyethylene glycol with a mole weight of about 500 as the water absorbing medium.

Ultrafiltration and reverse osmosis

In ultrafiltration and reverse osmosis the water diffuses from a concentrated solution through a selectively permeable membrane to an aqueous water absorbing phase with a much lower or almost zero solute concentration. The driving force for water transport is a pressure difference over the membrane exceeding the difference in osmotic pressure. The process is called ultrafiltration if the membranes allow the passage of water and other small molecules but impede the passage of molecules with a molecular weight of about 500 or higher. The term reverse osmosis is used when the membranes allow the passage of water but allow only a limited passage of small molecules, such as salts, monosaccharides and aromas. Reverse osmosis requires a much higher pressure, 50–75 atm, than ultrafiltration where pressures between 1 and 10 atm are usual. Most commonly, if not exclusively, used are the Loeb cellulose acetate membranes. An extensive review of membrane processes is given by Lacey (1972) and in the book of Sourirajan (1970). With direct osmosis they share the advantage of a low operating temperature. Because membranes with an acceptable permeability for water are not very selective for the lower molecular weight aromas, the commercial application for the dehydration of aromatic liquid foods will probably remain very limited. Ultrafiltration is very promising for the concentration of high molecular weight compounds such as protein solutions. From an energy point of view, these processes are superior to any other concentration process.

Freeze-concentration and clathration

Freeze-concentration and clathration involve a partial crystallisation of water and a subsequent separation of the crystals from the concentrate. Clathration processes are not in commercial use. Prohibitive for their application is the necessity of separating the clathrate former, such as freons, propane or butane, from the concentrated juice. Freeze-concentration is particularly suited for heat labile liquid foods containing volatile aromas. Since water is essentially withdrawn from the solution by the phase transformation from liquid to crystal, aroma losses by evaporation can be completely avoided. The low process temperature, being between -3 and -7°C, fully eliminates thermal decomposition reactions. Due to the high investment costs the dehydration costs are still high compared with the cost of evaporation.

PHASE EQUILIBRIA

In this section the thermodynamic relations are presented which describe the equilibrium between the phase to be concentrated (raffinate phase) and the water-extracting (extract) phase. The extract phase may be a gas, a solid or a liquid. Because preconcentration refers to liquids only the thermodynamics of solid mixtures will not be treated. Equilibrium always is established at the interface of the two phases. The rate of mass transfer between the phases is proportional to the deviations of the bulk concentrations of the two phases from their respective interface concentrations.

Thermodynamic equilibrium relations

The composition of a liquid or of a gaseous monophase mixture may be changed by contacting this phase with a second phase having a different chemical potential (partial Gibbs free energy) of its components. Mass will continue to be transferred until for all components the equilibrium conditions are satisfied. Necessary and sufficient conditions for equilibrium are (van Ness, 1964, p. 117):

$$\bar{\mu}_i^{\mathrm{I}} = \bar{\mu}_i^{\mathrm{II}} \qquad (i = 1, \ldots, n)$$
$$T^{\mathrm{I}} = T^{\mathrm{II}} \tag{1}$$

and if the phases are not separated by a selectively permeable membrane

$$P^{\mathrm{I}} = P^{\mathrm{II}}$$

where $\bar{\mu}_i$ is the chemical potential of component i, T is the temperature and P is the pressure. The superscripts I and II refer to phases I and II respectively. To relate the chemical potential to physical reality, the Lewis and Randall fugacity (f) is useful, which has the dimensions of pressure. For phase I

$$\bar{\mu}_i^{\mathrm{I}} = \mu_i^{\mathrm{I}} + RT \ln (\bar{f}_i^{\mathrm{I}}/f_i^{\mathrm{I}}) \tag{2}$$

and for phase II

$$\bar{\mu}_i^{\mathrm{II}} = \mu_i^{\mathrm{II}} + RT \ln (\bar{f}_i^{\mathrm{II}}/f_i^{\mathrm{II}}) \tag{3}$$

In eqns (2) and (3) R is the gas constant, $\bar{f}_i$ refers to the fugacity of component i in the mixture at the pressure and temperature of the system and f_i refers to the fugacity of that component in the pure state at the same temperature but at an arbitrary reference pressure P^r. For this reference pressure generally 1 atm is chosen. In a mixture of ideal gases the fugacity of each component is equal to its partial pressure. In real mixtures, the fugacity can be considered as a partial pressure which is corrected for non-ideal behaviour. Combination of eqns (1)–(3) yields (van Ness, 1964, p. 118):

$$\bar{f}_i^{\mathrm{I}} = \bar{f}_i^{\mathrm{II}} \tag{4}$$

Equation (4) is still of little practical use unless the fugacities can be related to the experimentally assessable quantities x_i, y_i, T and P, where x_i and y_i stand for the mole fraction of component i. The relation between fugacity and experimentally obtainable quantities is obtained by the introduction of the activity coefficient. The activity coefficient γ_i relates in a phase the fugacities $\bar{f}_i$ and f_i of component i. It is defined in each phase by

$$\gamma_i \equiv \frac{\bar{f}_i}{x_i f_i} \quad \text{or} \quad \gamma_i \equiv \frac{\bar{f}_i}{y_i f_i} \tag{5}$$

with the condition for condensable components that $\gamma_i = 1$ for $x_i = 1$ and $\gamma_i =$ constant for $x_i \to 0$. In this section, x refers to the mole fraction of a component in the liquid phase or membrane phase and y to the gas phase. The superscripts g, m and l will denote the gas, membrane and liquid phase respectively.

In a gas phase the standard state fugacity f_i is equal to the total pressure P, so

$$\gamma_i^{\mathrm{g}} = \frac{\bar{f}_i^{\mathrm{g}}}{y_i P} \tag{6}$$

In a liquid phase the standard fugacity at reference pressure $P^r = 1$ is related to the pure vapour pressure by

$$f_i^1 \equiv P_i^s \gamma_i^{gs} \exp\left(\int_{P_i^s}^{1} \frac{V_i^1}{RT}\, dP\right) \tag{7}$$

where P_i^s is the saturation vapour pressure of pure liquid or solid i at the temperature T of the system, V_i^1 is the molar liquid volume of pure i at the same temperature and γ_i^{gs} is the activity coefficient of pure saturated vapour of i at temperature T and pressure P_i^s. For water P_i^s is generally so far from critical conditions that γ_i^{gs} is very close to unity. At a reference pressure of one atmosphere the integral of eqn (7) is also close to unity and therefore f_i^1 is closely approximated by the saturation vapour pressure.

As is evident from eqn (7) the fugacities are dependent on pressure and consequently so are the activity coefficients. It is convenient to evaluate and compare all activity coefficients at the same pressure of say one atmosphere. For the relation between $\bar{f}_i^1$ at P, T, x_i and γ_i at T, x_i and $P = 1$ can be derived

$$\bar{f}_i^1 = \gamma_i^{P=1} x_i P_i^s \exp\left(\int_1^P \frac{\tilde{V}_i^1}{RT}\, dP\right) \tag{8}$$

In deriving eqn (8) f_i^1 is taken equal to P_i^s. In the integral $\tilde{V}_i^1$ denotes the partial molar liquid volume. In the following the superscript of the activity coefficient will be omitted and the pressure conditions will always refer to $P = 1$ atm. The fugacity of a pure solid (one component) is at moderate pressures closely approximated by the pure vapour pressure of that solid at the prevailing temperature:

$$\bar{f}_i^k \approx P_i^k \tag{9}$$

where k denotes the solid or crystal state.

On the basis of equilibrium condition (4) and eqns (5) through (8) we will now specify the equilibrium conditions for the phase systems

liquid–gas $P^g = P^I$
liquid–liquid $P^I = P^{II}$
liquid–liquid $P^I \neq P^{II}$
solid–liquid $P^k = P^I$

The first system is representative of evaporation and pervaporation, the second of liquid–liquid extraction and direct osmosis, the third system reverse osmosis and ultrafiltration and the fourth of freeze-concentration.

LIQUID–GAS, $P^g = P^I$

At the evaporation conditions suitable for heat labile liquid foods ($P \leq 1$ atm) the exponential term in eqn (8) can be neglected. Also the activity coefficient in the gas phase is very close to unity. Substitution for the fugacities from eqns (6) and (8) in (4) therefore yields the relation

$$y_i P = \gamma_i x_i P_i^{s} \tag{10}$$

The left-hand term in eqn (10) is the equilibrium vapour pressure of component i. For water the ratio of $y_w P/P_w^{s}$ is the so-called equilibrium humidity RH. So RH $= \gamma_w x_w$. If the mean mole weight of the dissolved solids in a liquid food is known, the activity coefficient of water can thus be calculated from water sorption isotherms.

From eqn (10) the partition coefficient or K factor is given by

$$K_i \equiv \frac{y_i}{x_i} = \gamma_i P_i^{s}/P \tag{11}$$

The relative volatility α_{aw} of volatile component i with respect to the volatile component water (w) in the mixture is defined by

$$\alpha_{iw} \equiv \frac{K_a}{K_w} = \frac{\gamma_a P_a^{s}}{\gamma_w P_w^{s}} \tag{12}$$

Here the subscript a denotes an aroma component. Because volatile aromas in foods are always present at very low concentrations γ_a is independent of the concentration of that aroma component, but of course strongly dependent on the other properties of the system, including water concentration, properties of dissolved solids, and T. At a low concentration of a and $x_w \to 1$ eqn (12) becomes

$$\alpha_{aw} = \frac{\gamma_a^{\infty} P_a^{s}}{P_w^{s}} \tag{13}$$

From this equation it is readily seen that the relative volatility of an aroma component at a given temperature is proportional to the product of activity coefficient and pure vapour pressure of that component.

LIQUID–LIQUID, $P^{I} = P^{II}$

For liquid–liquid extraction with immiscible phases or miscible phases separated by a selectively permeable membrane (direct osmosis) at or close to atmospheric pressure the exponential term in eqn (8) may again be neglected and substitution in eqn (4) yields

$$\gamma_i^{I} x_i^{II} = \gamma_i^{II} x_i^{II} \tag{14}$$

LIQUID–LIQUID, $P^{I} \neq P^{II}$

In order to obtain a pressure difference between the phases of any significance at or close to standard gravity by applying external pressure the phases have to be separated by a membrane that is impermeable to a part of the constituents. Substitution of (8) in (4) yields

$$\gamma_i^{\mathrm{I}} x_i^{\mathrm{I}} \exp\left(\int_1^{P_{\mathrm{I}}} \frac{\tilde{V}_i^{1,\mathrm{I}}}{RT}\,\mathrm{d}P\right) = \gamma_i^{\mathrm{II}} x_i^{\mathrm{II}} \exp\left(\int_1^{P_{\mathrm{II}}} \frac{\tilde{V}_i^{1,\mathrm{II}}}{RT}\,\mathrm{d}P\right) \quad (15)$$

If phase I is pure water at atmospheric pressure the equilibrium pressure in II is the so-called osmotic pressure π of the solution with a water concentration C_{w}, so

$$\pi = -\frac{RT \ln \gamma_{\mathrm{w}} x_{\mathrm{w}}}{\tilde{V}_{\mathrm{w}}^{1}} \quad (16)$$

In most cases $\overline{V}_i^1$ and $\tilde{V}_i^1$ do not differ and by good approximation $\overline{V}_i^1 = \tilde{V}_i^{1\mathrm{I}} = \tilde{V}_i^{1\mathrm{II}}$. From this and by substitution of eqn (15) in eqn (16) it follows

$$(P^{\mathrm{I}} - \pi^{\mathrm{I}}) = (P^{\mathrm{II}} - \pi^{\mathrm{II}}) \quad (17)$$

SOLID–LIQUID, $P^{\mathrm{k}} = P^{\mathrm{l}}$

Below the freezing temperature of an aqueous solution, a solid–liquid system tends to segregate a part of the water in the form of ice crystals. At equilibrium a substitution of the fugacities of the pure crystal phase and of the aqueous solution from eqns (8) and (9) respectively in (4) yields at moderate pressures

$$P_i^{\mathrm{k}} = \gamma_i^{1} x_i^{1} P_i^{\mathrm{s}} \quad (18)$$

The effect of process conditions and product properties on activity coefficients

The activity coefficients are dependent on pressure, temperature and the composition of the mixture. The effect of pressure has already been discussed. The temperature and composition effects will be treated in short.

TEMPERATURE EFFECT ON γ

The temperature dependence of an activity coefficient in the liquid phase is given by the exact thermodynamic relation

$$\frac{\partial \ln \gamma_i}{\partial(1/T)} = -\frac{H_i^0 - \overline{H}_i}{R} \quad (19)$$

where $\overline{H}_i$ is the partial molar enthalpy of component i in the solution and H_i^0 the enthalpy of component i in the pure standard state. The difference is the partial molar heat of mixing. It is evident from eqn (19) that if the heat of mixing is positive, so heat is evolved on mixing the pure component i with the other components, the activity coefficient increases with rising temperature. Unfortunately, values of $(H_i^0 - \overline{H}_i)$ vary considerably with temperature (Hougen and Watson, 1947) and are less available than the activity coefficients themselves. As can be deduced from water sorption isotherms, the activity coefficient of water in liquid foods increases with increasing temperature. In the water concentration range of practical interest for concentration processes (90–30 wt% water) γ_w is still close to unity. In a 50.6 wt% sucrose solution at 25°C, $\gamma_w = 0.983$ (Robinson and Stokes, 1955). Thus the variation of γ_w with temperature in concentration processes is hardly of any interest.

More important is the effect of temperature on the activity coefficients of aromas in the food. Activity coefficients of aroma can be as high as 350 000 (Chandrasekaran and King, 1971). Notwithstanding the high absolute values of aroma activity coefficients and the large variations with temperature (Glasstone and Pound, 1925) the relative volatilities α_{aw} vary only slightly with temperature. Therefore when γ_a of an aroma component is known at one temperature, the value of α_{aw} can be calculated for that temperature and used over the temperature range of interest.

CONCENTRATION EFFECT ON γ

The activity coefficient is a factor indicating the deviation from ideality of a component in a mixture. In ideal mixtures the activity coefficient is one; eqn (10) then becomes identical with the Raoult relation $y_i P = x_i P_i^s$. So the value of γ depends on the nature of the molecular interaction of component i with the other components in the mixture. From this it can be understood that a decreasing solubility, which results from increasing repulsive forces of a component in a solution, is accompanied by an increasing value of its activity coefficient. For components having a very low solubility in a solution it can be shown from eqn (14) that their activity coefficients are equal to the reciprocal of their molecular solubilities. Thus for *n*-alcohols for example the activity coefficients will because of decreasing solubility with increasing molecular weight strongly increase with the number of C atoms. From the above it will be clear that the water concentration affects the system properties and consequently also the activity coefficients of the constituents.

As mentioned already the activity coefficients of water in conditions prevailing during concentration are still quite close to unity and variations with water concentration and composition of dissolved solids are not yet important. Even at water concentrations of about 50 wt% the water molecules are in contact with, because of their low molecular weight with respect to those of the dissolved solids, mainly other water molecules (water concentration on a mole basis still >93%). Because of the very low concentrations of volatile aromas in liquid foods, being generally in the ppm range, the activity coefficients of these molecules are equal to those at infinitely low concentration of these components and are then fully controlled by the molecular properties of the dissolved solids–water mixture. Pierotti *et al.* (1959) and Wilson and Deal (1962) correlated activity coefficients at infinitely low concentrations with the molecular structure of both solute and solvent. Bomben gives the relative volatilities and activity coefficients of a number of aroma components at infinite dilution in water (Bomben and Merson, 1969). The pure vapour pressures, relative volatilities and activity coefficients of a number of aroma components at infinitely low concentration in water and in a 70 wt% sucrose solution are presented in Table I.

TABLE I

PURE VAPOUR PRESSURES, ACTIVITY COEFFICIENTS AND RELATIVE VOLATILITIES OF SOME AROMA COMPOUNDS AT INFINITE DILUTION IN AQUEOUS SOLUTIONS (CHANDRASEKARAN AND KING, 1971)

Aroma compound	*Temp.* (°C)	P_0 (mm Hg)	γ_0 0% *sucrose*	γ_0 70% *sucrose*	α_{aw} 0% *sucrose*	α_{aw} 70% *sucrose*
Methanol	25	95.11	1.53	—	7.84	—
Ethanol	25	58.76	3.37	6.7	7.79	15.5
Ethyl acetate	20	73.84	65	300	269	1 200
Isobutanol	25	17.50	42.90	—	22.50	—
Acetal	25	27.50	150	—	173	—
n-Hexanal	25	—	1 010	5 700	390	2 200
n-Butyl acetate	20	8.50	1 270	7 500	689	4 070
Ethyl-2-methyl butyrate	20	—	5 060	40 000	1 160	9 170
2-Hexanal	25	—	1 160	—	146	—
n-Hexanol	25	1.00	676	—	31.20	—
n-Hexyl acetate	20	—	23 100	350 000	1 580	23 900

For the homologous series the increase of the activity coefficient with increasing chain length is much greater than the corresponding decrease in pure vapour pressure and this results in an increase in relative volatility. It is also interesting to note that the activity coefficients of the aromas and their relative volatilities strongly increase with decreasing water concentration. The increase in relative volatility with increasing chain length appears to be greater when the water concentration is lower.

Effect of activity coefficients of aromas on their potential loss

The activity coefficients of aroma molecules strongly affect the retention of these aromas in a concentration process if there is no membrane present between the two phases or if the membrane is permeable to the aromas. We shall treat in short the different conditions of concentration.

EVAPORATION AND PERVAPORATION (LIQUID–GAS, $P^g = P^l$)

In the case of evaporation or pervaporation with a membrane permeable to the aromas the retention of a component after equilibrium evaporation (equilibrium between phases for the volatile aromas during the whole concentration process) is given by the Rayleigh relation

$$\frac{N_{ao}}{N_{at}} = \left(\frac{N_{wo}}{N_{wt}}\right)^{\alpha_{aw}} \qquad (20)$$

where N_{ao} and N_{wo} are the total mass of aroma component a and of water respectively in the original solution (feed) and N_{at} and N_{wt} the total mass of aroma component a and of water left after evaporation of $N_{wo} - N_{wt}$ mass of water. It can be calculated by eqn (20) that for a α_{aw} value 5, which is low compared with the α's of most aromas in foods, an increase of the dissolved solids concentration by equilibrium evaporation from 10 wt% to 50 wt%, so that $N_{wo}/N_{wt} = 9$, results in a loss of 99.998% of that aroma compound or a retention of only 0.002%.

DIRECT OSMOSIS (LIQUID–LIQUID, $P^{I} = P^{II}$)

If the membrane is only impermeable to the dissolved solids and permeable to water and aromas, the aroma loss after equilibrium concentration can be calculated with a relation identical to eqn (20). Instead of the relative volatility α_{aw}, however, the selectivity β_{aw} has to be used:

$$\beta_{aw} \equiv \frac{K_a}{K_w} = \frac{\gamma_a{}^{I}\gamma_w{}^{II}}{\gamma_a{}^{II}\gamma_w{}^{I}} \approx \frac{\gamma_a{}^{I}}{\gamma_a{}^{II}} \qquad (21)$$

where

$$K_a = \frac{x_a{}^{II}}{x_a{}^{I}} \quad \text{and} \quad K_w = \frac{x_w{}^{II}}{x_w{}^{I}}$$

If non-electrolytes such as polyethylene glycols are used as phase II the value of β_{aw} will be close to unity and the aroma loss will be proportional to the fraction of water removed. For electrolytes $\gamma_a{}^{II}$ will be higher than $\gamma_a{}^{I}$ and consequently the aroma retention will be better.

REVERSE OSMOSIS (LIQUID–LIQUID, $P^{I} > P^{II}$)

For a membrane permeable to the aromas, the aroma loss can again be calculated by eqn (20) but with the exponent β_{aw} instead of α_{aw}. As already noted the activity coefficient of an aroma in concentrated carbohydrate solutions is higher than in pure water, see Table I. But, moreover, the higher pressure at the side of the membrane where the dissolved solids concentration is higher results also in a higher value of the fugacity coefficient compared with that at the water side (see eqn 8). From this it can be concluded that β_{aw} is appreciably higher than one and that the aromas will be preferentially lost.

FREEZE-CONCENTRATION AND CLATHRATION (SOLID–LIQUID)

In freeze-concentration the ice phase is of a very high purity, the activity coefficient of the aromas in the ice phase $\gamma_a{}^{k}$ approaches infinity. A very high purity of the crystals is also observed in clathration.

THE RATE CONTROLLING FACTORS

One of the most commonly used equations to describe mass transfer including water and aroma transport in an isobaric and isothermal n-component system is the generalised Stefan–Maxwell equation, based on the hydrodynamic concept of diffusion. According to the model the unidirectional transport is given by the relation

$$C_i\left(\frac{\mathrm{d}\ln x_i\gamma_i}{\mathrm{d}z}\right) = \sum_{\substack{j=1\\ j\neq 1}}^{n} \frac{C_i\overline{N}_j - C_j\overline{N}_i}{\rho D_{ij}^{\mathrm{t}}} \qquad (i = 1, \ldots, n) \tag{22}$$

where C_i is the mass concentration of component i; C_j is the mass concentration of component j; ρ is the total concentration or density (=) g/cm^3; $\overline{N}_j$ is the mass flux of component j with respect to stationary co-ordinates; $\overline{N}_i$ is the mass flux of component i with respect to stationary co-ordinates; z is the co-ordinate of length; D_{ij}^{t} is the multi-component (turbulent or eddy) diffusivity or mobility of component i with respect to component j on a mass basis, and x_i is again the mole fraction. In eqn (22) the left-hand term is the driving force for transport of component i in that phase.

The water flux

Equations will be presented for the calculation of the water flux in the liquid phase, in the gas phase and in the membrane phase separately. Thereupon the interactive effect of the different phases upon mass transfer will be treated.

MASS TRANSFER COEFFICIENTS

Liquid phase

The molecular and turbulent diffusion of water and of dissolved solids in a liquid food can be treated as a binary system, one component being the water, the second the mixture of dissolved solids. Because of their extremely low concentration, the mass fluxes $\overrightarrow{N}_{\mathrm{a}}$ of the aromas do not influence the fluxes of water $\overrightarrow{N}_{\mathrm{w}}$ and of the dissolved solids $\overrightarrow{N}_{\mathrm{d}}$. In the case of no volume contraction on mixing and no transport of the dissolved solids through the interface, the net volume flow with respect to stationary co-ordinates of the liquid will be zero or $\overrightarrow{N}_{\mathrm{w}}/\rho_{\mathrm{w}} = -\overrightarrow{N}_{\mathrm{d}}/\rho_{\mathrm{d}}$, when ρ_{w} and ρ_{d} are the densities of pure water and of pure dissolved solids respectively. For these conditions it can be shown from eqn (22) that the flux F of water relative to the net velocity of the dissolved solids in the direction of diffusion z is given by

$$F_{\mathrm{w}} = -D_{\mathrm{w}} \frac{\dfrac{\partial \ln a_{\mathrm{w}}}{\partial \ln C_{\mathrm{w}}} \cdot \dfrac{\mathrm{d}C_{\mathrm{w}}}{\mathrm{d}z}}{1 - C_{\mathrm{w}}/\rho} \tag{23}$$

In this equation D_w is the binary eddy diffusion coefficient of water on a mass basis. At the interface between the drying phase and the extract phase or at the interface between the liquid phase and a membrane F_w is the water flux through that interface.

Both the binary eddy diffusion coefficient D_w and the activity coefficient are a function of the water concentration. We can therefore define a modified eddy diffusion coefficient $\mathscr{D}_w$

$$\mathscr{D}_w \equiv D_w \cdot \frac{\partial \ln a_w}{\partial \ln C_w} \tag{24}$$

By substitution of (24) in (23)

$$F_w = \frac{-\mathscr{D}_w \dfrac{dC_w}{dz}}{1 - C_w/\rho} \tag{25}$$

At the interface F_w is the mass flux of water through that interface and consequently the rate of dehydration per unit surface and unit time. In eqns (22)–(25) the diffusion coefficients are turbulent or eddy diffusivities and therefore dependent on the degree of turbulence and on the distance from the interface. In a still liquid or gas this diffusion coefficient becomes equal to the *molecular* diffusion coefficient on a weight basis.

We shall restrict the discussion to the concentration of liquids in a steady state. This means that at a certain place in the apparatus the conditions, including temperature, pressure and concentration, do not vary with time. The conductivity k of the liquid (or gas) for water between two planes parallel to the interface between drying and extractive phase or between gas or liquid phase and a membrane can be defined by

$$\frac{1}{k_w} \equiv \int_{z_1}^{z_2} \frac{dz}{\mathscr{D}_w} \tag{25a}$$

k in eqn (25a) is more generally called the mass transfer coefficient. By integrating eqn (25a) between the bulk concentration of the liquid, $C_w{}^b$, and the concentration at the interface $C_w{}^i$, and substituting in eqn (25), we get

$$F_w{}^i = k_w \rho \ln \frac{1 - C_w{}^b/\rho}{1 - C_w{}^i/\rho} \tag{26}$$

The superscripts i and b refer to the place at the interface and in the bulk respectively. By introduction of the logarithmic mean denoted by $(1 - C_w/\rho)$ ln, of $(1 - C_w{}^b/\rho)$ and $(1 - C_w{}^i/\rho)$, eqn (26) becomes

$$F_w{}^i = k_w \frac{(C_w{}^b - C_w{}^i)}{(1 - C_w/\rho)\ \ln} \tag{27}$$

Membrane phase

The eqns (23)–(25a) hold also for membranes. Equation (25a) is in this case integrated between the two interface concentrations C^{mI} and C^{mII} in the membrane in contact with gas or liquid phases I and II respectively (see Fig. 1):

$$F_w{}^m = k_w{}^m \frac{(C_w{}^{mI} - C_w^{mII})}{(1 - C_w{}^m/\rho^m) \ln} \tag{28}$$

where $(1 - C_w{}^m/\rho^m) \ln$ is again the logarithmic mean of $(1 - C_w{}^{mI}/\rho^m)$ and $(1 - C_w^{mII}/\rho^m)$. If the water concentration in the membrane is very low, as is the case in pervaporation where one side of the membrane is in

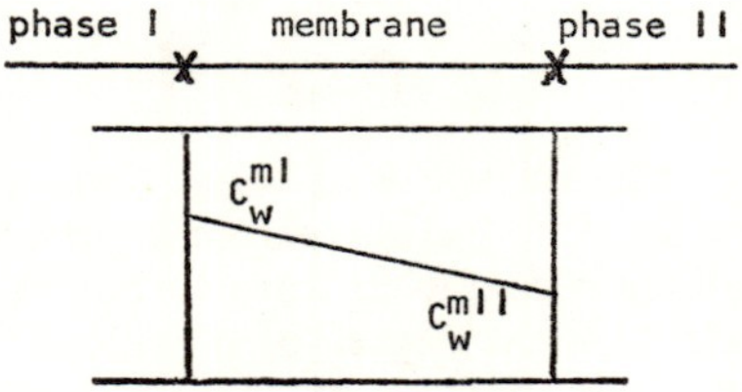

FIG. 1

contact with a dry gas atmosphere, the denominator in eqn (28) is about 1. If one side of the membrane is in contact with almost pure water, the water concentration in the membrane is independent of the driving force for reverse osmosis, thus

$$F_w = \bar{k}_w{}^m(C_w{}^{mI} - C_w^{mII}) \tag{29}$$

where

$$\bar{k}_w{}^m = \frac{k_w}{(1 - C_w{}^m/\rho^m)}$$

Although eqns (28) and (29) are correct, the driving force for water transfer through a membrane used for reverse osmosis or ultrafiltration is more commonly expressed in terms of pressures instead of concentrations. From eqns (15)–(17) it will be clear that the relation between the flux and pressure difference over the membrane can be written

$$F_w = k_w{}^P[(P^I - \pi^I) - (P^{II} - \pi^{II})] \tag{29a}$$

where $P^I - P^{II}$ is the pressure difference over the membrane and $\pi^I - \pi^{II}$ the osmotic pressure difference between either side of the membrane. In general phase II is almost pure water and thus $\pi^{II} = 0$.

Gas phase

In the gas phase the activity coefficient approaches unity, so $D_w = \mathscr{D}_w$. With this simplifying condition eqns (25)–(27) hold also for the gas phase. But here the flux F_w is negative, which means that the direction of the flux is away from the interface.

OVERALL TRANSFER COEFFICIENTS FOR WATER

In the most common situation the drying (raffinate) liquid phase I and the water withdrawing gas or liquid phase (extract phase) II are separated by a membrane m. In the case of direct contact between phases I and II

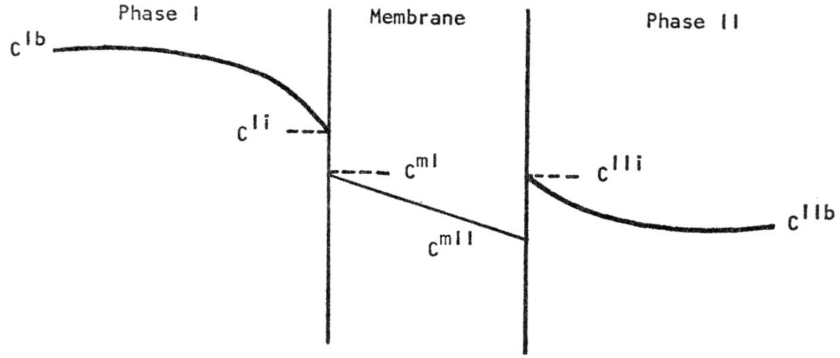

FIG. 2

the thickness of the separating membrane is zero. Under steady state conditions the water flux in the raffinate phase, the flux through the membrane (if present) and the flux in the extract phase are equal. The concentrations in these phases are presented in Fig. 2.

$$F_w = \frac{k_w^{\mathrm{I}}}{(1 - C_w^{\mathrm{I}}/\rho^{\mathrm{I}})\ \mathrm{ln}}(C^{\mathrm{Ib}} - C^{\mathrm{Ii}}) = \frac{k_w^{\mathrm{m}}}{(1 - C_w^{\mathrm{m}}/\rho^{\mathrm{I}})\ \mathrm{ln}}(C^{\mathrm{mI}} - C^{\mathrm{mII}})$$
$$= \frac{k_w^{\mathrm{II}}}{(1 - C_w^{\mathrm{II}}/\rho^{\mathrm{II}})\ \mathrm{ln}}(C^{\mathrm{IIi}} - C^{\mathrm{IIb}}) \tag{30}$$

An overall transfer coefficient k_0 can now be defined by the relation

$$F_w = k_{ow}^{\mathrm{I}}(C^{\mathrm{Ib}} - C^{\mathrm{I*(IIb)}}) \tag{31}$$

In eqn (31) $C^{\mathrm{I*(IIb)}}$ denotes the concentration that will exist in phase I if this phase is in equilibrium with the actual concentration C^{IIb} in phase II so

$$K_w^{\mathrm{I,II}} \equiv \frac{C_w^{\mathrm{I*(IIb)}}}{C_w^{\mathrm{IIb}}}$$

The term between brackets in eqn (31) is the overall driving force for mass transfer. A relationship can be derived between the overall coefficient and the partial mass transfer coefficient:

$$\frac{1}{k_{ow}^{I}} = \frac{(1 - C_w^I/\rho^I)\ln}{k_w^I} + \frac{(1 - C_w^m/\rho^m)\ln}{k_w^m} M_w^m + \frac{(1 - C_w^{II}/\rho^{II})\ln}{k_w^{II}} M_w^{II} \tag{32}$$

where

$$M_w^m \equiv \frac{C_w^{Ii} - C_w^{I*(mII)}}{C_w^{mI} - C_w^{mII}} \approx \frac{dC_w^{I*(m)}}{dC_w^m} \tag{33}$$

and

$$M_w^{II} \equiv \frac{C_w^{I*(IIi)} - C_w^{I*(IIb)}}{C_w^{IIi} - C_w^{IIb}} \approx \frac{dC_w^{I*(II)}}{dC_w^{II}} \tag{34}$$

$C_w^{I*(mII)}$ is the concentration in phase I that will exist if this phase is in equilibrium with the actual concentration C_w^{mII}; $C_w^{I*(m)}$ is the concentration in phase I that will exist if this phase is in equilibrium with the actual concentration C_w^m in the membrane, so

$$K_w^{Im} \equiv \frac{C_w^{I*(m)}}{C_w^m}$$

$C_w^{I*(IIi)}$ and $C_w^{I*(II)}$ are the concentrations of water in phase I if this phase is in equilibrium with the actual concentrations IIi and IIb, so

$$K_w^{I,II} = \frac{C_w^{I*(II)}}{C_w^{II}}$$

If the equilibrium concentration relations between phase I, the membrane and phase II are linear, $M_w^m = K_w^{Im}$ holds and so does $M_w^{II} = K_w^{I,II}$.

In the absence of a membrane (evaporation) relation eqn (32) reduces to

$$\frac{1}{k_{ow}^{I}} = \frac{(1 - C_w^I/\rho^I)\ln}{k_w^I} + \frac{(1 - C_w^{II}/\rho^{II})\ln}{k_w^{II}} \cdot M_w^{II} \tag{35}$$

In the case of evaporation the gas phase is almost 100% water, which means that $1 - C_w/\rho$ approaches zero and the partial mass transfer coefficient in the gas phase approaches infinity. So for evaporation the simplified relation holds:

$$\frac{1}{k_{ow}^{I}} = \frac{(1 - C_w^I/\rho^I)\ln}{k_w^I} \tag{36}$$

In the case of reverse osmosis or ultrafiltration with a low solute concentration in phase II, so that $\pi^{II} = 0$, eqns (30) and (31) become, in terms of pressure differences over the membrane,

$$F_w = \frac{k_w^I}{(1 - C_w^I/\rho^I)\ln}(C^{Ib} - C^{Ii}) = k_w^P(\Delta P - \pi_w^{Ii}) \tag{37}$$

The overall transfer coefficient, k_{ow}^{P}, can now be defined by

$$F_w = k_{ow}^{P}(\Delta P - \pi_w^{Ib}) \tag{38}$$

where π_w^{Ib} and π_w^{Ii} denote the osmotic pressures of solutions with water concentrations C_w^{Ib} and C_w^{Ii} respectively. The partial and overall transfer coefficients are related by

$$\frac{1}{k_{ow}^{P}} = \frac{1}{k_w^{P}} - \frac{(1 - C_w^{I}/\rho^{I}) \ln}{k_w^{I}} M_w^{P} \tag{39}$$

and

$$M_w^{P} \equiv \frac{\pi_w^{Ii} - \pi_w^{Ib}}{C_w^{Ii} - C_w^{Ib}} \approx \frac{d\pi_w}{dC_w} \tag{40}$$

The differential $d\pi_w/dC_w$ can be derived from experimental data or can be calculated theoretically from eqn (16).

It is obvious from eqns (38)–(40) that an increase in π_w^{Ib} resulting from a decrease in bulk water concentration, reduces the permeation rate. A similar result is obtained by a decrease in the mass transfer coefficient k_w^{I} in liquid phase I. This is because π_w^{Ii} of the liquid film adjacent to the membrane is higher than π_w^{Ib} in the bulk of the liquid. This phenomenon is known as the polarisation effect. With eqns (38)–(40) we have calculated the effect of ΔP and k_w^{I} upon the permeation rate of water from a sucrose solution. The calculations are made for cylindrical tubes with an inner diameter of 1 cm. The sucrose solution flows through the tubes. The physical data of sucrose are obtained from Sourirajan (1970). The value of k_w^{P} is obtained from Madsen (1973) for the DDS membrane type 995. The value of k_w^{I} is calculated by the empirical relation (58) with the constants $b_1 = 0.01$; $b_2 = 0.9$ and $b_3 = 0.35$. This relation will be given in a later section. In Fig. 3 the calculated flux is presented as a function of sucrose concentration with liquid velocity inside, and pressure difference over the membrane as parameter. It is clear from the figure that both the sugar concentration and the concentration polarisation strongly limit the permeation rate. At concentrations above about 35 wt% the flux is, at a liquid velocity of 25 cm/s, fully limited by the concentration polarisation. An increase in the k value of the membrane to infinity does not result in any increase in permeation rate. At a liquid velocity of 100 cm/s and above 40 wt% sucrose concentration in the bulk the flux becomes independent of the permeability of the membrane. Because a pressure difference of 100 atm already results in a reduced lifetime, it is obvious from the figure that even for very open membranes the permeation rate from a sucrose solution is limited to about 10 kg/cm^2 s at an exit sugar concentration of 35 wt%. For the retention of aromas in coffee, tea and fruit juices, very tight membranes have to be used. For these membranes k_w^{P} amounts to about 3.5×10^{-6} g/cm^2 s atm or about 0.13 kg/cm^2 h atm. With these low membrane permeabilities the practically attainable capacities for dehydrating aromatic liquid foods vary between 2 and 8 kg/m² h.

Water flux in crystallisation

The relations between partial and overall transfer coefficients for crystallisation are much more complex. In this case there is a resistance for mass

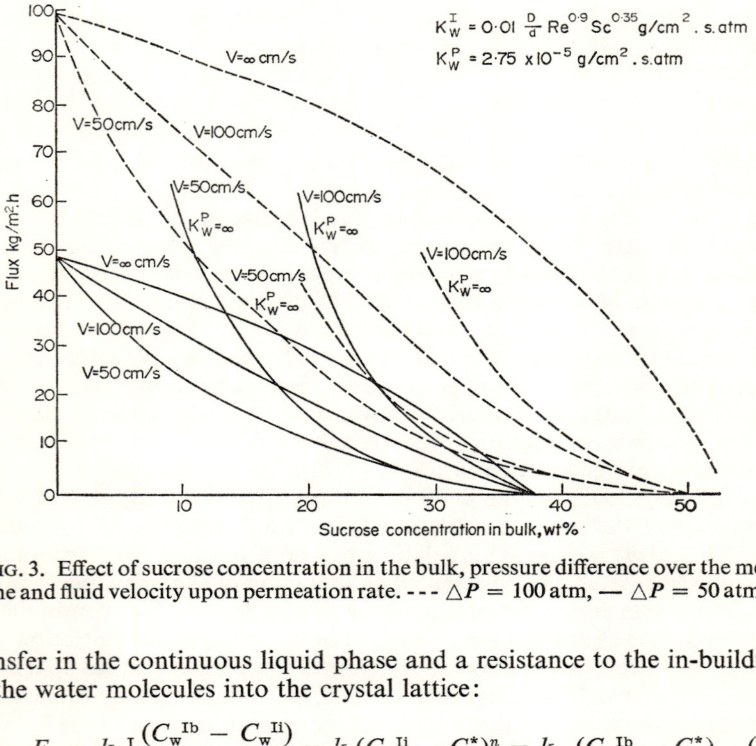

FIG. 3. Effect of sucrose concentration in the bulk, pressure difference over the membrane and fluid velocity upon permeation rate. --- $\triangle P = 100$ atm, — $\triangle P = 50$ atm.

transfer in the continuous liquid phase and a resistance to the in-building of the water molecules into the crystal lattice:

$$F_{\mathrm{w}} = k_{\mathrm{w}}^{\mathrm{I}} \frac{(C_{\mathrm{w}}^{\mathrm{Ib}} - C_{\mathrm{w}}^{\mathrm{Ii}})}{(1 - C_{\mathrm{w}}^{\mathrm{I}}) \ln} = k_{\mathrm{r}}(C_{\mathrm{w}}^{\mathrm{Ii}} - C_{\mathrm{w}}^{*})^{n} = k_{\mathrm{wo}}(C_{\mathrm{w}}^{\mathrm{Ib}} - C_{\mathrm{w}}^{*}) \quad (41)$$

where $C_{\mathrm{w}}^{\mathrm{Ib}}$ and $C_{\mathrm{w}}^{\mathrm{Ii}}$ are the water concentrations in the bulk and at the crystal surface respectively; C_{w}^{*} is the water concentration at the crystal temperature T^{k} and in equilibrium with the crystal; $k_{\mathrm{w}}^{\mathrm{I}}$ is the mass transfer coefficient in the liquid phase; k_{r} is the in-building or crystallisation rate constant; k_{wo} is the overall transfer coefficient; and n is an exponent.

Because the in-building rate does not have a first order dependence on the driving force for in-building ($n \neq 1$), a simple relationship between $k_{\mathrm{w}}^{\mathrm{I}}$, k_{r} and k_{wo} cannot be derived.

For detailed information about crystallisation kinetics the reader is referred to the paper on 'Freeze-concentration' (p. 115).

The dissolved solids flux

The flux of dissolved solids in the liquid phase and through membranes is calculated with the same eqns (23)–(29) derived for water. Instead of the subscript w the subscript d, indicating dissolved solids, has to be used. It is the objective that the membrane is only permeable to water and relatively impermeable to dissolved solids, thus $C_d^I \gg C_d^{II}$.

Due to the polarisation effect the dissolved solids concentration in the liquid phase I (raffinate) at the membrane interface is higher than in the bulk of that phase. For a constant value of ρ eqn (42) can be derived from eqn (37)

$$C_d^{Ii} = C_d^{Ib} + \frac{(1 - C_w^I/\rho^I)\ln}{k_w^I} \cdot k_{ow}^P(\Delta P - \pi_w^{Ib}) \qquad (42)$$

For the logarithmic mean of $(1 - C_w^{Ib}/\rho^I)$ and $(1 - C_w^{Ii}/\rho^I)$, $1 - C_w^I/\rho$ can be written by approximation, so

$$C_d^{Ii} = C_d^{Ib} + (1 - C_w^I/\tilde{\rho}^I)(\Delta P - \pi_w^{Ib}) \cdot \frac{k_{ow}}{k_w^I} \qquad (43)$$

For $C_d^{Ib} \gg C_d^{IIb}$ the flux of dissolved solids through the membrane becomes

$$F_d = \frac{k_d^m}{K_d^m}\left[C_d^{Ib} + (1 - C_w^I/\rho^I)(\Delta P - \pi_w^{Ib}) \cdot \frac{k_{ow}}{k_w^I}\right] \qquad (44)$$

where

$$K_d^m = \frac{C_w^{Ii}}{C_w^{mI}}$$

The aroma flux

THE PARTIAL TRANSFER COEFFICIENT FOR AROMA TRANSPORT

Liquid phase

Because of the very large gradients in activities of the aroma components in the liquid phase the effect of the water flux on the flux of aroma can be neglected. Under this condition, the flux of an aroma component becomes proportional to a simple pseudo-binary (molecular or turbulent) diffusion coefficient, the one component being the aroma, the other the mixture of water and dissolved solids. The binary aroma diffusion coefficient D_a is again strongly dependent on the water concentration, on the composition of dissolved solids, and on the properties of the dissolved solids. As can be seen from Table I, the activity coefficients of the aromas are also strongly dependent on water concentration. The aroma flux therefore is not only dependent on the aroma concentration gradient and on the value of D_a, but also on the water concentration gradient which by its influence on the activity coefficients affects the driving force for the aroma flux. The aroma flux is therefore proportional to the gradient in thermodynamic activity rather than in the concentration and can be expressed by (Bird *et al.*, 1963)

$$F_a = -D_a^1\left(\frac{dC_a}{dz} + C_a \frac{\partial \ln \gamma_a}{\partial C_w} \cdot \frac{dC_w}{dz}\right) \tag{45}$$

where F_a is again the flux of aroma component a relative to the interface on mass basis.

It can be expected that in all liquid foods the activity coefficients of the aromas increase with decreasing water concentration. This means that the value of $\partial \ln \gamma_a/\partial C_w$ is negative. At and close to the interface in the drying liquid dC_w/dz and dC_a/dz are both negative. From eqn (45) it is now evident that the flux F_a from the liquid to be concentrated decreases with increasing value of $\partial \ln \gamma_a/\partial C_w$. Fortunately eqn (45) contains only the logarithm of γ_a which varies much less than γ_a. According to the data on *n*-hexyl acetate in Table I the maximum value of $\Delta \ln \gamma_a = 2.7$ over the concentration range from 0 to 70 wt% sucrose. This means that in general it is justified to omit the second term between the brackets in eqn (45) and to use the simplified relation

$$F_a = -D_a^1 \frac{dC_a^1}{dz} \tag{46}$$

The mass transfer coefficient, k_a^1, of an aroma component can again be defined by

$$\frac{1}{k_a^1} \equiv \int_{z_1}^{z_2} \frac{dz}{D_a^1} \tag{47}$$

and

$$F_a = k_a^1(C_a^{1b} - C_a^{1i}) \tag{48}$$

Gas phase

The flux of aromas in the gas phase is the sum of the convective transport due to the water evaporation and the molecular or turbulent transport under the influence of an aroma concentration gradient

$$F_a^g = \frac{C_a}{\rho} \cdot N_w^g - D_a^g \frac{dC_a}{dz} \tag{49}$$

Because of the high relative volatilities of most aromas the first term in the right-hand side part of eqn (49) can generally be neglected. In terms of the mass transfer coefficient the relation becomes

$$F_a^g = k_a^g(C_a^{gi} - C_a^{gb}) \qquad (50)$$

Membrane phase

The aroma transport in the membrane is given by

$$F_a^m = k_a^m(C_a^{mI} - C_a^{mII}) \qquad (51)$$

in which k_a^m is defined by

$$\frac{1}{k_a^m} \equiv \int_{mI}^{mII} \frac{dz}{D_a^m} \qquad (52)$$

OVERALL TRANSFER COEFFICIENTS FOR AROMA

On an overall basis the flux is given by the relation

$$F_a = k_{oa}^I(C_a^{Ib} - C_a^{I*(IIb)}) \qquad (53)$$

where because of the low aroma concentrations the relation for k_o as given in eqn (32) reduces to

$$\frac{1}{k_{oa}^I} = \frac{1}{k_a^I} + \frac{M_a^m}{k_a^m} + \frac{M_a^{II}}{k_a^{II}} \qquad (54)$$

with

$$M_a^m \equiv \frac{dC_a^{I*(m)}}{dC_a^m} = K_a^m \equiv \frac{C_a^{I*(m)}}{C_a^m}$$

and

$$M_a^{II} \equiv \frac{dC_a^{I*(II)}}{dC_a^{II}} = K_a^{II} \equiv \frac{C_a^{I*(II)}}{C_a^{II}}$$

For evaporation and pervaporation processes the resistance to mass transfer in the gas phase can again be neglected. Due to the high relative volatility of most aromas K_a^{II} is very small. So eqns (53) and (54) reduce to

$$F_a = k_{oa}^I \cdot C_a^{Ib} \qquad (53a)$$

and for evaporation (without membrane)

$$k_{oa}^I = k_a^I \qquad (54a)$$

Relation between the mass transfer coefficients of water, dissolved solids and aromas

Liquid phase

For the pseudo-binary system water-dissolved solids the following relation holds:

$$k_w^l = k_d^l \qquad (55)$$

According to the Chilton–Colburn equation the relation between k_w and k_a is as follows:

$$k_a = k_w \left(\frac{D_a}{D_w}\right)^{2/3} \tag{56}$$

In this relation D_a and D_w are the molecular diffusion coefficients in the liquid phase on a weight basis of aroma and of water respectively.

Membrane phase

In the membrane the transport coefficients are controlled by the resistance of the molecules for diffusion. In general

$$k_w{}^m \neq k_d{}^m \neq k_a{}^m \tag{57}$$

Gas phase

For the gas phase, relation (56) is also applicable. The diffusion coefficients are in this case those for the gas phase.

Correlations for the mass transfer coefficients

The mass transfer coefficient for a component in a phase is of course dependent on the molecular diffusion coefficient of that component, on the physical properties of the solution including viscosity and density, and on the degree of mixing or turbulence in that phase. A general relation between these quantities can be written in the form

$$\frac{kd}{D} = b_1 Re^{b_2} Sc^{b_3} \tag{58}$$

where d is hydraulic diameter or length; Re is $\rho Vd/\mu$, the Reynolds number; Sc is $\mu/\rho D$, the Schmidt number; and b_1, b_2 and b_3 are constants. D in these relations is always the *molecular* diffusion coefficient on a weight basis.

Treybal (1955) presents eqn (58) for various conditions including flow inside pipes, flow transverse to cylinders, flow parallel to a flat surface. For the flow of gases parallel to plates, the constants in eqn (58) become $b_1 = 0.7$, $b_2 = 0.5$ and $b_3 = \frac{1}{3}$. For plates, d denotes the length of the plate in the direction of the flow. For flow of liquids between plates or inside tubes Timmins (1973) gives

$$b_1 = 0.01, \qquad b_2 = 0.9 \qquad \text{and } b_3 = 0.35$$

In evaporation processes the water flux is generally heat transfer controlled. For the effect of process conditions on heat transfer the reader is referred to the extensive literature on this subject. A concise review is given in the chapter on 'Heat transfer in evaporators' in Perry (1963). The equation for the overall heat transfer coefficient h_0 reads

$$\frac{1}{h_0} = \frac{1}{h_1} + \frac{\delta_w}{\lambda} + \frac{1}{h_2} + R_{f1} + R_{f2} \qquad (59)$$

where h is partial heat transfer coefficient; δ_w is wall thickness between heating medium and evaporating liquid; λ is heat conductivity of wall; R_f is fouling resistance. The subscripts 1 and 2 refer to inside and outside the separating wall respectively.

THE SELECTIVITY OF CONCENTRATION PROCESSES

The selectivity S of a concentration process can be defined by

$$S_{wi} = 1 - \frac{F_i/C_i}{F_w/C_w} \qquad (60)$$

For $S_{wi} = 0$ the ratio between the fluxes F_w and F_i is equal to the ratio between the initial concentrations C_w and C_i in the solution. This means that with respect to component i no concentration is obtained at all. For $S_{wi} < 0$ component i is withdrawn preferentially with respect to water. For $S_{wi} = 1$ component i is fully retained in the solution and the selectivity for that component is absolute.

Relations for the selectivity will now be given for dissolved solids and aromas for evaporation, pervaporation, direct osmosis and reverse osmosis.

Selectivity in evaporation

Dissolved solids: $$S_{wd} = 1 \qquad (61)$$

Aromas: $\alpha_{aw} > 5$ which means $C_a^{1*gb} \to 0$

$$S_{wa} = 1 - \frac{k_a^1}{k_w^1} \cdot \frac{(1 - C_w^1/\rho^1)\ln}{(1 - C_w^{1*(gb)}/C_w^{1b})} \qquad (62)$$

In eqn (62) $C_w^{1*(gb)}$ is the water concentration in the liquid phase in equilibrium with the water concentration in the bulk of the gas phase, C_w^{gb}. It is obvious from this equation that the selectivity increases with increasing value of C_w^1 and decreasing water concentration in the gas phase. For a rather dry gas atmosphere and after substitution of eqn (56) it follows by approximation

$$S_{wa} = 1 - \left(\frac{D_a^1}{D_w^1}\right)^{2/3} \approx 0 \qquad (62a)$$

Selectivity in pervaporation

Dissolved solids: $S_{wd} = 1$

Aromas: $\alpha_{aw} > 5$ and dry gas atmosphere

$$S_{wa} = 1 - \frac{\dfrac{(1 - C_w^{\,l}/\rho^l)\ln}{k_w^{\,l}} + \dfrac{(1 - C_w^{\,m}/\rho^m)\ln}{k_w^{\,m}} \cdot M_w^{\,m}}{\dfrac{1}{k_a^{\,l}} + \dfrac{M_a^{\,m}}{k_a^{\,m}}} \tag{63}$$

Also for pervaporation the selectivity increases with increasing water concentration in the liquid phase and decreasing water concentration in the gas phase. At a high value of M^m/k^m, eqn (63) can be approximated by

$$S_{wa} = 1 - \frac{k_a^{\,m}/M_a^{\,m}}{k_w^{\,m}/M_w^{\,m}} \tag{63a}$$

Selectivity in direct osmosis

The exact relation can be derived by substituting the complete relations for F_d/C_d and F_w/C_w in eqn (60). For high values of k^m/M^m the selectivity becomes for

Dissolved solids

$$S_{wd} = 1 - \frac{k_d^{\,m}/M_d^{\,m}}{k_w^{\,m}/M_w^{\,m}} \tag{63b}$$

Aromas

The selectivity for aromas is given by relation (63a).

Selectivity in reverse osmosis and ultrafiltration

Dissolved solids

$$S_{wd} = \frac{1 - \dfrac{k_d^{\,m}}{K_d^{\,m}}\left[1 + (1 - C_w^{\,I}/\rho^I)\dfrac{\Delta P - \pi_w^{\,Ib}}{C_d^{\,Ib}} \cdot \dfrac{k_{ow}^{\,P}}{k_w^{\,I}}\right]}{1 - k_{ow}^{\,P}\left[\dfrac{(\Delta P - \pi_w^{\,Ib})}{C_w^{\,Ib}}\right]} \tag{64}$$

From eqn (64) it is evident that the selectivity for dissolved solids increases with decreasing value of

$$\frac{k_d^{\,m}/M_d^{\,m}}{k_w^{\,m}/M_w^{\,m}}$$

with increasing value of C_w^{Ib} and especially with increasing pressure difference ΔP over the membrane.

Aromas

The selectivity for aromas is described by a relation similar to eqn (64).

CHEMICAL DETERIORATION

Loss of quality due to deteriorative chemical reactions is at low conversions almost directly proportional to the residence time of the reactants in the process apparatus and exponentially dependent on the process temperature. The conversion or quality loss is expressed by the relation

$$-\frac{dC_i}{dt} = k_r C_i \tag{65}$$

where k_r is the conversion rate constant, t is the residence time and C_i is the concentration of the reacting component. The temperature dependence of k can be expressed either by the Arrhenius equation

$$k_r = k_r^{\circ} \exp(-E_r/RT) \tag{66}$$

where k_r° is a constant; E_r the activation energy of the reaction; R the gas constant and T the absolute temperature, or by an exponential function of the form

$$k_r = k_r^{\circ} \exp aT \tag{67}$$

where a is the modified activation energy. It depends on the reaction mechanism whether eqn (66) or eqn (67) fits the experiments better. Equation (67) is preferred for the inactivation of micro-organisms and the thermal inactivation of thiamine and phosphatase. The activation energies are related by

$$a = \frac{E}{RT_1T_2} \tag{68}$$

So over a limited temperature range both relations show about the same temperature dependence.

The E values for enzymes are very high. From the experimental data of Verhey (1973) on phosphatase inactivation in raw skim-milk an E value of 104 kcal/mole has been calculated. Second to the enzymes is the physico-chemical decomposition of proteins. They have values which are generally between 40 and 70 kcal/mol. According to Burton and Jayne-Williams (1962), for non-enzymatic browning reactions E amounts to about 20 kcal/mol and for vitamin (riboflavin) destruction to about 15. Generally E lies between 10 and 30 for thermally sensitive liquid foods.

Because of the high values of E the chemical decomposition in freeze-concentration is negligible. For $E = 20$ kcal/mol the residence time of the liquid to be concentrated at -3°C may be a factor 500 longer compared with the residence time at a temperature of 50°C. Still quite low if not negligible are the thermal effects in membrane concentration processes where generally a temperature of 30°C is not exceeded. The effect may become important in evaporation processes.

Thermal degradation is always the result of a number of consecutive reaction steps. The further the reaction proceeds the stronger its adverse effect upon quality. In order to reduce these consecutive reactions to a minimum the residence time distribution of the liquid in the concentration apparatus has to be as small as possible. To this end plug flow must be aimed at. In order to reduce the absolute conversion, the mean residence time τ and the process temperature have also to be minimised. Because according to eqn (65) the thermal deterioration is directly proportional to τ and the conversion rate constant k_r, the effect of temperature on quality loss, ϕ_l, is for low conversions given by

$$E_{\phi l} = E_\tau \cdot E_r \qquad (69)$$

For a given feed composition, a constant concentration factor and a given concentration apparatus the residence time τ of the liquid in the apparatus at temperature T is inversely proportional to the dehydration capacity Q. The capacity in turn is a function of temperature and liquid viscosity.

Membrane processes

The capacity of a membrane apparatus is mainly influenced by the viscosity of the liquid. If the membrane itself is the controlling factor, the capacity will be inversely proportional to the viscosity of water. If due to strong concentration polarisation the liquid is the controlling factor, the capacity is according to eqn (58) at $b_2 = 0.9$ and $b_3 = 0.35$ about inversely proportional to the viscosity in the concentrated film. For the viscosity of water the energy of activation is about -4 kcal/mol in the temperature range 10–40°C. The energy of activation of a 60 wt% sucrose solution over the same temperature range is about -10. Because E_r varies between -10 and 30, a value in the range 0–25 follows from eqn (69) for the activation energy of quality loss. Consequently a decrease in temperature will, notwithstanding a decrease in capacity, always result in a decrease in quality loss by chemical deteriorative reactions.

Evaporation processes

For a given feed, constant concentration factor and given evaporator the residence time is inversely proportional to the overall heat transfer coefficient h_0. To the activation energy of h_0, values can be attributed between 1.5 and 6.7. So for the activation energy of the quality loss a value in the range 5–25 kcal/mol follows. Consequently also for evaporators a decrease of the process temperature or a decrease of the operating pressure also results in a decrease of thermal degradation.

PERFORMANCE AND ENERGY CONSUMPTION

Maximum attainable concentration

Drying is generally more expensive than concentration. Consequently the economics of preconcentration plus drying are improved if the product can be preconcentrated to as high a value as possible. The maximum concentrations attainable increase in the sequence reverse osmosis (ultrafiltration), freeze-concentration, pervaporation, evaporation.

Reverse osmosis

Depending on the mean molecular weight of the dissolved solids the maximum solute concentrate lies in the range 20–35 wt%. The maximum concentration for depectinised fruit juices is 30–35 wt%.

Ultrafiltration

Maximum concentration between 20 and 30 wt%.

Freeze-concentration

Depending on the viscosity of the concentrated product the maximum concentration is in the range 35–50 wt%, being about 40 wt% for fruit juices.

Pervaporation

The maximum concentration is strongly dependent on evaporation temperature. At about 50°C the range will be about 35–40 wt%.

Evaporation

By far the highest concentrations can be obtained in evaporation. The maximum concentration is in the range 50–80 wt%.

Energy consumption

The energy consumption will be calculated for the concentration of a 10 wt% solution to 35 wt% at a dehydration rate of 1000 kg/h. In order to compare the different processes, the energy consumption will be expressed in ton steam equivalents (TSE) needed for the withdrawal of 1000 kg water:

$$1\ \text{TSE} = \frac{\text{energy costs per 1000 kg water removal}}{\text{costs of 1000 kg steam}}$$

The steam equivalent is almost independent of heat energy prices and

TABLE II

ENERGY CONSUMPTIONS OF CONCENTRATION PROCESSES EXPRESSED IN TON STEAM EQUIVALENTS PER TON WATER REMOVAL

Process	*Steam equivalents* (tons)
Ultrafiltration	
efficiency[a] 75%	
pressure 2.5 atm	0.001
Reverse osmosis	
efficiency 75%	
presure 75 atm	0.028
Freeze-concentration	
efficiency 80%	
ΔT (difference between condensor and	
evaporator) = 20°C	0.090
ΔT = 40°C	0.196
ΔT = 60°C	0.386
Pervaporation	
efficiency 90%	1.111
Evaporation	
efficiency 90%	
single effect (without aroma recovery)	1.111
double effect (without aroma recovery)	0.555
triple effect (without aroma recovery)	0.370
single effect (with 90% aroma recovery by distill. α_{aw} = 4)	1.257
double effect (with 90% aroma recovery by distill. α_{aw} = 4)	0.701
triple effect (with 90% aroma recovery by distill. α_{aw} = 4)	0.510

[a] The efficiency is defined as the ratio of theoretical energy consumption to the actual energy consumption.

of currencies. For the conversion of electrical energy costs to thermal energy (steam) costs we used the ratio

$$\frac{\text{costs of 1 ton steam}}{\text{costs of 1 kWh}} = 143$$

The energy consumptions of the different processes are presented in Table II.

It is evident from Table II that by far the lowest energy consumption is attained in ultrafiltration and after ultrafiltration in reverse osmosis. The second best appears to be freeze-concentration. The highest consumption is needed for pervaporation and single effect evaporation.

REFERENCES

Ball, C. O. and Olson, F. C. W. (1957). *Sterilization in Food Technology*, McGraw-Hill, p. 512.

Bird, R. B., Stewart, W. E. and Lightfoot, E. N. (1963). *Transport Phenomena*, Wiley, New York.

Bomben, J. L. and Merson, R. L. (1969). Symposium on Flavors for Processed Foods. Sixty-second annual meeting of the Am. Inst. of Chem. Eng., Washington D.C.

Burton, H. and Jayne-Williams, D. J. (1962). *Recent Advances in Food Science*, Vol. 2, Butterworths, London, p. 106–116.

Chandrasekaran, S. K. and King, C. J. (1971). *Chem. Eng. Progr. Symp.*, Ser. 67, No. 107.

Chilton, T. H. and Colburn, A. P. (1934). *Ind. Eng. Chem.*, **26**, 1183.

Fisch, B. P. (1958). *Fundamental Aspects of the Dehydration of Foodstuffs*, Society of Chemical Industry, London, p. 143.

Glasstone, S. and Pound, A. (1925). *J. Chem. Soc.*, **127**, 2660.

Hougen, O. A. and Watson, K. M. (1947). *Chemical Process Principles*, Wiley, New York.

Lacey, R. E. (1972). *Chemical Engineering*, **4**, 56.

Lightfoot, E. N., Cussler, E. L. and Rettig, R. L. (1962). *A.I.Ch.E.J.*, **8**, 708.

Madsen, R. F. (1973). Sixth International Course on Freeze-drying and Advanced Food Technology, Bürgenstock, Lucerne, Switzerland, June, 1973.

Van Ness, H. C. (1964). *Chemical Thermodynamics of Non-electrolyte Solutions*, Pergamon Press, Oxford.

Perry, J. H. (1963). *Chemical Engineers Handbook*, 4th edn., McGraw-Hill, New York, pp. 11–35.

Pierotti, G. J., Deal, C. A. and Derr, E. L. (1959). *Ind. Eng. Chem.*, **51**, 95.

Robinson, R. A. and Stokes, R. H. (1955). *Electrolyte Solutions*, Butterworths, London.

Sourirajan, S. (1970). *Reverse Osmosis*, Logos Press, London, Table A-22.

Thijssen, H. A. C. (1966). Annex 1966-3 bulletin of the Int. Inst. of Refrigeration.

Thijssen, H. A. C. (1970). *J. Food Technology*, **5**, 211.

Timmins, R. (1973). Sixth International Course on Freeze-drying and Advanced Food Technology, Organized by Int. Inst. of Refrigeration, Bürgenstock, Switzerland, June.

Treybal, R. E. (1955). *Mass Transfer Operations*, McGraw-Hill, New York.

Verhey, J. G. P. (1973). *Neth. Milk Dairy J.*, **27**, 3.

Wilson, G. M. and Deal, C. A. (1962). *Ind. Eng. Chem. Fundamentals*, **1**, 20.

DISCUSSION

R. J. Clarke (General Foods Ltd., U.K.): Could you tell us a little more about the relative volatility at infinite dilution and its determination? It is very necessary to know it in order to determine the aroma retentions in evaporation.

H. A. C. Thijssen: The experimental determination of activity coefficients and of relative volatilities is not too difficult. If you are interested in the retention of say, methyl mercaptan, ppm quantities of it are introduced into the solution in a tube with a small gas head space. The tube is placed in a thermostat and shaken for some time, and after a short period phase equilibrium between the gas and liquid phases is obtained. Thereupon the concentration of methyl mercaptan is determined gas chromatographically in the gas head space. From this the relative volatility and activity coefficient can be calculated as a function of concentration. Does that answer your question?

R. J. Clarke: Yes, thank you, basically it does, except one easily finds published figures of the relative volatility for rather high concentrations of the aromatic components but not for infinite dilution.

H. A. C. Thijssen: You are absolutely right. Most data are given for high aromatic concentrations which are not relevant for practical conditions, but there are figures which I showed in Table I for extremely low concentrations in the ppm range. Professor King tells us that the concentrations are so low that a further decrease in concentration does not affect activity coefficient values.

A. S. Michaels (Alza Research): Further to this question, are you trying to establish some sort of correlation between relative and absolute volatilities and molecular weight of the aroma component? I think a more relevant quantity to use for predicting relative volatility is the solubility limit of the aroma component in the aqueous phase, because your activity coefficient is in inverse proportion to solubility by definition; and if the solubilities are low enough, then you can assume almost proportionality between activity and concentration up to the limit of solubility.

Another point is that relative volatility depends on the concentration of the other components in the aqueous solution, for instance sugar, and is directly dependent upon the effect of sugar on the solubility of the aromatic component in water. I would expect precise measurements of solubility limits could be helpful.

H. A. C. Thijssen: In fact my student, Dr. Bruin, measured the solubility limits of aroma components so as to calculate from them the activity coefficients. In my manuscript you will see that I have written: 'For components having a very low solubility in solution it can be derived from eqn. (14) that their activity coefficients are equal to the reciprocal of their molecular solubilities . . .'.

Norman Wookey (Tenstar): In some cases the situation is more complicated because the volatile component may be generated as well as removed during evaporation. I think that happens in concentrating glucose syrups where you get reactions generating undesirable components during the evaporation.

H. A. C. Thijssen: Yes, if you know the value of your reaction rate constant, the production rate of the component and the relative volatility of the component in the solution, you can calculate what the effect is of the process conditions upon the final concentration of the solution.

C. Judson King (University of California): Estimates of the solubility of these aroma compounds at low levels can be difficult. Measuring the relative volatility is a good way of finding the solubility which is otherwise difficult to come by.

We have done some measurements on limonene acetate which is an important compound in 'Surprise' juices, and it is interesting that the relative volatility becomes lower as the amount of dissolved solids increases. This is opposite to what your table shows. Apparently the volatilities of the more polar volatile compounds go up as the sugar content increases and with the non-polar compounds they go down.

H. A. C. Thijssen: Our measurements on coffee extract also differed from the data referred to; they showed a decrease in relative volatilities of the non-polar and polar components in coffee extract with increased solute concentration.

G. D. Saravacos (National Technical University, Greece): I should like to confirm what Dr. King has said: the high molecular compounds exhibit higher volatilities in water and as a general rule, if you add sugar, it increases the volatility. However, some experiments give opposite results. For instance, methyl anthranilate, the flavour component in grape juice, where by adding sugar (sucrose or glucose) or increasing the concentration increased volatility was not obtained. It may also be due to molecular structure and this compound is not very volatile, so volatility is not increased at all by adding more sugar.

H. A. C. Thijssen: This is again what Dr. King said. It is dependent upon the properties of the volatile components and in Table I we indicated the behaviour of rather polar aroma components.

E. Seltzer (Rutgers University): I would go further; there is a manufacturing process that has been in use for 35 years in the United States for aroma retention, in which hydrophillic materials are encapsulated in gums like gum arabic or gelatine or pectin. The hydrophilic materials are relatively insoluble in an aqueous matrix of the gum and by deliberately taking a high concentration of gums a low volatility is obtained. Generally this is about 30% on a weight basis of the gum solution, and one can spray dry or slab dry and get 90–92% retentions.

H. A. C. Thijssen: In these encapsulations you are aiming at forming little droplets with an impermeable skin so that this evaporation is diffusion controlled. Inside the skin the relative volatility of the aroma components can still be very high, 10 000 or so, and still good retentions can be achieved in spray drying. There are two different aspects: one, the tendency to evaporation indicated by relative volatility and, the other, the mobility of that component in the system to the evaporating surface; and in these gelatine or pectin systems there is a very rapid dry skin formation behaving like a selectively permeable membrane. Diffusion to the surface is thus very much restricted.

E. Seltzer: However, in determining the optimal amounts of the dispersed phase, the break point is around 30–35% for limonene acetate.

H. A. C. Thijssen: The membrane swells to a more open structure due to this limonene acetate or other components. The loss is not primarily due to a change in relative volatility, but a decrease in selectivity or an increase in permeability of the membrane to aromas like limonene.

S. A. Goldblith (M.I.T.): You have given a very useful summary on energy consumption for the different concentration processes. Have you done a

similar thing for different types of concentration plant in terms of capital investment cost and equipment capacities and then combined the two?
H. A. C. Thijssen: We did it only for freeze-concentration, and to some extent for a couple of evaporation processes, comparing the effect of different conditions upon the capital cost on the one hand and energy consumption on the other.

Fundamentals of Dehydration Processes

M. KAREL

Department of Nutrition and Food Science,
Massachusetts Institute of Technology,
Cambridge, Massachusetts, U.S.A.

NOTATION

(For specific units for each symbol consult equations in text.)

A	Area.
$\bar{B}$	Degree of browning calculated using average conditions.
B_{Σ}	Degree of browning calculated by integration of point conditions.
C, C_1	Constants.
D	Diffusivity.
D_1, D', D'', D_0	Diffusivities for specific conditions defined in text.
E, E_{D}	Energy of activation.
G_{I}	Rate of drying in constant rate period.
ΔH_{V}, $\Delta \bar{H}_{\mathrm{V}}$	Latent heats of vaporisation.
ΔH_{s}	Latent heat of sublimation.
K	Constant.
L	Thickness of slab.
L_{e}	Lewis number.
P^0	Vapour pressure.
R	Gas constant.
R_{d}, R_{s}, R_0	Resistances to sublimation.
T	Temperature.
T_{s}	Surface temperature.
T_{a}	Air temperature.
T_{w}	Wet bulb temperature.
T_{H}	Platen temperature.
a	Water activity.
b	Permeability of food to water vapour.
h	Film coefficient for heat transfer.
k_{d}	Thermal conductivity of dry layer.

k_i	Thermal conductivity of frozen layer.
k_g	Film coefficient of mass transfer.
m	Moisture content.
$m_c, m_e, m', m_l, m_h, m_i$	Specified moisture content.
n	Constant.
p	Partial pressure of water.
p_a	Partial pressure of water in air.
p_s	Partial pressure of water at slab surface.
p_i	Partial pressure of water at ice surface.
$r(t)$	Location of interface of 'dry' layer as a function of time.
t	Time.
w	Weight.
x	Distance from surface.
z	Thickness of 'dry' layer, in air drying.
α	Thermal diffusivity.
β	Constant.
ε	Emittance of food surface.
ε''	Loss factor.
μ	Slope of moisture content gradient with time.
ρ_s	Bulk density of solids.
ν	Frequency.
σ	Stefan–Boltzmann constant.

INTRODUCTION

Food preservation processes have in common their goal of extending the shelf life of foods to allow storage and convenient distribution. The first and most dangerous limitation of shelf life is due to activity of micro-organisms, hence the food preservation processes are designed to eliminate the danger of microbial spoilage, or at least to control it to avoid potential for health-threatening activities of bacteria, and other micro-organisms. Dehydration processes achieve this aim by lowering the availability of water to micro-organisms. In addition to microbial growth, chemical and physical processes may occur in foods during storage, with a potential for deterioration of quality. Control of water content affects the rate of these processes, and may contribute to extension of storage life.

A fundamental aspect of dehydration processes is therefore the relation between the amount and state of water in foods, and the effects of this water on biological, chemical and physical processes limiting the shelf life of foods. The next section of the present paper deals with this fundamental aspect of dehydration.

Removal of water from foods is the essence of dehydration processes, and this operation is a *combined heat* and *mass transfer operation*. A fundamental aspect of dehydration deals therefore with the relations governing the transport of mass and of heat during the various types of specific processes available for the drying of foods. In the present paper we present these relations in two idealised situations, which represent major types of dehydration: in the third section (p. 55) we deal with the *drying of a slab of food in air* at constant temperature and humidity, and in the fourth section (p. 60) we discuss the heat and mass transfer in several types of *freeze dehydration* situations.

An important fundamental aspect of food dehydration is the deterioration of food quality during the dehydration process itself. Dehydration, like other food preservation processes, is conducted under conditions which are often compromises between the desire to maximise the retention of desirable food attributes, such as colour, taste and nutrition, and on the other hand an equally justified desire to conduct the operation economically and efficiently. In the final section we deal with the influence of the process of dehydration on several food properties, including flavour retention. In particular, we present also an example of an approach to a quantitative analysis of quality losses during processing which may lead to improved process optimisation.

THE STATE OF WATER IN FOOD

The fundamental purpose of food dehydration is to lower the availability of water in the food to a level at which there is no danger of growth by undesirable micro-organisms. A secondary purpose is the lowering of the water content in order to minimise rates of chemical reactions, and to facilitate distribution and storage.

The availability of water for microbial growth and chemical activity is determined not only by the total water content but also by the nature of its binding to foods. The first subject to be considered therefore in an analysis of the fundamentals of dehydration is the nature of water binding in the food.

There is substantial evidence that of the total water present in food a portion is strongly bound to individual sites, and that an additional amount is bound less firmly, but is still not available as a solvent for various soluble food components. There is a variety of methods by which the degree of binding of water within the food can be established, and among these are:

(1) Determination of unfrozen water.
(2) Nuclear magnetic resonance.
(3) Dielectric properties of food.
(4) Sorption behaviour.

Additional methods of determination of water binding have in fact been proposed, but the methods listed above provide adequate insight into the pertinent properties of water in the food.

A number of authors have established that upon cooling water-containing foods to low temperatures, well below the freezing point of the food, a portion of the total water remains unfrozen. The most successful methods of analysis of unfrozen water have utilised differential thermal analysis. Food samples with different total water contents have been cooled to temperatures below $-60°C$ and then rewarmed in the DTA apparatus. At water contents at which only 'unfreezable water' is present there is an absence of DTA peaks, which in 'free water' are due to the absorption of the latent heat of fusion. By using a sufficient number of different water contents it is possible to determine quite precisely the point at which all is 'unfreezable'. Recent estimates by Duckworth (1971, 1972) were in the range of 0.13–0.46 g water per g of solids for food components. Typical foods had values between 0.2 and 0.3 g water per g of solids.

A view has often been expressed that the non-freezing fraction of water is in a sense already frozen by being ordered, having binding energies as great or greater than those in ice, and by having a low mobility. Recent research, however, with infra-red spectroscopy and nuclear magnetic resonance (Walter and Hope, 1971) indicates that the water remaining unfrozen does in fact possess substantial mobility and does not show the orderly structure expected of ice-like aggregates.

Wide line proton magnetic resonance is used routinely to determine water (and other liquids having a high proton content) in foods. The basis of its application is as follows: protons of molecules in liquid state, as a result of the more or less random motion of these molecules, experience a similar net magnetic field when placed in a magnet. They therefore give sharp and large signals in NMR determination, because they absorb radio energy at about the same frequency. In solids, however, interactions with neighbouring atoms change the proton response, and the absorptions by the many protons are spread out over a large frequency. Hence NMR signals are broad and shallow, and within the 'window' used to record the signal they may be negligible compared to liquid signal. A portion of the total water in foods shows signals intermediate in character between liquid and solid water, and if the total water content is known, NMR may be used to estimate 'bound water content' (Shanbhag *et al.*, 1970). The amount of water bound to hydrophilic food components is estimated at about 0.1–0.3 g/g of solids.

The amount of water bound by proteins, polymers, and foods can also be made by dielectric measurements. Dielectric properties of water molecules depend on the molecular environment of these molecules, and 'bound' water shows properties intermediate between the rigidly held dipoles of ice and the much more mobile liquid water molecules. The amount of bound water is, however, difficult to estimate by this method. Brey *et al.* (1969), working with proteins, reported 0.05–0.1 g/g solids; Roebuck *et al.* (1972) reported 0.3–0.4 g/g bound in starch.

The most successful method for studying properties of water is the study of sorption isotherms, that is curves relating the partial pressure of water in the food to the water content. In some cases it is more convenient to study the curves relating *water activity* (a) to the water content. Water activity is defined as the ratio of partial pressure of water in the food (p) to the vapour pressure of water (P^0) at the given temperature.

Typical isotherms in foods are S-shaped, and they may be approximated by a variety of mathematical relationships, some of which are listed in Table I.

TABLE I

SOME PROPOSED FOOD-WATER ISOTHERMS

Linear	$m = c_1 a + c_2$
Oswin	$m = c(a/1 - a)^n$
Kuhn	$m = c_1(\ln a)^n + c_2$
Fugassi	$m = \dfrac{c_1 a}{c_2 a(1 - a) + c_3 a}$

A portion of the total water content present in food is held strongly bound to specific sites. Foods contain many polar sites including, for instance, the hydroxyl groups of polysaccharides, carbonyl and amino groups of proteins, and others on which water can be held by hydrogen bonding or by ion–dipole bonds, or by other strong interactions. The most effective way of estimating the contribution of adsorption of specific sites to total water binding is the use of the B.E.T. isotherm. This mathematical relation, shown in eqn (1), is based on over-simplified assumptions but is extremely useful as an estimate of the so-called 'monolayer value', which we shall consider as equivalent to the amount of water held adsorbed on specific sites.

$$\frac{a}{m(1 - a)} = \frac{1}{m_1 C} + \frac{C - 1}{m_1 C} a \qquad (1)$$

where a = water activity; m = water content g H_2O/g solids; m_1 = monolayer value; C = constant. By obtaining isotherms at several different temperatures and then plotting the partial pressure of water in a food at a given water content as a function of the reciprocal of absolute temperature, it is possible to estimate the total heat of adsorption of water in food. (We define this heat of adsorption, $\Delta \overline{H}$, as the amount of heat required to remove 1 mol of water *at the given water content* from the food into the vapour phase.) This heat of adsorption includes the latent heat of vaporisation, but as shown in Fig. 1, at low water contents there is an additional contribution of adsorption. Reported maximum latent heats for water in foods range from about 11 kcal/mol to as much as 20 kcal/mol. Monolayer values and maximum heats of adsorption for typical food systems are shown in Table II.

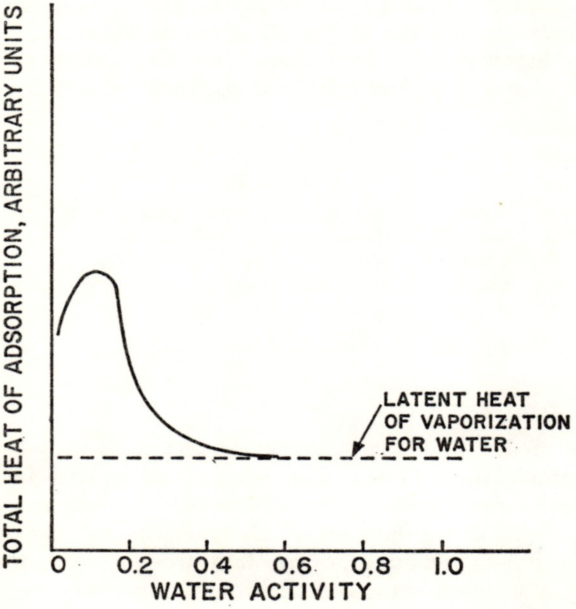

FIG. 1. Typical dependence of the total heat of sorption (heat of vaporisation + heat of adsorption) on water activity.

In addition to water adsorbed on specific sites there is a portion of the total water in which the vapour pressure is depressed by its being present in small capillaries. The relationship between water activity and radius of capillary is shown in eqn (2):

$$\ln (a) = \frac{-2\gamma}{r} \cdot \cos \theta \cdot C_1 \tag{2}$$

where γ = free surface energy of water; r = capillary radius; θ = contact angle; C_1 = constant.

Finally, water capable of acting as a solvent is subject to depression of vapour pressure due to solute. Ideally this vapour depression is given by

Raoult's law, but in foods there are usually substantial deviations from the ideal relationship. Table III presents data relating the water activity of water to concentrations of various solutes. If ideal relations are followed, then water activity should be numerically equal to mole fraction of water

TABLE II

SELECTED B.E.T. MONOLAYER VALUES (m_1) AND MAXIMUM HEATS OF ADSORPTION ($\Delta\bar{H}$) FOR FOOD AND FOOD COMPONENTS

Substance	m_1(*g/g solids*)	$\Delta\bar{H}$(*kcal/mol*)
Lactose amorphous	0.06	11.6
Dextran	0.09	12
Starch	0.11	14
Ribonuclease	0.06	16
Egg albumin	0.06	17
Spray dried skim milk	0.05	16
Potato flakes	0.06	
Shelled corn	0.05	12.5

TABLE III

WATER ACTIVITY OF AQUEOUS SOLUTIONS OF SOME POTENTIAL FOOD HUMECTANTS (22°C)

Solute	*Concentration (% weight/weight)*	*Approximate water activity*
Sucrose	58.4	0.90
	67 [a]	0.86
Glucose	47 [a]	0.92
Invert sugar	63 [a]	0.82
Sucrose 37.6% and invert sugar 62.4%	75 [a]	0.71
	9.3	0.94
NaCl	19.1	0.85
	27 [a]	0.74
Lactose	14.5 [a]	0.99
Maltose	48.8 [a]	0.95

[a] Saturated solution.

in the solution. Deviations from ideality may arise from several causes and are especially significant at high concentrations.

Not all of the water present in food acts as a solvent and recent work by Duckworth (1972) has demonstrated that not only all of the water bound on specific sites (monolayer water), but an additional fraction of the total water as well, do not dissolve solutes.

He used NMR techniques and noted that proton-containing solutes such as sugars give in solution NMR signals which differ from those given in the precipitated state. Duckworth (1972) found that each solute had a specific water activity at which it went into solution, and that the presence of other insoluble components in the food did not alter this activity. Thus he found that sucrose went into solution at an activity of 0.82 in several sucrose–polymer–water systems. The amount of water to which this activity corresponds varied from 0.11 to 0.34 g/g solids, depending on what polymer was used in the system. The monolayer value was well below this amount, corresponding to 20–45% of the non-solvent water. The monolayer value was about 55–75% of water not capable of dissolving urea, which went into solution at a lower activity than that at which sucrose became soluble.

Soluble substances present in foods may also undergo phase transformations and the occurrence of these phase changes may depend on the presence of water. Sorption of water in such systems is complicated, and may require consideration of kinetics as well as equilibrium of sorption. The considerations involved are illustrated in Fig. 2, which shows the sorption behaviour of sucrose in several states. Crystalline sucrose sorbs very little water until the water activity reaches approximately 0.8 and the sucrose begins to dissolve. When the drying procedures are sufficiently rapid to produce amorphous sucrose, however, exposure to increasing humidities results in water uptake, reaching sorption levels far higher than that of crystalline sucrose. This difference in behaviour is due to the higher internal area available for water sorption in the amorphous material, and greater ability of water to penetrate into the H-bonded structure, which is less regular than in sucrose crystals. However, the very adsorption of water results in the capability for breaking some H-bonds and imparting a mobility to the sugar molecules; and this mobility results eventually in the sucrose transforming from the metastable amorphous state to the more stable crystalline state. In this process the sugar loses water, often as a result of exposure to increasing relative humidity, because the recrystallisation that occurred extremely slowly at low humidities speeded up with transient water pickup at higher humidities (Fig. 2).

The knowledge of sorption behaviour of water in foods is pertinent to dehydration in the following respects.

(1) It determines the optimal water content and water activity to which the food must be dehydrated in order to assure adequate stability.
(2) It relates the concentration of water in food to its partial pressure, and this relation is of key importance in the analysis of mass and heat transfer during dehydration.
(3) As noted above the thermodynamic relationships involved in sorption may be used to estimate the amount of heat required to remove the water from food.

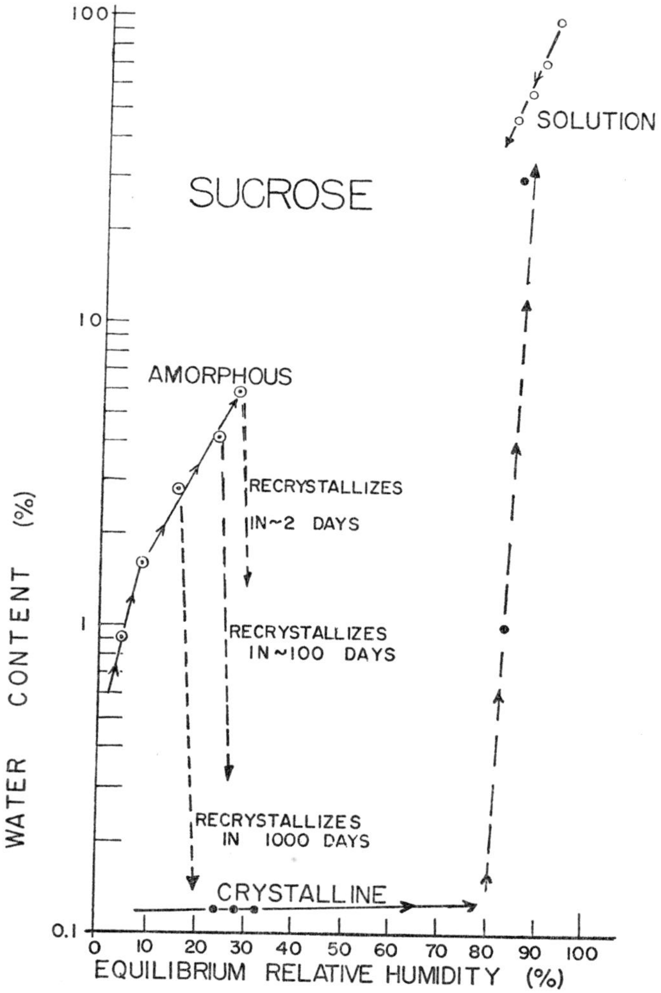

FIG. 2. Sorption behaviour of water in sucrose.

The various food deterioration mechanisms depend on water activity in different ways. Figure 3 shows schematically the dependence on water activity of bacterial and mould growth, of enzyme activity, of non-enzymatic browning, and of lipid oxidation. The growth of micro-organisms can often be considered directly and simply related to water activity. While there are complicating factors, it is usually feasible to accept

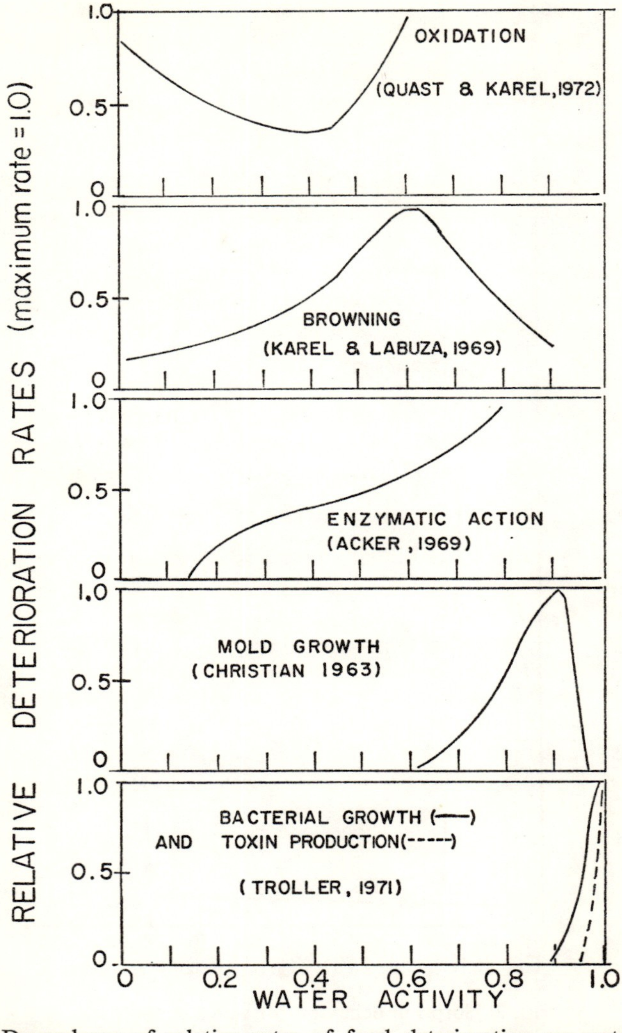

FIG. 3. Dependence of relative rates of food deterioration on water activity. *Bacterial growth:* S. aureus; *Toxin production:* Enterotoxin B (after Troller, J. A., *Appl. Microbiol.*, **21**, 435 (1971)). *Mould growth:* Xeromyces bisporus (after Christian, J. H. B., in *Recent Advances in Food Science* (ed. Leitch and Rhodes), Butterworths, London, 1963, p. 248). *Enzymatic activity:* Oat lipase acting on monoolein in ground oats (after Acker, L., *Z.f. Ernährungswissensch.*, Suppl. 8, 45 (1969)). *Browning:* Non-enzymatic browning of pork bites, after Karel, M. and Labuza, T. P., M.I.T. Report on Contract Research with U.S. Air Force School of Aerospace Medicine, Brooks AF Base, Texas, Contract No. F41-609-68-C-0015, 1969. *Oxidation:* Lipid oxidation in potato chips (after Quast, D. G. and Karel, M., *J. Food Sci.*, **37**, 584 (1972)).

this general rule which is due to the classical work of Scott and his co-workers (Scott, 1957). Chemical stability of foods is related to water content in more complicated ways (Karel, 1973).

DEHYDRATION OF A SLAB BY AIR-DRYING

Dehydration of foods requires simultaneous transfer of mass and of heat, and this operation can be achieved in different types of equipment. Equipment may be classified in a variety of ways including by method of heat transfer, by pressure level at which the dehydration is conducted, or by the method by which the food is conveyed. The present paper, however, is limited to fundamentals, and consideration will be given only to two idealised cases.

(1) Dehydration of a slab in air, with heat transfer by convection, and with the temperature and relative humidity of the air kept constant.
(2) Freeze dehydration of a slab, with heat transport occurring in several ways.

In air drying under constant environmental conditions the process of drying may be divided into a 'constant rate' period and one or more 'falling rate' periods. In this respect we may divide materials to be dehydrated into two categories, non-hygroscopic and hygroscopic. Our definition of a non-hygroscopic material shall be that the partial pressure of water in the material is equal to the vapour pressure of water. In hygroscopic materials, however, the partial pressure of water becomes *less* than the vapour pressure of water at some critical level of moisture which we shall refer to as m_h. We can define therefore (m_i = initial moisture content) for non-hygroscopic foods:

$$p = P^0 \quad \text{for} \quad 0 < m \leq m_i$$

for hygroscopic foods:

$$p = P^0 \quad \text{for} \quad m_h < m \leq m_i$$
$$0 < p < P^0 \quad \text{for} \quad 0 < m \leq m_h$$

We can describe the 'constant rate' period as one in which the surface of the slab is maintained at a moisture level assuring that $p = P^0$ at the wet bulb temperature T_w. This will be true as long as the moisture content at the surface is greater than zero for non-hygroscopic materials, or greater than m_h for hygroscopic ones.

Under these conditions, the resistance to heat and mass transfer is located solely in the air stream, and since in our idealised system the conditions of the air are constant, this resistance does not change with time. The driving force for mass transfer ($p - p_a$) and that for heat transfer ($T_a - T_w$) also remain constant, hence the constant rate period

$$-\frac{dw}{dt} = Ah(T_a - T_w)(\Delta H_v)^{-1} \tag{3}$$

where w = weight of slab (lb); h = film coefficient of heat transfer (Btu/h/ft/°F); T_a = air temperature (°F); T_w = wet bulb temperature of air (°F); ΔH_v = latent heat of vaporisation of water (Btu/lb); t = time (h); A = area of slab (ft^2).

$$-\frac{dw}{dt} = Ak_g(P^0 - p_a) \tag{4}$$

where k_g = gas coefficient of mass transfer ($lb/h/torr/ft^2$); p_a = partial pressure of water in air (torr). The constant rate period continues as long as the supply of water to the surface suffices to maintain the saturation of the surface to give a constant surface temperature, and level of water pressure equal to P^0 at the wet bulb temperature.

The constant rate period ends when this condition no longer applies. One explanation of the situation at the end of this period is that at this point 'dry' patches appear at the surface, and thus the area for the evaporative process decreases (Harmathy, 1969). Our approach, however, shall assume that in our infinite slab geometry the conditions are uniform at any given slab depth, including the depth zero, corresponding to the surface level.

In our idealised system geometry we can divide the drying of hygroscopic foods into three periods, as represented schematically in Fig. 4, and that of non-hygroscopic into two periods.

The first period is the constant rate period with constant drying rate G_I and during this period the water removed from the surface is constantly resupplied by capillary flow to the surface, as shown by eqn (5).

$$G_I = -\frac{dw}{dt} = k_g A(P^0 - p_a) = -D_L \rho_s A \frac{dm}{dx} \tag{5}$$

where D_L = overall internal liquid diffusivity in slab (ft^2/h); ρ_s = bulk density of dry solids in slab (lb/ft^3); m = moisture content (lb H_2O/lb solids); x = distance (ft).

The end of the constant rate period occurs when the internal water supply is inadequate to saturate the surface. Krischer and Kröll (1963) and Görling (1958) present the end of 'constant rate' period data in terms of 'break point' curves. Under idealised conditions (Krischer and Kröll, 1963) the average moisture content (critical moisture) of the slab at which the constant rate period ends is given by

$$m_c = \frac{1}{3}\frac{k_g(P^0 - p_a)L}{D_L \rho_s} \tag{6}$$

where L = slab thickness (ft); m_c = critical moisture content (lb/lb). During the first falling period we assume that the transport from a level in the slab at which the saturation conditions still prevail ($p = P^0$) to the surface, occurs by transport in the gas phase, and from the surface to bulk

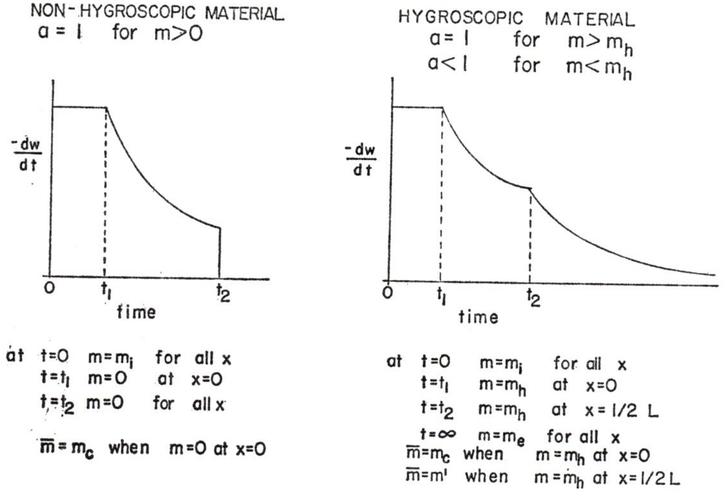

FIG. 4. Schematic representation of drying periods in non-hygroscopic and hygroscopic materials dried in air with constant temperature and humidity.

of the air we have the transport through the gas film in the air. Hence the drying rate is given by

$$-\frac{\mathrm{d}w}{\mathrm{d}t} = \frac{A}{\frac{z}{b} + \frac{1}{k_g}} \cdot (P^0 - p_a) \tag{7}$$

where b = permeability of the food material to vapour flow (lb ft/ft²/torr/h). The rate is assumed to fall during this period due to the increasing internal resistance, the cause of which is the increase in distance z from the saturation surface within the slab to the slab surface.

Finally we reach the end of the second period when the centre of the slab no longer has enough water to maintain a partial pressure equal to P^0. In non-hygroscopic foods, this corresponds to the end of the drying; in hygroscopic foods this point corresponds to the reaching in the centre of the slab of the local moisture content m_h.

In the case of non-hygroscopic foods, the drying is completed at the end of the second period (the first falling rate period), but in the case of hygroscopic foods we encounter the third period during which the partial pressure of water in the food is everywhere below the saturation level P^0. In this situation the drying occurs by desorption throughout the food and the usual idealised assumption is to consider that at the beginning of the third period the moisture content is uniform throughout the slab. Of course this assumption is patently untrue, but it aids in evaluation of the differential eqn (8), which governs the desorption in this third period.

$$\frac{\mathrm{d}m}{\mathrm{d}t} = \frac{b}{\rho_s}\frac{\mathrm{d}^2 p}{\mathrm{d}x^2} \tag{8}$$

If an assumption is made that the relationship between p and m is linear (another patently untrue, but convenient oversimplification), we can write eqn (9):

$$\frac{\mathrm{d}m}{\mathrm{d}t} = \frac{b\beta}{\rho_s}\frac{\mathrm{d}^2 m}{\mathrm{d}x^2} = D\frac{\mathrm{d}^2 m}{\mathrm{d}x^2} \tag{9}$$

where β = a linear isotherm constant relating p to m; D = effective diffusion coefficient for water vapour in food during the third drying period.

The desorption process which occurs during the third period of drying has been treated thoroughly by previous authors, including King (1968). Assuming that mass transfer occurs only in the vapour phase, and no shrinkage, as well as a constant total latent heat of vaporisation, but taking into consideration the dependence of D on process variables, King (1968) suggests the relation

$$\frac{\delta m}{\delta t} = \frac{\delta}{\delta x}\left(D\frac{\delta m}{\delta x}\right) \tag{10}$$

which is of course equivalent to eqn (9) when D is constant. King (1968), however, notes that D depends on process variables, and presents a series of equations representing different cases of desorption including:

(1) internal resistance controlling;
(2) external resistance controlling;
(2) both internal and external resistances important.

The usual practice in the literature, however, has been to assume that the effective diffusion coefficient D is constant over the entire falling rate period, or at least over significant portions of this period, so that the drying process can be represented by one or more straight lines when the quantity $\ln(m - m_e)$ is plotted against time (Jason, 1958) (m_e being the equilibrium moisture content). These straight lines are due to the assumed validity of eqn (11) which represents the solution of eqn (10) using simplified assumptions including: internal resistance control, isotropic medium, D constant over each segment of the drying period.

$$\frac{m - m_e}{m' - m_e} = \frac{8}{\pi^2} \exp\left(-\frac{\pi^2}{4}\frac{D}{L^2}t\right) \tag{11}$$

where m = average moisture content; m' = moisture content at the beginning of the drying segment. The problem of applying eqn (11) lies in lack of reliable data on effective diffusion coefficients for foods. Typical values of D observed in the literature have ranged from 10^{-8} cm²/s to 10^{-5} cm²/s, often for very similar materials. It should be noted, however, that often the values of D have been obtained from drying data in which the assumptions underlying eqn (11) were not realised. Thus King (1968) states that the values of D reported by Saravacos and Charm (1962) and by Fish (1958) have been apparently derived from experiments in which external resistances were of significance.

Dehydration in the falling rate period is not only an unsteady state process with respect to moisture gradients, but also with respect to temperature. The temperature at the beginning of the falling rate period is equal to T_w, and rises through the falling rate period toward the dry bulb temperature of air T_a. The relative effects of temperature and moisture gradients inside the food may as a first approximation be related to the Lewis number ($L_e \equiv \alpha/D$), where α is the thermal diffusivity in units consistent with those of D. Young (1969) noted that when L_e is greater than 60 the thermal gradients can be neglected. In fact a number of investigators, including Jason (1958), Chirife (1971) and Vaccarezza *et al.* (1971), have measured temperature distribution in foods during drying and found little or no temperature gradients. Harmathy (1969) studied theoretical distribution of moisture and temperature during drying of a porous slab, and found negligible temperature gradients. It is therefore feasible in many cases to treat dehydration of a slab in air having constant wet and dry bulb temperatures as if the temperature of the slab varied with time, but not with position within the slab.

A more serious and significant deviation from the assumed idealised situation lies in the changes of geometry and structure of drying food materials.

In dehydration processes other than freeze-drying structural changes in the food materials occur freely, since there is a possibility for liquid flow, solute redistribution and shrinkage. As a result the mass transport properties change drastically during the dehydration process. Fish (1958) reports that in scalded potatoes the diffusion coefficient for water is 10^{-8}–10^{-7} cm²/s when the moisture is still at a level of 15–20% (dry basis), but that it drops to less than 10^{-10} when moisture is less than 1%. Similar data were reported for other materials, but in some foods diffusion coefficients do not fall off as sharply with decreasing water content as in the case of potatoes noted by Fish (1958). Thus Jason (1958) observed that fish muscle dehydration behaviour could be described by the use of two diffusion coefficients, the first (D') valid down to about 5% free water and the second (D'') down to almost complete dehydration. Jason (1958) found that D' was about 2–4 × 10^{-6} cm²/s for various species of fish, and D'' 0.1–1.0 × 10^{-6} cm²/s.

However, mass transport in air drying depends on drying conditions and cannot be described adequately by constant diffusion coefficients.

FREEZE-DRYING OF A SLAB

Principal types of transport

Freeze-drying, like dehydration, is a coupled mass transfer and heat transfer process. The rate of sublimation is given by

$$G = \frac{A(p_i - p_c)}{R_d + R_s + R_0} \tag{12}$$

where p_i = partial pressure of ice inside the food; p_c = partial pressure of water in the condenser; G = rate of sublimation; R_d = resistance of the 'dry' layer in the food; R_s = resistance of space between food and condenser; R_0 = constant. At the same time, the heat of sublimation ΔH_s must be supplied and therefore

$$G = q/\Delta H_s \tag{13}$$

We shall consider three cases which represent the three basic types of possibilities in vacuum freeze-drying (Fig. 5).

(1) Heat transfer and mass transfer pass through the same path (dry layer), but in opposite directions.
(2) Heat transfer through the frozen layer, mass transfer through the dry layer.
(3) Heat generation within the ice (by microwaves), mass transfer through the dry layer.

An additional possibility closely related to the first case above is freeze-drying at atmospheric pressure, rather than in a vacuum.

Heat and mass transfer through the dry layer

Consider the following case which is simplified, but nevertheless typical of most vacuum freeze-drying operations. The material to be dried is heated by radiation to the dry surface, and its internal frozen layer temperature is determined by the balance between heat and mass transfer. For simplicity's sake, we shall consider a slab geometry with negligible end effects. We shall assume that the maximum allowable surface temperature T_s is reached instantaneously and that the heat output of the external heat supply is adjusted in such a manner as to maintain T_s constant throughout the drying cycle. We also assume that the partial pressure of water in the drying chamber, p_s, is constant, and that all of the heat is used for sublimation of water vapour.

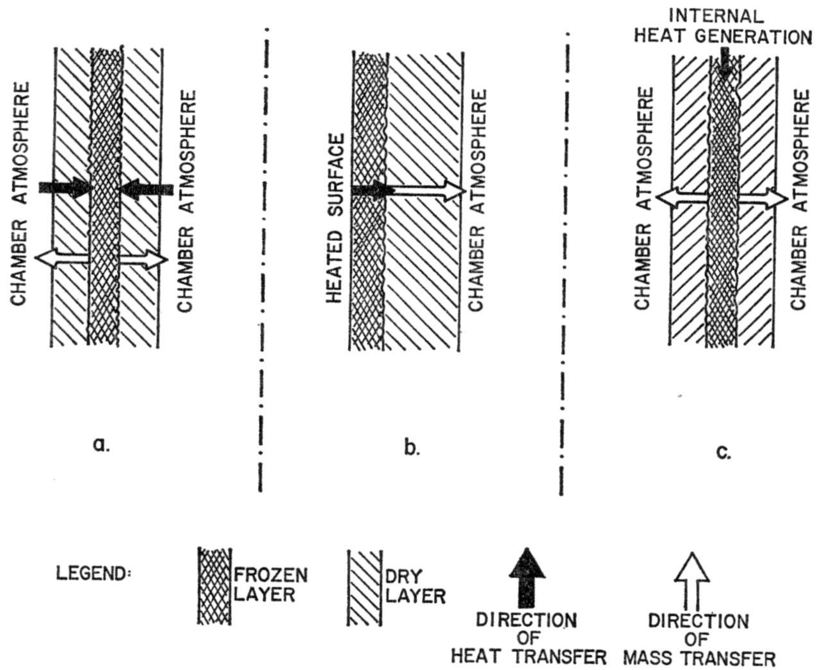

FIG. 5. Types of heat transfer defining different types of freeze-drying: (a) Heat transfer by conduction through dry layer; (b) heat transfer through frozen layer; (c) internal heat generation by microwaves.

Under these conditions, the heat transfer at any instant is given by:

$$q = Ak_d(T_s - T_i)x_d^{-1} \tag{14}$$

where x_d = thickness of dry layer (ft); k_d = thermal conductivity of dry layer (Btu/ft/h/°F); T_i = temperature of ice surface within the food (°F). The sublimation rate is given by

$$G = -\frac{dw}{dt} = Ab(p_i - p_s)x_d^{-1} \tag{15}$$

where w = weight of water in slab (lb); t = time; b = permeability of dry layer (lb/ft/h/torr); p_s = partial pressure of water at dry layer surface (torr). At the same time, given slab geometry, and assuming that at the ice–dry layer interface the moisture content drops from the initial value of m_i to the final value m_f, we obtain a relation between loss of weight and the rate of recession of the interface.

$$-\frac{\mathrm{d}w}{\mathrm{d}t} = A\rho(m_i - m_f)\frac{\mathrm{d}x_d}{\mathrm{d}t} \tag{16}$$

where ρ = bulk density of solids in the drying slab (lb/ft^3); m_i = initial moisture content (lb water/lb solids); m_f = final moisture content (lb water/lb solids). By combining eqns (15) and (16) we obtain eqn (17).

$$x_d\,\mathrm{d}x_d = \left(\frac{b}{\rho(m_i - m_f)}\right)\cdot(p_i - p_s)\,\mathrm{d}t \tag{17}$$

We can make the assumption that the rate of heat transfer is equal to the rate of sublimation multiplied by the latent heat of sublimation:

$$Ak_d(T_s - T_i)x_d^{-1} = Ab(p_i - p_s)x_d^{-1}\Delta H_s \tag{18}$$

and after simplification we have a relation between the pressure and temperature:

$$p_i = p_s + \frac{k_d}{b\Delta H_s}T_s - \frac{k_d}{b\Delta H_s}T_i \tag{19}$$

Since we assume that p_s, b, ΔH_s, k_d and T_s are all constant, we have a linear equation relating p_i with T_i as shown in Fig. 6. In that same figure, we also show the thermodynamic relation between p_i and T_i. We see that there is only one point at which curves representing the two equations meet. This means that if the assumptions inherent in our analysis are true, the frozen layer temperature T_i will remain constant throughout the drying cycle.

As a consequence, eqn (19) can be readily integrated to calculate the freeze-drying time, because it contains only the two variables x_d and t.

$$\int_0^{L/2} x_d\,\mathrm{d}x_d = \frac{b(p_i - p_s)}{\rho(m_i - m_f)}\int_0^{t_d}\mathrm{d}t \tag{20}$$

and the drying time is given by

$$t_d = \frac{L^2\rho(m_i - m_f)}{8b(p_i - p_s)} \tag{21}$$

where L = thickness of slab (ft); t_d = drying time. Because of the equivalence of heat and mass transfer we could equally well integrate the heat transfer equation (eqn 14) and have a corresponding equation for drying time (eqn 22).

$$t = \frac{L^2 \rho (m_i - m_f) \Delta H_s}{8 k_d (T_s - T_i)} \tag{22}$$

We see therefore that drying time depends on the following variables:

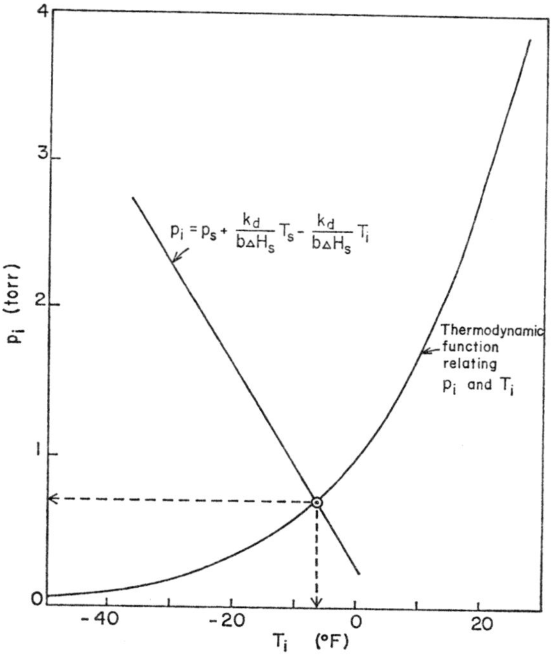

FIG. 6. Graphical determination of temperature and partial pressure of ice at the frozen layer interface.

maximum permissible surface temperature (T_s);
initial and final moisture contents (m_i, m_f);
bulk density of solids (ρ);
latent heat of sublimation (ΔH_s);
thickness of the slab (L);
thermal conductivity of the dry layer (k_d);
permeability of the dry layer (b).

Frozen layer temperature depends neither on the overall thickness of the drying material, nor on the thickness of the layer already dry.

During freeze-drying with heat transfer by radiation to the dry layer surface, and by conduction through the dry layer, the surface temperature is ideally brought rapidly to the maximum permissible level, and maintained at this level by changing the radiator temperature in accordance with a suitable programme. In practice, the changes are based on an experimentally determined empirical programme, but the changes could be made by feedback devices. The maximum surface temperature is usually dictated by quality considerations, especially flavour and colour changes.

The ice temperature is determined by interactions between surface temperature, chamber vapour pressure p_s and the dry layer properties

TABLE IV

SOME TYPICAL FROZEN LAYER TEMPERATURES AND SURFACE TEMPERATURES IN FREEZE-DRYING

	Temperature (°C)	
Food	*Frozen layer*	*Surface*
Chicken dice	−20	60
Orange juice	−45	45
Shrimp	−20	75
Beef	−15	60

k_d and b. The frozen layer temperature must be maintained below a critical level which depends on the nature of the product, and on its thermal history. Ideally, it should also be below the eutectic melting point, which may be in some cases 50°F or more below the melting point of ice. Typical ice temperatures existing in freeze-drying of foods under conditions in which the total pressure was primarily due to water vapour, and the heat transfer took place via the dry layer, are shown in Table IV.

The above idealised situation may be brought closer to reality by assuming that the freeze-drying time may be divided into three periods. During the first, the platen temperature remains constant, while the surface temperature increases to the maximum permissible level. The interface between the ice and the dry layer is assumed to be at a constant level.

During the second period the surface is maintained at a constant temperature (maximum permissible surface temperature, T_s^*) by gradually decreasing the platen temperature. The interface temperature is assumed to remain constant. In the third period the ice disappears and the internal temperature rises, an assumption being made that the heat supply goes only into raising of the dry layer temperature. (Heat of desorption for any water which is desorbed from the dry layer during this period is not considered.)

During the first period, given specified platen temperatures the solution to be provided is usually the rate of recession of the interface, and the rate of increase of the surface temperature. For the case of an infinite slab with constant dry layer properties the heat flow is governed by the following equations.

The temperature in the dry region satisfies eqn (23):

$$\alpha \frac{\delta^2 T}{\delta x^2} = \frac{\delta T}{\delta t} \quad \text{for} \quad 0 \leq x \leq r(t) \tag{23}$$

where α = thermal diffusivity; x = position in the dry layer (surface defined as 0); $r(t)$ = location of the ice–dry layer interface as a function of time. At the surface:

$$-k_{\mathrm{d}} \frac{\delta T}{\delta x} = \varepsilon\sigma(T_{\mathrm{H}}^4 - T_{\mathrm{s}}^4) \tag{24}$$

where ε = emittance of surface; σ = Stefan–Boltzmann constant; T_{H} = platen temperature; T_{s} = surface temperature (varying with time). It is further assumed that the interface recession can be approximated by eqn (25):

$$-k_{\mathrm{d}} \frac{\delta T}{\delta x} = \Delta H_{\mathrm{s}}(m_{\mathrm{i}} - m_{\mathrm{f}})\rho \frac{\mathrm{d}r}{\mathrm{d}t} \tag{25}$$

Cho and Sunderland (1970) obtained the following useful quasi-steady state expressions relating platen temperature and interface temperature to drying rate:

$$r = \frac{T_{\mathrm{s}} - T_{\mathrm{i}}}{(T_{\mathrm{H}}^4 - T_{\mathrm{s}}^4)} \frac{k_{\mathrm{d}}}{\varepsilon\sigma} \tag{26}$$

$$\frac{\mathrm{d}r}{\mathrm{d}t} = \frac{k_{\mathrm{d}}}{\rho(m_{\mathrm{i}} - m_{\mathrm{f}})\Delta H_{\mathrm{s}}} \frac{T_{\mathrm{s}} - T_{\mathrm{i}}}{r} \tag{27}$$

The second period can be analysed using either of the following assumptions.

(1) Ice disappears, before the maximum surface temperature is reached while an ice interface is still present.
(2) The maximum surface temperature is reached while an ice interface is still present.

This second case is more likely in industrial practise.

Equations (24), (25) and (26) apply here also, but now the surface temperature is constant at the maximum permissible level T_{s}^*, and the platen temperature T_H varies. Analytical solutions for this case are available, and are given by eqns (28) and (29).

$$t = t_1 + \frac{(m_i - m_f)\rho\Delta H_s}{2k_d(T_s^* - T_i)}(r^2 - r_1^2) \quad (28)$$

where t_1 = duration of the first period; r_1 = interface position at end of first period.

$$T_H = \left(T_s^* + \frac{k_d}{\varepsilon\sigma r}(T_s^* - T_i)\right)^{1/4} \quad (29)$$

Analytical or numerical solutions were also developed for shapes other than slabs. Thus Dyer and Sunderland (1971) published solutions for slabs, cylinders and spheres. The problem of cyclically varying chamber pressure was analysed recently by Mellor and Greenfield (1972).

TABLE V

TYPICAL THERMAL CONDUCTIVITIES OF FREEZE-DRIED FOODS

Food material	*Gas and pressure* (torr)	k_d [Btu/(°F ft h)]	*Reference*
Turkey meat	Air or He, 0.1	0.02	Triebes and King (1966)
Turkey meat	Air, 10	0.045	Triebes and King (1966)
Turkey meat	He, 10	0.07	Triebes and King (1966)
Haddock	Air, 0.5	0.013	Lusk et al. (1964)
Beef	Air, 1	0.013	Bralsford (1967)
Apple	Air, 0.1	0.01	Harper (1962)
Coffee	Air, 1	0.02	Fito et al. (1972)

Recent work has also confirmed that the assumption of a uniformly receding interface, and of constant dry layer properties, does not lead to serious errors. In this respect we should mention the work of Hatcher and Sunderland (1971), who conducted an experimental study of moisture and temperature distribution during the freeze-drying of beef.

In order to apply the relations listed above information is needed on the various specific transport properties of the food materials. Some data are available in the literature, including work by Margaritis and King (1971) on water transport in turkey meat, and recent data by Petree and Sunderland (1972) on thermal radiation properties of freeze-dried meats.

The transport properties k_d and b in themselves depend on pressure, and on the kind of gas filling the pores of the dry layer. The thermal conductivity dependence on pressure is shown in Table V which lists some selected values of k_d. The permeability may be considered to be inversely proportional to total pressure, at least for pressures in excess of one torr. Table VI gives some typical values of permeability of foods.

It should be noted that the thermal conductivities of dry layers of foods are extremely low, and compare with conductivities of insulators such as cork and styrofoam. As a consequence, the temperature drop across the dry layer is large, and with surface temperatures often limited to values below 150°F because of danger of discoloration, and in some cases to values below 100°F because of danger of denaturation, the resultant ice temperature is usually well below 0°F. Except for foods with very low eutectic melting points, it is the surface temperature that limits the drying rate.

TABLE VI

TYPICAL PERMEABILITIES OF FREEZE-DRIED FOODS

Food material	*Total pressure* (torr)	*Permeability* [1 lb/(torr ft h) × 10^3]	*Reference*
Beef	1.6	4.6	Hill (1967)
Turkey breast	0.1	5.1	Sandall *et al.* (1968)
Turkey breast	100	0.25	Sandall *et al.* (1968)
Coffee (20% initial solids)	1	0.01	Quast and Karel (1968)
Coffee (10% initial solids)	0.15	0.02	Lambert and Marshall (1962)
Beef, slow frozen	1	0.01	Bralsford (1967)
Beef, quick frozen	1	0.003	Bralsford (1967)

As a consequence of the above limitation, drying rates attainable in practise are very much below the maximum rates attainable with ice. Thus, for materials loaded into the freeze-drier at about 2–4 lb/ft^2 of tray surface, which corresponds to industrial practise reported in the literature, average drying rates are of the order of 0.30 lb of water removed per square foot. The corresponding drying times are 6–10 h.

A much more rapid rate can of course be achieved by decreasing the particle size and loading rates. This corresponds to reduction of the average thickness of the dry layer and thus of mass and heat transfer resistances. This approach is, however, limited only to selected products, since efficient operation requires specialised equipment such as continuous freeze-driers.

In general, operating conditions in freeze-drying of foods include maximum surface temperatures of 100–180°F, and chamber pressures of 0.1–2 torr. Freeze-drying of biological specimens, of vaccines and of micro-organisms is usually conducted with maximum surface temperatures of 70–90°F, and chamber pressures below 0.1 torr. It is possible to conduct freeze-drying at atmospheric pressure, provided the gas in which the drying is conducted is very dry. All of the theoretical relations discussed above apply in this case, but in addition to mass and heat transfer resistances in the food, one must also consider those in the gas phase.

The heat transfer is improved, but the mass transfer deteriorates as the pressure increases, and it is mass transfer that becomes limiting in atmospheric freeze-drying. As a consequence, for all but very small particles, the drying rates are very slow.

Heat transfer through the frozen layer, with mass transfer through the dry layer

Freeze dehydration of liquids, and of solids capable of intimate contact with a heating surface, can be conducted with heat transfer through the frozen layer as indicated in Fig. 5b. Practical systems in which we can observe situations approximating the idealised scheme of Fig. 5b include, for instance, the tubular drier invented by Seffinga, as well as several other patented procedures for freeze-drying liquids (Noyes, 1968). Experimental results in a pilot scale drier were obtained by Lambert and Marshall (1962). The mass and heat transfer equations are given below for the case of a slab with negligible end effects.

The mass transport of water vapour is given by eqn (15) just as in the previous case:

$$-\frac{\mathrm{d}w}{\mathrm{d}t} = Ab(p_\mathrm{i} - p_\mathrm{s})x_\mathrm{d}^{-1} \tag{15}$$

Similarly the rate of recession of the frozen layer and increase of the dry layer is given by eqn (17):

$$x_\mathrm{d}\,\mathrm{d}x_\mathrm{d} = \left(\frac{b}{\rho(m_\mathrm{i} - m_\mathrm{f})}\right)\cdot(p_\mathrm{i} - p_\mathrm{s})\,\mathrm{d}t \tag{17}$$

The heat transfer rate, however, is now given by

$$q = Ak_\mathrm{i}(T_\mathrm{H} - T_\mathrm{i})x_\mathrm{i}^{-1} \tag{30}$$

where T_H = temperature of wall in contact with frozen layer (°F); k_i = thermal conductivity of the frozen layer (Btu/ft/h/°F); x_i = thickness of the frozen layer (ft). As a consequence the relation between pressure and temperature of the ice interface becomes more complicated. In the case of heat transport through the dry layer, a linear equation relating p_i and T_i contained no additional variables. In the present case, however, eqn (31) contains an additional variable:

$$p_\mathrm{i} = p_\mathrm{s} + \left(\frac{k_\mathrm{i}}{b\Delta H_\mathrm{s}}\right)\left(\frac{x_\mathrm{d}}{x_\mathrm{i}}\right)(T_\mathrm{H} - T_\mathrm{i}) \tag{31}$$

It may appear that two additional variables are contained in eqn (31), x_d and x_i, but these two variables are interrelated by eqn (32), and one can be converted into a function of the other as shown in eqn (33).

$$x_d = L - x_i \tag{32}$$

$$p_i = p_s + \left(\frac{k_i}{b\Delta H_s}\right)\left(\frac{x_d}{L - x_d}\right)(T_H - T_i) \tag{33}$$

The interface temperature and pressure are therefore no longer independent of time of drying, even if k_i, ΔH_s, x_d, b, p_s and T_w remain constant. In fact, p_i must be evaluated as a function of x_d. This can be achieved readily by assuming an x_d and then determining the corresponding p_i as shown in Fig. 6. This is then repeated for another assumed value of x_d and if enough points are taken, one obtains p_i as a function of x_d:

$$p_i = \mathrm{f}(x_d) \tag{34}$$

Substituting this function into eqn (20) allows its integration by analytical or, if $\mathrm{f}(x_d)$ is too complicated, by numerical methods.

$$\int_0^L x_d \, dx_d = \frac{b(p_i - p_s)}{\rho(m_i - m_f)} \int_0^{t_d} dt \tag{20}$$

$$\int_0^L \frac{x_d}{\mathrm{f}(x_d) - p_s} \, dx_d = \frac{b}{\rho(m_i - m_f)} \int_0^{t_d} dt \tag{35}$$

The most convenient way to solve the above equations is through the use of computer, either digital or analogue.

In the case described here, the balance between ease of heat transfer and of mass transfer changes continuously, with mass transfer becoming more difficult as drying progresses (longer path through the dry layer) and heat transfer becoming progressively easier (shorter path through the frozen layer). The resistance of the frozen layer to the transport of heat is in any case not too formidable, with the thermal conductivities often as high as 1.0 Btu/ft/h/°F^{-1} and therefore up to two orders of magnitude higher than the thermal conductivities of the dry layer. As a consequence, theoretically calculated averaged drying rates with heat input through the frozen layer often exceed those attained in drying with heat transfer through the dry layer. The drying rate, for materials dried with heat transfer to the dry layer surface, can be increased very significantly by continuously removing most of the dry layer. This could be achieved, at least in theory, by a rotating knife or other scraping device, as suggested by Greaves (1960). The improvement theoretically achievable by this method of minimising mass transfer resistance is evident from the calculated drying rates of a hypothetical slab of one inch thickness, with the following properties and drying conditions:

$$b = 2 \times 10^{-2} \text{ lb h/ft/torr};$$
$$p_s = 0.005 \text{ torr};$$
$$k_d = 5 \times 10^{-2} \text{ Btu/h/ft/torr};$$
$$k_i = 1.0 \text{ Btu/h/ft/torr};$$
$$\rho = 20 \text{ lb/ft}^3;$$
$$m_i = 2.5 \text{ lb water/lb solids};$$
$$m_f = 0.05 \text{ lb water/lb solids};$$
$$\Delta H_s = 1200 \text{ Btu/lb}.$$

Maximum permissible surface temperature: 125°F.

(1) Heat transfer by radiation to two faces of slab: 8.75 h.
(2) Heat transfer by conduction to 'backface' of slab (wall at 10°F): 13.5 h.
(3) Heat transfer by conduction to 'backface' of slab (wall at 28°F): 7.2 h.
(4) Heat transfer by conduction to 'backface' of slab (wall at 10°F, dry layer removed continuously): 4.0 h.

Freeze-drying with heat input by microwaves

The limitations on heat transfer rates in conventionally conducted freeze-drying operations have led early to the attempt to provide internal heat generation through the use of microwave power. Work on this problem was started in the 1950s and the early attempts were reviewed by Burke and Decareau (1966) and by Copson (1962).

The generation of heat by microwaves depends on the presence of dipoles (in the case of foods, primarily water) which when placed in a rapidly changing electric field undergo changes in orientation which can result in 'friction' and consequently generation of heat. Frequencies available for industrial applications such as freeze-drying are limited by Federal authorities to two: 915 MHz and 2450 MHz. Equation (36) gives the amount of power generated in the material in the electric field:

$$\text{Power}\left(\frac{\text{watts}}{\text{cm}^3}\right) = E^2 \cdot \nu \cdot \varepsilon'' 55 \times 10^{-14} \tag{36}$$

where E = electric field strength (V/cm); ν = frequency (Hz); ε'' = loss factor.

The loss factor ε'' is an intrinsic property of foods which depends strongly on temperature and composition. In particular, liquid water absorbs much more energy than ice or dry food components. This is evident for instance from data shown in Table VII.

Theoretically, the use of microwaves should result in a very accelerated rate of drying, because the heat transfer does not require internal temperature gradients, and the temperature of ice could be maintained close to the maximum permissible temperature for the frozen layer, without the need for excessive surface temperatures. If, for instance, it is permissible to maintain the frozen layer at 10°F, then the drying time for an ideal process using microwaves for a one-inch slab with properties identical to those described in the preceding section would be 1.37 h.

We should note that this drying time compares very favourably with the 8.75 h required for the case of heat input through the dry layer, 13.5 h for heat input through the frozen layer without dry layer removal, and even

TABLE VII

DIELECTRIC LOSS FACTOR OF SELECTED FOODS AND FOOD COMPONENTS AT 2000–3000 MHz

Material	*Temperature* (°C)	*Loss factor*	*Reference*
Raw beef	−40	0.083	Burke and Decareau (1966)
Raw beef	5	10.56	Burke and Decareau (1966)
Freeze-dried beef	5	0.122	Burke and Decareau (1966)
Ice	−12	0.003	von Hippel (1954)
Water	25	12	von Hippel (1954)
Cooked beef	−20	0.65	Bengtsson and Risman (1971)
Cooked beef	25	12	Bengtsson and Risman (1971)

with the relatively short drying time of 4 h for the case in which the dry layer was continuously removed. In laboratory tests on the freeze-drying time of a one-inch thick slab of beef, an actual drying time of slightly over 2 h was achieved, compared with about 15 h for conventionally dried slabs (Hoover *et al.*, 1966).

In spite of these apparent advantages, the application of microwaves to freeze-drying has not been successful. The major reasons for the failures are the following.

(1) Energy supplied in the form of microwaves is very expensive. A recent review estimates that it may cost 10–20 times more to supply one Btu from microwaves than it does from steam.
(2) A major problem in the application of microwaves is the tendency to glow discharge, which can cause ionisation of gases in the chamber, and deleterious changes in the food, as well as loss of useful power. The tendency to glow discharge is greatest in the pressure range of

0.1–5 torr and can be minimised by operating the freeze-driers at pressures below 50 μm. Operation at these low pressures, however, has a double drawback: (a) It is quite expensive, primarily because of the need for condensors operating at a very low temperature, and (b) the drying rate at these low pressures is much slower.

(3) Microwave freeze-drying is a process which is very difficult to control. Since water has an inherently higher dielectric loss factor than ice, any localised melting produces a rapid chain reaction which results in 'runaway' overheating.

Effect of structure and structural changes on heat and mass transfer in freeze-drying

The temperature and moisture gradients in the freeze-drying materials depend strongly on the properties of the dry layer, and these properties are set in part during the freezing. For instance, if solute migration is possible during freezing, an impermeable film which impedes drying can form at the surface (Quast and Karel, 1968). Slush freezing can prevent the formation of such a film or it may be removed mechanically. Slow freezing produces bigger crystals, hence usually bigger pores and better mass flow during drying and reconstitution, if the solute migration and film formation are prevented. The beneficial effect of slow freezing is, however, often offset by the structure's collapse if the frozen layer temperature is too high.

Ideally, the vapour flows through the pores and channels left by ice crystals. However, if freezing produces isolated ice crystals surrounded by a solid matrix, then the vapour must diffuse through the solids.

A similar situation results if the matrix collapses at the ice front and seals the channels. Collapse often occurs at a fixed temperature similar to the recrystallisation temperature (MacKenzie, 1966; Ito, 1971) when the matrix is sufficiently mobile to allow flow under the influence of various forces (Table VIII). Freezing causes a separation of the aqueous solutions present in foods into a two-phase mixture of ice crystals and concentrated aqueous solution. The properties of the concentrated aqueous solution depend on temperature, concentration and composition. If drying is conducted at a very low temperature, then mobility in the extremely viscous concentrated phase is so low that no structural changes occur during drying, and resultant structure consists of pores in the locations which contained ice crystals, surrounded by a dry matrix of insoluble components and precipitate compounds originally in solution. If, on the other hand, the temperature is above a critical level, mobility in the concentrated aqueous solution may be sufficiently high to result in flow and loss of the original separation and structure.

During freeze-drying there exist both temperature and moisture gradients in the drying materials, and the mobility and therefore collapse of the concentrated solutions forming the matrix may vary from location to location. Mobility of an amorphous matrix depends on moisture content as well as on temperature, hence collapse can occur at areas other than ice surface (Karel and Flink, 1973a).

TABLE VIII

TYPICAL 'COLLAPSE' TEMPERATURES

Substance	*Collapse temperature*	*Reference*
Sucrose	−25	Ito (1971)
Glucose	−40	Ito (1971)
Dextran	−10	MacKenzie (1966)
Coffee extract (25%)	−20	Bellows and King (1972)
Apple juice (22%)	−41.5	Bellows and King (1972)
Grape juice (16%)	−46	Bellows and King (1972)
Sucrose (25%)	−24	Bellows and King (1972)
Sucrose (20%) + NaCl (5%)	−43	Bellows and King (1972)

QUALITY CHANGES DURING DEHYDRATION

Introduction

As discussed above, dehydration has as its aim the lowering of water activity to a level at which deterioration of food quality proceeds at a rate slow enough to allow long-term storage, and convenient distribution. However, during the dehydration process itself, some deterioration of quality is possible. Quality deterioration is due to the following factors related to the dehydration process.

(1) High temperatures used in some dehydration processes.
(2) Rearrangement of components, and internal flow during dehydration.
(3) Use of vacuum in freeze-drying and other vacuum processes.
(4) The process of water removal *per se*.

Among changes most commonly associated with the dehydration process are: *textural changes* in the food due to removal of water and subsequent cross-linking of polymeric constituents, *flavour losses*, and *nutrient losses* or *quality changes* due to chemical reactions occurring during the process of dehydration, in particular those due to non-enzymatic browning. The textural changes are a very complex subject, and in many cases are directly related to the process of water removal. Flavour loss is controlled by the structural changes within the food. Reactions producing nutrient losses are dependent on the moisture and temperature gradients during the dehydration process.

Changes in texture and in rehydration capability of dehydrated foods

The quality of dehydrated foods is often limited by changes in their texture and their rehydration capability. Tough, 'woody' texture; slow and incomplete rehydration and loss of the typical fresh food juiciness are most common quality defects of such foods. The physicochemical basis for these changes is as yet not fully understood. In the case of plant materials loss of cellular integrity and crystallisation of polysaccharides are thought to play a key role in these changes, and it is known that crystallisation of polysaccharides such as starch and cellulose is in fact promoted by removal of water. In baked products and in various other starch-based products the retrogradation of starch is an obvious mechanism and is in fact closely tied to the very process of dehydration.

In the case of animal-derived tissues used as dehydrated foods tenderness losses are due to aggregation of muscle proteins, and in particular of the actomyosin fraction. Details of these aggregative reactions are still unclear. Textural changes may be due to one or all of the following events in the actomyosin complex: aggregation or cross-linking of undenatured protein; denaturation of proteins followed by aggregation; or interaction of the native or denatured proteins with lipids or carbohydrates.

Myosin is considered the most likely site for the formation of cross-links in the actomyosin complex, and it is probable that S–S bond formation is involved. It is known that these bonds can be formed in the frozen state, and are stabilised or made irreversible by drying (Buttkus, 1970; Karel and Flink, 1973a). It is possible therefore to observe textural changes in freeze-dehydrated foods as well as in air-dehydrated foods in which the temperature during drying is much higher. Temperature of drying is, however, of key importance to the rate of formation of irreversible aggregates.

It is interesting to note that, while freezing prior to dehydration may have some deleterious effects in promoting actomyosin aggregation, the same processes, namely freezing, thawing and then rehydration, are used to improve rehydration characteristics of dehydrated vegetables. In this case the effect is due to increased internal porosity caused by the cavities left by large ice crystals. The flow of water during rehydration is facilitated, but the water-binding capability of the food material is not necessarily enhanced. Thus in many cases foods which have been made porous by slow freezing (large crystals), may imbibe more water during rehydration, but this water is loosely held.

A related method, designed to prevent shrinkage of the food tissues during dehydration, has been patented by Haas (1971). He proposed freezing foods that have been previously pressurised with 500–1500 psig of methane, nitrogen, CO, air, freon or ethane (CO_2 is not effective). The pressurised, frozen substances are then air-dried and retain their shape.

Other remedies to the texture deterioration problem have included additives (in particular phosphates, chelating agents and buffers for proteinaceous foods, and plasticisers such as glycerol for vegetable tissues), and physical disruption of the cross-linked structure. In this last respect an interesting, but as yet not approved method, is based on irradiation of dehydrated vegetables with ionising radiations, preferably at a low temperature (Gardner and Wadsworth, 1969).

Flavour retention in dehydrated foods

Retention of aroma compounds and flavours in dehydrated foods is an important attribute of quality. Conversely, it is often important to remove, during drying, traces of solvents used during preparation for drying of some types of foods (for instance, solvent-defatted oilseeds, or fish protein concentrate prepared by extraction with organic solvents). The factors controlling the removal of volatile organic compounds from drying foods are therefore of fundamental importance to dehydration processing.

In general it is recognised that freeze dehydration results in the most retention of flavours, and that some other dehydration processes produce satisfactory retentions, especially when conducted under optimal conditions.

In spite of the vacuum used in freeze-drying, retentions of many highly volatile compounds are high. It is also observed that compounds of different vapour pressures have similar and often high retention, as is evident from the data shown in Table IX, in which retention of pyridine in coffee powder and in dehydrated model systems is compared with prediction of retention based on various models using vapour pressure of pyridine and of water as criteria of retention.

The retention of organic volatiles has been considered to result from surface adsorption of the volatile on the dry layer of the freeze-drying sample (Rey and Bastien, 1962) or from an entrapment mechanism which immobilises the volatile compounds within the amorphous solute matrix (Thijssen, 1971; King, 1970; Flink and Karel, 1970a; Karel and Flink, 1973a). A simple experiment shows that in freeze-dried carbohydrate solutions the retention phenomena depend on local entrapment rather than adsorption. Maltose solutions were frozen in layers, some of which were volatile-containing, others volatile-free. After freeze-drying the layers were separated and analysed separately for the volatile. The results (Table X) show that volatile is retained in those areas where it was initially present. This experiment demonstrates that adsorption is not the major retention mechanism in freeze-dried sugar solutions since volatile escaping from the lowest layer of Sample A was not retained in large amounts on the upper dry layers. Further, volatile retention is unaffected by the passage of water vapour through the volatile-free maltose layers, as shown for Sample B. In a companion experiment, sections of the freeze-dried cake were cut

TABLE IX

RETENTION OF PYRIDINE IN COFFEE POWDER

	Retention (%)
A. Predicted from theoretical relations	
Raoult's law	7.5
Volatility in infinitely dilute solutions, data of Thijssen[a]	0.01
Calculated using volatility determined experimentally for 0.01% aqueous solution at −20°C[b]	$<10^{-5}$
B. Observed retention	
Coffee powder	50
Freeze-dried model (1% glucose)[b]	85
Freeze-dried model (1% starch)[b]	50

[a] Thijssen, 1970.
[b] Fritsch *et al.*, 1971.

TABLE X

RETENTION OF 2-PROPANOL IN SPECIFIED LAYERS OF FREEZE-DRIED MALTOSE SOLUTIONS[a]

	Sample A		*Sample B*	
	Before freeze-drying	*After freeze-drying*	*Before freeze-drying*	*After freeze-drying*
Top layer	0	0	4	2.52
Middle layer	0	0.05	0	0.05
Bottom layer	4	2.73	0	0.02

[a] 2-Propanol content is g/100 g solids.

from a sample perpendicular to the mass transfer axis. Essentially uniform retention was observed for the whole sample, supporting observations that volatile retention is determined locally in the food and not by surface adsorption. The gross structure of the freeze-dried material is freely permeable to the flow of volatile from the lower freeze-drying levels, and the retained volatile is not located on the surface of the dry layer but within the amorphous solute matrix.

The physical aspects by which the volatile is entrapped within the amorphous solute matrix are only partially understood. Perhaps two mechanisms, selective diffusion (Thijssen, 1971) and micro-regions (Flink and Karel, 1970a), represent macro- and micro-views of the same basic phenomenon. The size of the entrapments is small since grinding and evacuation of the dry material does not release any volatile. Recently the size of the micro-regions has been shown to vary with, among other things,

TABLE XI

TYPICAL LEVELS OF RETENTION OF SPECIFIC ORGANIC VOLATILES IN SOLUTIONS CONTAINING HIGH CONCENTRATIONS (OVER 18%) OF SPECIFIED SOLIDS, AND 0.75% OF THE VOLATILE

Organic volatile	*Volatile vapour pressure at* −30°C (torr)	*Volatile retention for specified solid* (%) *maltose*	*sucrose*	*glucose*
Acetone	11	50	55	25
2-Propanol	0.7	65	75	50
1-Butanol	0.1	55	70	31
Tert-butanol	0.44	75	80	50
Ethanol	1.1	45	—	55

the solubility of the organic volatile in the aqueous solution. Retained hexanal (about 1 g hexanal per 100 g maltodextrin) freeze-dried from an aqueous 20% maltodextrin solution appeared under the optical microscope within the amorphous solute matrix as 2–6 μm droplets. More soluble alcohols showed smaller droplets down to the limits of resolution.

The degree of retention in low molecular weight carbohydrates is a function of the chemical natures of the non-volatile solute and of the entrapped volatile, and is also dependent on processing conditions. Table XI shows typical retentions observed when various volatiles were freeze dried in aqueous solutions containing carbohydrates. Table XII shows the influence of processing conditions on retention.

The formation of volatile-retaining micro-regions in freeze-dried foods is controlled by interactions of the volatiles with non-volatile solids and water during freezing, drying and storage. Flink and Karel (1970a) studied volatile retention in freeze-dried carbohydrate solutions, and postulated that crystallisation of water during freezing results in formation of micro-regions containing highly concentrated solutions of carbohydrates and volatiles. As the local moisture content within these regions decreases, first due to freezing and then to sublimation, there occur associations between the molecules of solute. In the case of carbohydrates these associations are caused by hydrogen bonds formed between hydroxyl groups of carbohydrate molecules (Flink and Karel, 1972; Karel and Flink, 1973a). We have recently observed (Chirife *et al.*, 1973) that molecular associations entrapping volatiles within microregions seem to occur also in polar polymers containing no hydroxyl group.

The structure of the microregions and the permeability of these regions to water and to organic vapours depend strongly on local water content.

TABLE XII

TYPICAL EFFECTS OF CHANGES IN PROCESS VARIABLES ON FLAVOUR RETENTION IN FREEZE-DRYING, AND AN EXAMPLE OF THESE EFFECTS IN FREEZE-DRYING OF COFFEE

Change in process variable	*Effect on flavour retention*
Increased temperature	Decrease
Faster freezing	Decrease
Increased sample thickness	Decrease
Increased chamber pressure	Decrease
Increased solid content	Increase

Example: Relative retention of coffee volatiles in freeze-drying (after Ettrup Petersen *et al.*, 1973.)

Process conditions	*Relative retention* (%)
Slow frozen, 5 mm thick, 0.2 torr	87
Slow frozen, 15 mm thick, 0.2 torr	100
Quick frozen, 3 mm thick, 0.2 torr	47
Slow frozen, 5 mm thick, 0.8 torr	65
Slow frozen, 15 mm thick, 0.8 torr	35
Quick frozen, 3 mm thick, 0.8 torr	36

As this content decreases, the ease of loss of organic volatiles decreases until at some critical moisture level there is no further loss (Flink and Karel, 1970a, 1970b). Exposure of freeze-dried carbohydrate solutions containing entrapped volatiles to water vapour shows the following pattern: at low humidities there is no volatile loss even after evacuation for prolonged periods; at higher humidities a rapid volatile loss occurs until a new level of retention is reached, which is again stable unless the humidity is increased further.

It was determined that the critical point for initiation of volatile loss corresponds to sorption of water to levels above the calculated B.E.T. monolayer value (Flink and Karel, 1972). Below this level water is sorbed on those hydroxyl groups of the carbohydrates which do not participate in the structure-forming hydrogen bonds. Adsorption of water in amounts below the monolayer value does not therefore disrupt micro-region structure, and the volatile retention is not diminished. At moisture contents above this level, however, the sorbed water competes for hydroxyl groups involved in structure forming, the micro-region structure is disrupted, and a volatile loss occurs. Humidification resulted in a new level of retention, as long as the original structure of the freeze-dried materials was not destroyed by sufficient water to cause either dissolution or crystallisation.

These observations confirmed the existence of a microstructure which undergoes partial collapse upon humidification to a level above the monolayer value, this partial collapse becoming complete only upon dissolution

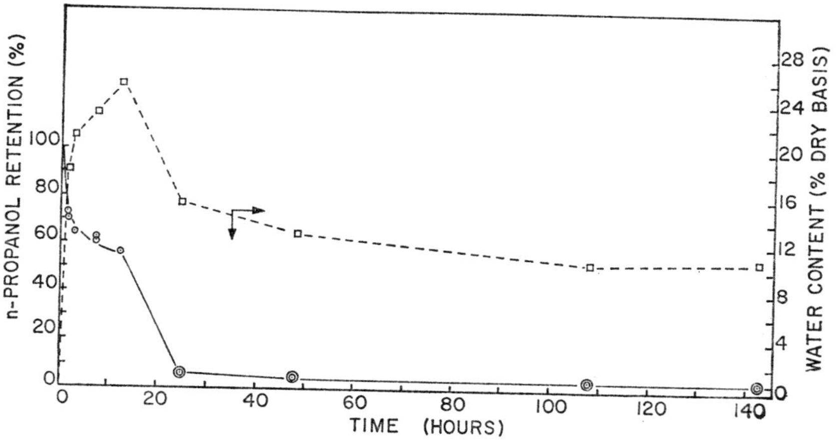

FIG. 7. Adsorption of water (dashed line) and retention of *n*-propanol (solid line) in freeze-dried maltose humidified to 75% relative humidity at 25°C.

or crystallisation (Flink and Karel, 1972; Chirife and Karel, 1973a). Figure 7 shows the loss of entrapped *n*-propanol during recrystallisation of freeze-dried maltose solutions. Like other sugars, maltose upon freeze-drying gives an amorphous cake which recrystallises upon humidification (Guilbot and Drapron, 1969). The results of humidification of maltose at 75% relative humidity are presented in Fig. 7. Humidification causes the moisture content to increase to approximately 25% (dry basis) and then to fall, indicating recrystallisation. The humidification had a pronounced effect on retention of entrapped *n*-propanol. After freeze-drying and before humidification the amorphous cake of maltose contained 3.47 g *n*-propanol/100 g maltose. As the water content of the maltose increased towards 25%, the propanol content was decreasing slowly towards a new value of about 50% of the original. Then, as recrystallisation began, propanol was lost rapidly; the final retention was very low. The rate of propanol loss during this period parallelled the progress of crystallisation. The high rate of propanol loss is a consequence of the extensive disruption of the carbohydrate–carbohydrate bonds present in the amorphous cake. The results obtained with maltose thus are very similar to those found by Flink and Karel (1972) in crystallisation of lactose after exposure to 61% relative humidity.

The microstructure developed in freeze-dried carbohydrates and responsible for volatile retention can be disrupted by influences other than sorption of water. Polar molecules capable of structure disruption should

TABLE XIII

RELEASE OF *n*-PROPANOL ENTRAPPED IN FREEZE-DRIED MALTOSE SOLUTIONS, AFTER EXPOSURE TO SPECIFIED VAPOURS

(*n*-propanol content prior to exposure ≅ 3.5 g propanol/100 g maltose)

Vapour to which maltose was exposed		*Temperature* (°C)	*Propanol released* (%)
Water, vapour activity	0.61	25	62
Water, vapour activity	0.75	25	95
Methanol, vapour activity	0.75	25	68
Ethanol, vapour activity	0.75	25	0
Ethyl ether, saturated vapour		25	0
Acetic anhydride, saturated vapour		25	5
Benzene, saturated vapour		25	0
Aniline, saturated vapour		25	32
Ethyl ether, saturated vapour		37	0
Benzene, saturated vapour		37	0

release entrapped volatiles, and high temperatures may cause thermal disruption of structure and volatile release.

Table XIII presents results of exposure of maltose containing entrapped propanol to various organic vapours. The results in Table XIII represent 'pseudo-equilibrium' values in that at each condition of exposure a new level of retention was established which did not change with additional time of exposure. In the same study we also observed that in the absence of water or other polar vapours maltose did not release entrapped propanol when it was subjected to vacuum at elevated temperatures up to and including 82°C. At 100°C, however, a partial release of propanol occurred during evacuation (Chirife and Karel, 1973a).

The most recent work in our group in the field of retention of organic volatiles during dehydration has produced the following advances.

(1) The microscopic studies were continued using a freezing/freeze-drying microscope, and have confirmed the importance of solubility relationships in the distribution of entrapped volatiles within the freeze-dried materials. Volatiles having a low solubility readily condense into droplets visible under the microscope, whereas highly soluble volatile compounds are entrapped in the form of very small droplets, and perhaps also as molecular species, distributed within the amorphous carbohydrate structure.

(2) Using ^{14}C-labelled alcohols we were able to extend our studies to retention of organic volatiles present in very low concentrations, which are in the range of those typical of flavour compounds in foods. Table XIV summarises results obtained at low concentrations using several different entrapping solids, including polymers. We have found that the microregion concept applies fully at these low, and therefore practically important, concentrations.

TABLE XIV

RETENTION OF PROPANOL IN FREEZE-DRIED SOLUTIONS CONTAINING INITIALLY 20% OF SPECIFIED SOLIDS

Solid	*Initial volatile content* (ppm)	*Type of freezing*	*Retention* (%)
Dextran-10	100	rapid	56
Dextran-10	100	slow	97
Maltose	100	rapid	70
Maltose	100	slow	88
PVP	100	slow	58
Starch	100	rapid	38
Cellulose	250	rapid	8

(3) We have extended our work to the study of polymers, including polyvinylpyrrolidine (PVP), starch, cellulose, dextran and proteins. We established that in polymers, in addition to retention by entrapment or by inclusion within the polymer chains, there is a small but significant contribution of adsorption. The contribution of adsorption which we found in the case of cellulose, and of PVP, is between 5 and 35% of the retention due to entrapment, whereas in the case of sugars the contribution of adsorption was negligible.

Nutrient destruction and other deteriorative reactions occurring during drying

Reactions occurring during drying can result in quality losses, in particular in nutrient losses and in other deleterious changes due to non-enzymatic browning. Attempts have been made to analyse these changes in a quantitative manner and to relate them to process conditions. Such quantitative analysis would facilitate the determination of optimal drying conditions with respect to nutrient retention and organoleptic quality.

Kluge and Heiss (1967) measured rates of non-enzymatic browning in a model food system (glucose, glycine and cellulose) and used these rates to develop relations for the calculation of browning occurring in drying. Labuza (1972) suggested the use of computers for prediction of the extent of deterioration occurring during storage. The principle is as follows.

Given the rates of deterioration as functions of moisture content and temperature, and the knowledge of the distribution of moisture and of temperature within the food throughout the dehydration process, a suitable integration procedure with respect to time and space is carried out to obtain the total deterioration occurring in the process.

Actually, two limiting cases may be recognised which make the analysis less cumbersome than is implied by the general statement.

(1) In *air dehydration* it is usually possible to assume that the temperature gradient within the food is negligible. Thus it is possible to calculate the moisture–temperature profiles by an assumption that moisture content varies with time and location, but the temperature varies with time and not with location in the food.

(2) In *freeze-dehydration*, within the 'ice-free' layer (deterioration in the frozen layer during drying may often be neglected), it may be assumed that the temperature varies with time and location, but that the moisture gradients are small. In fact, in freeze-drying it is often permissible to simplify the situation even further and to assume that, as long as *any ice* is present, the *temperature difference across the dry layer* remains unchanged.

We have recently evaluated the extent of non-enzymatic browning in dehydration of potato slabs using data on kinetics of non-enzymatic browning in potatoes published by Hendel *et al.* (1955) and some simplifying assumptions concerning moisture and temperature distribution in the potato slab. The rate of non-enzymatic browning depends, of course, on both moisture content and temperature. Hendel *et al.* (1955) observed that at any moisture and temperature the browning rate was constant, and the degree of browning increased linearly with time. Similar observations were made by Mizrahi *et al.* (1970) for freeze-dried cabbage, by Karel and Labuza (1968) for model systems, and by Karel and Nickerson (1964) for orange juice powder. Karel and Flink (1973b) observed linear increases in the browning of freeze-dried model systems at each of several temperatures.

Figure 8 presents typical data derived from the results of Hendel *et al.*

(1955). Using this data it is possible to calculate browning rates in potato at any combination of moisture content and temperature.

The moisture–temperature distribution for any drying condition was derived using idealised assumptions, including the following:

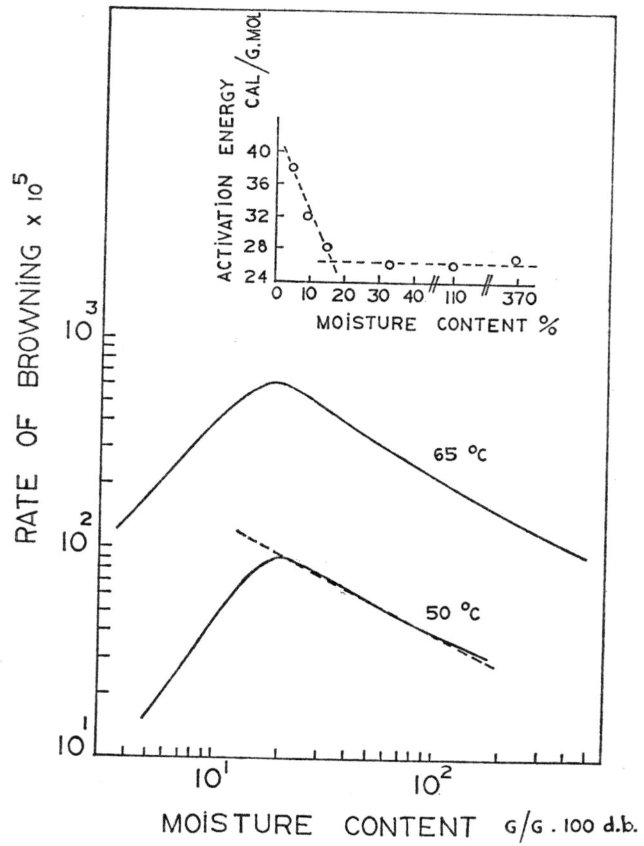

FIG. 8. Effects of moisture content and of temperature on non-enzymatic browning in potatoes (after Hendel *et al.*, 1955).

(1) Slab geometry.
(2) No shrinkage.
(3) Constant condition of air temperature, humidity and velocity.
(4) Drying in the falling rate period is considered with moisture content = 3.5 g/g at the beginning of that period.

(5) Moisture distribution in potato slabs given by eqn (37) and the average moisture given by eqn (38).
(6) A diffusion coefficient independent of concentration but depending on temperature as shown in eqn (39).

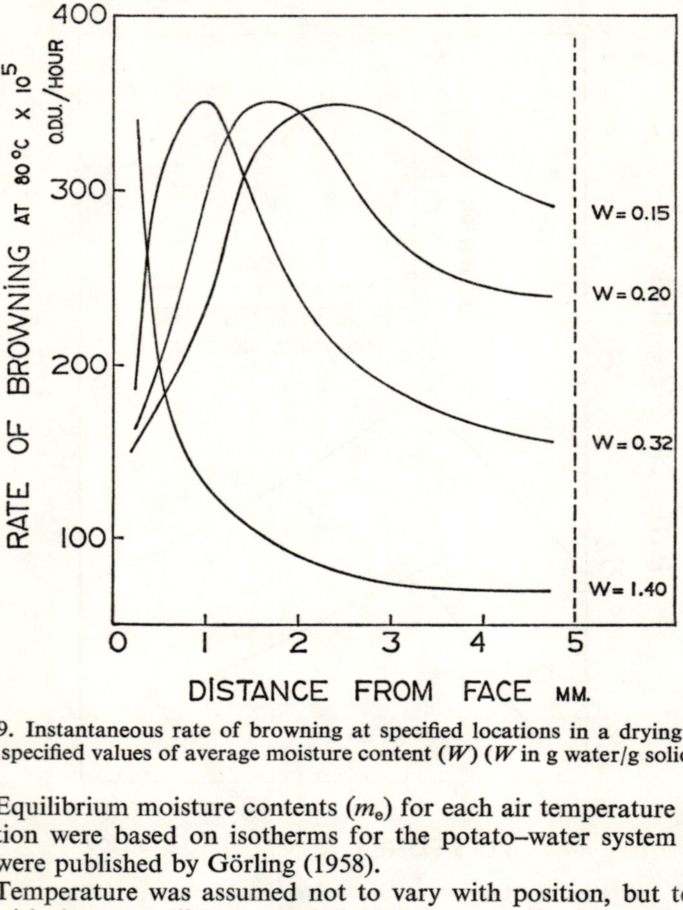

FIG. 9. Instantaneous rate of browning at specified locations in a drying potato slab, at specified values of average moisture content (W) (W in g water/g solids).

(7) Equilibrium moisture contents (m_e) for each air temperature condition were based on isotherms for the potato–water system which were published by Görling (1958).
(8) Temperature was assumed not to vary with position, but to vary with time according to eqn (40).

The moisture distribution throughout the potato slab is given by:

$$m(x) = m_e + \frac{4}{\pi}(m_i - m_e) \sum_{n=0}^{\infty} \frac{(-1)^n}{(2n+1)} \cos \frac{(2n+1)}{2L} \pi x \times \exp\left[-\frac{\pi^2}{4} \frac{(2n+1)^2}{L^2} Dt\right] \tag{37}$$

The mean moisture content is given by

$$\frac{m - m_e}{m_i - m_e} = \frac{8}{\pi^2} \sum_{n=0}^{\infty} \frac{1}{(2n+1)^2} \exp\left(-\frac{(2n+1)^2}{L^2} \frac{\pi^2}{4} Dt\right) \tag{38}$$

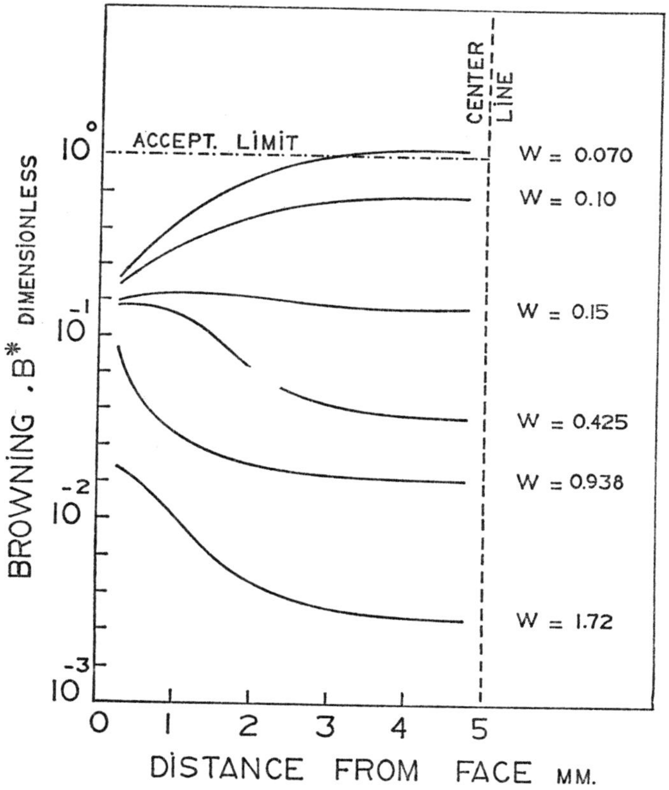

FIG. 10. Browning gradients in a potato slab dried in air at 71.2°C to specified values of average moisture content (W).

where: $m(x)$ = moisture content at location x in the slab; L = half-thickness of the slab. The dependence of diffusivity on temperature is given by

$$D = D_0 \, e^{-E_D/RT} \tag{39}$$

where: D = diffusivity (cm^2/min); D_0 = constant (assumed = 66 cm^2/min); E_D = activation energy for diffusion (assumed equal to 7500 cal/mol). The temperature depends on time:

$$\ln\left(\frac{T_a - T_w}{T_a - T}\right) = \mu t \tag{40}$$

where μ = slope of the line ln (m/m_i) v. time.

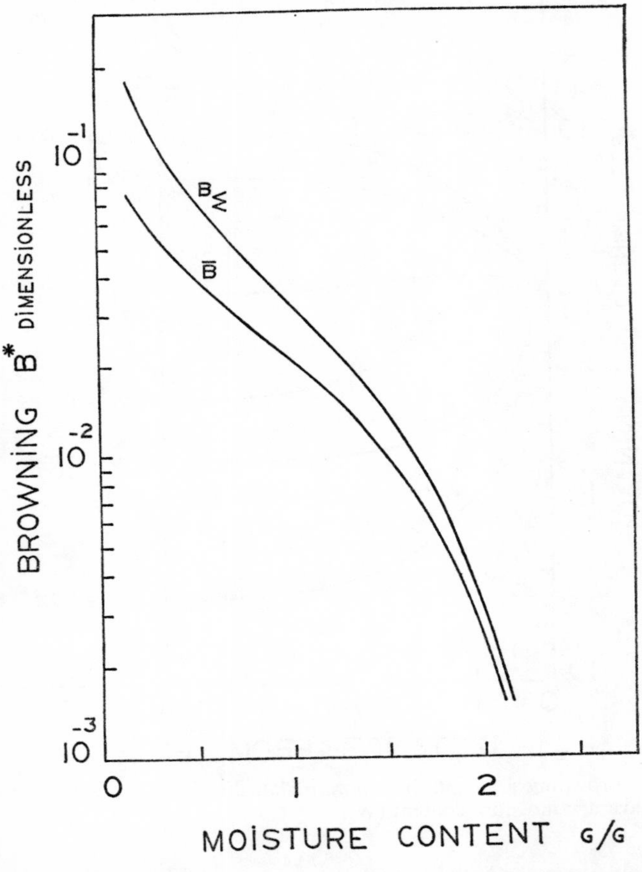

FIG. 11. Comparison of browning in a potato slab calculated by using average moisture contents ($\bar{B}$) with browning calculated with consideration of moisture gradients existing during drying at 71.2°C. (B_Σ).

A computer programme was used to calculate the rate of browning, by first calculating the moisture and temperature distribution in the potato slab and then computing the corresponding browning rate.

Figure 9 shows the instantaneous browning rates within the potato slab as a function of position and degree of drying.

Figure 10 shows the browning that occurs in the drying slab at various locations when drying is conducted at an air temperature of 71.2°C.

Finally, Fig. 11 shows the predicted amount of browning for different final moisture contents for the slab dried at 71.2°C. It is evident that there is a difference between the total browning calculated by summing up browning at each location (B_Σ) compared with browning computed by using an average value of moisture for each drying time ($\bar{B}$).

An advantage of the approach discussed here is that once a computer programme is written and constants calculated, it is possible to simulate rapidly various conditions of drying and to determine optimum conditions. The major limitations on the approach arise from a great paucity of kinetic data on nutrient deterioration reactions, and on physicochemical properties of foods.

REFERENCES

Bellows, R. J. and King, C. J. (1972). 'Freeze-drying of aqueous solutions: maximum allowable operating temperatures', *Cryobiology*, **9**, 559.

Bengtsson, N. E. and Risman, P. O. (1971). 'Dielectric properties of food at 3 GHz as determined by a cavity perturbation, *J. Microwave Power*, **6**(2) 101.

Bralsford, R. (1967). 'Freeze-drying of beef. I, Theoretical freeze-drying rates of beef', *J. Fd. Technol.* (*London*), **2**, 339.

Brey, W. S., Heeb, M. A. and Ward, T. M. (1969). 'Dielectric measurements of water sorbed on ovalbumin and lysozyme', *J. Colloid Interface Sci.*, **30**, 13.

Burke, R. F. and Decareau, R. V. (1966). 'Freeze dehydration of foods', *Advances in Food Research*, Vol. XIII, 1.

Buttkus, H. (1970) 'Accelerated denaturation of myosin in frozen solutions', *J. Food Sci.*, **35**, 558.

Chirife, J. (1971). 'Diffusional process in the drying of tapioca root', *J. Food Sci.*, **36**, 327.

Chirife, J., Karel, M. and Flink, J. (1973). 'Studies on mechanisms of retention of volatiles in freeze-dried food models: The system PVP: *n*-propanol', *J. Food Sci.*, **38**, 671.

Chirife, J. and Karel, M. (1973a). 'Contribution of adsorption to volatile retention in a freeze-dried food model containing PVP', *J. Food Sci.* (in press).

Chirife, J. and Karel, M. (1973b). 'Effect of structure disrupting treatments, on volatile release from freeze-dried maltose', *J. Fd. Technol.* (*London*) (submitted for publication).

Cho, S. H. and Sunderland, J. E. (1970). 'Approximate solution for rate of sublimation-dehydration of foods', *Trans. ASAE*, **13**(5), 559.

Copson, D. A., (1962). *Microwave Heating*, Avi Publishing Company, Westport, Conn.

Duckworth, R. B. (1971). 'Differential thermal analysis of frozen food sys-

tems. I. The determination of unfreezable water', *J. Fd. Technol.* (*London*). **6**, 317.
Duckworth, R. B. (1972). 'The properties of water around the surfaces of food colloids', *Proc. Inst. Food Sci. Technol.* (*U.K.*), **5**(2), 60.
Dyer, D. F. and Sunderland, J. E. (1969). 'The influence of varying interface temperature on freeze-drying', *Bull. l'Inst. Internat. du Froid.*, Annexe 1969-4, 37.
Dyer, D. F. and Sunderland, J. E. (1971). 'Freeze-drying of bodies subject to radiation boundary conditions', *Trans. ASME, J. Heat Transfer*, **93**, 427.
Ettrup Petersen, E., Lorentzen, J. and Flink, J. (1973). 'Influence of freeze-drying parameters, on the retention of flavor compounds of coffee', *J. Food Sci.*, **38**, 119.
Fish, B. P., (1958). 'Diffusion and thermodynamics of water in potato starch gel', in *Fundamental Aspects of the Dehydration of Foodstuffs*, Society of the Chemical Industry, London, 143.
Fito, P. J., Pinaga, F. and Aranda, V. (1972). 'Heat transfer properties in freeze-dried foods, Proceedings, Internat. Sympos. on Heat and Mass Transfer Problems in Food Eng., 24–27 October, 1972, Wageningen, Netherlands, Vol. 2, F6-1.
Flink, J. and Karel, M. (1970a). 'Retention of organic volatiles in freeze-dried solution of carbohydrates', *J. Agr. Fd. Chem.*, **18**, 295–297.
Flink, J. and Karel, M. (1970b). 'Effect of process variables on retention of volatiles in freeze-drying', *J. Food Sci.*, **35**, 444–447.
Flink, J. M. and Karel, M. (1972). 'Mechanisms of retention of organic volatiles in freeze-dried systems', *J. Fd. Technol.* (*London*), **7**, 199.
Flink, J. M. and Labuza, T. P. (1972). 'Retention of 2-propanol at low concentrations by freeze-drying carbohydrate solutions', *J. Food Sci.*, **37**, 617–618 (1972).
Fritsch, R., Mohr, W. and Heiss, R. (1971). 'Untersuchungen über die Aroma-Erhaltung bei der Trocknung, von Lebensmitteln nach vershiedenen Verfahren', *Chem. Ing. Technik*, **43**, 445.
Gardner, D. S. and Wadsworth, C. K. (1969). 'Low-temperature irradiation Treatment of Dehydrated Potatoes', U.S. Patent 3-463-643.
Görling, P. (1958). 'Physical phenomena during the drying of foodstuffs', in *Fundamental Aspects of the Dehydration of Foodstuffs*, Society of the Chemical Industry, London, 42.
Greaves, R. I. N. (1960). 'The application of heat to freeze-drying systems', *Annals N.Y. Acad. Sci.*, **85** (ART. 2), 682–688.
Guilbot, A. and Drapron, R. (1969). 'Evolution, en fonction de l'humidité relative, de l'état d'organisation et de l'affinité pour l'eau, de divers oligosides cryodeshydrates', *Bull. Intern. Inst. Refriger.*, Annexe 1969-9, 191.
Haas, G. J. (1971). 'New drying technique', *Food Eng.*, **43**(11), 58.
Harmathy, T. Z. (1969). 'Simultaneous moisture and heat transfer in porous systems with particular reference to drying', *I & EC Funda.*, **8**, 92.
Harper, J. C. (1962). 'Transport properties of gases in porous media at reduced pressures with reference to freeze-drying', *A.I.Ch.E.J.*, **8**(3), 298.
Hatcher, J. D. and Sunderland, J. E. (1971). 'Spiked-plate freeze-drying', *J. Food Sci.*, **36**, 899.

Hendel, C. E., Silveira, V. G. and Harrington, W. O. (1955). 'Rates of non-enzymatic browning of white potato during dehydration', *Food Technol.*, **9**, 433.

Hill, J. E. (1967). 'Sublimation dehydration in the continuum, transition and free molecule flow regimes', Ph.D. Thesis. Georgia Inst. Technol. Sept., 1967.

Hoover, M. W., Markantonatos, A. and Parker, W. N. (1966). 'UHF dielectric heating in experimental acceleration of freeze-drying of foods', *Food Technol.*, **20**, 807.

Ito, K. (1971). 'Freeze-drying of pharmaceuticals. Eutectic temperature and collapse temperature of solute matrix upon freeze-drying of three-component systems', *Chem. Pharm. Bull.* (*Japan*), **19**(6), 1095.

Jason, A. C. (1958). 'A study of evaporation and diffusion processes in the drying of fish muscle', in *Fundamental Aspects of the Dehydration of Foodstuffs*, Society of the Chemical Industry, London, 103.

Karel, M. (1973). 'Recent research and development in the field of low moisture and intermediate moisture foods', *CRC Critical Rev. of Food Technol.*, **3**, 329.

Karel, M. and Flink, J. M. (1973a). 'Influence of frozen state reactions on freeze-dried foods', *J. Agr. Food Chem.*, **21**, 16–21.

Karel, M. and Flink, J. M. (1973b). 'Mechanisms of deterioration of Nutrients', M.I.T. Report on Contract Research Project No. 9-12485 with Manned Spacecraft Center, N.A.S.A.

Karel, M. and Labuza, T. P. (1968). 'Optimization of Protective Packaging of Space Foods', M.I.T. Report on Contract Research Project No. F41-609-68-C-0015 with Aerospace Medical Division, U.S. Air Force.

Karel, M. and Labuza, T. P. (1968). 'Non-enzymatic browning in model systems containing sucrose', *J. Agr. Food Chem.*, **16**, 717–719.

Karel, M. and Nickerson, J. T. R. (1964). 'Effects of relative humidity, air, and vacuum on browning of dehydrated orange juice', *Food Technol.*, **18**(8), 104–108.

King, C. J. (1968). 'Rates of moisture sorption and desorption in porous dried foodstuffs', *Food Technol.*, **22**, 509.

King, C. J. (1970). 'Recent Developments in Food Dehydration Technology', Proc. 3rd Internat. Congress Food Sci. and Technol., Washington, D.C., 565.

Kluge, G. and Heiss, R. (1967). 'Untersuchungen zur besseren Beherrschung der Qualität von Getrockneten Lebensmitteln unter besonderer Berücksichtigung der Gefriertrocknung', *Verfahrentechnik*, **6**, 251.

Krischer, O. and Kröll, K. (1963). *Trocknungstechnik. I. Die wissenschaftlichen Grundlagen der Trocknungstechnik*, 2nd edn, Springer-Verlag, Berlin.

Labuza, T. P. (1972). 'Nutrient losses during drying and storage of dehydrated foods', *CRC Crit. Rev. Food Technol.*, **3**, 217.

Lambert, J. B. and Marshall, W. R., Jr. (1962). 'Heat and mass transfer in freeze-drying', in *Freeze Drying of Foods* (ed. F. R., Fisher), National Research Council, National Academy of Sciences, Washington, D.C., 1962.

Lusk, G., Karel, M. and Goldblith, S. A. (1964). 'Thermal conductivity of some freeze-dried fish', *Food Technol.*, **18**(10), 121.

MacKenzie, A. P. (1966). 'Basic principles of freeze-drying for pharmaceuticals', *Bull. Parenteral Drug. Assoc.*, **20**, 101 (1966).

Margaritis, A. and King, C. J. (1971). 'Measurement of rates of moisture transport within the solid matrix of hygroscopic porous materials', *I & EC Funda.*, **10**, 510.

Mellor, J. D. and Greenfield, P. F. (1972). 'Heat and Vapour Transfer Problems in Cyclic-pressure Freeze-drying', in Proceedings, Internat. Sympos. on Heat and Mass Transfer Problems in Food Eng., 24–27 October 1972, Wageningen, Netherlands, Vol. 2, p. E8-1.

Mizrahi, S., Labuza, T. P. and Karel, M. (1970). 'Computer-aided predictions of extent of browning in dehydrated cabbage', *J. Food Sci.*, **35**, 799–803.

Noyes (1968). *Freeze-drying of Foods and Biologicals*, Noyes Dev. Corp., Park Ridge, N.J., 1968.

Petree, D. A. and Sunderland, J. E. (1972). 'Thermal radiation properties of freeze-dried meats', *J. Food Sci.*, **37**, 209.

Quast, D. and Karel, M. (1968). 'Dry layer permeability and freeze-drying rates in concentrated fluid systems', *J. Food Sci.*, **33**, 170.

Rey, L. and Bastien, M. C. (1962). 'Biophysical aspects of freeze-drying. Importance of the preliminary freezing and sublimation periods', in *Freeze-Drying of Foods* (ed. F. R. Fisher), National Academy of Sciences, National Research Council, Washington, D.C., 25.

Roebuck, B. D., Goldblith, S. A. and Westphal, W. B. (1972). 'Dielectric properties of carbohydrate–water mixtures at microwave frequencies', *J. Food Sci.*, **37**, 199.

Sandall, O. C., King, C. J. and Wilke, C. R. (1968). 'The relationship between transport properties and rates of freeze-drying poultry meat', *Chem. Eng. Progress Symp. Series*, **64**(86), p. 43.

Saravacos, G. D. and Charm, S. E. (1962). 'A study of the mechanisms of fruit and vegetable dehydration', *Food Technol.*, **16**, 78.

Scott, W. J. (1957). 'Water relations of food spoilage microorganisms', *Advan. Food Res.*, **7**, 83.

Shanbhag, S., Steinberg, M. P. and Nelson, A. I. (1970). 'Bound water defined and determined at constant temperature by wide-line NMR', *J. Food Sci.*, **35**, 612.

Thijssen, H. A. C. (1970). 'Concentration processes for liquid foods containing volatile flavours and aromas', *J. Fd. Technol.* (*London*), **5**, 211.

Thijssen, H. A. C. (1971). 'Flavor retention in drying preconcentrated food liquids', *J. Appl. Chem. Biotechnol.*, **21**, 372 (1971).

Triebes, T. A. and King, C. J. (1966). 'Factors influencing the rate of heat conduction in freeze-drying', *I & EC Process Res. and Dev.*, **5**(4), 430.

Vaccarezza, L., Lombardi, J. L. and Chirife, J. (1971). 'Mecanismos de Transporte en el Secado de Remolacha Azucarera', Paper presented at VIII. Jornadas de Inv. en Ciencias de las Ing. Quimica y Quimica Aplicada, M. del Plata, Argentina, 23 September, 1971.

Von Hippel, A. (1954). *Dielectric Materials and Applications*, M.I.T. Press, Cambridge, Mass.

Walter, J. A. and Hope, A. B. (1971). 'Nuclear magnetic resonance and the state of water in cells', *Progr. Biophys. Mol. Biol.*, **23**, 3.

Young, J. H. (1969). 'Simultaneous heat and mass transfer in a porous hygroscopic solid', *Trans. ASAE*, **12**, 720.

DISCUSSION

J. D. Mellor (C.S.I.R.O., Australia): In your second method, transferring heat through the ice phase for freeze-drying, it might be possible to use a granulated frozen product such that the ice crystal size exposed would be greater than the size of the physical structure of the substrate, then the exposed ice crystals would sublime at a faster rate and possibly with a better flavour retention in the substrate. Could you comment on this idea?

M. Karel: I apologise for ignoring completely the physical aspect of freeze-dried products. Nowadays practically all the important applications of freeze-drying concern granulated frozen liquids with fine particle sizes drying in beds or as moving layers. I anticipate, however, some problems: the dry powder produced might be difficult to collect or have an undesirable colour because of the fine particle size or such a bulk density that it would require agglomeration. My main concern is the quality of the product. Flavour retention would depend on the entrapment of flavour components in the substrate. You may then have for this proposed treatment a material for freeze-drying with no mass transfer resistance to water vapour which you want, against a high mass transfer resistance to flavour components. Perhaps Prof. Thijssen would like to comment?

H. A. C. Thijssen (Eindhoven University): You can expect an improvement in aroma retention by finely grinding frozen materials and drying them as single particles themselves in thin layers, and the trend is now in this direction.

M. Karel: Very simply, the idea may mean a fantastic iceberg effect for freeze-drying.

J. D. Mellor: Aroma retention also means selective evaporation rates.

M. Karel: I think both Prof. Thijssen and I agree that, except for the situations where the limit would be exceeded, this might be the case with thinner layers.

S. A. Goldblith (M.I.T.): May I just make a pragmatic point: how do you get the right bulk density for coffee you want to sell?

E. Seltzer (Rutgers University): The coffee most commonly sold in the United States is freeze-dried in 7 min by using a 'fluidised' bed drying by flowing down an inclined pipe, granules screened for size which have been frozen fairly slowly. I have shown some slides at M.I.T. of some commercial equipment which is the largest use of freeze-drying in the United States.

Could you also show again Fig. 3, the slide on the browning reaction, because I want, for the purposes of this Symposium, to update information by pointing out an important anomaly in the range of browning.

M. Karel: Figure 3 shows the point at which one specific example of browning occurs rather than a typical range. It in fact refers to freeze dehydrated pork.

E. Seltzer: Well, what I am going to say concerns freeze-dehydrated pork in which the monolayer moisture content is 3.9–4.2. The military want dried pork for rations; they are, next to the coffee industry, the largest users of dehydrated foods in the United States. They are preoccupied at present with compressing products to save space. In the case of pork one tries to freeze-dry to 2% average moisture but ends up with a range of 5–10% moistures, then re-humidification brings the moistures higher in order to compress the dried pork so that it will not fragment and fall apart. After compression it is re-dried again to 2% moisture in the hope that it will brown nicely (of course there is possible lipid oxidation). We found the compressed 2% moisture product rehydrates more quickly than if it remained at 10–12% moisture. However, it also means a fairly steep increase in browning with storage time whereas the product left at 11% moisture browned much more slowly. Lipid oxidation may also be involved here.

M. Karel: Browning is a very complicated reaction involving the condensation of carbonyl group compounds. The groups come from a number of sources, with carbohydrates the most common. However, one excellent source involves lipid oxidations and, in the case of beef and pork which have already been dried once, then re-humidified and compressed, you have already a source of lipid derived carbonyl compounds for protein lipid interaction. I have dealt with it in a recent paper.

E. Seltzer: Thank you. In fact it was explained in 'Advances in Food Research' some twelve years ago.

A. S. Michaels (Alza Research): I am prompted to seek parallels between your observations on water and volatile components and studies on transport processes in hydrophilic polymeric membranes. You focus on the transport of small molecules through a rather ill-defined matrix, whereas the membranes we focus on the matrix as the focal point for understanding the molecular scale transport through it. In this way, transport for small molecules through say, polymer ethyl, might give a lead in controlling what is going on with volatile loss in food dehydration.

There are two points: first, the water diffusion coefficient is concentration dependent in a water-soaked polymer like polymer ethyl. In bone dry polymer alcohol the diffusion coefficient changes 8 or 10 orders of magnitude with concentration. Small molecules moving through hydrous polymers show an even stronger concentration dependence than a solvating substance such as water. Second, the established time for a particular water content is a critical variable; the longer the polymeric material stays in a given state of hydration the lower, as a rule, the permeabilities for water and other substances tend to go. It is due to re-arrangement in the matrices because chain polymers and big molecules move slowly. I think these points may be helpful in explaining volatile loss and retention. One other point, you were puzzled by the difference in flavour retention between pepsin polymer and bovine serum, here macro-molecules have an enormous effect upon the morphology of the resulting matrix. If corpuscular protein which has good stability and rigidity is dried it leaves voidage that is probably the same order of magnitude as the molecular size that you are trying to retain—like flavour molecules. Macromolecules packed tightly end up as dry structures with a permeability like that of inorganic gases.

M. Karel: I think I do not overlook these points and I have published some work on polymer diffusion coefficients and their concentration dependence and so has Prof. Thijssen. The same thing that breaks up the hydrogen bonds in polyvinyl alcohol breaks up the hydrogen bonds in maltose and maltodextrose. The problem is complicated by the poor definition of the state of components of foods, so I would much rather work with individual polymers. Most of us are aware of the polymer work, but we are unable to characterise food materials well enough. We have studied various diffusion theories, Michaels' work on xylene and other problems.

M. Gianturco (Coca Cola, U.S.A.): Prof. Thijssen has said aroma retention in pervaporation of coffee is 50% under unspecified conditions. I think under good conditions of spray drying higher retentions are possible, as high as 90% of almost any volatile; therefore, in your view what are the advantages of freeze-drying over spray drying?

M. Karel: If this were true, were there no other factors, such as control of bulk density, I do not see the reason for freeze-drying for some specific liquids. I have been indoctrinated with the idea that for a number of liquids the thermal conditions pertaining in freeze-drying, and the flavour retention combined result in a superior product.

H. A. C. Thijssen: Theoretically, it can be shown that optimal spray drying conditions produce less thermal deterioration than freeze-drying where the product remains for many hours at temperatures of at least 50°C with a temperature effect on activation, as against optimal spray drying conditions where the product rises above 100°C for only a few seconds. This has been proved, for example, by spray drying skim-milk with the enzyme phosphatase. As for aroma retention in spray drying it is possible to achieve about 99%. It may be that because freeze-drying conditions can be easily studied on a bench scale cheaply whereas spray drying requires industrial-size units for optimal conditions, more experimental work has been carried out on freeze-drying. For the optimum you can spray dry a liquid product for about one quarter of the price with almost a full aroma retention and no chemical deterioration whatsoever.

M. Karel: I am worried by such substances as phosphatase because a bit of off flavour at the product surface may be enough to deteriorate it. Your approach by taking the effect of temperature on activation energy is a good overall guide as to whether you need freeze-drying or not, but sometimes you need to be more specific.

S. A. Goldblith: As a comment for Mr. Gianturco and all concerned, while this is a scientific discussion, freeze-drying of coffee came about in part as a marketing approach.

A. I. Morgan, Jr. (Western Region Research Laboratory, California): May I raise a point about entrapping molecules by varying the surface matrix. Adding the correct surface active agent to the product to be freeze-dried can vary the behaviour of the matrix without changing the bulk composition being processed. For example, freeze-drying films of orange juice containing a soluble protein gives very good volatile retention for some proteins and poor for others.

M. Karel: P. McNulty and I have discussed the transport of flavour components from emulsions across various surface active layers. I think you cannot really expect a monolayer of surface active agents to retard mass transport of polar organic compounds.

A. I. Morgan, Jr.: I said orange juice and limonene.

M. Karel: Limonene is not the only polar component in orange juice, and it is a complicated one. We found that cholesterol, which is non-polar, is retarded.

N. Meisel (Les Micro Ondes Industrielles): Why have you not said anything about simple vacuum dehydration which can be used with microwave heating?

M. Karel: I have chosen to concentrate on those fundamentals which were basic to products rather than to equipment. I have worked with vacuum dehydration and I agree it is an important and potentially under-utilised method.

Essence Recovery on Citrus Evaporators

D. T. SHORE

A.P.V. Ltd., Crawley, Sussex, U.K.

INTRODUCTION

An important part of the evaporation process for concentrating fruit juices is essence recovery. In the recovery of essences in fruit juices, single components are not dealt with. Thijssen (this volume) has shown that the volatile flavour compounds are very readily removed as soon as they enter the evaporator and will come off in the first effect. The question is, what comes off? Taking a flavour, which is a subtle balance of many compounds, how can we determine with certainty from the relative volatilities of the compounds concerned what flavour compound is collected.

PROBLEMS WITH ESSENCES

It is proven by experience, particularly in Israel and Brazil, that in the citrus fruit juice industry any product resulting from a combination of concentration and an essence recovery system, taking as its feed the flash from one of the stages, or a pre-stripping stage of an evaporator, followed either by condensation or fractionation, cannot simulate the natural fresh flavour of the juice. Primarily for this reason the purchasers of these plants still use in many cases a cut-back technique whereby putting some of the fresh juice with the concentrate produces a more natural flavour. But they still buy these plants because some of their customers like to know that there is an essence plant and that they can have essences which can be sold!

Concord grape has often been referred to, where it is easy to recover methyl anthranilate using a suitable distillation column. Following very close criteria as to how many plates to put in the column or the packing to put in to get a number of equivalent plates, it is certain what the overhead fraction is going to be. But in the case of citrus fruits it is impossible to forecast.

The other problem is that neither the essence nor the concentrate produced is stable on storage. If the juice is not stored under very cold conditions the essence in the juice deteriorates. And from a straight equilibrium flash, e.g. from the first effect of an evaporator, the condensate, whether or not distilled to produce a higher concentration, contains two very distinctive phases: the oil and water phase compounds, and if these two phases themselves are not separated then the essence deteriorates very quickly indeed. The water phase can be made reasonably stable, but the oil phase contains unwanted substances in the flavour spectrum which really ought to be removed. Removing these, in order to improve the stability, makes a very complicated plant which seems to be never paid for and becomes obsolete three years later.

There is no single process which is generally acceptable in the industry. For depending upon the quality of the juice prepared and depending on the customer to whom the juice manufacturer is supplying, he will sell: either, short-time evaporated juice with no essence recovery; he will sell juice with some cut-back material; or he will sell juice with some oil phase material added back; or he will sell juice with some oil phase and water phase essence added back; or, he may indeed remove some of the peel oil components and add back the water phase and supply separately the oil phase material. All these things can happen with the same plant on different days of the week.

EVAPORATOR PLANTS AND EQUIPMENT

In a double-effect evaporator with the feed to the first effect, the vapour which is coming off the first effect and condensing on the shell side of the second effect will contain a great proportion of the volatile organic material. If vapour is bled so that stage two of the evaporator is only a partial condenser, a rich fraction of essence in the vapour leaving the evaporator is obtained. If this is brought to a long counter-current condenser, this acts as a wetted-wall distillation column, which means the condensate leaving the bottom is in near equilibrium with the dilute essence that is coming in at the base of the condenser in the vapour and there is a high concentration in the overhead fraction. The condensate fraction is discarded via a drain line. The overhead fraction passes to a chiller and a scrubber and if it is a long enough condenser a 200-fold essence is obtained without the complication of a distillation plant. A refrigerated chiller filled with liquid ammonia is often used—although a cold glycol circuit is preferred for convenience in engineering, particularly for somewhere like Israel.

A double-effect evaporator normally has live steam fed to the first effect, but is of course using the vapour boiled off in that first effect to do the evaporation in the second effect. That means using half a pound of steam for every pound of water evaporated. If a thermocompressor supplied with high pressure steam is incorporated in the circuit, vapour can be drawn from between the two effects to produce twice the evaporation because of the re-entrainment of one part of steam with one part of the vapour. Triple-effect operation and double effect with thermocompression give about the same thermal efficiency. However, the capital cost is lower because the cost of a thermocompressor is small compared with that of an additional effect, i.e. an extra stainless steel evaporator.

On applying the essence recovery system to the double-effect thermo-compression concentrator, there will be a 50% loss of vapours into the recompression cycle; therefore, it is not good from the point of view of applying essence recovery, since on a partial condensation system only 50% approximately of the essences are recovered anyway, giving only 25% yield overall.

In pasteurising fruit juices it has been common practice to use a rather high temperature and a holding time considerably longer than for most other food products, causing quite a lot of thermal damage to sensitive volatile flavour compounds before the evaporation, so the final quality of the essence or the juice being produced may be impaired.

The next system I wish to describe feeds the juice at a temperature lower than the pasteurising temperature into the second effect of the evaporator, then through the pasteuriser before feeding it to the first effect. The strip fraction containing the essences is taken from the steam chest of the third effect of the evaporator. If this is done, a higher quality product ensues. In this 'mixed feed' process a thermocompressor may be added to the first effect to improve the thermal economy to that of a quadruple-effect evaporator without the loss of volatiles referred to when a 'forward feed' double-effect evaporator is employed.

This concept may be applied to existing evaporators employing a thermocompressor on the first effect, i.e. there is a clear advantage in feeding what is called a mixed feed evaporator (that is, feeding raw juice into the second effect before the first effect, and then into the third) since two problems are avoided: first, pasteurisation heat damage to the essence fraction and, secondly, loss of the essence fraction in the thermocompressor loop.

PROCESS WITH CONCENTRATE FEED TO THIRD EFFECT

Finally, there is one other process which has become very interesting in Israel. It was previously mentioned that when the essence is put straight back into the concentrate it does not keep very well; and after one or two months at, say, 20°C, there is some deterioration in flavour, hence the practice of storing these essences separately. It is believed that the problem is the oil phase; but if the oil phase is separated and only the water phase put back some flavour components are lacking. A very interesting process tested this season in Israel uses a triple-effect evaporator operated in the mode described above, feeding 2, 1 and 3, and taking off from the steam side of the third effect to an essence plant consisting of a partial condensor followed by a scrubber. An essence concentrate of about 200-fold is obtained. Instead of putting that back into the final concentrate out of the third effect, it is put back into the *feed* to the third effect stage of evaporation. Although some essence is then lost in the last stage of evaporation, a more stable product results. The reasons cannot be explained.

DISCUSSION

S. A. Goldblith (Chairman introducing the speaker): We live in an era where, in America at least, many of our children are unfamiliar with freshly squeezed orange juice or freshly brewed coffee. In the case of the former it is, in large measure, due to the success of the evaporators and essence recovery. Therefore, it is particularly fitting that we have a contribution concerning this subject.

H. A. C. Thijssen: Can you say a little more about the oil phase problem?

D. T. Shore: There are a number of laboratories who, on our behalf, have been studying samples. Unfortunately, the factor I am looking for is the storage life. This last season in Israel was a very poor one for fruit—and everybody says 'Well you can't believe anything you've learned this year—we'll have to do it all again next year!' Of course some studies are being made, but we have no answer to the problem at the moment. There is obviously fractionation and removal of some undesirable factors, but we do not know what.

C. H. Mannheim: I was not really aware of the process mentioned in the last statement you made of sending the essence fraction back to the third effect of the concentration and that there is an undesirable fraction in the so-called aroma that you remove, either in the water or the oil phase. There is a belief that the oil phase is unstable and, therefore, it has to be separated and discarded, because compared with cold pressed oil obtained from the peel, it is actually a heated oil which some people call a stripped oil similar to the oil recovered during peel concentration. Originally, after separation of the oil phase from the water phase, it was discarded—this was about two or three years ago. Later on our studies showed that, despite the fact that it had been heated, this oil phase has quite a good flavour and a very good aroma and it is not so unstable and does not produce, to the best of my knowledge, anything objectionable, when stored at 0°C or a little less. No water should be present, and oxidation prevented.

To try and answer your main problem 'Why is oil stable in a citrus concentrate—properly treated, not mistreated—and does not deteriorate provided you store it at 0°C'. The stability, we believe, is because in citrus there are several natural anti-oxidants. For instance, if you take vitamin C—extremely stable in citrus—and you put the same amount of vitamin C in apple juice it will deteriorate in no time. The reason, I believe, is that in a natural state the oil is adsorbed on the pulp, because citrus oil is found there. You can easily prove this by centrifuging and separating any kind of fruit juice to obtain a kind of serum and also a pulp. This pulp is colourful and smells and tastes good with all the aroma and oil in it. This is a proper adsorption. Therefore, if you mix this oily fraction with the more highly concentrated juices the volatile losses are less. Considering everything, I was more in agreement with the beginning of your talk, rather than the end.

S. A. Goldblith: Thank you, I would like to add one thing, Prof. Mannheim. They are building in Florida today 'Tank farms' to store 80–90 000 gal citrus concentrate at about $-8°C$ for an extended period of time—up to a year—assuming, of course, that the microbiological problems, the cloud formation, the pectimethylestorase and the other enzymes have all been taken care of.

M. Gianturco: I would add that in this type of bulk storage the oil is only added at the time the product is packed for market. There are a number of companies, however, that use only the oil phase and do not use the water phase essence, and they find it quite stable. There is an unstable fraction in the essence, but it is neither the oil phase nor the water phase, if your collection system is efficient enough you will collect this fraction: it is a gaseous fraction that will impart an off aroma to the juice if it is stored in contact with the concentrate and when reconstituted the juice will have an off flavour.

D. T. Shore: Really confirming what I said: that the juice manufacturer does everything! He supplies it with the oil phase, without the oil and with both the oil and water phases.

Rheological Aspects of Fruit Juice Evaporation

G. D. SARAVACOS

Department of Chemical Engineering,
National Technical University,
Athens, Greece

INTRODUCTION

Large quantities of fruit and vegetable juices are concentrated by thermal evaporation. Evaporation is a reliable and economical method of concentration which is applied widely in the food and the process industries.

The design of evaporators is based on the principles of heat transfer and on practical experience from the operation of pilot plant units and industrial installations. Evaporation is a physical separation process in which heat and mass transfer take place simultaneously. The evaporation rate is controlled mainly by heat transfer and all the effort to improve it is directed towards accelerating the heat transfer rate.

The quality of fruit juices may be damaged by high temperatures, as a result of undesirable chemical reactions affecting the flavour or the colour of the product. The deterioration of quality is a time–temperature process and, in this respect, evaporators may be considered as chemical reactors in which operating temperature and residence time are the most important parameters (Meffert, 1964).

Short residence time and low temperature are necessary for heat-sensitive products like orange juice. These conditions are met in single-pass evaporators, such as falling or rising film (tube or plate) and agitated film types. Some products are relatively heat stable (e.g. tomato and grape juices) and they can stand relatively high temperatures and long residence times. These conditions prevail, for example, in forced circulation evaporators (Moore and Pinkel, 1968).

The rheological properties (viscosity, viscometric constants) are very important in the evaporation of fruit juices, because they are related directly to the transfer of heat and to the fouling of the evaporator surfaces.

In evaporation, heat transfer is controlled by the resistance of the juice side, since the resistances of the heating medium (condensing steam) and the metal wall are relatively low. The heat transfer coefficient of the juice is a function primarily of its flow rate and its rheological properties. The various flow patterns in evaporators (tubular, film, agitated film) aim at increasing the heat transfer coefficient of the product (McAdams, 1954).

Fouling in evaporators is caused by the formation of scale on the heated surfaces and it is a serious problem in the evaporation of some juices. Although the mechanism of fouling is quite complicated, the temperature of the heated surface and flow characteristics are considered as two important factors (Morgan, 1967).

RHEOLOGY OF FRUIT JUICES

A detailed study of the rheological properties of fruit juices, concentrates and purées was undertaken, because only limited information is available in the literature (Saravacos, 1968, 1970). The flow properties of these materials are also needed for fluid flow calculations and for quality considerations. The measurements were made with a tube viscometer and a coaxial–cylinder viscometer which permit various shear rates (van Wazer *et al.*, 1963).

Clear (depectinised) fruit juices are similar to sugar solutions, i.e. they are Newtonian fluids with a constant viscosity at a given concentration and temperature. Cloudy juices, containing suspended colloids and other solids, are non-Newtonian, usually pseudoplastic, and their apparent viscosity decreases at high shear rates. These juices become thinner at high flow rates or at high speeds of agitation.

The viscosity of all juices increases exponentially at higher concentrations. The viscosity of the Newtonian juices decreases significantly at higher temperatures. The effect of temperature is more pronounced the higher the concentration of the juice. Temperature has a relatively smaller effect on the apparent viscosity of pseudoplastic juices, and concentrates show the least dependence on temperature (Saravacos and Moyer, 1967; Saravacos, 1970).

EVAPORATION OF FRUIT JUICES

It is generally difficult to predict the evaporation behaviour of fruit juices based on their physical and chemical properties alone. Experimental data on the evaporation of a particular product are usually needed for the design of commercial units. Industrial evaporators are not suitable for experimentation, and pilot plant or laboratory units are normally used for research and development work.

A number of experiments were made using the pilot plant falling film evaporator, installed at the New York State Agricultural Experiment Station. The evaporator consisted of a vertical stainless steel tube 2 in.

in diameter by 10 ft long equipped with all the required pumping, heating and condensing accessories (Saravacos *et al.*, 1970).

Various types of apple and grape juices were evaporated, in order to establish the effect of rheological properties on the evaporation rate. The latter was expressed as the overall heat transfer coefficient (U), based on the internal surface of the evaporator tube. The boiling temperature was varied from 85 to 212°F and the flow rate was kept constant at a value resulting in complete coverage of the evaporation surface.

Clear apple juice was prepared by enzyme depectinisation and filtration. High heat transfer coefficients were obtained, ranging from 350 Btu/h ft^2 °F at 15° Brix to 210 Btu/h ft^2 °F at 65° Brix. The heat transfer coefficient, as expected on the basis of viscosity, increased significantly at higher boiling temperatures. No fouling of the evaporator surface was noticed even at 212°F, and it is apparent that clear juices can be concentrated in high-temperature evaporators with short residence times, if the product is very heat sensitive. In a multiple-effect evaporator the concentrated juice should be preferably at the highest temperature (counter-current flow of product/steam), because of its marked effect on viscosity.

Cloudy apple juice was concentrated to about 60° Brix, and the concentrate formed a gel upon cooling. The heat transfer coefficients of the cloudy juice were significantly lower than those of the clear juice (Fig. 1). The decrease of heat transfer coefficient due to concentration was more pronounced in the cloudy (pseudoplastic) juice. Evaporation of the cloudy apple juice at temperatures higher than 150°F had a negative effect on the heat transfer coefficient. At these high temperatures the evaporator surface was fouled considerably by deposits of the product, which reduced the evaporation rate.

Grape juices gave similar results. Depectinised Concord grape juice was evaporated at rates similar to the clear apple juice. Cloudy grape juices, containing colloids and other solids, were evaporated with more difficulty and some fouling was noticed at high boiling temperatures. High-extraction grape juices were found to be non-Newtonian fluids which were concentrated at much lower rates than the clear juices.

Fouling is affected strongly by the temperature of the heated surface (Morgan, 1967), and evaporation at low temperatures is the best method of avoiding it. Non-Newtonian juices have a fouling tendency which is accelerated at high temperatures. As explained earlier, temperature has a strong effect on fouling rate, resulting in a significant decrease of the evaporation rate. Fouling is associated with various chemical reactions, such as denaturation of proteins, which are strongly dependent on temperature.

The rheological properties of fruit juices have a significant effect on the heat transfer rate in various types of evaporators, as shown in the following examples.

(1) Forced circulation evaporators are advantageous for pseudoplastic products, because the reduction in apparent viscosity at high flow rates results in higher heat transfer coefficients. These units are suitable for products which can withstand long residence times (e.g. tomato juice).

(2) Agitated film evaporators are used for viscous pseudoplastic juices which require short residence times (e.g. orange juice).

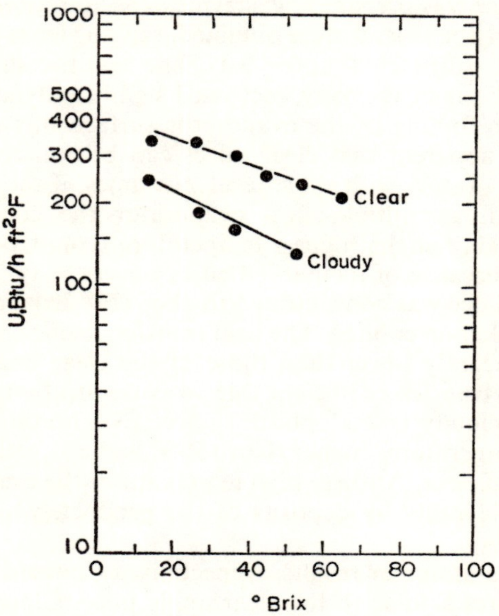

FIG. 1. Overall heat transfer coefficients in the evaporation of depectinised (clear) and cloudy apple juices (130°F).

(3) The rotary steam-coil evaporator (WURLING) has been found successful for the concentration of tomato paste. The high shear rates, developed in the vicinity of the rotating coil, reduce the apparent viscosity of the tomato paste, resulting in high heat transfer coefficients. Boiling takes place from a pool of tomato paste which can withstand the relatively long residence times.

CONCLUSIONS

The design and operation of evaporators can be improved by a knowledge of the rheological properties of the particular fruit juice. The type of flow, the operating temperature, and the residence time in the evaporator can be selected so that a high evaporation rate is achieved. Fouling of the evaporator surfaces is related to the temperature of the heated surface, the rheological properties, and the chemical composition of the juice.

ACKNOWLEDGEMENT

The experimental work was conducted at the Department of Food Science and Technology, New York State Agricultural Experiment Station, Cornell University, Geneva, N.Y., and all technical help is gratefully acknowledged.

REFERENCES

McAdams, W. H. (1954). *Heat Transmission*, McGraw-Hill, New York.

Meffert, E. F. Th. (1964). 'Residence Times and Changes in Fruit Juice in Evaporators', International Federation of Fruit Juice Producers, V.219, Juris Verlag, Zurich.

Moore, J. G. and Pinkel, E. B. (1968). 'When to use single pass evaporators', *Chem. Eng. Progr.*, **64**, 39.

Morgan, A. I., Jr. (1967). 'Evaporator concepts and evaporator design', *Food Technol.*, **21**, 1353.

Mutzenburg, A. B. (1965). 'Agitated film evaporators. I. Thin film technology', *Chem. Engng*, 175.

Saravacos, G. D. and Moyer, J. C. (1967). 'Heating rates of fruit products in an agitated kettle', *Food Technol.*, **21**, 54A.

Saravacos, G. D. (1968). 'Tube viscometry of fruit purées and juices', *Food Technol.*, **22**, 1585.

Saravacos, G. D. (1970). 'Effect of temperature on the viscosity of fruit juices and purées', *Food Sci.*, **35**, 122.

Saravacos, G. D., Moyer, J. C. and Wooster, G. D. (1970). 'Concentration of liquid foods in a pilot-scale falling film evaporator', *Food Sciences Bulletin*, No. 4, Cornell University, Geneva, New York.

van Wazer, J. R. *et al.* (1963). *Viscosity and Flow Measurement*, Interscience, New York.

DISCUSSION

D. T. Shore: When you referred to the difference in overall heat transfer coefficients of two boiling temperatures, 85° and 130°F, were those with the same temperature difference across the evaporator, and what size would that temperature difference be?

G. D. Saravacos: We experimented with the two temperatures, studying the whole curve of heat transfer coefficients at various boiling temperatures between 85° and 212°F, and we tried to keep the same temperature difference of about 50°F so as not to introduce another parameter. By increasing the boiling temperature, it was very clear in the depectinised juice that there was a very marked difference in heat transfer coefficients but, when we studied cloudy and pulpy juices, when we had a boiling temperature above 130°F, there was severe fouling and a reduction in heat transfer coefficients. At 212°F we were unable to get good heat transfer coefficients at all.

A. S. Michaels: I presume that you found the turbidity in these juices to be due to pectin-like materials which very often go through spontaneous coagulation and precipitation at elevated temperatures. This could very easily account for the fouling conditions you observed, because the higher temperatures are the places where the precipitation takes place most readily and that would obviously be at the transfer surfaces.

G. D. Saravacos: When you heat a cloudy apple juice serious coagulation does happen but it probably can be stabilised without difficulty. With fouling on a heat transfer surface, then you have a different situation. Pectin is suspected as one cause, but other quite minor constituents such as proteins which are more heat labile and which are denatured are involved.

A. I. Morgan, Jr.: If you use stainless steel tubes and measure the overall heat transfer coefficients it is very difficult to determine what additional resistance is being slowly added on the inside unless the heat flux is almost exactly equal at all times. By running a series of synthetic substances to which we added various components, e.g. fibre, pectin cloud, salt content, along with pH and a whole array of possibilities we got the impression that fouling seems almost always to go back to a gummy protein somewhere, even in a juice as free of protein as apple or grape. A false grape juice does not cause much fouling until there is a little glue in the form of a soluble protein. The damage it does is then accentuated by the presence of fibre. Nothing seems to be able to get a hold of the wall and stay there unless there is a little protein glue to stick the fibre on. If you stick trypsin in, for example, you almost eliminate the phenomenon.

G. D. Saravacos: From the little evidence we have, it most probably is protein. We could not complete the whole experiment when we used cloudy apple juice because the fouling was so severe even at very low boiling temperatures of 80°F. When the temperature of the jacket was very low at about 120–130°F we could get a good evaporation. I would say it is a matter of wall temperatures, not boiling temperatures, because once the wall is above a certain temperature then you get immediate fouling due to protein coagulation.

As far as the heat transfer coefficients are concerned, I agree with you that you have to know precisely those for the steam and wall sides of the metal. But in this case I would say that the controlling factor was the juice side; we had rather low heat transfer coefficients, so I would say the controlling factor was in the juices.

D. T. Shore: You had a particular length of tube in your evaporator, and I do not really see how you can draw too many conclusions for different concentration ranges within the same tube unless you are going to recirculate some of the juices. Of course, this affects the fouling problems not only in changing the liquid wetting of the wall but with any re-cycling of the material, such as protein material.

G. D. Saravacos: Perhaps I have not explained clearly, but the experiments were done as a series of measurements with a step by step increase in concentration but with a very high flow rate. You cannot work in an evaporator without wetting the whole inside. Experiments were conducted to see what the flow rate would be by increasing flow rate versus the evaporation rate; and we observed a certain minimum or optimum flow rate when we were sure that the tube was completely covered with a film of juice.

Session II

NON-MEMBRANE CONCENTRATION

Introduction

ARTHUR I. MORGAN, JR.

Director, Western Regional Research Laboratory,
U.S. Dept. of Agriculture,
California, U.S.A.

Workers in our field have a rather strong, almost intuitive notion that preconcentration can offer practical advantages over straight evaporative dehydration of foodstuffs from the single strength. We believe this for at least three reasons:

(1) Nearly all dehydrators consume several times more energy than can be accounted for in the latent heat of the water evaporated (order of 2–5 times) whereas the energy consumption of multiple-effect evaporators can be less than the latent heat of the water evaporated. In my country, which enjoys relatively cheap power, the term 'energy crisis' keeps coming up more and more often, so it appears that energy costs may become a more important factor with us than previously.

(2) The overall heat transfer rate that can be realised in dehydration equipment is usually less than can be realised in concentrators, per unit area of installed heat-transfer surface. Put in engineer's language, heat transfer coefficients in dehydrators are generally smaller than in evaporators, and the amount of installed heat-transfer surface in a dehydrator must therefore be correspondingly greater than in an evaporator, for a given duty. Thus, hardware costs may be minimised by preconcentration.

(3) It may be necessary to concentrate the foodstuff simply in order to get it into a physical condition in which it can be conveniently handled in a dehydrator (e.g. the dehydration of a citrus juice).

These considerations strongly suggest that some fraction of the water in a foodstuff should be removed by preconcentration, and the remainder in a dehydrator. This conclusion holds with particular force when preconcentration can be accomplished by use of multiple-effect evaporation, but it is not necessarily limited to such instances. Even single-effect evaporation can offer enough advantage to be worth while.

The use of single- or multiple-effect evaporators is a familiar practice, and we will welcome any discussion of recent improvements in that field. Nevertheless, we shall, I hope, be dealing with other forms of preconcentration as well; including those which may not involve evaporation at all.

I would like to mention several possibilities to you, some of which I am certain will be given specific mention by the speakers, and others which may be worth considering during the discussion.

The most obvious is, of course, freeze-concentration, which will be discussed by the first speaker. The method, as you probably know, can, in principle, remove water at an energy consumption of only about 15–30% of single-effect evaporation, and without risk of heat damage.

In the past, this method has been plagued by the problem of yield loss due to entrainment of solutes in the ice phase. Perhaps today's speaker will reveal to us how this problem has been solved.

My own laboratory has been quite active both in non-thermal concentration and in evaporation. Our non-thermal concentration work has encompassed both membrane and non-membrane methods. An example of the latter is our Osmovac process for making dehydrated fruit slices. In this process, the fruit is partially dewatered osmotically by immersion in strong sugar solution, and then vacuum dehydrated.

We also supported a study of factors affecting flavour retention in low-temperature evaporation of fruit juices, particularly evaporation from a semi-frozen slush.

In the thermal evaporation field, our attention has focused primarily on two areas: first, the fouling problem, that is preventing burning and formation of deposits on heat-transfer surfaces, and second, flavour retention by means of essence stripping and recovery.

These fouling studies encompassed both evaporator design and control. One result of this work was the invention of a rotary-steam-coil vacuum evaporator, called the Wurling evaporator, which has enjoyed considerable commercial success. The basic principle consists of moving the heat-transfer surface through the fluid, rather than attempting to move the fluid, as is done conventionally. The Wurling evaporator permits juices and purees to be concentrated to an exceptionally high degree.

We also conducted a study of feed-back and feed-forward control of tubular evaporators, with the objective of minimising the fouling of heat-transfer surfaces through correct programming of steam temperature.

We investigated a method of concentrating juices by spraying into superheated steam under partial vacuum, in order to eliminate heat-transfer surfaces altogether.

And last, but not least, we developed the WURVAC process of essence stripping and recovery. This is a method in which aroma volatiles are flash-evaporated from a juice or puree, the organics stripped from the condensate by means of an inert gas, collected in a small amount of aqueous phase, and returned to the juices.

There are many other potential methods of concentrating liquid foods while retaining the flavour and aroma. One that we have not investigated is the separation of water by means of solid clathrates, a scheme once proposed for the desalination of water, but which might prove applicable to concentrating foods.

So I come before you as one who has over the years taken an active part in non-membrane concentration. I maintain a keen interest in the subject, and I share your eagerness to hear whatever good news is revealed by today's speakers, and by the discussion that will follow.

Freeze-concentration

H. A. C. THIJSSEN

Physical Technology Laboratory,
Department of Chemical Engineering,
Eindhoven University of Technology,
Eindhoven, The Netherlands

INTRODUCTION

In freeze-concentration water is first partly segregated from the aqueous solution by crystallisation and thereupon the ice is separated from the concentrated liquid phase. The crystallisation and separation operations can be performed in the same apparatus. In actual practice, however, these operations are performed in two separate apparatuses.

The process is particularly suited for the concentration of heat labile liquid foods. Since water is withdrawn from the solution by the phase transformation from liquid to crystal, aroma losses by evaporation can be completely avoided. If high local supercoolings during crystallisation are prevented, ice crystals of a very high purity are produced. The loss of solute is then completely controlled by the degree of perfection of the ice–liquid separation. With the right crystallisation apparatus and a complete ice–liquid separation, freeze-concentration therefore exhibits unique qualities. Except for the economics freeze-concentration is superior to evaporation and reverse osmosis for the concentration of liquid foods containing volatile aromas.

In evaporators volatile aromas in the feed are almost quantitatively lost with the water vapour. The quality can partly be restored by separating the aromas from the vapour leaving the evaporator in a distillation column and by feeding them back to the concentrated liquid. Very volatile aromas, however, are lost with the inert gases and aromas with a volatility equal to—or less than—the volatility of water in the solution cannot be recovered from the vapour. Partly because of the incomplete aroma recovery, being generally not more than 60%, and partly because of degradative thermal reactions, the original quality of these liquid foods, which include coffee and tea extracts and most fruit juices, is never fully restored. Technically, however, evaporation is the best developed concentration process. Moreover, reverse osmosis does not exhibit the same selectivity as freeze-concentration with respect to small aroma molecules. Because membranes of the required selectivity have low permeabilities, their application in dewatering aromatic liquid foods will probably remain restricted to a few special products.

Disadvantages of freeze-concentration which prevented this technique from becoming competitive in the past were the loss of dissolved solids with the ice, the loss of a part of the volatile aromas during the ice–liquid separation and above all the high processing costs. Capital costs were 3–10 times higher than those of evaporators with aroma recovery. However, recent developments have shown that solute losses can be reduced to a negligibly low level provided the liquids are clear or of a colloidal nature.

The object of this paper is to describe the present state of freeze-concentration with special emphasis on aromatic liquid foods. This is because of the fact that along with the low process temperature it is the possibility of complete aroma retention that puts this technique in a favourable position. In addition, future developments will be touched upon. In order to make it possible to compare the performance of the different systems the main fundamental aspects of crystallisation and of crystal–liquid separation will be given first. Special attention will be paid to the effect of the mean crystal size upon the separation. The relatively high investment costs or high solute losses of some commercial systems can partly be attributed to a lack of understanding of these fundamentals. The paper will be concluded with an economic analysis and a cost comparison with evaporation. However, these evaluations should be interpreted with extreme caution and should be used only as a means of comparing different processes and different combinations of techniques.

CRYSTALLISATION

Water can be crystallised from a solution by cooling and removal of the heat of crystallisation in any type of apparatus, including pipe heat exchangers, plate heat exchangers, stirred jacketed vessels, scraped heat exchangers, vacuum crystallisers, internally cooled rotating drums, and cooled endless belts. Provided the solution is seeded with active nuclei, crystallisation will always take place, irrespective of the process conditions. Critical in the process is the separation of the crystals from the concentrate and also the velocity at which this separation can be performed (capacity). The cost of separation and the solute losses with the ice strongly increase with decreasing crystal size. The cost of crystallisation in general increases with increasing crystal size. The optimum crystal size is determined by the type of crystallisation process, the crystallisation conditions, the type of separator and especially by the value of the concentrate. A high value of the concentrate demands a low allowable loss and this in turn requires large crystals. In each case, the process conditions that yield a certain crystal size at minimum cost or that result in the maximum crystal size at constant cost must be known.

Fundamentals of crystallisation

Water forms a eutectic mixture with the other components of liquid foods. Down to the eutectic temperature, therefore, water segregates upon cooling in the form of ice crystals, leaving a concentrated liquid phase. Figure 1 shows experimental freezing curves of apple juice, coffee extract and some sugar solutions. It can be seen that in freezing apple juice of 11 wt% to −5.8°C the solute content of the liquid phase becomes 40 wt% under equilibrium conditions and that 81.5% of the water segregates as ice crystals.

Crystallisation may take place either in layers, as in pipes, in plate heat exchangers, on discs, on drums, on endless belts, or in stirred suspensions of ice crystals. Because of the difference in crystal growth in layer freezing and in suspension freezing these two main types of crystallisation will be treated separately.

LAYER FREEZING

Layer freezing or regular freezing is the name given to processes in which a crystal layer is deposited upon one which has previously been formed from the same solution. By this unidirectional freezing, the ice grows in the form of needles or bars with an irregular cross section perpendicular to the cooled surface. The rate of freezing or the growth velocity of the tips of the ice crystals decreases with increasing solute concentration and decreasing bulk supercooling at the crystal tips (Huige, 1972). A decrease in freezing rate results in an increase in diameter of the crystals (Koonz and Ramsbottom, 1939; Kramers, 1958; Luyet, 1962; King *et al.*, 1968). The mean diameter of the crystals decreases with increasing solute concentration and at a constant concentration decreases with a decreasing value of the molecular diffusion coefficient of water and with increasing viscosity (Rulkens and Thijssen, 1972; Rohatgi *et al.*, 1969). The viscosity and molecular diffusion coefficient are of course coupled. The mean thickness of the liquid separating parallel crystals increases with increasing solids concentration up to about 20% and remains constant at higher concentrations (Rulkens and Thijssen, 1972). Solute rejection in the direction of the growth of the crystals is only appreciable at a very low solute concentration, say below 1% (Dschu, 1967; Kepner *et al.*, 1969). In unidirectional freezing of solutions containing more than 10% dissolved solids, where the non-frozen liquid layer is at about freezing temperature or somewhat undercooled, the solute concentration upon thawing of the frozen layer is equal to that of the non-frozen liquid. This means that by merely separating the frozen layer from the non-frozen liquid outside that layer, no concentration effect is obtained. Only under extremely slow freezing conditions, say growth velocities of one centimetre per day or less, can rejection be expected. These very slow freezing rates were in former days employed for the preparation of 'apple jack'. When the non-frozen liquid is vigorously

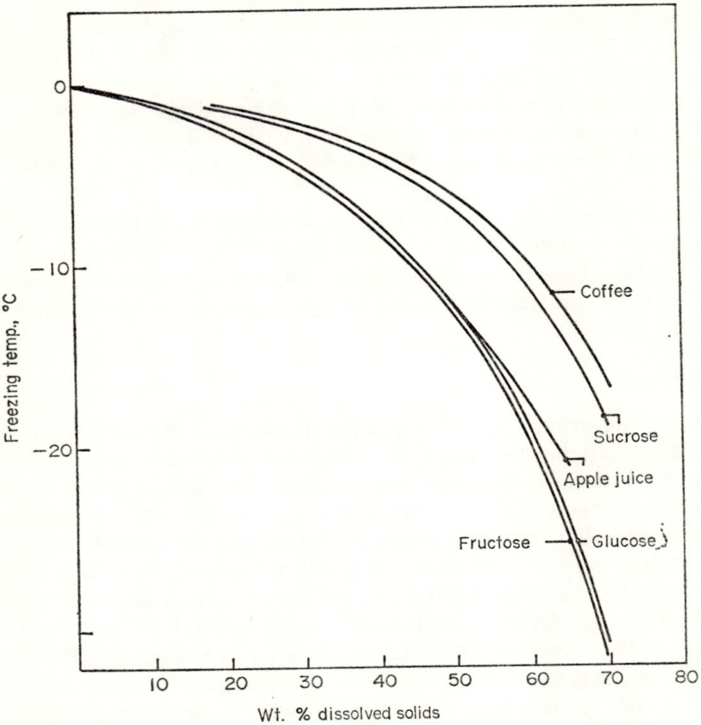

FIG. 1. Freezing curves of coffee extract, apple juice and of some sugars (Chandrasekaran and King, 1971, Heisz and Schachinger, 1951).

stirred and heated, slow freezing by equi-directional cooling via the frozen layer results in some solute rejection. The effect, however, is still so weak that also in this case separation of the porous solid layer without washing cannot be used as an effective concentration technique.

BULK CRYSTALLISATION

The kinetics of nucleation and growth of freely suspended crystals in stirred suspensions have been studied much more extensively.

Crystal growth rate

The mean crystal size in continuously operating crystallisers is, in sequence of decreasing influence, dependent on solute concentration, bulk supercooling, mean residence time of the crystals in the crystalliser and degree of turbulence in the crystal suspension. The experimentally observed effect of the solute concentration in a glucose solution upon growth rate is

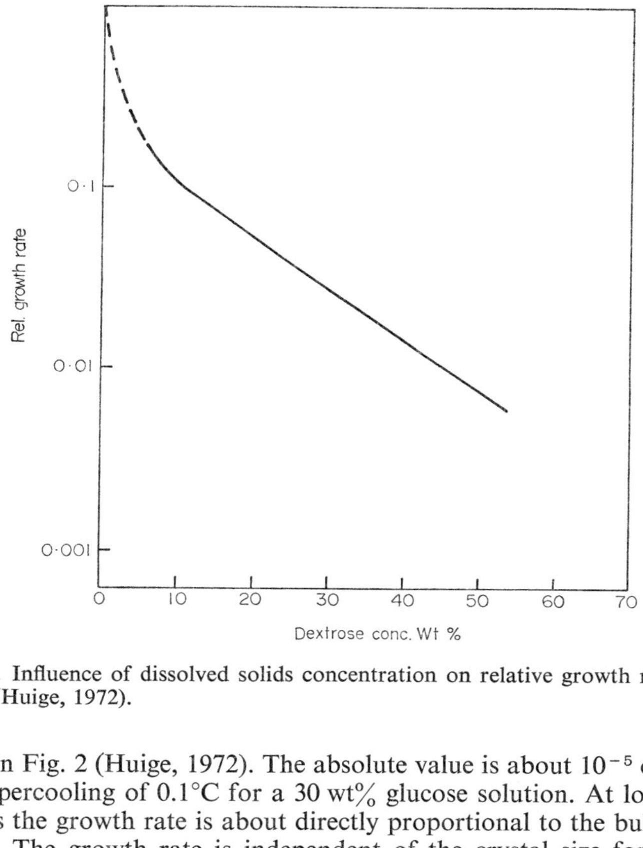

FIG. 2. Influence of dissolved solids concentration on relative growth rate of ice crystals (Huige, 1972).

shown in Fig. 2 (Huige, 1972). The absolute value is about 10^{-5} cm/s at a bulk supercooling of 0.1°C for a 30 wt% glucose solution. At low supercoolings the growth rate is about directly proportional to the bulk supercooling. The growth rate is independent of the crystal size for crystals larger than about 50 μm. This means that at constant bulk supercooling and constant solute concentration, the mean crystal size produced in a continuous stirred tank crystalliser is proportional to the crystal residence time. The effect of the power input of the stirrer per unit mass of suspension upon the growth rate appears to be small (Huige, 1972). The effect of the mean residence time of the crystals in stirred crystallisers with a constant crystallisation capacity but at varying crystal concentration on mean crystal diameter is illustrated in Fig. 3. The figure clearly demonstrates that the method of cooling also strongly influences the crystal growth rate and mean crystal size. The less than linear effect of residence time on crystal size is here caused by the decrease of the bulk supercooling with increasing crystal concentration at constant capacity. At a given rate of heat withdrawal per unit mass of suspension the bulk supercooling is

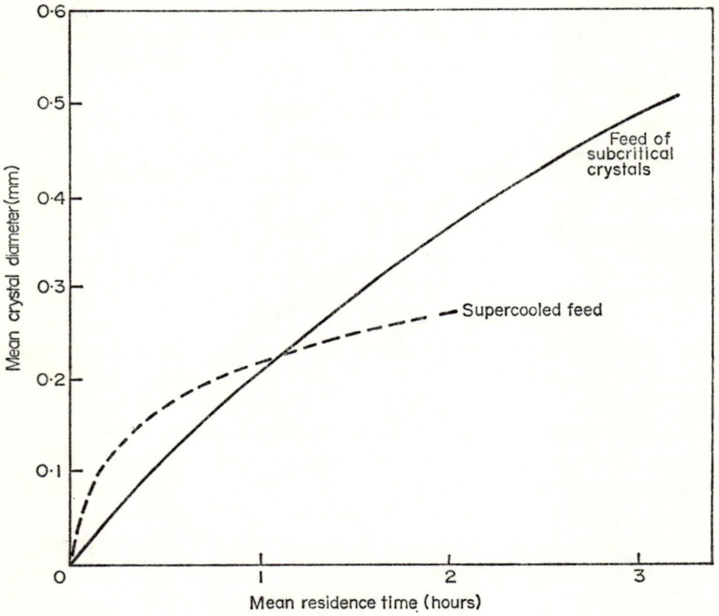

FIG. 3. Effect of mean residence time of the crystals on mean crystal diameter in a 30 wt% dextrose solution (Huige, 1972).

inversely proportional to the total area of the crystals per unit mass of suspension. This total crystal surface area per unit suspension volume is equal to the product of the crystal mass concentration (crystal mass per unit volume of suspension) and the specific surface area of the crystals. At a given crystallisation capacity a high supercooling can thus only be obtained if the number of crystals per unit volume of suspension is low. The same conclusion can, of course, be drawn from a mass balance over the crystals produced. To obtain the largest possible crystals at an acceptable residence time, the above dictates a low net nucleation rate of crystals.

Nucleation rate

The nucleation rate increases with increasing solute concentration and is proportional to the square of the bulk supercooling (Huige, 1972). Because the heat of crystallisation can generally not be withdrawn uniformly from the whole crystal suspension, there will always be spots with a higher supercooling than the mean bulk supercooling. Because of the second order dependence of the nucleation on supercooling, these colder spots contribute much more to the nucleation and less to the crystal growth than the bulk. Crystal growth exhibits only a first order dependence. When the stirring rate in the crystalliser is increased, the volume fraction and degree of supercooling of the cold spots will decrease and consequently so will the nucleation rate.

However, not all nuclei born in the cold spots will survive the mixing with the bulk of the suspension. This is because smaller crystals have a lower equilibrium temperature in a solution than larger ones. The effect of the diameter of spherical isotropic crystals upon the lowering of the equilibrium temperature is expressed by the Gibbs–Thomson equation

$$\Delta T^* = \frac{4\sigma T^*}{d\rho_s \Delta H} \quad (1)$$

where ΔT^* is the lowering of the equilibrium temperature of a crystal with diameter d in a solution with respect to the equilibrium temperature T^* of that same solution for a very large crystal, σ is the surface free energy, ρ_s is the density of the crystal and ΔH the heat of crystallisation. The calculated effect of the crystal diameter on ΔT^* is presented in Fig. 4. The crystal diameter in equilibrium with a certain bulk supercooling we will call the critical diameter. Upon mixing of a suspension of small crystals with a suspension of large crystals, the bulk temperature will be established somewhere in between the lower equilibrium temperature of the smaller crystals and the higher equilibrium temperature of the large crystals. Because this bulk temperature is higher than the equilibrium temperature of the small crystals the latter will melt. The large crystals on the contrary will grow. This growth of larger crystals at the expense of smaller crystals is known as the ripening effect. The melting rate of the smaller subcritical crystals and the growing rate of the larger supercritical crystals increase with an increase in the difference between the respective values of the sizes of the small and large crystals. So if small nuclei produced in cold spots are removed from there immediately and are homogeneously mixed with the bulk of larger crystals almost all of these nuclei will melt.

Energy consumption

The energy consumption of crystallisation is dependent on the energy input by stirring and pumping, the heat influx from the environment, the degree of heat exchange between the ice and concentrate stream leaving the freeze-concentrator and the warmer feed, the temperature of the feed and the method of heat extraction. First we shall calculate the energy consumption of a single-stage heat pump per unit calorie of heat withdrawal. Thereupon the energy consumption will be calculated per unit mass of ice.

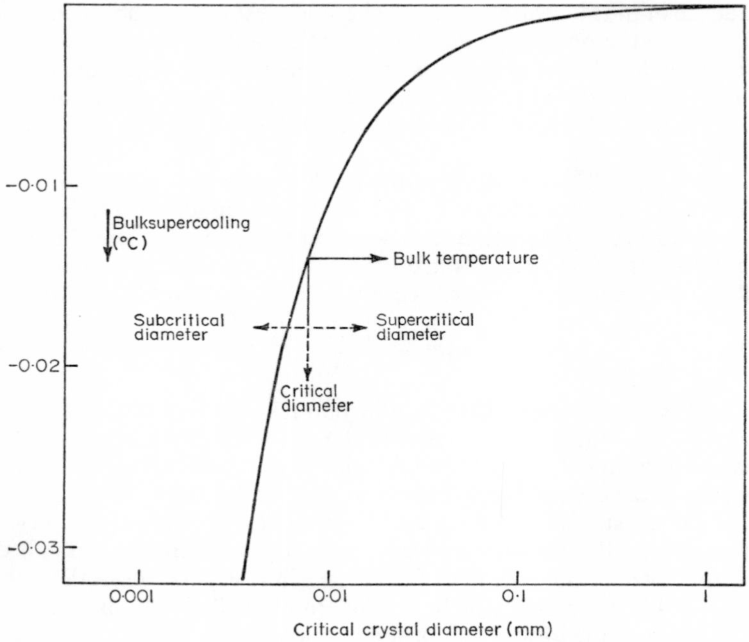

FIG. 4. Relation between bulk supercooling of a liquid and the critical diameter of ice crystals (Huige, Senden and Thijssen, 1972).

Energy consumption of heat pump

The energy requirements of a mechanical heat pump hardly depend on the refrigerant used in the compressor. We shall calculate the energy consumption of an electromotor driving a single-stage ammonia compressor. The brake power consumption for compressors with a capacity of about 100 000 kcal/h has been obtained from data from compressor manufacturers. For the electromotor we have taken an efficiency of 93%. Figure 5 gives the relation between the energy consumption of the motor in kilowatt-hours per 100 000 kcal heat extraction and the temperature difference between the compressor and evaporator of the refrigeration unit. The condensor temperature is taken as parameter. It is obvious from the figure that at small temperature differences between compressor and evaporator, the energy consumption hardly depends on the absolute value of compressor- or of evaporator-temperature.

Relative energy consumption of crystallisation

The thermal efficiency of the crystalliser can be defined as

$$\beta \equiv \frac{\text{kg ice removal} \times 80}{\text{heat extraction by compressor}}$$

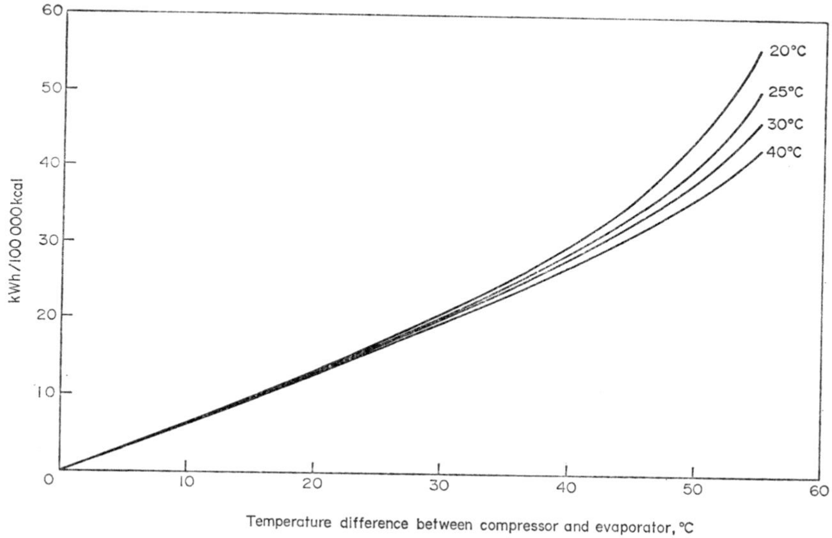

FIG. 5. Effect of temperature difference between condensor and evaporator for an ammonia single-stage compressor on electricity consumption per 100 000 kcal heat extraction. The condensor temperature is taken as parameter. Capacity of the compressor is about 100 000 kcal/h.

in which the value 80 is the heat of crystallisation per kg water. β is a dimensionless quantity. A value of 1 means that the crystalliser functions ideally, with 100% heat exchange between the separated ice and concentrate and the feed, perfect insulation and no energy consumption for pumping and stirring.

In order to make cost comparisons that are independent of the rapidly changing values of currencies we have expressed the energy consumptions in steam equivalents. We may assume that the ratio of the price of electricity to the price of steam is fairly stable. For steam we have taken a price of Dfl.10 (Dutch florins) per metric ton and for electricity a price of

Dfl.0.07 per kWh. By conversion the price of 1 kWh is equivalent to 0.007 tons of steam. Using these steam equivalents, we can also directly compare the energy consumption of freeze-concentration with the steam consumption of evaporators. For a condensor temperature of 30°C, Fig. 6 gives the energy consumption in tons steam equivalents per ton ice production as a function of the temperature difference ΔT between condensor and evaporator with the thermal efficiency as parameter.

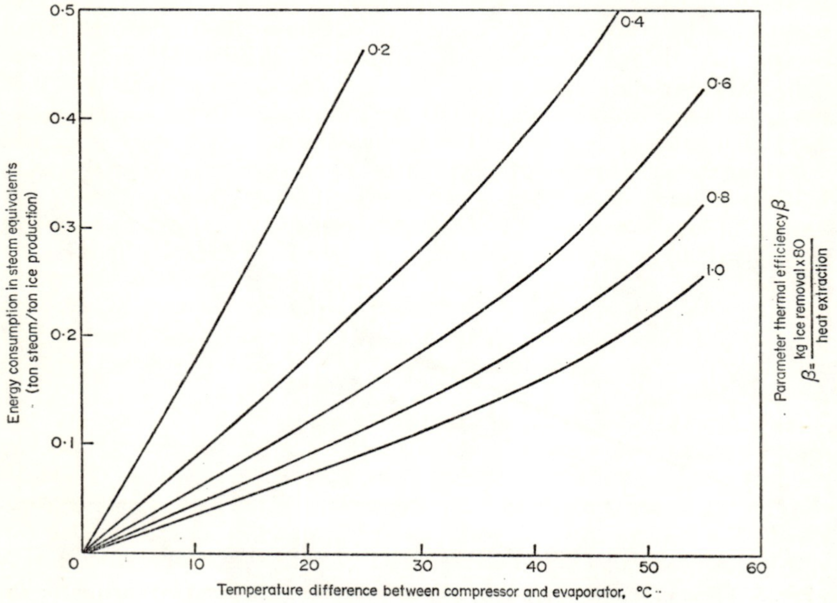

FIG. 6. Relation between the energy consumption of a crystalliser expressed in steam equivalents and the temperature difference between compressor and evaporator of the refrigerator. The parameter is the thermal efficiency of the crystalliser.

It is clear that to reduce the energy consumption a low temperature difference between the compressor and the condensor must be aimed at. This means also that the temperature difference between the evaporating refrigerant and the crystallising liquid has to be kept small. The steam consumptions of single-, double- and triple-effect evaporators with a thermal efficiency of 1.0 are 1.0, 0.5, and 0.333 tons steam per ton water evaporation respectively. Because for the concentration of liquid foods more than triple effects are never used it is evident that the theoretical energy consumption of crystallisation compares very favourably with that of evaporation.

Technical crystallisers

The heat of crystallisation can be withdrawn directly both by vacuum evaporation of a part of the water and by evaporation of a secondary refrigerant such as butane or indirectly by means of a refrigerant that is separated from the process liquid by a solid wall.

CRYSTALLISERS WITH DIRECT HEAT REMOVAL

In vacuum freezers the crystallising solution boils at an absolute pressure of about 2 mm of mercury. At a liquid temperature of -3°C, 140 kg of water must be evaporated for the crystallisation of 1000 kg of ice. Direct cooling has distinct advantages but also some disadvantages compared with indirect cooling. A disadvantage is that due to evaporation, a part of the aromas will be lost together with the inert gases. Direct cooling, however, eliminates a cooling wall and consequently the use of expensive scraped heat exchanging surfaces such as are necessary in indirectly cooled bulk crystallisers. Moreover, by compressing the water vapour from 2 mm to about 7 mm of mercury and using the separated ice crystals as a coolant to condense the compressed vapour, very low energy consumptions can be attained. For large vacuum freezers with vapour compression by steam boosting (Aackter and Barak, 1967) energy consumptions as low as 8 kWh per 1000 kg water removal are reported. Direct cooling has been studied extensively for sea water desalination. The process is not yet applied commercially for liquid foods.

As mentioned already the advantages of simplicity and low energy costs must be paid for by a lower quality of the concentrate compared with the product obtained from an indirectly cooled crystalliser. At least theoretically, aroma losses can substantially be reduced if the gas–vapour mixture leaving the condensor is stripped countercurrently in an aroma absorber with the concentrated liquid leaving the ice–concentrate separator. A process example of aroma recovery is given by the flowsheet in Fig. 7. If the temperature of the condensor is not too low and a part of the non-condensables leaving the first section of the aroma absorber are recirculated over the condensor, the aromas are not, or only to a small extent, condensed in the condensor and are only removed from the gas–vapour mixture by the concentrated liquid in the aroma absorber.

CRYSTALLISERS WITH INDIRECT HEAT REMOVAL

Only indirectly cooled crystallisers have, so far, found application in the food industry. There are two classes of crystallisers using indirect heat transfer, viz. internally and externally cooled crystallisers. The first class involves the extraction of heat through the wall surrounding the crystallising solution. The second class is characterised by the adiabatic operation of the crystalliser: the heat is externally transferred from the feed to the crystalliser.

Internally cooled crystallisers

This class can be divided into crystallisers producing a solidified or almost solidified suspension, and a group of crystallisers producing pumpable slurries.

The first group belongs to the category of layer freezing. Upon solidification of a layer of the desired thickness this layer can either be washed *in situ*

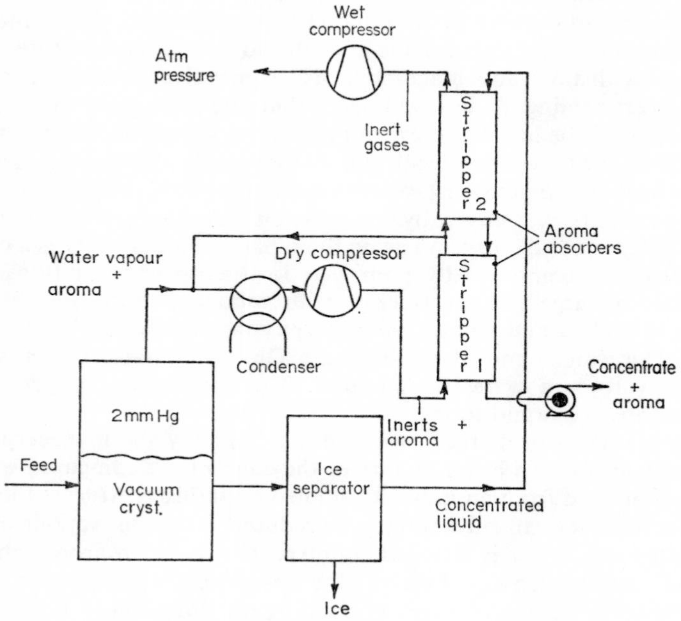

FIG. 7. Flowsheet of vaccuum crystalliser with aroma recovery.

or mechanically removed as solid slabs or as flakes and be separated elsewhere. By partial solidification, even dilute solutions can be concentrated in one step to 40% or more. The washing *in situ* of frozen layers inside the pipes or between the plates of a heat exchanger has the advantage of simplicity; this method, however, is not yet in use on a commercial scale. The method of separated crystallisation and separation operations is employed in the Linde–Krause crystalliser (Thijssen, 1968) which consists of a large stainless steel drum, internally refrigerated, rotating in a trough of the liquid to be concentrated. The 'solid' layer is removed continuously by a scraper knife. Because of the tiny ice crystals the liquid–ice separation is difficult. The process is used commercially for orange juice. In another version (Urban, 1969) the liquid is sprayed onto a slowly revolving drum or on internally refrigerated discs, and is removed as flaked ice.

Most crystallisers of this class used for freeze-concentration belong to the second group and produce pumpable suspensions. In the more 'classic' types, the crystal suspension remains in the crystalliser only a few minutes. Because of the short residence time the product crystals are very small, being less than 50 μm. In the votator process (Anon., 1965) the crystals are produced in horizontal scraped surface heat exchangers. Short residence times are also employed in the Gasquet process (Gasquet). Their horizontal scraped surface tubular heat exchanger has an inside diameter of 14 in. Much longer residence times, up to several hours, are taken in the Union Carbide process (Wenzelberger, 1969; Toulmin, 1959; Anon., 1964) by Struthers (Muller, 1966; Muller, 1967), by Daubron and by Phillips Petroleum Comp. Union Carbide has acquired several patents on a method of 'controlled' ice nucleation and crystal growth as developed by the Commonwealth Engineering Co., Dayton, Ohio. They claim the production of large crystals by using a cascade array of crystallisers–separators. Struthers advocate the use of stirred but non-scraped indirectly cooled crystallisers. To obtain large crystals they closely control temperature, driving force, turbulence and crystal concentration. Daubron (Daubron) employs vertical ammonia-jacketed scraped cylindrical vessels. The residence time of the crystals is about 3 h. The nuclei are formed on the cooled wall, and the crystals grow in the slowly stirred bulk of the suspension. An advantage of this process is its simplicity. One of the patents of Phillips Petroleum takes advantage of the ripening effect. The crystals formed in the heat exchanger are recirculated through a holding or ripening tank. At a total residence time of 1.3 h spherical crystals ranging in size from 0.4 to 2 mm are formed while concentrating beer from 3.2 to 6.0 wt% alcohol.

Externally cooled crystallisers

The class of externally cooled crystallisers can be sub-divided into three main types. The first type involves supercooling of the feed stream. The supercooled crystal-free liquid releases its cold in the crystalliser. High supercoolings of up to 6°C appear to be feasible. To minimise the change of primary nucleation and crystallisation in the heat exchanger, which may cause a complete blockage of the liquid path, the wall of the heat exchanger contacting the process liquid must be highly polished (Thijssen *et al.*, 1968). With this type, local supercoolings in the crystalliser can virtually be suppressed. In 30 wt% glucose solutions, crystals of about 0.2 mm are obtained at a mean residence time in the crystalliser of only about half an hour. The liquid from the crystalliser can be recycled to the heat exchanger. The crystals are retained in the crystalliser by means of a filter in the suction line. This type of crystalliser, however, is not commercially available.

The second type is characterised by recirculation of the whole suspension from the crystalliser to the heat exchanger. Residence time of the crystals in the heat exchanger is generally short compared with their residence time in the crystalliser. Consequently the crystal growth occurs mainly in the crystalliser. The process has been developed by Lurgi (Döge and Vêlebil, 1969) and is marketed by Krauss Maffei (Krauss Maffei). They use a scraped surface heat exchanger. The slurry leaving the heat exchanger is gently mixed with the suspension in the crystalliser. The ripening effect in the suspension leaving the heat exchanger is optimal if the flow through the crystalliser before it returns in part to the heat exchanger is a plug flow.

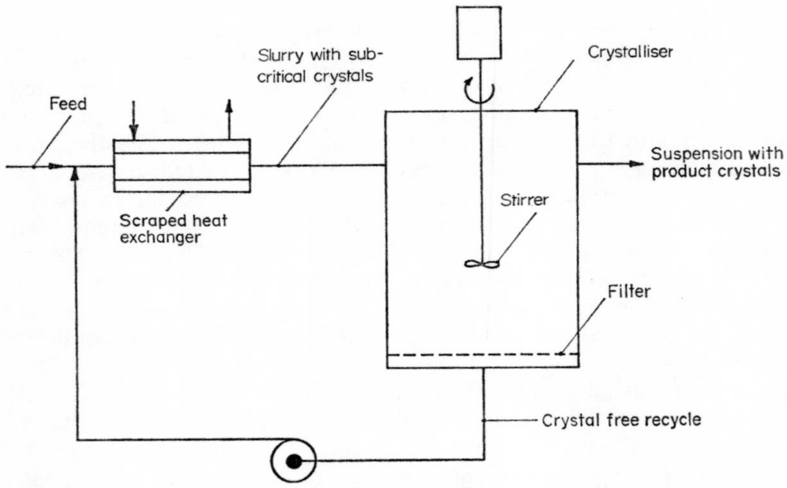

FIG. 8. Schematic representation of crystallisers externally cooled by sub-critical crystals.

Sepial uses a cascade of two internally cooled crystallisers. The majority of the ice crystals are formed in the first crystalliser, and the crystals are grown to the desired dimensions in the second stage.

The last type of crystalliser in this class is that producing sub-critical crystals in an external heat exchanger. Crystal-free liquid is partly recirculated from the crystalliser to the heat exchanger (see Fig. 8). The heat flux through the wall of the scraped heat exchanger is taken very high. This results in a very strong nucleation. The residence time of the crystals in the heat exchanger is only a few seconds. The extremely small crystals produced in the heat exchanger are fed continuously to the crystalliser, where the residence time is at least half an hour. Because of the relatively long residence time in the crystalliser the bulk supercooling is small, amounting to less than 0.02°C. The small crystals in the feed stream to the crystalliser appear to be sub-critical under these circumstances and they melt when mixed with the suspension of large crystals. The heat of melting of the small feed crystals is consumed by the growing large crystals. An increase in nucleation rate in the heat exchanger causes an increase of the mean diameter of the product crystals. The calculated effect of the mean

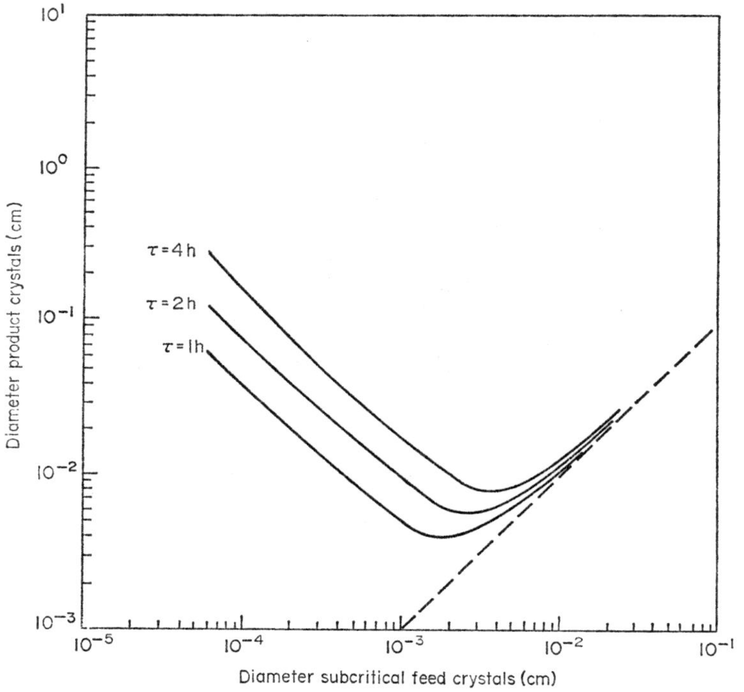

FIG. 9. Calculated effect of diameter of sub-critical crystals fed to stirred ripening tank upon diameter of product crystals in ripening tank. Parameter is mean residence time of crystals.

diameter of the nuclei leaving the heat exchanger on the mean size of the crystals in the crystalliser is presented in Fig. 9. The process has been developed at the Eindhoven University of Technology (Huige, Senden and Thijssen, 1972; Huige, 1972) and further developed to a commercial scale by Grenco (Grenco, 1972). At a residence time of about 2 h, spherical crystals of about 1 mm are grown from a 40 wt% sucrose solution. Figure 10 is a photograph of ice crystals grown from 40 wt% coffee extract.

FIG. 10. Photograph of ice crystals produced in a crystalliser externally cooled by sub-critical crystals. Enlargement ×60.

ICE–CONCENTRATE SEPARATION

The commercial applicability of freeze-concentration stands or falls with the effectiveness of the separation. The separation can be performed either batchwise or continuously in presses, filtering centrifuges, wash columns, or a combination of these devices.

In all separators the capacity is inversely proportional to the viscosity of the concentrate and directly proportional to the square of the mean diameter of the crystals as expressed by the relation

$$Q = \frac{\Delta P g d_e^2}{0.2\mu l} \cdot \frac{\varepsilon^3}{(1-\varepsilon)^2} \; (=) \; \text{cm}^3/\text{cm}^2 \text{ s} \qquad (2)$$

where Q = draining rate from the crystal bed (cm^3 liquid per second and cm^2 cross section of the bed); ΔP = pressure difference exerted over the bed by compression, centrifugal force or pressure drop of the filtrate (kg/cm^2); d_e = effective diameter of the crystals (cm); μ = viscosity of liquid (poise); l = thickness of the bed (cm); g = acceleration due to gravity (cm/s^2); and ε = the volume fraction in the bed filled by the liquid phase. The combined properties of the bed are represented by the bed permeability K

$$K \equiv \frac{g d_e^2}{0.2} \cdot \frac{\varepsilon^2}{(1 - \varepsilon)^2} \tag{3}$$

The fractional loss of solute with the ice is given by the relation

$$\gamma = Z \left| \frac{x_c}{x_F} - 1 \right| \tag{4}$$

where γ = the fraction of the solute in the feed to the freeze-concentrator that is lost with the ice; Z = the amount of concentrated liquid that remains occluded per kg of ice (kg/kg); x_c = the weight fraction of dissolved solids in the concentrated liquid phase leaving the crystalliser; and x_F = the weight fraction of dissolved solids in the feed to the freeze-concentrator. It is apparent from eqn (4) that the effect of an incomplete separation strongly increases with an increase of the concentration factor x_c/x_F.

Presses

Both hydraulic piston presses and screw presses are used. The loss of solute is fully determined by the amount of liquid that remains occluded in the compressed ice cake. After compression the solids are so heavily occluded that they cannot be removed by washing the compressed ice cake. At pressures up to 10 kg/cm^2 about 0.6 kg liquid remains occluded per kg of ice. At pressures of the order of 100 kg/cm^2 and long compression times, the amount of occluded liquid can be reduced to about 0.05 kg/kg ice. Because of these high residual liquid contents, presses can, according to eqn (4), only be used at values of the concentration factor x_c/x_F approaching 1.

Filtering centrifugals

In basket centrifuges, an ice bed is obtained with a porosity of between 0.4 and 0.7. Values of 0.4 are only obtained for more or less spherical crystals. Dendrites give high porosities. In contrast to presses, however, a part of the pores are emptied in the centrifugal field. The remaining liquid in the cake remains as:

(1) liquid held by viscous and capillary forces in the interstices between the crystals;
(2) liquid held to the crystal surfaces by viscous forces.

According to Brownell and Katz (1947) the final residual fractional filling

S_r of the voids between the crystals after centrifugation until constant liquid content of the cake is

$$S_r = c\left(\frac{\sigma g}{Krw^2\rho}\right)^{0.264} \tag{5}$$

where c = a constant dimensionless factor; K = permeability as defined by eqn (3) (cm^3/s^2); w = angular velocity of the centrifuge (rad/s); ρ = the liquid density (g/cm^3); and r = radius of basket (cm). It is evident from eqns (3) and (5) that the residual liquid content is inversely proportional to the root of the mean crystal size. At constant whizzing times of the cake and constant speed of the centrifuges the residual liquid content is according to Silverblatt and Dahlstrom (1954) also inversely proportional to the viscosity of the liquid. Pankovic (1962) has studied the effect of the speed of rotation upon the residual liquid content in a push centrifuge. With snow he found a strong decrease in the range from 1 to 1000 times gravity. For g values from 1000 to 2000 the residual water content remains more or less constant at a level of 0.1 kg liquid/kg ice. Okabe (1962) has studied the separation of a brine from an ice slurry in an under-driven basket centrifuge operating at 750 g. From his experiments a residual value of 0.19 kg liquid/kg ice can be calculated.

It is, however, the possibility of washing the cake with water or with melted ice that renders separation by centrifuges more effective than by presses. The wash water of course will dilute the concentrate. Depending on crystal size and liquid viscosity the fractional loss γ can, however, still be as high as 10%. An even more serious disadvantage of centrifuges is the loss of volatile aromas. During spinning of the liquid from the filter cake the liquid is brought into intimate contact with an excess of air.

Wash columns

A perfect separation of ice and liquid without any dilution can be obtained in wash columns. Because wash columns are completely closed and operate without a gas headspace, aroma losses are virtually zero. The term wash columns applies to any separation process in which the separation is mainly the result of a displacement of the concentrated liquid between the crystals by the melt of the purified crystals or by water. The washing can be performed batchwise or continuously. The batch process only serves a useful purpose when washing *in situ* grown crystals inside pipes or between plates. The continuously operating wash columns are also known under the somewhat misleading name of 'column crystallisers'. In continuous columns, crystal and liquid phase are moved countercurrently in intimate contact. This process was patented by Arnold (1951). The crystal suspension from the crystalliser, which can be directly connected to the wash column, is fed to one end of the column. The concentrated liquid is removed from that same column end through a filter. The crystals are moved to the other end of the column where they are melted. In the column the crystals are washed countercurrently with a wash liquid obtained by melting part of the purified crystals leaving the column. This reflux liquid flow is conveyed through the column by applying a higher pressure to the liquid phase at the column end where the crystals are melted. Thus the crystals move in a direction of decreasing solute concentration in the liquid phase. If the feed to be concentrated is fed directly to the crystalliser, the column is fed with a suspension in which the concentration of the liquid phase is equal to that of the concentrated product. Higher separation capacities and better separations are obtained if the liquid feed is introduced between the crystalliser and the wash column. In that case the solute concentration of the liquid phase of the suspension fed to the column becomes now more diluted and less viscous. The liquid leaving the column from that same column end serves as feed to the crystalliser.

Several modes of operating wash columns or column crystallisers have been described in the literature (Albertins *et al.*, 1967; Arkenbout *et al.*, 1973). They differ in the way the crystals are forced to move through the column and can be grouped into columns in which the driving force for countercurrent transport is the difference in density between the crystals and the liquid phase, columns with a screw or spiral conveyor, and columns in which the ice is moved as a porous cylinder by means of a reciprocating piston at one column end.

Buoyancy bed columns

In buoyancy bed columns the mean driving force for the countercurrent movement of crystals and liquid is the density difference. Buoyancy bed columns have been extensively tested for the brine–ice separation in sea-water desalination by crystallisation studies. Arkenbout *et al.* (1973) describe a column which is divided into a number of sections. The sections are separated by perforated plates. Crystals and liquid can pass through the perforations. On each perforated separation wall a number of metal balls are kept in a dancing motion by vibrating the whole column. By the action of the shaking balls a radial mixing of the suspension is obtained and the passage of the crystals through the perforations is facilitated. The capacity of buoyancy bed columns is of course limited by the hindered rising velocity of the crystals with respect to the velocity of the liquid phase.

Screw conveyor columns

The screw conveyor column has been developed by Schildknecht (1961) and Schildknecht and Breiter (1970a, 1970b). The crystal suspension is fed to the annulus between two concentric cylinders. In the annulus a spiral which is lenticular in cross section rotates. A schematic diagram of the column is presented in Fig. 11. The spiral is used to convey the crystals through the column as well as to agitate the crystal mass. For this purpose both rotary and oscillatory motions are imparted to the spiral. The column has been extensively tested for organic systems.

Piston bed columns

In piston bed columns the crystal suspension is fed in at one end of the column and compacted by expressing to a solid but still porous ice cylinder.

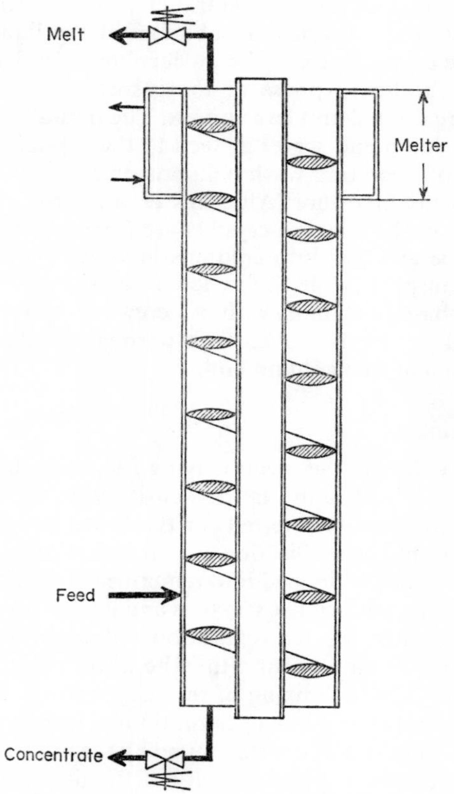

FIG. 11. Schematic representation of the 'Schildknecht' wash column.

The concentrated liquid leaves the column through a filter. By means of a reciprocating piston the ice bed is forced to move towards the other end of the column in countercurrent to the wash liquid. A wash column is depicted schematically in Fig. 12.

The Phillips Petroleum Company (1965, 1967) (McKay and Dale, 1963; Malick and Dale, 1964; McKay and Goard, 1965; U.S. pat., 1967; Dutch pat. appl., 1965) has studied the application of wash columns for the freeze-concentration of liquid foods. Wash columns in which an almost flat wash front perpendicular to the direction of movement of the crystals can be obtained have been the subject of detailed studies at the Eindhoven

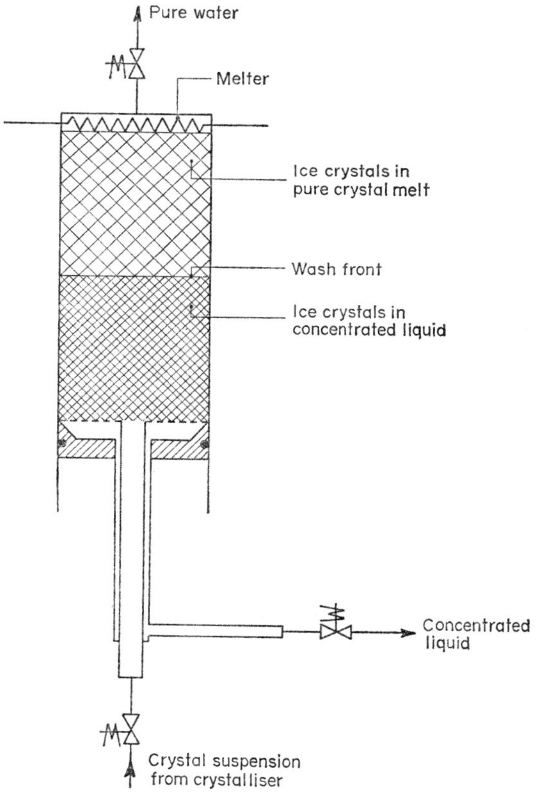

FIG. 12. Schematic representation of a wash column.

University of Technology (Vorstman and Thijssen, 1972, 1973). The distance between the part of the bed with still undiluted liquid and the part of the bed with purified crystals is only a few centimetres. In the case of stable displacement of a concentrated liquid the concentration of dissolved solids in the melt of the ice crystals leaving the columns is generally considerably less than 10 ppm. Figure 13 is a photograph of a stable displacement of a 40 wt% (model) fruit juice from an ice suspension with crystals having a mean diameter of 0.3 mm. The capacity of the column for this suspension is about 10 000 kg/h ice removal per m^2 column cross section.

FIG. 13. Photograph of a stable displacement in a wash column.

The displacement, however, is only perfect and stable if the following criterion is satisfied:

$$\frac{d_{\text{mean}}^2}{\mu_{\text{ml}}} > 10^{-3}\ \text{cm}^2/\text{P}$$

in which d_{mean} is the mean diameter of the crystals in centimetres and μ_{ml} the viscosity in poises of the liquid to be displaced by the wash water. The capacity of the wash column is approximately directly proportional to the ratio $d_{\text{mean}}^2/\mu_{\text{ml}}$.

Combination of press and wash column

The most economical process is obtained by combining a press and a wash column. A typical arrangement is depicted schematically in Fig. 14. The concentrated slurry leaving the crystalliser is partly separated in a press. The ice cake, still containing about 40 wt% occluded concentrate, is thereupon dispersed in the feed to the system. The diluted ice suspension is completely separated in the wash column. The liquid separated from the ice in the wash column is fed to the crystalliser.

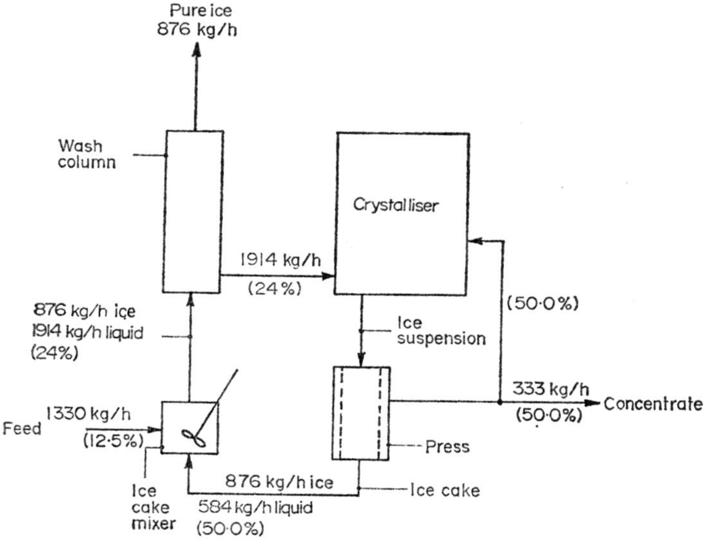

FIG. 14. Schematic diagram of a typical combination of a press and a wash column.

A simple press for compressing the ice suspension to a residual amount of occluded liquid of about 40 wt% is considerably cheaper than a much more sophisticated wash column. In the typical example shown in the figure the concentration of the liquid to be separated from the ice in the wash column becomes 24 wt% instead of 50 wt%. The effect of the concentration on viscosity at freezing temperature of the solution is shown for sucrose–water in Fig. 15. From the figure it can be seen that at 50 wt% the viscosity is 0.5 P whereas at 24 wt% it amounts to only 0.06 P. With the arrangement in Fig. 14 the capacity of the wash column can consequently be increased by a factor of 8.3. Another even more important advantage is that crystal suspensions leaving the crystalliser which either because of a too small mean crystal diameter or because of a too high viscosity of the liquid or both do not satisfy the stability criterion of a wash column can still be completely separated with the combination of press and wash column.

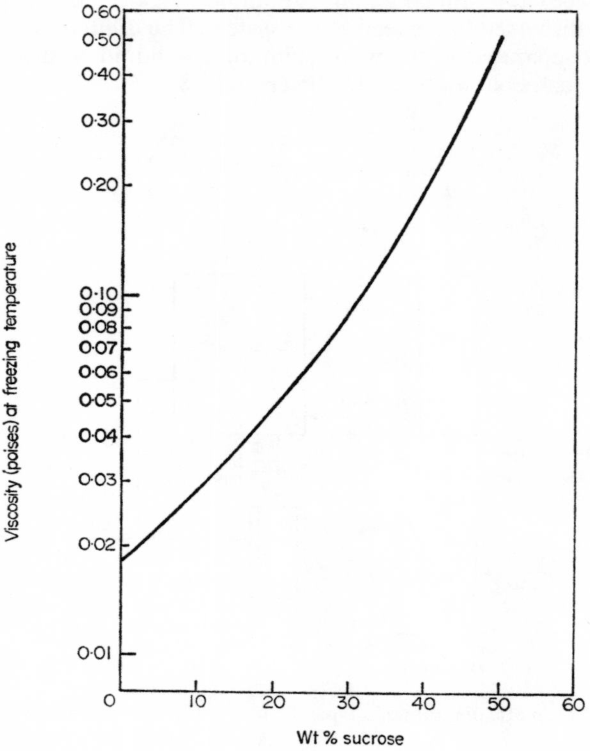

FIG. 15. Effect of sucrose concentration on viscosity at the freezing temperature of the solution.

ECONOMICS

The total costs of concentration excluding labour and costs of building are the sum of the process costs plus the value of solute that is lost in the separation process or

$$J_t = IR/hC_h + U + CP_c \tag{6}$$

where J_t = the total concentration costs per metric ton of water removal, I = total installed costs excluding building; R = the sum of annual depreciation, interest on invested capital, and maintenance; C_h = hourly dewatering capacity (tons); h = number of operational hours per year; U = cost of utilities per ton of water removal; C = kg of concentrated liquid that are lost per ton dewatered; P_c = the value per kg of concentrate.

The capital costs, the utility costs and the product losses vary greatly with the concentration system. It can be put as a rule of thumb that the more sophisticated the system is, the higher the capital costs become and the lower the product losses. The economics of freeze-concentration are given by Van Pelt (1973) and Thijssen (1973) for the system of a crystalliser producing large crystals (sub-critical crystals fed to ripening tank) with a wash column. However, the data given by these authors make comparisons difficult. Van Pelt calculates the dewatering costs for a cascade array of crystallisers and Thijssen uses a single-stage crystalliser with a press as pre-separator and a wash column for the complete separation. Just to make comparisons possible we shall calculate the costs of four single-stage freeze-concentration systems with very different types of crystallisers and separators. It must be borne in mind, however, that crystallisation to the desired concentration in more than one step is generally more economical.

System I

Crystalliser is a scraped heat exchanger, residence time of crystals in heat exchanger less than 5 min. The suspension with small crystals, diameter $< 100\ \mu m$, is separated in a press.

System II

Same crystalliser as I, the crystal suspension is separated in a basket centrifuge. The ice cake is washed with 0.15 kg water per kg of ice before unloading.

System III

The crystalliser consists of a scraped heat exchanger in series with a ripening tank. The crystal suspension is recirculated between ripening tank and heat exchanger. Residence time of crystals in heat exchanger per passage about 1 min, residence time in stirred ripening tank about 3 h. The ripening tank acts as an ideal mixer for the crystals. The crystal suspension (crystals $> 150\ \mu m$) is separated in a basket centrifuge identical with that in system II.

System IV

The crystalliser consists of a heat exchanger (producing very small sub-critical crystals) and a stirred ripening tank. Crystal-free liquid is recirculated from the ripening tank to the heat exchanger. Residence time liquid in heat exchanger < 10 s, residence time in ripening tank 3 h. The crystal suspension (crystals $> 300\ \mu m$) is separated in a wash column.

The freeze concentrators are calculated for the concentration in one step of a 25 wt% sucrose solution to 45 wt% at dewatering capacities of 500 and 1000 kg/hour. The number of operational hours per year is 2000 and 5000. The costs of the utilities are: electricity 0.07 Dfl/kWh; steam 10 Dfl/ton. Depreciation of invested capital 10 years; interest on capital 8%. Total annual capital costs 14.9%. Maintenance 2.1%. Cost of maintenance and capital costs 18%. The costs of the freeze-concentrators are based on installed costs, excluding the building. Labour costs are not included. The capital costs of the units are of course very dependent on the number of units that the manufacturer can produce annually. We shall calculate the installed costs for a small series—annual production by one manufacturer of about 2 units—and for larger series—annual production of about 5 units. The price will become appreciably lower for large series of 10 or more. The costs of the small series are close to the price of comparable units that are available on the market. Because the costs of the four systems are calculated according to the same rules, the relative costs are quite accurate.

TABLE I

SOLUTE LOSSES WITH THE ICE IN FRACTION OF SOLUTE IN FEED, $c_c/c_F = 0.45/0.25 = 1.8$

System	*Size of crystals from crystalliser*	*Residence time crystals* (h)	*Type of separator*	γ
I	Small, $<100\ \mu m$	<0.1	Press	0.13
II	Small, $<100\ \mu m$	<0.1	Basket Centrifuge	0.08
III	Large, $>150\ \mu m$	3	Basket Centrifuge	0.02
IV	Large, $>300\ \mu m$	3	Wash column	0.002

The solute losses are strongly dependent on the size of the crystals and the type of separator. From our own experiments and from information from users of freeze-concentration units we have calculated solute losses based on solute present in the feed. The results are collected in Table I. The effect of solute losses on processing costs will be calculated for solute values of 1, 5 and 10 Dfl/kg dry solids.

The installed costs of a freeze-concentration unit are split up into the refrigerator, including single-stage compressor, heat exchanger, piping and instruments, the crystalliser, the separator and the rest which includes storage vessels, piping, pumps, instruments and miscellaneous. The costs of the four systems for a capacity of 1000 kg water removal are presented in Table II.

The dewatering costs expressed in Dutch florins per 1000 kg water removal for capacities of 500 and 1000 kg/h dewatering are presented in Table III. These costs do not include the losses of dissolved solids with the ice in the separator. It is interesting to note that even for 5000 production hours per year, the capital costs contribute more than 75% to the total costs. It is obvious from the table that the more sophisticated the freeze-concentration process is, the higher the dewatering costs will be.

The total dewatering costs for the four systems, including the value of

TABLE II

INVESTMENT COSTS OF FREEZE-CONCENTRATION UNITS IN THOUSANDS (DUTCH FLORINS) FOR 1000 kg WATER REMOVAL PER HOUR. LARGER SERIES. SINGLE-STAGE CONCENTRATION FROM 25 TO 45%

	Part of apparatus	*Concentration system* I	II	III	IV
Small series	Refrigerator	170	170	182	218
	Crystalliser	323	323	467	647
	Separator	85	391	391	306
	Rest	170	170	170	170
	Total	748	1 054	1 210	1 376
Larger series	Refrigerator	135	135	145	174
	Crystalliser	256	256	371	514
	Separator	67	310	310	243
	Rest	135	135	135	135
	Total	593	836	961	1 093

TABLE III

COSTS OF CONCENTRATION PER 1000 kg WATER REMOVAL IN DUTCH FLORINS, SOLUTE LOSSES NOT INCLUDED. SINGLE-STAGE CONCENTRATION FROM 25% TO 45%

	Dewatering capacity	500 kg/h				1 000 kg/h			
	Small/larger series	*Larger*		*Small*		*Larger*		*Small*	
System	*Operational hours*	2 000	5 000	2 000	5 000	2 000	5 000	2 000	5 000
I	Costs, Dfl/1 000 kg	72	34	88	40	61	29	75	35
	% capital costs	88	75	90	79	87	73	90	78
II	Costs, Dfl/1 000 kg	49	45	120	54	84	39	103	46
	% capital costs	91	80	93	84	90	78	92	82
III	Costs, Dfl/1 000 kg	114	51	141	62	97	45	119	54
	% capital costs	91	81	93	84	89	77	91	81
IV	Costs, Dfl/1 000 kg	129	61	158	73	113	54	138	64
	% capital costs	88	74	89	78	87	73	89	77

the solute losses, are collected in Table IV. The effect of the value of the dry solute on dewatering costs is presented graphically in Fig. 16 for 5000 production hours per year and a dewatering capacity of 1000 kg/h. The value of dry orange juice is about 2.50 Dfl/kg (value ex factory) and that of coffee extract between 5 and 10 Dfl/kg. Because the curves relating solute value with total cost price intersect at about 0.3 Dfl/kg it may be concluded that irrespective of the capital costs the freeze-concentration system with the lowest solute losses also gives the lowest total dewatering

TABLE IV

COSTS OF CONCENTRATION, INCLUDING SOLUTE LOSSES, PER 1000 kg WATER REMOVAL. FOR SINGLE-STAGE CONCENTRATION FROM 25% TO 45%

Value of dry solute (Dfl/kg)	*Dewat. cap.*	500 kg/h				1 000 kg/h			
	Small/larger series	*Larger*		*Small*		*Larger*		*Small*	
	Operational hours	2 000	5 000	2 000	5 000	2 000	5 000	2 000	5 000
10	System I	800	762	816	768	789	757	803	763
	System II	547	493	568	502	532	487	551	494
	System III	226	164	253	174	209	157	231	166
	System IV	140	72	170	84	124	65	150	75
5	System I	436	398	452	404	425	393	439	399
	System II	323	269	344	278	307	262	327	270
	System III	170	107	197	118	153	101	175	110
	System IV	135	67	164	79	119	59	144	70
1	System I	145	106	161	113	134	102	148	108
	System II	144	90	165	99	128	83	148	91
	System III	125	63	152	74	108	56	130	65
	System IV	130	62	159	74	114	55	140	65

costs. This is about 60 Dfl/1000 kg water removal for System IV. Figure 17 shows the dewatering costs for freeze-concentration System IV and a single-effect centrifugal film evaporator with aroma recovery by distillation. The costs of freeze-concentration of the solution from 25% to 45% by single-stage concentration for 2000 and 7500 operational hours per year appear to be higher by factors of 2.2 and 1.6 respectively than the costs of evaporation.

Effect of concentration

Both the capacity of the crystalliser and the capacity of the separator strongly decrease with increasing product concentration. The factor controlling the capacity is the viscosity of the product at freezing temperature.

The effect of the concentration on the relative growth rate and therefore also on the capacity of the crystalliser can be deduced from Fig. 2. The effect of the concentration upon the capacity of the separator can be deduced from its effect on viscosity as shown in Fig. 15 and eqn (2). From these figures the capacity can be calculated for sugar solutions as a function of product concentration. The freeze-concentration System IV having a capacity of 1000 kg dewatering per hour for a 45 wt% sugar solution has

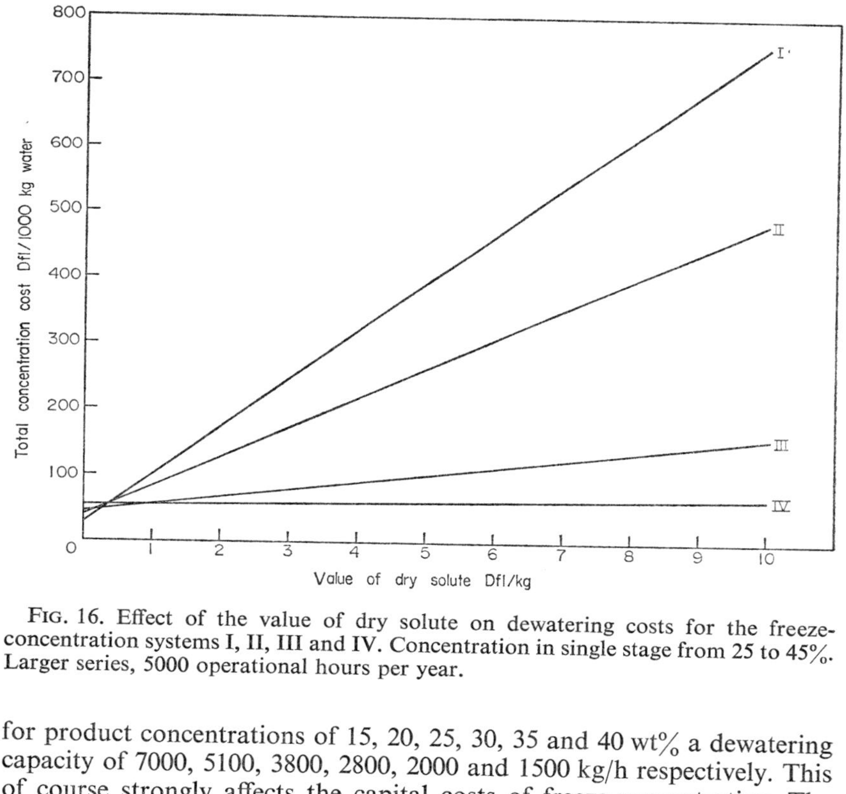

FIG. 16. Effect of the value of dry solute on dewatering costs for the freeze-concentration systems I, II, III and IV. Concentration in single stage from 25 to 45%. Larger series, 5000 operational hours per year.

for product concentrations of 15, 20, 25, 30, 35 and 40 wt% a dewatering capacity of 7000, 5100, 3800, 2800, 2000 and 1500 kg/h respectively. This of course strongly affects the capital costs of freeze-concentration. The utility costs, on the other hand, are almost independent of the product concentration.

Table V presents the investment costs for larger series with a capacity of 1000 kg water removal per hour and at product concentrations of 15, 20, 25, 30, 35, 40 and 45 wt%. The total dewatering costs for System IV in dependence on the product concentration are given in Table VI. Because of the very small losses of solute in this system the additional costs of losses

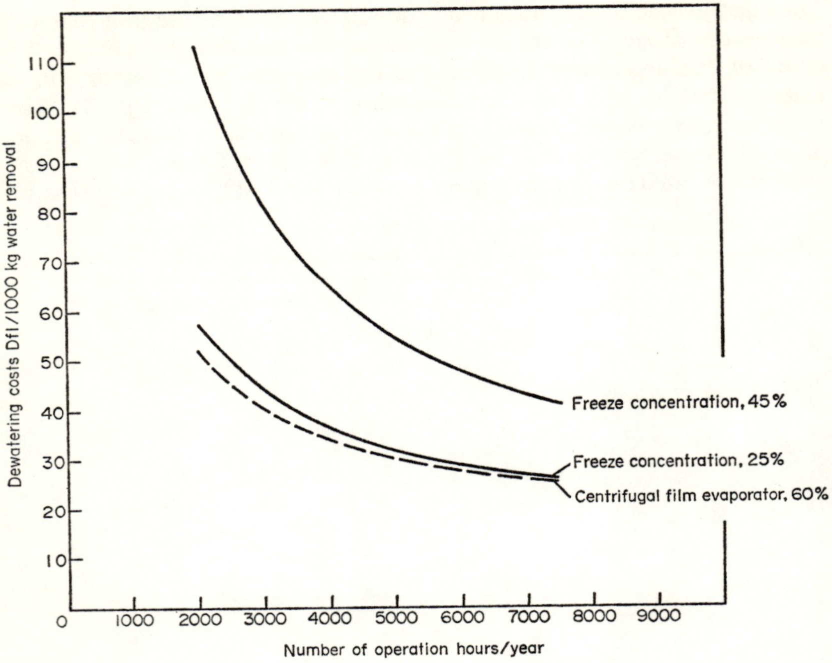

FIG. 17. Effect of number of operational hours and of product concentration on dewatering costs for freeze-concentration System IV in single stage and for centrifugal film evaporation with aroma recovery by distillation. Capacity 1000 kg dewatering per hour.

TABLE V

INVESTMENT COSTS FOR LARGER SERIES OF FREEZE-CONCENTRATION UNITS IN THOUSANDS (DUTCH FLORINS) FOR 1000 kg WATER REMOVAL PER HOUR FOR PRODUCT CONCENTRATIONS OF 15, 20, 25, 30, 35, 40 and 45 wt%

Product concentration wt %	*Concentration system*			
	I	II	III	IV
15	260	295	321	365
20	280	328	361	405
25	308	371	412	468
30	344	430	481	547
35	398	521	588	668
40	463	625	712	810
45	593	836	961	1 093

are neglected here. The costs for 25 wt% product concentration are also presented in Fig. 17. In the low-concentration range the freeze-concentration costs are about equal to those of a centrifugal film evaporator with aroma recovery.

TABLE VI

EFFECT OF PRODUCT CONCENTRATION ON TOTAL DEWATERING COSTS FOR SYSTEM IV. DEWATERING CAPACITY 1000 kg/h. LARGER SERIES. SINGLE-STAGE CONCENTRATION

Product concentration wt %	*Dewatering costs* (Dfl/1 000 kg)	
	2 000 h	5 000 h
15	48	28
20	51	29
25	57	31
30	64	34
35	75	39
40	88	44
45	113	73

CONCLUSIONS

Heat labile and aromatic liquid foods can be concentrated without any loss in quality if the separation of the concentrated liquid from the ice is performed in a closed apparatus without a gas headspace, such as a press or a wash column. A complete separation without any solute losses is possible in wash columns. The capital costs increase with increasing sophistication of the system and consequently decreasing solute losses. Nevertheless total dewatering costs are lowest for the system with the lowest solute losses. The maximum concentration that can be attached in freeze-concentration is about 45 wt%. The maximum concentration is controlled by the viscosity of the concentrate at freezing temperature; the maximum viscosity being about about 100 cP. A decrease in viscosity results in an increase in the capacity of the concentration unit. The processing costs therefore sharply increase with increasing product concentration. From this it can be concluded that for a concentration factor of more than about 2 concentration has to take place in two or more stages.

Future developments

The energy consumption of freeze-concentration is about inversely proportional to the difference between condensor temperature and evaporator temperature of the compressor. At a thermal efficiency of 60% and with a temperature difference smaller than 48°C the energy consumption becomes better than that of a triple-effect evaporator. Future developments therefore will have to be directed at crystallisers operating at a small temperature difference between refrigerant and crystal suspension. Because of the high capital costs the main reduction in dewatering costs has to come from a further reduction in the price of freeze-concentration units. A reduction in investment costs may, however, never be accompanied by an increase in solute losses.

REFERENCES

Aackter, M. and Barak, A. (1967). 2nd Eur. Symp. on Fresh Water from the Sea, Athens, May, pp. 89–91.

Albertins, R., Gates, W. C. and Powers, J. E. (1967). *Fractional Solidification* (ed. M. Zief and W. R. Wilcox), Marcel Dekker, New York, 1967.

Anon. (1964). Union Carbide Corp. Dutch Patent Application No. 263,468.

Anon. (1965). 'Votator "Cleanwall" Scraped Surface Heat Exchanger', Techn. Bull. of Votator Division, Chemetron Corp., p. 7.

Arkenbout, G. J., Kuyk v., A. and Smit, W. M. (1973). *Chemistry and Industry*, 3 February, 139.

Arnold, P. M. (1951). U.S. Pat. No. 2,540,977.

Brownell, L. E. and Katz, D. L. (1947). *Chem. Eng. Progress*, **43**, 601.

Chandrasekaran, S. K. and King, D. J. (1971). *J. Food Science*, **36**, 699.

Daubron; Notice Concentration, No. 2, 254, Daubron S.A., France.

Daubron; Special Brochure No. 23,867, Daubron S.A., France.

Döge, F. and Vêlebil, Q. (1969). Germany, Informationblatt G.K. 101, 102, 103, 104 and 105.

Dschu, I. L. (1967). *Kältetechnik–Klimatisierung*, **19**, 279.

Gasquet Catalogue No. 1685, Bordeaux, France.

Grasso's Koninklijke Machinefabrieken (1969). Dutch Patent Application No. 69,18686.

Grenco Pamphlet (1972). 'Freeze concentration of food liquids', Grenco N.V., Netherlands.

Heisz, R. and Schachinger, L. (1951). *Food Technol.*, **5**, 211.

Huige, N. J. J. (1972). 'Nucleation and growth of ice crystals from water and sugar solutions in continuous stirred tank crystallizers', Ph.D. thesis, Eindhoven University of Technology, Netherlands.

Huige, N. J. J., Senden, M. M. G. and Thijssen, H. A. C. (1972). Chisa Symp., Prague, 11–15 September, 1972.

Krauss Maffei pamphlet B4721, 3rd edn, Gefrierkonzentration G.K.

Kepner, R. E., Straten v., S. en Weurman, C. (1969), *J. Agr. Food Chem.*, **17**, 1123.

King, C. J., Lam, W. H. and Sandall, O. C. (1968). *Food Technol.*, **22**, 100, 1302.

Koonz, C. H. and Ramsbottom, J. M. (1939). *Food Research*, **4**, 120.

Kramers, H. (1958), 'Fundamental aspects of the dehydration of foods', Aberdeen Conference of the Soc. of Chem. Industry, 1958.

Luyet, B. J. (1962). *Freeze-drying of Foods* (ed. F. R. Fishes), Nat. Acad. Sci., Washington, D.C.

Malick, E. A. and Dale, H. (1964). 'Study of concentrated and reconstituted beers', Ann. Conv. of the Am. Soc. of Brewing Chemists, New York, 6 May, 1964.

McKay, D. L. and Dale, G. H. (1963). 'The Phillips Fract. Cryst. Process Applied to Beer Concentration', Ann. Conv. of the Master Brewers Assoc. of America, Baltimore, 26 September.

McKay, D. L. and Goard, H. W. (1965), *Chem. Eng. Progress*, **61**, 99.

Muller, J. G. (1966), Annexe 3 au Bullt. de l'Inst. Int. du Froid.

Muller, J. G. (1967). *Food Technol.*, **21**, 49.

Okabe, T. (1962). *Dechema-Monografien*, Vol. 47, p. 815.

Pankovic, Z. (1962). *Dechema-Monografien*, Vol. 47, p. 805.

Pelt van, W. H. J. M. (1973). Sixth Int. Course on Freeze-Drying and Advanced Food Technology, Int. Inst. of Refrigeration, Bürgenstock, Lucerne, Switzerland.

Phillips Petroleum Co. (1965). Dutch Patent Application No. 651,051,6.

Phillips Petroleum Co. (1967). 'Fractional Crystallization', U.S. Patent No. 3,339,372.

Phillips Petroleum Co. (1965). Dutch Patent Application No. 640,060,8.

Rohatgi, P. K., Jain, S. M., Frenck, D. N. and Adams, C. M. (1969). *Trans. Metallurgical Soc. of AIME*, **245**, 267.

Rulkens, W. H. and Thijssen, H. A. C. (1972). *J. Food Technol.*, **7**, 79.

Schildknecht, H. (1961). *Anal. Chem.*, **181**, 254.

Schildknecht, H. and Breiter, J. (1970a). *Chemiker Zeitung*, **94**, 3.

Schildknecht, H. and Breiter, J. (1970b) *Chemiker Zeitung*, **94**, 81.

Silverblatt, C. E. and Dahlstrom, D. A. (1954). *Ind. Eng. Chem.*, **46**, 1201.

Thijssen, H. A. C., Vorstman, M. A. G. and Roels, J. A. (1968). *J. Crystal Growth*, **3**, 4, 355.

Thijssen, H. A. C. (1968). *Dechema-Monografien*, Vol. 63, p. 153.

Thijssen, H. A. C. (1973). Sixth Int. Course on Freeze-drying and Advanced Food Technology, Int. Inst. of Refrigeration, Bürgenstock, Lucerne, Switzerland, 1973.

Toulmin, H. A. (1959). Union Carbide Corp., U.S. Patent No. 2,877,851.

Urban, B. (1969). Annexe 3 au Bull. de l'Inst. Int. du Froid, 169.

Vorstman, M. A. G. and Thijssen, H. A. C. (1972). Int. Symp. on Heat and Mass Transfer Problems in Food Eng., October, Wageningen, Netherlands, 1972.

Vorstman, M. A. G. and Thijssen, H. A. C. (1973). *De Ingenieur*, **84**(45), 65–69.

Wenzelberger, E. P. (1969). Union Carbide Corp., U.S. Patent No. 2,977,234.

DISCUSSION

M. Karel (M.I.T.): I would like to say a word about the effect of proteins and solid particles on losses during the separation when comparing sucrose with orange juice or milk at the same viscosity. Experiments with orange juice indicate that the same crystal growth rate is obtained in orange juice solutions as in coffee extracts and some other solutions. Stable displacement is still achieved in the wash column, the only point is that fibrous and other solid particles behave in this washing process as crystals and a separation occurs between the solubles and the solid particles. The fibrous or cellular materials are withdrawn with the ice. If the ice stream containing these fibres is melted, the fibres can easily be separated and fed back to the concentrate. However, care should be taken that the concentration of cellular material is not too high, otherwise the permeability of the ice bed becomes too low, problems of displacement instability will arise, and the column will cease to give a satisfactory separation.

You spoke only about the wash column. How would the separation be affected in a centrifuge?

H. A. C. Thijssen: Under the conditions prevailing in centrifuges, stable displacement of the concentrated liquid by the wash liquid is never obtained, and thus appreciable solute losses occur for large crystals of the order of 1–2%. This is too high. Figure 16 shows that even for orange juice costing 2½ Dfl/dry kg, the total concentration cost is nearly doubled, and it might be necessary to evaporate the melted ice fraction to recover the solute.

C. H. Mannheim (Technion, Israel): I agree that the most important factor in freeze-concentration is the solute loss. This is emphasised by the fact that your figure of 2½ guilders for dry orange solids is very low in the context of today's rising prices.

Returning to Professor Karel's comments, did you start with fresh orange juice in these experiments?

Karel: No, It was frozen, concentrated and rediluted orange juice.

J. D. Mellor (CSIRO, Australia): Would you like to discuss the economics of the basket centrifuge which results in 2% losses, compared with the wash column giving 0.2%, in a little bit more detail. The basket centrifuge may possibly take up one quarter or one fifth of the area occupied by the column, and space is often an important economic factor in industrial processes.

Thijssen: Referring to the same series of equipment which I described in the paper, the cost contributions of the various separation systems are as follows: basket centrifuge is about 310 000 guilders installed, the press 57 000 and the wash column 243 000. So the wash column is somewhat cheaper compared with the separator. In fact the wash column requires somewhat less space and the losses are negligible. On the other hand a centrifuge always operates regardless of the size of the crystal. A decrease in crystal size results in a longer residence time in the centrifuge and in higher losses. With a wash column, if you are below the stability criterion, no separation at all will be achieved. The wash column either operates perfectly, or it doesn't operate at all. Does this answer your question?

Mellor: Partly. Do the figures for space requirement include all the associated equipment that is needed?

Thijssen: For the wash column the only equipment needed is a scraper knife at the top which requires very little space and the air cylinder which is mounted in the same vertical plane as the wash column. The space required for a column of capacity 500 kg/h is about 1½ m^2. The column diameter is 50 cm.

A. L. Moeller (Ecal Nateko AB, Sweden): If I may refer to Table I in your paper, you compare product losses for the various systems using different crystal sizes. If, for example, System III was using the same crystal size as System IV then the losses should be reduced by a factor of 4 to 0.5%, if the inverse square relationship suggested by eqns (3) and (5) in the paper holds. Would this be correct?

Thijssen: We did some experiments with centrifuges and collected data from industry and all of these are brought into the figures, but the data is only rough. It also depends on the density.

Moeller: But it is dependent on the sizes of the crystals?

Thijssen: Yes, if by size you mean the crystal diameter. We use the effective crystal diameter measured in the following way. A crystal bed is formed from the crystal suspension and its permeability and porosity measured. From the permeability of the bed and the porosity we calculate effective diameter, because this gives a real indication of the behaviour of the crystals in a bed such as in a centrifuge or in a wash column.

Moeller: There's another point too. Could there be a possibility of lowering the costs in your wash system by using the heat evolved during the freezing to melt the ice crystals?

Thijssen: Yes, sir. But the controlling factor in the cost price is the capital cost which makes up about 75% of the total cost. The energy costs can easily be reduced by a factor of two but this has little effect upon the total cost. First a cheaper crystallisation system or a cheaper separating system must be found, then the energy problem may become more interesting.

Chairman: A further inducement to select high solute recovery systems is the need to avoid high B.O.D.'s in the effluent stream.

C. Varsel (Coca Cola Company, U.S.A.): Dr. Thijssen, do you get quantitative recovery of aroma components in the freeze-concentration of orange juice?

Thijssen: The aroma recovery is absolutely 100%. I do not know too much about orange juice, but if we look again at the model solution and at coffee extract, we get water from the wash column without any taste or odour and indeed it could be used for distilled water.

Chairman: Could we have your comments on the quality say of concentrates produced in this way versus product concentrated in a double or triple effect evaporator with an essence recovery?

Thijssen: Yes—in fact there is no comparison because in freeze-concentration you have used the selective process and water is withdrawn by the phase transformation to the crystalline state. If you have, moreover, a complete separation of the crystals from the liquid phase in the wash column, there are no solute losses. There is also no chemical deterioration, because the residence time may be 2 or 3 hours, but the temperature does not rise above $-5°C$. The quality is thus comparable to that of the fresh product. This could never be the case for a process like evaporation where the separation is performed at a higher temperature. In evaporation the recovery of the aroma from the vapour leaving the evaporator is generally not more than 60–70%. In addition there is a change in the bouquet because not all components are collected at their original levels relative to each other.

Non-membrane Concentration

C. H. MANNHEIM and MRS. N. PASSY

Department of Food Engineering and Biotechnology, Technion, Israel Institute of Technology, Haifa, Israel

INTRODUCTION

The major processes used in the food industry for food preservation include: pasteurisation, sterilisation, freezing, dehydration and concentration.

Evaporation for purposes of concentration is one of the most important unit operations in the food industry. In addition to serving as a method of preservation it offers the advantages of lowering packaging, storage and transportation costs due to the large reduction in weight and volume. Very large evaporation facilities are used in the fruit and vegetable juice industries for producing concentrates, in dairies for the production of milk and whey concentrates, in sugar and syrup manufacturing and in alcohol distilleries. In addition, evaporators are often used prior to drying, such as in instant coffee production, milk powder manufacturing, etc., with the aim of reducing overall production costs.

However, it is also well known that despite the widespread use of thermal concentration processes there are several problems to be overcome and these include:

(1) Deposit formation on heat transfer surfaces of a burnt layer.
(2) Loss of volatile aromas.
(3) Heat damage to sensitive products.
(4) Browning during evaporation or storage of final product.

There does not exist a universal evaporator for all purposes and numerous designs and makes are available. In choosing the specific evaporator one must pay close attention to initial investment, operating costs and ease of maintenance before making the final decision.

Ways and means to overcome the above-mentioned problems and the merits of the various basic designs of evaporators will be discussed.

TYPES OF FLUIDS TO BE CONCENTRATED

Liquid foods are concentrated for a variety of reasons.

(1) The operation may be one stage in a process such as preconcentration of milk or coffee prior to spray or freeze-drying.
(2) It may be used to reduce bulk in order to save on packaging, shipping and storage costs.
(3) Removal of water offers great savings in the cost of freezing concentrates.
(4) It may prevent microbial spoilage due to high osmotic pressure.

The majority of fluids in the food industry fall, for evaporation purposes, into one of the following types; moderately heat sensitive, very heat sensitive and viscous products.

Moderately heat sensitive products

This group includes products such as glucose and sucrose solutions, malt and wort extracts, clear soups, brines and some fruit juices such as apple and pineapple juice. The products tolerate evaporation temperatures of 50°C and above and relatively high Δt between steam side and product side in the evaporator.

Very heat sensitive products

The damage which can occur in these products may be caused by decomposition, polymerisation and coagulation of some constituents. Dairy products and juices such as citrus are included in this group. The presence of protein is often the controlling factor in the design of evaporation equipment. Since protein is sensitive to heat above 50°C, where denaturation is quite rapid, flavour changes and accumulation of denatured protein scale on heating surface may occur. Evaporators for this group should have high turbulence to minimise scale formation, short contact time to limit degradation and flavour changes, and low temperatures. They must have facilities for easy cleaning.

Another problem in this group of products is aroma and flavour changes. For example, in citrus juices a so-called 'cooked flavour' is formed. Blair *et al.* (1952) attributed these off-flavours to changes in the peel oil and lipid fractions. By a catalytic reaction in the acid solution followed by hydration and dehydration reactions, a series of compounds such as 1,8-terpine, 1,4-cineole and alpha-terpinene is formed. Some of these have pungent odours responsible for the off-flavour in canned orange juice.

In previous years the approach to solve the problem of these products was to use low temperature evaporators but relatively long contact times. However, in recent years most of these evaporators have been replaced by new ones using higher evaporation temperatures but much shorter contact times and include aroma recovery systems in the pre-evaporation step.

In the old low temperature evaporators, using ammonia condensors, evaporation temperatures were between 20 and 35°C and residence times in the system were 20–35 min or even longer. The new evaporators use 30–45°C evaporation temperatures with residence times of 0.5–5 min.

Among this group of heat sensitive products enzyme extracts and other pharmaceutical products such as antibiotics, insulin and blood products should be mentioned.

Viscous products

For evaporation purposes these products should be sub-divided into three types, namely: non-heat sensitive, heat sensitive, and pseudo-plastic (non-Newtonian). As an upper limit of final product viscosity a figure of 20 000 cP is normally stated.

For the non-heat sensitive products, of high viscosity, evaporation is facilitated by using higher evaporation temperatures, such as 70–90°C. Examples of such products are agar-agar, gelatine, glues and meat extract.

Heat sensitive viscous products are more difficult to concentrate, since with the decrease in temperature required to prevent product deterioration the viscosity increases. Normally, temperatures of about 50°C and thin film evaporators are applied here. An example of such products is yeast extract.

Pseudo-plastic products can best be concentrated in agitated evaporators since viscosity decreases with application of mechanical force. These products include high pulp containing juices such as tomato, peach and apricot.

EVAPORATOR SYSTEMS

Introduction

Basically an evaporation system consists of a heat exchanger, supplying the sensible heat to raise the product to its boiling point and provide the latent heat of vaporisation, a vapour and liquid separator, a condensor to remove the resultant vapour and if concentration is carried out under reduced pressure a vacuum system.

The separator is usually of a ‘cyclone’ type, to facilitate separation of vapour from the concentrated liquid and it is designed in such a way that entrained droplets are precipitated by centrifugal action. Removing the concentrate and the condensate is done by pumps or a barometric leg. The vacuum system consists either of a vacuum pump or a steam ejector.

The steam heated evaporator is the most common one found in the food industry. The rate of heat transfer is governed by the movement of liquid over the heating surface. Previously, when it was thought that high temperatures should be eliminated during concentration, refrigerant heat-pump cycle evaporators like Mojonnier and Buflovak types were used.

Heat transfer—definitions

For energy, in the form of heat, to pass between two points there must be a temperature gradient. Heat transfer may occur by conduction, convection and radiation, which are briefly described as follows.

CONDUCTION

This is the transfer of heat by molecular kinetic energy alone where no motion takes place, and it occurs only in solids. In fluid flow, heat transfer is by pure conduction in the laminar boundary layer only.

CONVECTION

This is the transport of heat by currents within a fluid, generated either by density differences—natural convection, or by mechanical assistance—forced convection.

Heat transfer from fluid to wall is described by the equation:

$$Q = h \cdot A \cdot \Delta t$$

where: Q = amount of heat (kcal/h); A = area (m^2); Δt = temperature gradient (°C); h = film heat transfer coefficient (kcal/m^2/h/°C). Typical values of h for air, water and steam are given in Table I.

TABLE I

FILM HEAT TRANSFER COEFFICIENTS

Fluid	*Type of heat transfer*	h(kcal/m^2/h/°C)
Air	Natural convection	1–50
Air	Forced convection	10–250
Water	Heating/cooling	250–20 000
Water	Boiling	1 500–45 000
Steam	Film condensation	5 000–20 000
Steam	Drop-wise condensation	25 000–100 000

RADIATION

This is emitted from all bodies and will be negligible in this discussion.

During evaporation, all types of heat transfer occur simultaneously. In the evaporator, heat is transferred from a heating medium through a separating wall to the medium to be evaporated. The heating medium is generally steam, which condenses on this wall, and releases its latent heat which is taken up by the product.

The overall heat transfer equation is:

$$Q = K \cdot A \cdot \Delta t$$

where K is the overall heat transfer coefficient taking in consideration convection of liquid to the wall, conduction through the wall and convection from the wall to the liquid.

In industrial plants it is important that the heat flux $q = Q/A$ will be as high as possible since this involves lower plant costs and less space. Since $q = K \cdot \Delta t$, the overall heat transfer coefficient, and the temperature differences are the limiting factors in the evaporation process.

Heat transfer in evaporators

The factors affecting heat transfer in evaporators are:

(1) temperature difference between the steam and the product,
(2) type of heat transfer between wall and product,
(3) type of heat transfer between condensing steam and wall,
(4) product viscosity, velocity and film thickness.

TEMPERATURE DIFFERENCE

A larger Δt can be obtained by lowering the product temperature by operating under vacuum, and raising steam temperature. The limiting factors for Δt are the allowable stresses in the evaporator due to pressure differences and the sensitivity of the product to heat damage which affects maximum allowable wall temperature.

HEAT TRANSFER—WALL TO PRODUCT

Studies show that when the surface reaches the local boiling temperature of the liquid, small bubbles are formed which detach from the surface in increasing numbers, as the surface temperature increases. At a certain critical heat flux and Δt, the bubbles coalesce to form a continuous isolating film of vapour between the wall and the product. At heat flux and Δt's above this point the heat transfer coefficient decreases.

The surface tension between the wall and the product determines the tendency of the film towards boiling. At low surface tension, the bubbles can easily detach from the surface so that a higher critical heat flux is obtained. If the product has a high surface tension, the bubbles will coalesce to a film at a lower heat flux, thus reducing heat transfer coefficients.

HEAT TRANSFER—CONDENSING STEAM TO WALL

The two condensation regimes which give the highest heat transfer coefficients are film condensation and drop-wise condensation (Table I). It should be mentioned here that in most evaporator types, film condensation occurs, while in the centrifugal evaporator (Centri-Therm) condensation is drop-wise. When vapour condenses on a surface, a low condensate surface tension produces wetting of the surface, so that a continuous film is formed and the transfer coefficient is reduced. A high surface tension, on the other hand, causes the droplets to coalesce rather than to spread out, exposing a larger fraction of the heat transfer surface, and thereby increasing the heat transfer coefficient.

TABLE II

SPECIFIC VOLUME OF VAPOUR *v.* PRESSURE

Abs. pressure (kg/cm^2)	*Steam specific volume* (m^3/kg)
0.1	15.3
0.5	3.4
1.0	1.7
1.5	1.2
2.0	0.9

PRODUCT VISCOSITY, FILM THICKNESS AND VELOCITY

With increasing viscosity, due to evaporation, the film thickness becomes greater, the velocity lower, and the heat transfer coefficients are reduced.

Vacuum in evaporators

In concentration of food liquors and especially those that are heat sensitive, the process is usually carried out under vacuum. At reduced pressures the specific volume of vapour increases with decreasing pressure (Table II). The increase in volume is very rapid at the low pressures and this is a limiting factor in the construction of evaporation units. In addition, it is necessary at low pressure to incorporate adequate vapour pipe-work to the condenser in order to limit the vapour velocity and prevent excessive pressure drops.

Thermal economy

Evaporation plants can be classified according to their thermal economy.

SINGLE-EFFECT EVAPORATION

In a single-effect evaporator, the steam is used only once for evaporation. The concentrated liquid is recirculated for further concentration or, alternatively, leaves the plant.

MULTIPLE-EFFECT EVAPORATION

In multiple-effect evaporation a series of evaporator bodies are connected so that the vapour from one body is the heating medium for the next body. Passing from single effect to multiple effect does not necessarily alter in any way the features and type of construction. It affects only the interconnecting piping and the operation. The purpose of multiple effect is to reduce the steam consumption for a given amount of evaporation. Although multiple effect involves greater steam economy, the plant becomes considerably more expensive and complicated to operate. Furthermore, multiple-effect evaporation involves a very high temperature in the first effect, and the total liquid hold-up volume increases considerably. This is a disadvantage for heat sensitive products, since residence times become long.

USE OF FURNACE GASES

Another way to reduce evaporating costs was suggested by Vincent (1967). He proposes to use hot gases from a citrus peel dehydrator. These gases are a mixture of furnace flue gases and vapours from the drying peels. The hot gases are used to heat the first and second effect of a tubular falling film evaporator. This system is being successfully applied to the concentration of citrus peel waste liquor and is now being investigated, by the author, for citrus juice evaporation.

EVAPORATOR TYPES

Over the past 50 years a wide range of equipment for concentration has been designed, ranging from simple boiling pans to modern and efficient agitated film and centrifugal evaporators.

The classification offered here is based on the nature of heat transfer surface.

(1) Tubular surface with natural or forced circulation.
- (a) Vertical tubes with climbing film.
- (b) Vertical tubes with falling film.
- (c) Inclined tubes.
- (d) Horizontal tubes.

(2) Flat heating surfaces—plate evaporators.
(3) Stationary cylindrical surfaces with agitated film.
(4) Stationary conical surfaces.
(5) Rotating conical surfaces.

Tubular surface

The climbing or falling film tubular evaporator has in the past dominated the industry and is still extensively used for the very largest capacities. Units of natural and forced circulation are to be distinguished.

NATURAL CIRCULATION UNITS

In these evaporators, also called thermosyphons, circulation of the liquid is brought about by convection currents arising from the heating surface. In this group the tubes may be horizontal, vertical or inclined with the steam being inside or outside.

The absence of forced circulation in horizontal evaporators results in low heat transfer coefficients, especially for viscous concentrates. Therefore, these evaporators are better suited for non-viscous solutions that do not deposit any scale.

The vertical type evaporator is much more versatile and is suited for solutions that deposit salt or scale, entrainment losses in it are low.

Evaporators in which the spreading of the film is vapour induced include the types which are called rising and falling film evaporators.

Rising film

In a rising or climbing film evaporator the fresh liquid to be concentrated is fed in at the bottom of the heat exchanger and rises in it. A mixture of liquid and vapour comes out at the top of the tubes and passes to an external separator, thus the evaporator is continuous in operation. Within the tubes there are three distinct regions. At the bottom, under the static head of liquid, no boiling takes place, only simple heating. In the centre region the temperature rises sufficiently for boiling and vapour is produced; heat transfer rates are still low. In the upper region the volume of vapour increases and the remaining liquid is being wiped into a film on the tube surfaces resulting in good heat transfer conditions. The disadvantages of this type are the relatively large hold up of liquid in the lower regions of the tubes giving long contact times (15–30 min) (Bedford, 1972). Also, evaporation ratio in a single pass is usually not sufficient to reach the required concentration, so that recycling is necessary, extending residence time. In the central portion of the tube, formation of scale, protein deposits and other fouling is often found to be most severe. Therefore, only for protein free materials may long runs be achieved. Due to low heat rates obtained in this equipment the capital cost may be quite high.

Falling film

To eliminate the restrictions caused by the static head of liquid, and to take advantage of the high heat transfer rates obtained with a thin film of liquid, the falling film evaporator was designed. Higher efficiencies and thermal economy are achieved and residence times are reduced to 1–5 min.

The evaporator is constructed of a vertical calandria and a separator. Even distribution of liquid over the heat transfer surface becomes an important factor in this design in order to avoid local hot spots and burning. In the most common system the feed is distributed at the top and sprayed through a nozzle plate into each tube, forming a relatively thin film on the tube wall. The film falls under gravity down the tube. The vapour formed together with concentrated liquid is passed to a separator. High heat transfer rates permit shorter tubes in the calandria and the lower temperature differences give higher thermal economy. Due to these advantages this evaporator is widely used.

FORCED CIRCULATION EVAPORATORS

Increased velocity of flow of the liquid through the tubes will bring about a marked increase in the liquid film heat transfer coefficients. The circulation is achieved with the aid of an external circulating pump which forces the liquid over the heating surface. The circulation pump is most frequently a centrifugal pump, but for extremely viscous liquids a positive pump is used. Forced circulation enables higher degrees of concentration to be accomplished since the heat transfer rate can be kept up in spite of the increased viscosity of the liquid. The three basic components of the forced circulation evaporators, namely, the heat exchanger, the separator and the pump, may be arranged in several forms.

VACUUM PANS

These are one of the simplest types of evaporator and exist in many forms. Shell and tube calandria may be used for heating the larger pans with natural circulation of the liquid. Smaller pans are heated by double jackets and mechanical agitation is maintained by means of a stirrer.

Operation of pans is usually batch wise with high residence times thus requiring operation under vacuum. Due to their simplicity and adaptability to a wide range of products, even at high viscosities, vacuum pans are widely used especially in small plants in jam manufacture and final concentration of tomato paste. Vacuum pans may be set up in two or three effects for better steam economy.

Flat heating surfaces—plate evaporators

To meet the ever-increasing demand for shorter residence times and lower operating temperatures the plate evaporator was developed and is widely used in the food and especially in the dairy industry.

Principally, the plate evaporator differs from the conventional unit in that the tubular calandria is replaced by a compact arrangement of gasketed plates held together in a frame. Steam economy is possible by using multiple effects with vapour recompression for heating. High heat transfer coefficients are obtained and viscous materials are handled at relatively high temperatures but short contact times. In this evaporator, the designer attempted to secure the best operating conditions by combining the climbing film and falling film principles (Fig. 1).

The mechanism of boiling within two parallel plates is quite different from the one occurring within a tube. The heat transfer area is much larger per unit height, and retention times are much lower. By varying the plate gap and the width of the plates, it is possible to control the vapour velocity for efficient heat transfer but still ensuring proper wetting. Dimensional adjustments achieve the optimum operating conditions needed for different viscosities, concentration ratios and boiling temperatures.

The main design features of the plate evaporator are as follows. Evaporation is carried out in sets of four plates in which there are rising and falling film product passages with steam passages in between. Rubber gaskets define the flow of product and steam through suitable parts cut in the plate as well as providing the necessary seals between one plate and the next. In the APV evaporator the inlet of the feed is at the bottom of the plate (Fig. 1). The liquid film is carried up the plate, heat being applied on the other side. At the top of the climbing film plate the material weirs over and then enters a falling film section. The material arriving at the top of the up plate is still associated with a high vapour velocity, there is thus no distribution problem in entering the down plate, where the second stage of evaporation takes place and viscosity of the product becomes high.

A number of plate units, each embodying four individual plates, may be mounted in a frame to give the necessary heat transfer area. All plate units in the frame are fed in parallel with the material to be evaporated, and the concentrated product together with the vapour produced leave from the head of the frame for separation in a centrifugal separator. This separator may be of the horizontal or, alternatively, of the vertical cyclone type, depending upon the duty and the space available. In a single-effect unit, the vapour is removed through a condenser. Multiple effect can be used to achieve the most efficient operation in terms of steam consumption per unit of evaporation. A typical arrangement of a double-effect plant with thermal recompression is shown in Fig. 2. Various arrangements of preheating are used making full use of effluents from the evaporator such as condensate and also using inter-effect vapours. Triple-effect or double-effect with thermocompression is common, resulting in about 31 lb of evaporation per lb of steam supplied (according to APV).

A new plate evaporator based on the falling film principle only was

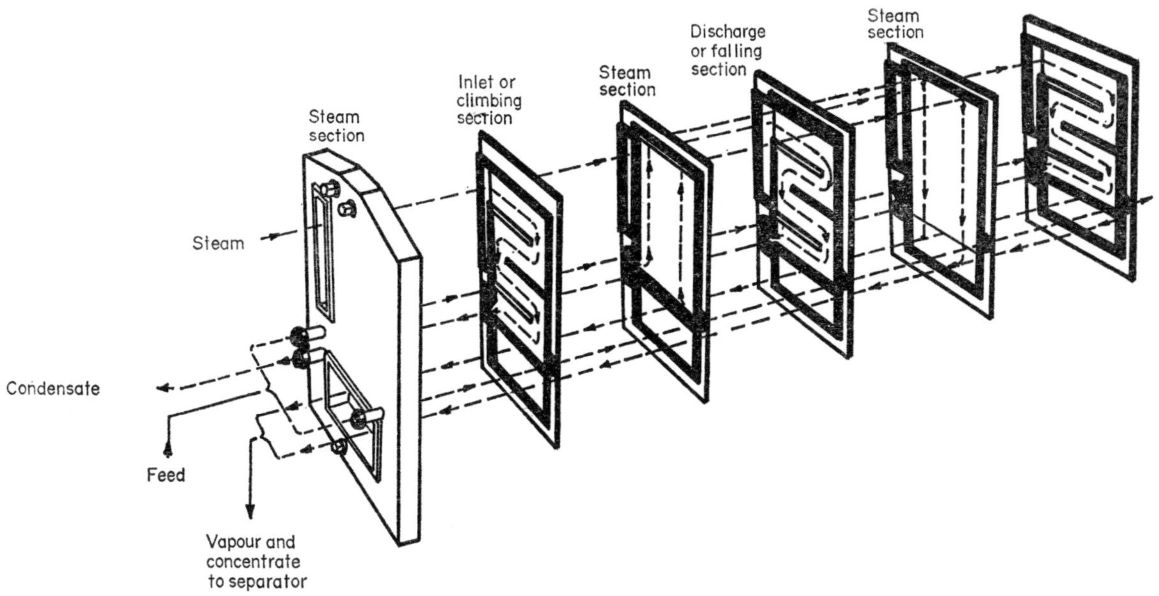

FIG. 1. APV Plate Evaporator, diagrammatic arrangement of plates for one feed pass. (By courtesy of APV International Ltd., England.)

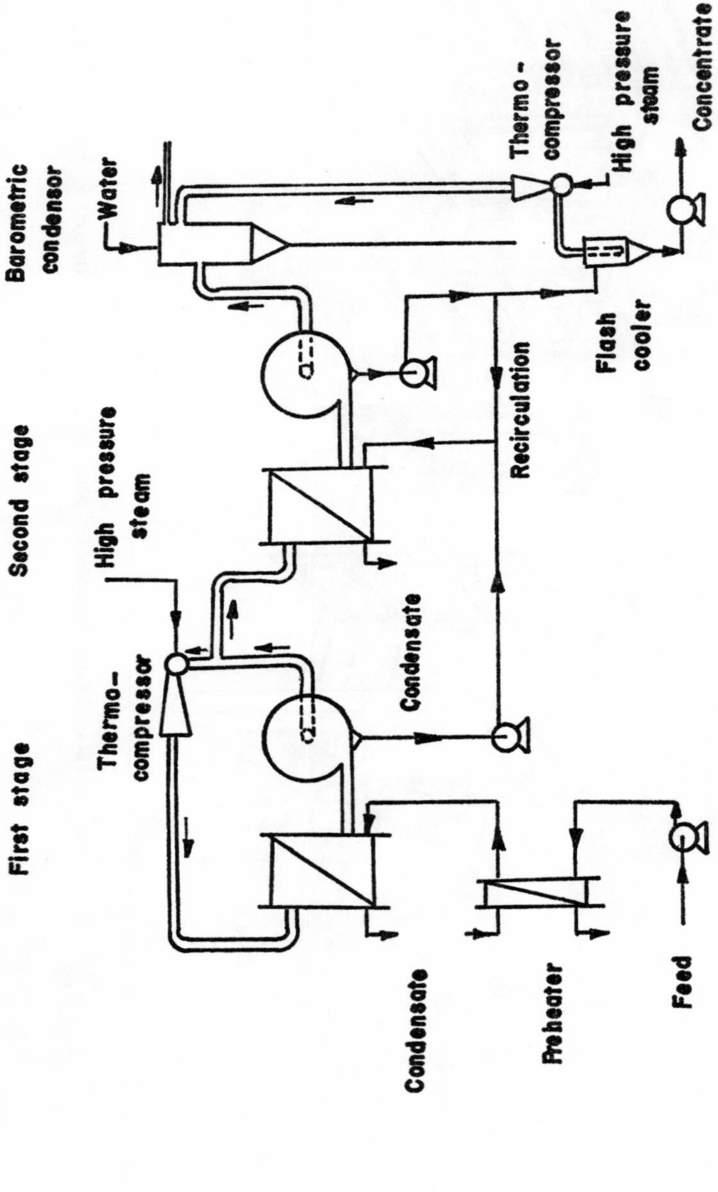

FIG. 2. APV Plate Evaporator, flow diagram for two-stage plant. (By courtesy of APV International Ltd., England.)

designed recently by the APV Company. A prototype of this design is presently being tested in Israel with citrus juices. The new design comprises a falling film product plate together with adjacent steam plate, i.e. two plates per unit as compared with four on the previous design (Fig. 3). The length of the plate is equal to the length of combined up and down plates and the plates are 50% wider, hence fewer plates and a shorter pack are required for the same duty. The distribution system here is unique and works by feeding each plate through a series of carefully graded orifices which causes a metered mixture of liquid and flash vapour to be injected at a number of points across the top of the plate to ensure that the liquid flowing down the entire width of the plate is an even unbroken film. This arrangement, together with a special design which divides the plate into two identical halves in order to avoid recirculation which is undesirable when processing citrus juice, gives a very short residence time of the product in the system. In a typical fruit juice application the company claims to obtain vapour velocities of 45 and 110 ft/s in the first and second effect respectively.

The main advantages of the plate evaporator may be summarised as follows.

(1) It is flexible, so that surface area can be added or removed as required.
(2) Very low headroom is needed for such an installation.
(3) There is good sanitary construction and easy accessibility to heat transfer surfaces.
(4) Multiple-effect operation may be obtained.
(5) Flexibility for multiple duties including heat sensitive products due to short residence time.

The limitations of the plate evaporator are:

(1) The need for rubber gaskets for sealing rules out any duties involving the use of organic solvents and places an upper limit on the operating temperature.
(2) Liquids containing suspended solids constitute a problem due to the narrow gap between the plates needed to ensure conditions for proper climbing—falling film operation. It is necessary to place filters in the line to protect the evaporation from large particles.
(3) For even distribution, and to ensure good wetting of the plates, orifice pieces between the header ports formed in each plate and the evaporative surface are needed. These must be removed for cleaning.
(4) Since the evaporator is based on standard pressed plates, which have steam and vapour ports of a fixed size, the maximum amount of vapour which can be handled by the machine is restricted.

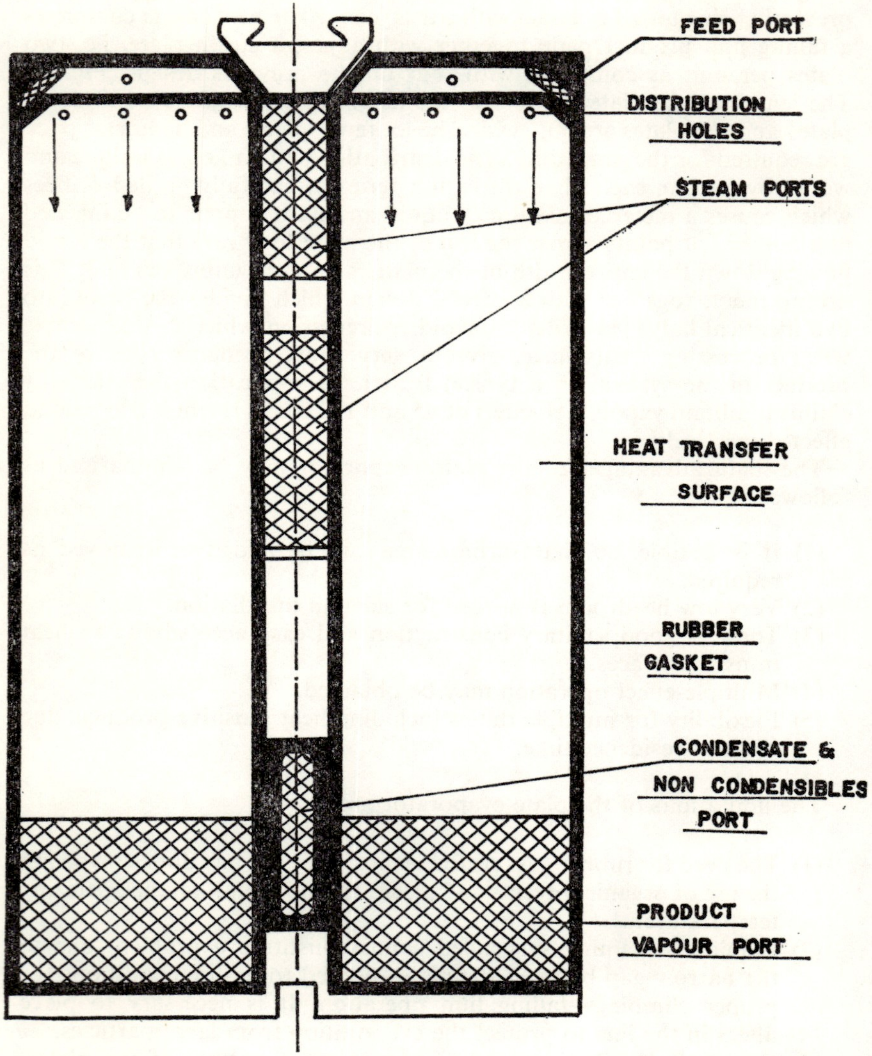

FIG. 3. Schematic diagram of new plate of APV Falling Film Evaporator. (By courtesy of APV International Ltd., England.)

(5) To maintain proper wetting usually recirculation is required in this system. Attempts to overcome this limitation are being made in the new design by dividing individual effects into a number of stages in series.

Stationary cylindrical surfaces with agitated film—thin film evaporator

While evaporating heat sensitive viscous fluids containing suspended matter, the problems of maintaining adequate rates of heat transfer and proper distribution of product over heat transfer area are intensified. Also liquids with low thermal conductivities, and liquids containing solids that may crystallise during the evaporation process pose a problem. The application of a mechanical method for spreading a thin film on the heat transfer surface has major advantages (Fig. 4). In an evaporator of this type the product is spread by an internal rotor. The rotor may have a fixed clearance of 0.2–2.0 mm, or fixed blades with adjustable clearance, or blades which actually wipe the heat exchange surface. The unique feature of this equipment is not the thin film itself, but the mechanical agitator device for producing and agitating the film. Two main types of such evaporators are to be found. One type comprises a vertical cylindrical jacketed tube containing a concentric rotor whose blades have a small clearance from the heated wall (Mutzenberg, 1965). In the other type, the jacketed tube is horizontally disposed and conical in shape. Within this tube rotates a concentric rotor with blades having a clearance from the heated wall.

Comparison of advantages and disadvantages of the vertical and horizontal machines may be summarised.

(1) The vertical position guarantees the maximum possible uniformity of film flow, while the horizontal machine permits back mixing of the film which results in an unfavourable residence time distribution.
(2) The vertical machine operates in a countercurrent manner as regards liquid and vapour flow. This may result in fractionation which is helpful in distillation but less desired in fruit juice concentration. In the horizontal machine the flow is co-current and the problem of entrainment and carry over droplets into the vapour is great.
(3) The vertical machine requires less floorspace, but more headroom.
(4) The manufacture of a horizontally mounted rotor with sufficient stiffness to varying temperatures to ensure small wall clearances is difficult especially in large machines.

Therefore for evaporation purposes the vertical machines are preferred. Design of blades has a significant importance on evaporator perform-

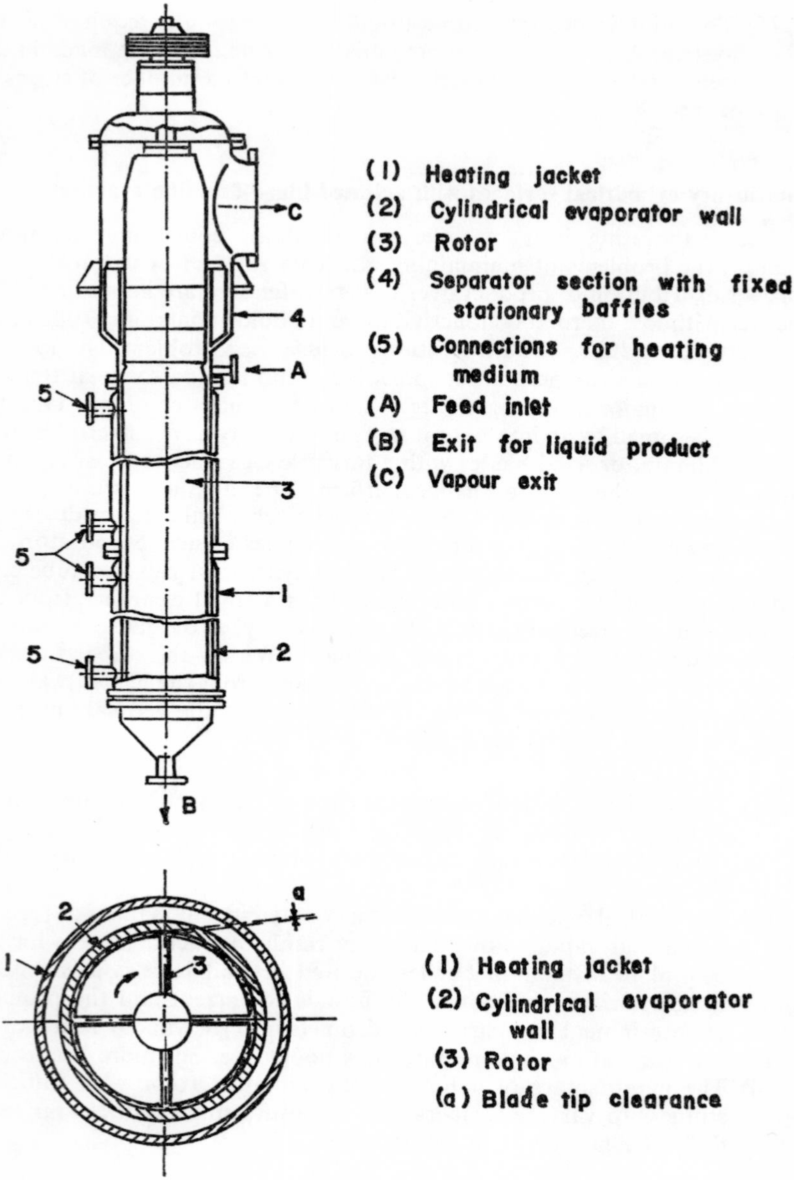

FIG. 4. Schematic section of Agitated Film Evaporator. (From *Industrial Chemist*, June, 1963, courtesy of Chemical Processing, England.)

ance. The rotor baldes, or wipers, force the liquid into a film flowing downwards under the influence of gravity; the blades dip into the film, so keeping it in violent agitation. The blade tip clearance should be as small as possible and should almost wipe the wall, for best performance (Reay, 1963). Obviously the film thickness differs from one liquid to another depending on its physical properties. Rotor speed also has some effect on heat transfer. Besides the beneficial effect on heat transfer, high rotor speeds are useful in enhancing the action of the centrifugal de-entrainment device in the upper part of this type.

The object in processing heat sensitive materials in agitated film evaporators is to expose them to heat for the shortest possible time, which in these machines is usually in the order of seconds. However, in some applications larger residence times are required. This can be achieved by means of a Flow Control Ring (Reay, 1963) which can quickly be installed at the lower end of the rotor thus increasing hold-up. Mean residence time values are dependent on width of Flow Control Ring and rotor speed. Values between 3 and 100 s can be obtained. One important advantage of this residence control is that it offers another means of control over evaporation ratio. In thin liquids high evaporation ratios would have been difficult to obtain without the risk of producing dry spots on the wall. This means that unless a Flow Control Ring is used in these cases, recirculation would be required.

The advantages of this type of evaporator are as follows.

(1) It can handle highly viscous products.
(2) Pulpy and foaming materials can be concentrated.
(3) Residence times are short.
(4) High evaporation rates.
(5) It can operate at low Δt between wall and product, thus eliminating fouling.

Disadvantages of this equipment are as follows.

(1) There is a moving rotor with small blade clearence which requires precise alignment resulting in expensive construction. If the blades are in direct contact with the wall, due to unalignment, undesirable wear will occur and the product may be contaminated with metal parts.
(2) It is difficult to clean.
(3) High headroom required for demounting rotor for inspection and cleaning.
(4) Only available in single effect therefore poor steam economy.
(5) High capital and operating costs.

The agitated thin film evaporator should only be used in cases where the problem of steam economy is unimportant, such as in the final stages of evaporation of a highly viscous product. In this case the amounts of water to be removed are small but potential heat damage due to fouling or burning on is large.

Agitated film evaporators, for example, are most suitable for final concentration of hot-break tomato paste. According to Jacobi (1971) the evaporation rates in an agitated film evaporator with this product were 195 kg/h/m^2 as compared with 103 kg/h/m^2 in a horizontal thin film and 93 kg/h/m^2 in the rotary-coil evaporator.

Stationary conical surface—Expanding Flow Evaporator

Figure 5 shows an Expanding Flow Evaporator. The liquid to be evaporated is fed to alternate conical interspaces through nozzles in the central spindle. As the liquid passes upwards in a channel of increasing flow area (hence the name 'Expanding Flow'), it evaporates. The mixture of concentrated liquid and vapour leaves the cones at their outer peripheries. A special design of shell induces a cyclone effect for effective vapour–liquid separation. Primary steam is introduced at the peripheries of alternate conical channels. The flow area is decreasing as the steam condenses, thus maintaining a high condensing film coefficient, and condensate is removed through holes at the cone centres. The expanding flow evaporator has a number of advantages: a hold-up time of 1–2 min, low head room, and a high degree of flexibility. By changing the number of cones the evaporator capacity can easily be adapted to various requirements.

Rotating conical surfaces—Centri-Therm Evaporator

The main feature of the Centri-Therm Evaporator is the application of rotating conical evaporating surfaces. In conventional evaporators the liquid film is generally transported over the evaporating surface by means of the vapour and the final concentration of the product is therefore limited. The transporting force can be very much increased with mechanical assistance, which means considerably higher velocity and therefore large heat transfer coefficient and the possibility of obtaining a higher final concentration. A higher velocity will also mean shorter hold up time on the heat surfaces and this naturally has a great influence on final product quality, notably on heat sensitive liquids. In the Centri-Therm the mechanical assistance is the centrifugal force, which, at the speed employed, is more than a hundred times as big as the force of gravity.

A cross section of a Centri-Therm machine is shown in Fig. 6. In the centre there is the rotating bowl in which the double-walled cones are

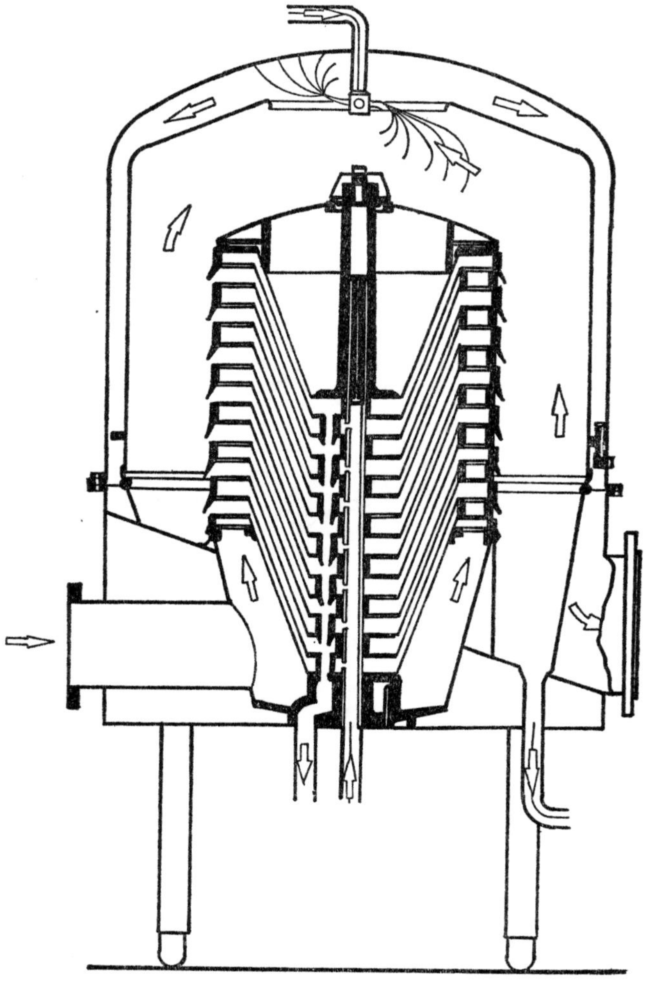

FIG. 5. Schematic section of Expanding Flow Evaporator. (Courtesy of the Alfa Laval/De Laval Group, Sweden.)

nesting. This bowl or rotor body is fastened to a tubular shaft. The steam and vapour chambers are separated in the rotor body by an intermediate partition.

The cross section of the Centri-Therm resembles that of an expanding flow evaporator; however, the evaporators are completely different in

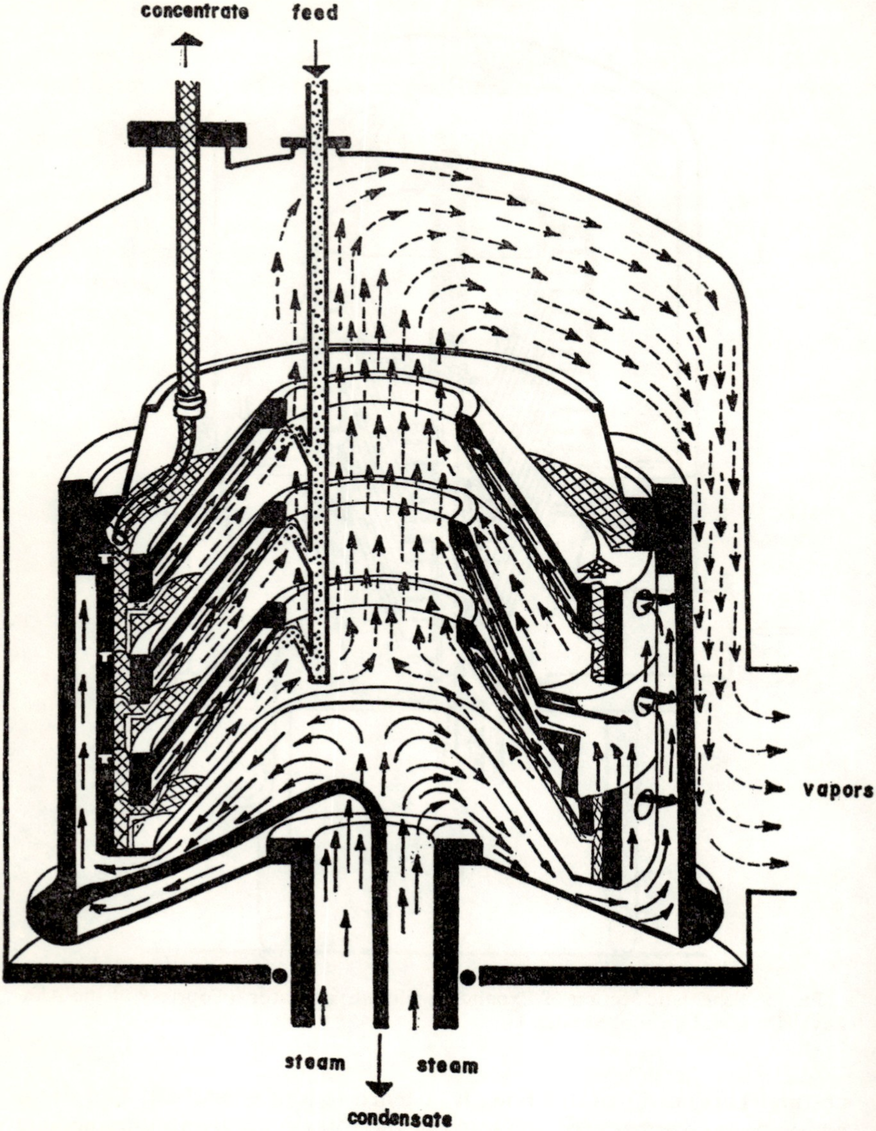

FIG. 6. Schematic section of Centri-Therm Evaporator. (Courtesy of the Alfa-Laval/De Laval Group, Sweden.)

their structure. In the Expanding Conical Flow single-walled stationary cones are employed instead of double-walled rotating ones as in the Centri-Therm. The liquid rises in the former evaporator in the same way as in any climbing film evaporator while in the Centri-Therm it is spread over the surface by centrifugal force.

In the Centri-Therm the liquid is fed in from the top of the machine (Fig. 6), via a distribution pipe which is fixed to the shell and passes through the cone-stack's centre part, and is injected onto the underside of the rotating cones. The liquid is immediately spread out in a film of about 0.1 m thickness and passes over the heat transfer surface in about 1 s. The liquid is immediately heated to evaporation temperature and begins to boil. Since the film is so thin, vapour is released very rapidly and passes out through the centre of the cone-stack to the surrounding mantle. From here, the vapour is fed to the condenser via the vapour outlet.

The concentrated product collects in a series of rings round the insides of the cones' peripheries and is led via vertical holes to a paring chamber in the cone-stack's upper part, from which it is removed by a stationary paring tube fixed to the mantle.

Primary steam is fed to the evaporator from beneath via a hollow spindle to the steam chamber at the periphery of the cone-stack. The steam passes to the spaces between the conical elements via holes in the rings to which the cones are welded. The steam is condensed on the cone surface, on the other side of which the product is being evaporated.

As soon as the condensate forms droplets on the cone surface, they are thrown off by centrifugal force against the outer cone of the element. The condensate passes out via the same holes through which the steam entered and collects in the steam chamber, in the bottom of which is a paring chamber. Condensate is finally removed from the evaporator by stationary paring tube.

Since the droplets of condensate are 'centrifuged' away from the heat transfer surface as fast as they are formed, no condensate film, which would offer resistance to heat transfer, is formed. The drop-wise condensation which consequently occurs, gives rise to very high heat transfer coefficients—k values of up to 7000 kcal/m^2/h/°C have been observed.

In principle, the k value on the product side is increased if the product film thickness is reduced under otherwise constant conditions. The same applies if the film velocity is increased. A higher final product viscosity leads to low product velocity and a thicker film.

Centrifugal force, to which the liquid is subjected by the rotation of the heat transfer surface, affects film thickness and velocity such that increased force gives a thinner film and a higher velocity. The magnitude of the centrifugal force is dependent, amongst other things, on the cone's rotational speed n, $F \propto n^2$. From the above statements it is clear that a higher rotational speed produces a higher *k* value. The speed cannot, however, be increased indefinitely, since the liquid film can become discontinuous. This would mean that the heat transfer surface would be only partly utilised, and also that burning-on would result.

The cone elements are stacked up on each other and can easily be removed for inspection and cleaning. They are sealed from each other by *O*-rings, for which the desired quality can be selected to suit the solvent involved. The liquid is spread on the cones by means of a distributor which consists of a main pipe with a number of short branches fitted with nozzles.

As previously mentioned, the liquid is spread out by centrifugal force into a very thin film as soon as it makes contact with the rotating cone surface. The product takes about 1 s to pass over the surface, but this short time is sufficient to evaporate the liquid to the required concentration. The residence time within the evaporator is, as a result, so short that no noticeable deterioration occurs of the product's characteristic taste, colour, protein and vitamin content. In addition, decomposition of organic materials and burning-on is eliminated.

SEALS, LUBRICATION AND BEARINGS

To a system working under vacuum, seals and packings naturally are important details. To seal off the rotating part, expensive carbon seals are used while stationary seals consist of *O*-rings.

The main advantage of the Centri-Therm is the high final product quality obtainable in one pass in a very short time (1–10 s). The viscosity of the end product can be up to 20 000 cP. The evaporator is therefore suitable for concentrating very heat sensitive and viscous materials such as citrus juices, coffee extracts and pharmaceuticals. Juice concentrates made in Centri-Therm are usually of high aromatic quality making aroma recovery unnecessary. The equipment is compact, requires low headroom and is of sanitary construction. On the other hand, about 1.1 kg steam is needed per kg water evaporated since the high Δt required in Centri-Therm does not permit double-effect operation. Capacities of Centri-Therm range from 50 kg/h in a pilot plant machine to 2400 kg/h evaporated water in the larger industrial equipment. At present an evaporator with a capacity of about 5000 kg/h water evaporated is being tested. This means that for large capacities several evaporators are needed, i.e. very high investment. Since this evaporator is rotating, the seals require close attention to assure proper operation and they must be replaced periodically.

CHOICE OF EVAPORATORS

Choice of the suitable evaporator for a given duty demands a thorough knowledge both of different types of commercial evaporation plants and of the product to be evaporated. Considerations in specifying an evaporation plant include:

(1) Capacity expressed as kg/h evaporated water.
(2) Concentration degree, usually expressed as percent dry solids in the evaporated product.
(3) Heat sensitivity of the product, related to hold-up time and temperature.
(4) Recovery of volatile substances.
(5) Sanitary construction and ease of cleaning in place.
(6) Reliability and simplicity of control.
(7) Plant dimensions, related to available floorspace and headroom.
(8) Investment and operating costs related to evaporation rate per unit heat transfer area.

While the technologist has as a primary objective the highest product quality, management is often more interested in economics. Jacobi (1971) notes that in the food industry, contrary to all expectations, the cost problem is often weighed heavier than the quality factor. The choice is therefore usually a compromise between technical performance and economic feasibility.

Following are some guidelines for choice of evaporator type.

Heat sensitivity of products

If the product to be evaporated is very heat sensitive, and there are risks of it being damaged if exposed to excessive hold up times at elevated temperatures, an evaporator giving ultra-short contact time and low evaporation temperature is to be chosen. Table III is a comparison of residence times in different evaporator types.

Viscosity of products

Evaporators for viscous products must be equipped with arrangements to spread the liquid in thin film and to transport it with high velocity over the heat transfer surface. This can be done either by mechanical assistance, i.e. agitated or wiped film, or by centrifugal force. If the product, besides being viscous, is temperature sensitive, the demand under 'heat sensitive products' must also be met. The maximum concentrate viscosity which can be handled in different types of evaporator equipment is approximately as follows:

Climbing film	100 cP
Falling film	200 cP
Plate	300–400 cP
Agitated film	20 000 cP
Centri-Therm	20 000 cP

TABLE III

COMPARISON OF RESIDENCE TIMES IN VARIOUS EVAPORATORS

Type of evaporator	*No. of stages*	*Residence time*
Vacuum pan	One stage	One to several hours
Climbing film	One stage with recirculation	$\frac{1}{2}$–1 h
Climbing film	One stage and single pass	About 1 min
Falling film	One stage and single pass	About 1 min
Falling film	Five stages and single pass	About 4 min
Plate	Three stages and single pass	About 4 min
Expanding flow	Two stages and single pass	1–2 min
Agitated film	One stage	20–30 s
Centri-Therm	One stage	1–10 s

In addition to viscosity, as mentioned, pulp content is another limiting factor in choosing an evaporator for a given duty.

Temperature difference

A high temperature difference between the primary and secondary sides enables reduction in heat transfer surfaces and a higher degree of concentration without the need for recirculation. Residence time is therefore shorter and heat treatment more gentle.

In order to be able to use a high temperature difference the evaporator must be so constructed that the film thickness and velocity over the heat transfer surface can be accurately controlled during evaporation. In an evaporator of conventional (tubular) type, a high temperature difference involves a risk of burning-on, especially if the product contains protein.

Space requirements

Maintenance and service are facilitated if the dimensions of the evaporator plant are small. Sufficient headroom must be available when the evaporator is to be disassembled for inspection and cleaning—this is especially important for vertical climbing film evaporators.

Cleaning, servicing, etc.

Modern evaporators operate with a minimum of attention and often with a varying degree of automatic control. The evaporator should be so constructed that it can be cleaned with the minimum of labour, and preferably by automatic In Place Cleaning Techniques. It should be easy for inspection and the heat transfer surface should be easily accessible for removal and cleaning.

Steam consumption

Especially for plants of high capacity, steam consumption is a significant factor in the choice of evaporator. As a rule those types of evaporators especially designed for low steam consumption operate on products which tolerate high temperatures and longer hold-up times.

Following are examples of steam consumption for plants having one- two- or three-effect.

Steam consumption (kg/h *evap. water*)

No. of effects	*Without compression*	*With compression*
1	1.1	0.6
2	0.6	0.4
3	0.4	0.3

Water consumption

This has recently become a very important factor due to the acute shortage of water in some regions and effluent problems in all places. Water consumption is affected primarily by condenser efficiency, temperature in the final effect and cooling water supply temperature. Cooling water consumption is reduced with increase in the number of effects. Cooling water can be saved by using a portion of the product vapour for product preheating, as is usually the case when surface condensers are used. With a cold feed, which is common in the dairy industry, very considerable cooling water savings can be made.

Summary of recommendations

Table IV summarises the recommendations for the choice of evaporators. In every case a few evaporator types are recommended. The more sophisticated evaporators like agitated and wiped film, and rotary conical may be used in most cases; however, due to their low capacity, high investment and operating cost, these should be chosen judiciously.

TABLE IV

SUMMARY OF RECOMMENDATIONS FOR CHOICE OF EVAPORATOR

Product description		*Suitable evaporator type*	*Comments*
Viscosity	*Heat sensitivity*		
Low or medium	None	A. Tubular B. Plate C. Stationary conical	Occurs within the chemical and cellulose industries. Types with horizontal tubes less suitable for scalebuilding products
High	None or low	A. Agitated or wiped film B. Rotating conical C. Vacuum pan	Batch evaporators may be used for products such as: agar-agar, gelatine, meat extracts
Low or medium	Sensitive	A. Tubular B. Plate C. Stationary conical	Includes products like milk and fruit juices to medium solid content. Horizontal tubes not suitable when fouling risks exist
High	Sensitive	A. Agitated film B. Rotating conical	Includes most fruit juice concentrates, yeast extracts and some pharmaceuticals. For pulpy products only agitated film should be used
Low	Very sensitive	A. Tubular B. Plate C. Stationary conical	Single pass desired
High	Very sensitive	A. Rotating conical B. Plate	Only single pass desired. Includes citrus concentrates and some pharmaceuticals and egg white

MAIN COMMERCIAL EVAPORATOR TYPES

The main commercial evaporators used in the food industry, and especially in the fruit juice industry, according to types described above, are as follows.

Tubular evaporators

(1) T.A.S.T.E., falling film.
(2) Wiegand, falling film.
(3) Holvrieka, falling film.
(4) Laguilharre, falling film.
(5) Kestner, climbing film.
(6) Jensen & Andersen, climbing film.
(7) Unipektin, horizontal film.
(8) Bergs Maskin, climbing and falling film.
(9) Blaw-Knox, falling film.
(10) Buflovak, falling and rising film and horizontal tubes.
(11) Henszey, climbing film with mechanical vapour compression.
(12) Rossi & Catelli, falling film.

Most of above makes of evaporators work on a similar principle of operation. Mention will be made of those evaporators which are of special importance in the food industry.

THE T.A.S.T.E. (Gulf, Tampa, Florida, IMC in Italy and others)

One of the most important falling film evaporators is the so-called T.A.S.T.E. evaporator. This is in principle a three-effect, five-stage evaporator especially designed for the citrus industry. The T.A.S.T.E. evaporator operates on the high temperature short time principle. Preheating, concentration from a 12°Bx juice to a 60°Bx concentrate and flash cooling are obtained in about 3–4 min. In this evaporator preheated juice enters a distribution cone and goes through a 10°C flash, which atomises the juice, thus feeding the tubes with a descending mist rather than a film. This mist is fed into the tubes at a velocity of 7 m/s and is accelerated due to thermal expansion, to about 200 m/s at the end of the 12 m long tubes. This combination of mist and high velocity gives very high heat transfer coefficients and prevents localised heating and burning of the product. The concentrate is flash cooled, and the manufacturers claim this cooler removes also any off-flavour produced during the concentration process.

The evaporator can be equipped with an aroma recovery unit which may be needed due to removal of aroma during flashing.

The advantages of the T.A.S.T.E. evaporator are:

(1) Good steam economy (about 0.35 kg steam/kg water).
(2) Ease of installation since it comes in one or maximum two pieces.
(3) Ease of operation and cleaning.
(4) Wide range of capacities—from 1300 kg/h to 27 000 kg/h water evaporated.
(5) Product quality is good especially if used in connection with aroma recovery or cut back juice.

The disadvantages are:

(1) High head room required.
(2) For best quality, aroma recovery and/or cut back are necessary.
(3) Limited mainly to citrus juice concentration.

ROSSI & CATELLI (Parma, Italy)

This company makes a 'downward forced flow' evaporator with circulation. In the double-effect version the feed juice enters the second effect which consists of an inclined tubular heat exchanger and a cyclone type vapour separator. From the second effect juice is delivered by gravity into the first effect which is a falling film forced circulation evaporator. This double-effect evaporator is especially suited for concentrating tomato juice to solid contents over 40%. Residence times in this equipment are fairly long and range from 30 to 90 min. Evaporation capacities range from 3500 kg/h to about 20 000 kg/h water evaporated.

Flat heating surfaces—plate evaporators

(1) APV plate evaporator.
(2) Schmidt/Bretten plate evaporator.

The plate evaporators work according to the climbing and falling film principle as described previously (Figs. 1 and 2). Residence time in single-effect units is claimed to be about 3 min, and in double-effect evaporators with recirculation about 4 min. At present APV is testing a new plate evaporator based solely on the falling film principle.

Stationary cylindrical thin-film evaporators with agitation

(1) LUWA agitated film.
(2) Sambay (LUWA) wiped film.
(3) Rotafilm wiped film with condenser in centre of machine.
(4) Pfaudler wiped film.
(5) Kontro tapered agitated film.

Other evaporators

Stationary and rotating surface evaporators called Expanding Flow and Centri-Therm are made only by Alfa-Laval and were described previously.

TITANO—Continuous rotary coil evaporator (Manzini, Parma, Italy)

The Titano is a double-effect evaporator with reverse feed. It consists of a preconcentration stage which is a climbing film second-effect evaporator with natural circulation. The first effect, but final stage concentrator, is a trough-shaped chamber with a multiple-coil rotating in it. Equipment was especially designed for tomato juice concentration and manufacturers claim to be able to achieve above 40°Bx hot-break concentrates. Residence time is about 1 h or more, final evaporation temperature is about 60°C. Capacities range from about 3000 kg/h to 23 000 kg/h of evaporated water.

PROBLEMS IN EVAPORATION

Fouling

The deposition of a burnt layer of organic matter on the hot surface of an evaporator constitutes a severe problem during evaporation. This phenomenon is called fouling and is caused by destruction of temperature sensitive components in the food. Fouling causes a decrease in conduction of heat to the fluid through heating surfaces, decreases heat transfer rates, i.e. plant capacity, and reduces product quality and storage stability. In the evaporation of highly viscous products like tomato juice, it is a dominant factor, and therefore most studies concerning this problem were conducted on this medium (Adams *et al.*, 1955; Morgan and Wasserman 1959; Morgan and Carlson, 1960; Morgan, 1967).

Morgan and Wasserman (1959) investigated the composition of foul in forced and natural circulation evaporators during tomato juice concentration. They found a significant accumulation of protein in the foul (33% in the foul as compared to 13% in the product). Rupture of tomato cells seemed to increase fouling rates by release of protein. Morgan and Carlson (1960) found that a standard cleaning procedure could restore the surface to the same 'clean value' (h_0) heat transfer coefficient after each fouling experience. h_0 was best indicated by extrapolation of the heat transfer coefficient in the liquid (h_L) to zero time (Fig. 7). At any time since cleaning, the decrease in liquid heat transfer coefficient h_L may be interpreted as an increase in the fouling coefficient—h_F (Morgan, 1967).

$$\frac{1}{h_L} = \frac{1}{h_0} - \frac{1}{h_F}$$

This implies that heat transfer into the liquid from the fouled layer is the same as from the clean surface except for a loss of heat in passing through the fouled layer. The rate of change of this coefficient with time is referred to as the fouling rate.

Operating variables were studied (Carlson and Morgan, 1962) for their relative effect on fouling rates of fruit and vegetable pulps. One of the most important factors found to influence fouling was the temperature of the heat transfer surface (Fig. 7).

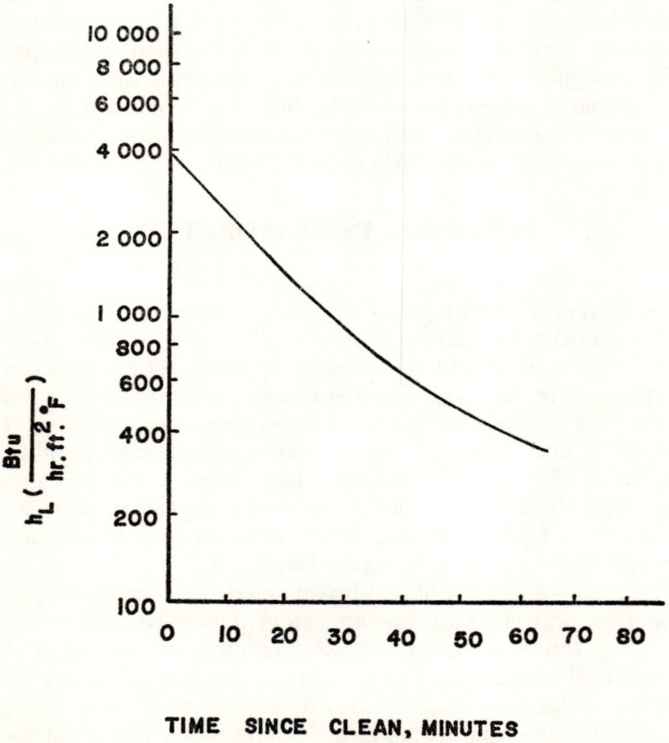

FIG. 7. Variation of liquid heat transfer coefficient with time during concentration of tomato juice. (From *Food Technology*, **14**, 525, courtesy of the Institute of Food Technologists.)

Adams *et al.* (1955) found a decrease in film deposition due to pectin degradation, lower coil surface temperature, coating the heat transfer surface with silicone and the addition of some chelating agents to tomato juice. These investigators also found that a pre-evaporation treatment or injecting direct steam at 40 psig into blanched hot broken tomatoes prevented film deposition. According to Adams *et al.* (1955) particle size had little effect on fouling and only drastic particle reduction, by means of homogenisation or colloid milling, increased film deposition.

Agitation of the product film in mechanically agitated evaporators was claimed also to prevent scaling and produce a highly uniform concentrate (Fisher, 1965). However, Adams *et al.* (1955) claimed that heating

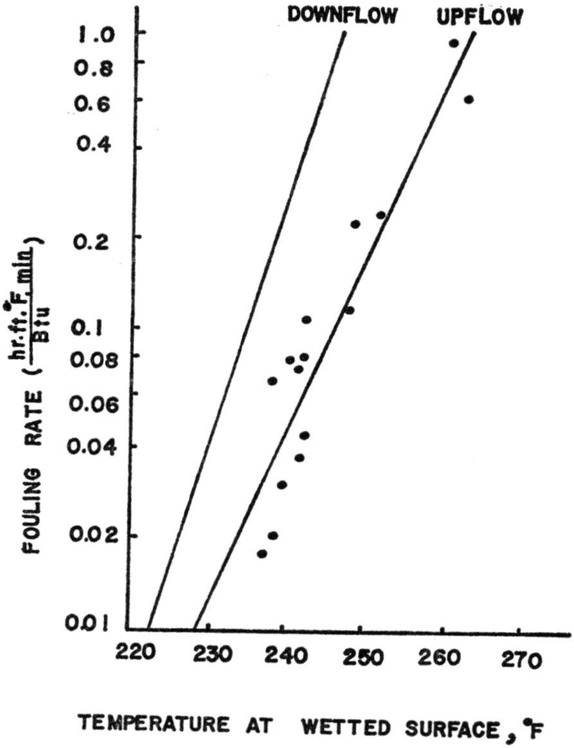

FIG. 8. Effect of flow direction on fouling rate of boiling tomato paste (20% solids). (From *Food Technology*, **21**, 1354, courtesy of the Institute of Food Technologists.)

coil temperatures above a certain critical level promoted film deposition despite high turbulence in the boiling medium.

Other operating variables found to influence fouling were direction of fluid flow in evaporator, boiling or warming juice on heat transfer surface and amount of vapour in fluid. Figure 8 shows that a rising film evaporator causes less fouling than a falling film evaporator. This is probably due to improper distribution in the latter case which causes local hot spots and thus burning on.

Another parameter is the presence or absence of boiling on the heat transfer surface. Fouling was much more rapid when the juice was warmed on the surface as compared with being boiled for the same wall temperature (Fig. 9). Due to this effect fouling can begin at the entrance to an evaporator where boiling has not yet begun and can slowly spread to the remainder of the surface.

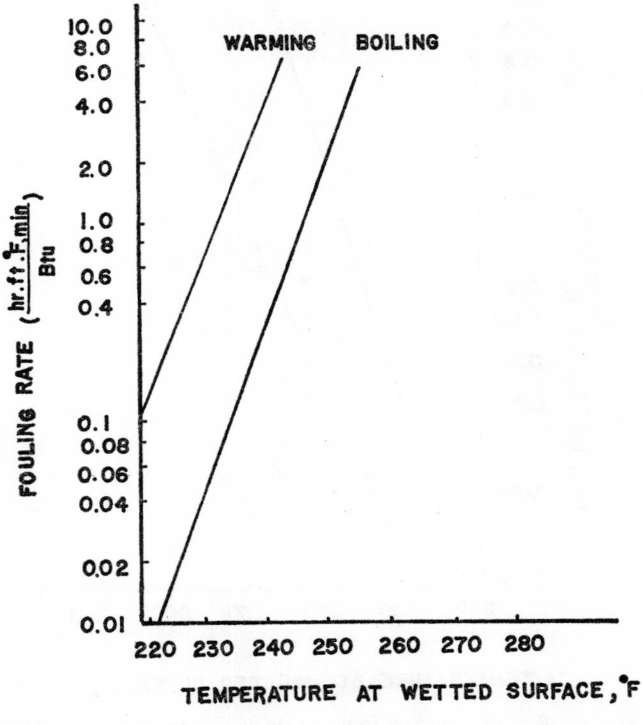

FIG. 9. Effect of boiling on fouling rate of tomato paste. (From *Food Technology*, **15**, 596, courtesy of the Institute of Food Technologists.)

An increase in weight fraction of vapour in liquid on the heat transfer surface increased the rate of fouling (Fig. 10). The effect becomes pronounced when a vapour fraction above 25% is reached.

Chemical analysis of fouled layers of various foods showed (Morgan, 1967) that their sugar, fibre, pectin and ash values were similar to those of the liquid being separated. However, in all cases the Kjeldahl nitrogen values were much higher in the fouled layer than in the liquid food.

Furthermore, when temperature dependence on fouling rates was treated in an Arrhenius plot, activation energies between 50 and 70 kcal/g mol were obtained. These values are in the range of known activation energies for protein denaturation.

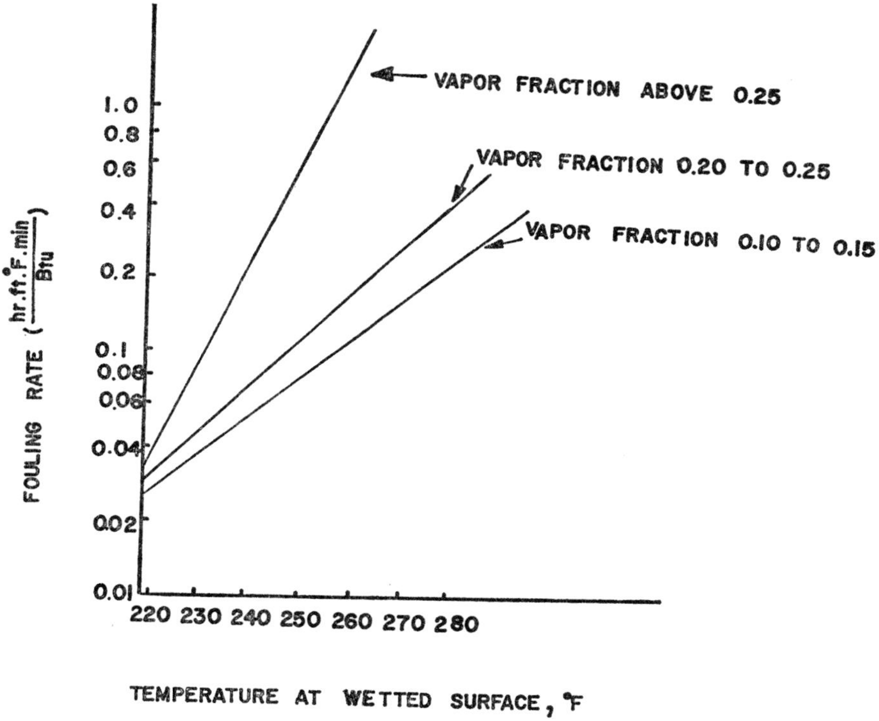

FIG. 10. Effect of vapour fraction on fouling rate of tomato juice (From *Food Technology*, **15**, 596, courtesy of the Institute of Food Technologists.)

An investigation of the influence of components of the feed liquid (Morgan and Wasserman, 1959) showed that pectin and fibre affected fouling markedly and this effect was explained by the influence of these components on viscosity. However, protein does not have a marked influence on consistency and its fouling effect is apparently by chemical reaction. Morgan and Wasserman (1959) also showed that once protein was denatured by evaporation or by preheat treatment as mentioned by Adams *et al.* (1955), fouling was minimised in subsequent evaporation.

In order to prevent product deterioration Kopelman and Mannheim

(1964) suggested separating tomato juice centrifugally, into serum and pulp, prior to evaporation. They found that by concentrating the low viscosity serum alone, fouling was avoided, better heat transfer coefficients were obtained, enabling the attainment of higher concentrations.

Based on the above studies, in order to achieve good evaporator operation with minimum fouling, the following points should be observed when designing an evaporator:

(1) Sufficient feed preheaters should be supplied so that boiling occurs throughout the tubes, thus preventing local boiling.
(2) The vapour fraction on the heat transfer surface should be minimised by avoiding flashing, high vapour velocities and providing efficient means of vapour removal.
(3) Viscosity of liquid should be as low as possible.
(4) The heat transfer surface must be as cool as possible.
(5) The surface should be completely wetted.
(6) The stagnant layer of liquid on the surface should be kept as thin as possible.

A rotary coil evaporator was suggested by Randall *et al.* (1966) for use on viscous liquids. This evaporator consists of a rapidly rotating, steam filled coil under the surface of a pool of liquid boiling in vacuum. Surface wetting is maintained by the pool of liquid. The pool volume is maintained constant with dilute feed. The stagnant layer is kept thin by the rapid rotation of the coil.

The evaporator is used commercially for producing tomato paste. Another application of this evaporator is the manufacture of fruit jams, jellies and preserves.

Aroma retention and recovery

In the process of concentrating liquid food many volatile and odorous components are removed with the water vapours. The typical fresh aroma bouquet of liquid foods can be preserved in the final product by two principal methods. They can be either retained in the product or fraction of it which is not evaporated, or recovered in the evaporation process and reincorporated at a later stage. A comprehensive review of the literature on this subject was presented by Mannheim and Passy (1972).

Briefly, in the conventional systems for recovering volatile flavours, the feed material, which is an aqueous solution of volatiles, is subjected to heat in order to vaporise part of the juice. The resulting vapour fraction, containing water vapour, volatile flavour components and non-condensable gases, is separated from the residue (the stripped juice) and condensed.

The vapour fraction is then passed to a recovery system for concentration of the volatile flavours. Fractional distillation, stripping or extraction are forms of aroma concentration.

Processes for preconcentration and aroma recovery may be divided into the following groups.

(1) Processes performed entirely under atmospheric pressure.
(2) Processes performed partially or wholly under reduced and low temperature to preserve the aromatic quality of the heat sensitive concentrate and/or essence.
(3) Processes based on adsorption of the aromatic substances on charcoal and their back extraction; however, none are used commercially for foods at present.
(4) Processes based on stripping aromatic substances with an inert gas.
(5) Processes based on extraction of aromatic substances with a condensed gas such as carbon dioxide.

Pre-evaporation under atmospheric conditions, followed by fractional distillation is the most suitable method for the recovery of water soluble essences such as those of apples, pears, peach and berry juices and is widely accepted in these industries today. For heat sensitive juices, such as citrus, pre-evaporation must be carried out under vacuum and it is best followed by stripping of vapours to obtain the concentrated aroma. The WURVAC system (Bomben *et al.*, 1966) employing a liquid sealed vacuum pump seems suitable, provided losses in vent gases are minimised by using sufficient cold traps or other means. In the case of citrus juices the aroma should be stripped from the juice prior to any other heat treatment such as pasteurisation.

Use of liquid carbon dioxide at ambient temperature and pressures of about 70 atm to extract aroma compounds has been suggested by Schultz and Randall (1970). Liquid carbon dioxide is a selective solvent for aroma constituents such as esters, aldehydes, ketones and alcohols and therefore may find a wide range of applications. Sugars, acids, salts, amino acids, oils and water are insoluble in it. Highly concentrated essence extracts in water free oil form were obtained from apples, pears, orange juice and roasted ground coffee. This process is not yet used commercially probably due to the technical difficulties caused by working at higher pressures.

Water based aroma preparations are relatively unstable in storage and care must be taken to keep them at low temperatures, and in the absence of light and oxygen. If reincorporated into products they lose their intensity relatively fast and should therefore be added to the final product as close to the time of consumption as possible. This essence instability makes flavour retention in the product or fraction of it with the avoidance of any heat treatment of aromatic substances rather attractive. A method of separating orange juice into pulp and serum prior to evaporation, concentrating serum alone and using unheated pulp to reinforce concentrate was shown to be very effective (Peleg and Mannheim, 1970) (Fig. 11). Compared with similar processes, the pulp was not subjected to any heat

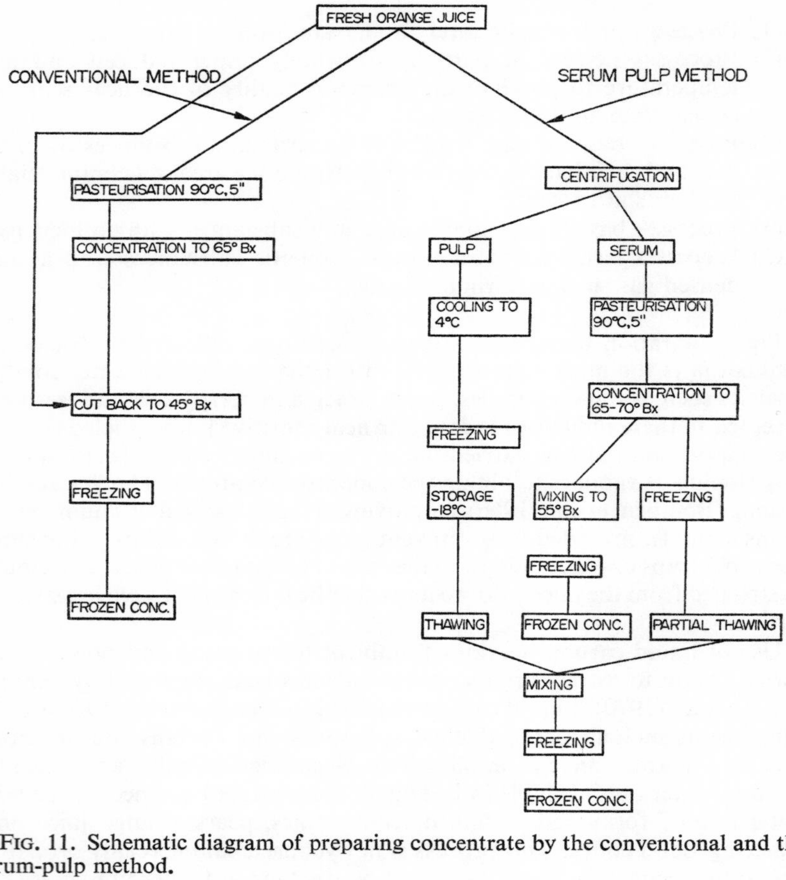

FIG. 11. Schematic diagram of preparing concentrate by the conventional and the serum-pulp method.

treatment and therefore all the aromatic compounds in it were retained undamaged. The concentrate was stable, despite the active enzymes in it, due to the fact that the final product was about 55°Bx, at which concentration enzymatic activity at low temperatures was negligible. Care must be taken to keep the separated pulp chilled or frozen until reincorporation into the concentrate. The final product must be preserved in the frozen state. In addition, when concentrating only low viscosity serum, heat transfer rates are improved and fouling is prevented as mentioned previously.

Reverse osmosis has also been suggested for aroma retention. In order to retain all water soluble components very tight membranes are required giving very low permeation rates. Furthermore, the limit of final concentration using reverse osmosis is about 40°Bx. However, since aroma losses decrease markedly with a decrease in water concentration, reverse osmosis could be considered as an initial concentration step followed by any other method (Merson and Morgan, 1968).

Preconcentration treatments

All liquids receive some treatment prior to concentration. Most liquids such as fruit juices are screened to obtain even particle size in order to achieve even evaporation rates and uniform products. Other liquids, such as sugar and salt solutions, are filtered to remove undesirable and foreign particles. Some fruit purees such as berries are pretreated with pectic enzymes to facilitate juice extraction, increase yields and obtain juices rich in colour. Other juices, such as apple and grape, are treated with the same enzymes in order to obtain clear juices (Charley, 1969). Endo (1965) provides a detailed examination of enzymes needed for apple juice clarification. He found that they must be a mixture of the endo polygalacturanoases (PG) and pectinesterases (PE). In addition to clarifying the juices the above enzymes reduce the viscosity of liquids significantly and thus facilitate the concentration process.

While it is possible to make high quality cloud free concentrates, such as apple, the production of pulpy concentrates, with high final viscosity, still poses a problem. This is especially true if final concentration is carried out at relatively low temperatures (30–45°C) in order to preserve aroma and prevent browning. In order to be able to concentrate pulpy liquids Wucherpfennig (1964) suggested separating centrifugally the coarse pulp from plum nectar and concentrating the juice with fine pulp alone. A similar process was suggested by Pimazzoni (1961) and Kopelman and Mannheim (1964) for tomato juice concentration. This process gave good results with tomato juice and was reported as being used commercially for tomato paste manufacture in Italy. In this process care must be taken to achieve proper remixing of concentrated serum with pulp. Often, parts of the pulp must be discarded, as the viscosity and stability of final product is sometimes not satisfactory.

The serum–pulp separation process could not be applied to all fruit nectars. In some cases the separated serum contained too much pulp and the pulp too much water. Sulc and Ciric (1968) reported a new pretreatment of the pulp using a new pectolytic enzyme called Rohament P

(Gramp, 1969). This enzyme is rich in pectinglycosides and protopectinases and causes maceration of the cells, resulting in high juice yields with typical colour of the product. The pectic substances are hydrolysed only to a small extent, causing a sufficient reduction in viscosity to be able to treat the products. The cloud stability of the products, on the other hand, increased significantly and no separation occurred after one month's storage (Gramp, 1969).

Briefly, the preconcentration treatment consists of pasteurising the mashed product at about 85°C to inactivate any undesirable enzyme and reduce the microbial load. Pulp is cooled to 45°C and treated with 0.2% Rohament P at a pH range of 2.5–4.5 for 30 min (for berries) or 120 min (for carrots). After enzymatic hydrolysis, product is heated again to 85°C and then screened or separated in a decanter type centrifuge.

Mizrahi and Mannheim (1973) investigated the feasibility of using Rohament P prior to tomato juice concentration in order to obtain a concentrate which could be reconstituted into a cloud stable juice. While treated raw juice was very stable the concentration step caused a decrease in viscosity and resulted in an unstable reconstituted product. This phenomenon may be due to a shrinkage of the suspended particles owing to the increase in soluble solids and volume reduction during concentration. This shrinkage was irreversible and was explained by the formation of cross linkages in the pectin structure. Berk and Mizrahi (1965) suggested the ultrasonic radiation of orange juice in order to reduce its viscosity. This treatment reduced the viscosity of the concentrate to about $\frac{1}{3}$ its initial value and enabled achievement of final soluble solids contents of up to 80°Bx. However, the final product had a very undesirable off-flavour apparently due to oxidation.

SUMMARY

Choice of evaporator requires detailed information about available commercial plants as well as knowledge of product characteristics. The potential adverse effects which may occur during concentration must also be considered. Design of evaporator should ensure proper distribution of liquid over the heat transfer surface to prevent local hot spots, and burning on. The product should not be flashed into the evaporator, in order to prevent fouling as well as excessive aroma losses. For non-heat sensitive juices preconcentration under atmospheric conditions followed by fractional distillation is most suitable for aroma recovery. For heat sensitive juices pre-evaporation must be carried out under vacuum followed by nitrogen stripping of vapours to obtain an essence. Preconcentration treatments, such as separating the juice into pulp and serum, facilitate concentration and enable the use of unheated pulp to improve quality of final product.

REFERENCES

Adams, H. W., Nelson, A. I. and Legault, R. R. (1955). 'Film deposition on heat exchanger coils', *Food Technol.*, **9**, 354.

Bedford, B. P. (1972). 'Evaporator types and application', *Food Manuf.*, **3**, 24.

Berk, Z. and Mizrahi, S. (1965). 'A new method for the preparation of low viscosity orange juice concentrates by ultrasonic irradiation', *Fruchtsaft-industrie*, **10**, 71.

Blair, J. S., Godar, E. M., Masters, J. E. and Riester, D. W. (1952). 'Exploratory experiments to identify chemical reactions causing flavor deterioration during storage of canned orange juice', *Food Research*, **17**, 235.

Bomben, J. L., Kitson, J. A. and Morgan, A. I. (1966). 'Vacuum stripping of aromas', *Food Technol.*, **20**, 125.

Carlson, R. A. and Morgan, A. I., Jr. (1962). 'Fouling inside vertical evaporator tubes', *Food Technol.*, **16** (11), 112.

Charley, V. L. S. (1969). 'Some advances in food processing using pectic and other enzymes', *Chem. Ind.*, **20**, 635.

Endo, A. (1965). *Agr. Biol. Chem.*, **29** (2), 129.

Fisher, R. (1965). 'Agitated thin film evaporators. II. Process application', *Chem. Eng.*, **9**.

Gramp, E. (1969). 'Die Verwendung von Enzymen in der Labensmittelin-dustrie', *Deutsche Lebensmittel Rundschau*, **65** (11), 343.

Jacobi, E. (1971). 'Dunnschichtverdampfer in der Lebensmittelindustrie', *Gordian*, **71** (11), 347.

Kopelman, J. and Mannheim, C. H. (1964). 'Evaluation of two methods of tomato juice concentration. I. Heat transfer coefficients', *Food Technol.*, **18**, 907.

Mannheim, C. H. and Passy, N. (1972). 'Aroma Retention and Recovery during Concentration of Liquid Foods', 3rd Nordic Aroma Symposium Hemeelinna, Finland.

Merson, R. L. and Morgan, A. I. (1968). 'Juice concentration by reverse osmosis', *Food Technol.*, **22** (5), 97.

Mizrahi, S. and Mannheim, C. H. (1973). 'Cloud stability of reconstituted tomato juice', Annual Report to Min. of Agriculture (In Hebrew), Technion, Israel Inst. of Technol. Haifa, Israel.

Morgan, A. I. Jr. and Carlson, R. A. (1960). 'Fouling inside heat exchanger tubes', *Food Technol.*, **14** (11), 594.

Morgan, A. I., Jr. (1967). 'Evaporation concepts and evaporation design', *Food Technol.*, **21** (10), 63.

Morgan, A. I. and Wassermann, T. (1959). 'Fouling of evaporator tubes by tomato', *Food Technol.*, **13** (12), 691.

Mutzenberg, A. B. (1965). 'Agitated thin-film evaporators. Thin-film Technology', *Chem. Eng.*, Sept. 13.

Peleg, M. and Mannheim, C. H. (1970). 'Production of frozen orange-juice concentrate from centrifugally separated serum and pulp', *J. Food Science*, **35**, 649.

Pilnik, W. (1971). 'Die Bedeutung von Enzymen bei der Fruchtverarbeitung Dechema-Monographen', Symposium–Frankfurt, October.

Pimazzoni, O. (1961). *Industria Conserve.*, **36**, 35.
Randall, J. M., Carlson, R. A., Graham, R. P. and Morgan, A. I. (1966). 'Yield tomato paste with 50% solids', *Food Eng.*, **38** (3), 168.
Reay, W. H. (1963). 'Recent advances in thin-film evaporation', *Ind. Chemist*, June.
Schultz, W. G. and Randall, J. M. (1970). 'Liquid carbon dioxide for selective aroma extractions', *Food Technol.*, **24**, 1282.
Sulc, D. and Ciric, D. (1968). 'Herstellung von Frucht und Gemuesemark Konzentraten', *Flussinges Obst.*, **35** (6), 230.
Vincent, D. B. (1967). 'Waste heat evaporates waste', *Food Eng.*, **39** (2), 84.
Wucherpfennig, K. (1964). *Flussiges Obst.*, **4**, 186.

DISCUSSION

C. Varsel (Citrus Research & Development, Coca Cola Co. U.S.A.): In the serum-pulp method you claim retention of flavour with the pulp. Does this include the water soluble aroma components?
C. H. Mannheim: Yes definitely. Well, I will admit here we never did a mass balance, but we ran a lot of sensory evaluations and we found that we retained most of the aroma components.
Varsel: And do you have any idea of the enzyme content of the serum? If there are still enzymes in the serum these may cause product deterioration.
Mannheim: Yes, some enzymes remain in the serum but we inactivate them by pasteurisation.
F. Bramsnaes (Danmarks Tekniske, Højskole): In the serum-pulp method the pulp is not pasteurised but only kept cool. Do you think there will be any microbiological problems, particularly from yeasts?
Mannheim: There is a problem, but it can be overcome. In the citrus industry, for example, when they use the cut-back method several techniques are used. One technique is to use very good sanitary practices. First of all, the oranges are cleaned very carefully using a chlorine rinse and a final rinse. The equipment is kept very clean and the juice is chilled quickly. In a similar way, using good sanitary practice, the pulp is taken from the centrifuge to a Votator type cooler, or frozen immediately. In this way spoilage problems may be avoided.
H. A. C. Thijssen (Eindhoven University): Professor Mannheim, in your remarks you said that flashing into the evaporator should be prevented in order to limit excess aroma losses. Normally, in most types of evaporators, excluding possibly the Centri-Therm, there is more or less phase equilibrium between the gas phase and the liquid phase, and so in all cases there is about the same aroma loss. Why, then, should flashing lead to a higher aroma loss than normal evaporation?
Mannheim: Yes, I agree. From the thermodynamic point of view there shouldn't be any difference however the vapours are separated. But the fact is that if a liquid is flashed into the evaporator, apparently a much more efficient stripping of the aromas from the juice is obtained than if the liquid is boiled on the heat transfer surface. For example, we tried to use Centri-Therm as a pre-evaporator

for aroma recovery by passing juice very quickly through the unit without any flashing. This unit was an extremely poor stripper.

Chairman: That could be explained by the fact that in a film evaporator such as the Centri-Therm having laminar flow and a very short residence time, it might be expected that the loss of aromas from the liquid layer is diffusion controlled, and that equilibrium is not attained between the liquid and the gas phase.

Mannheim: Yes, but in the other evaporators, such as the plate evaporator, the liquid leaving the evaporator is much more concentrated. The separation takes place from a much more concentrated liquid than when the feed is flashed. Professor Thijssen showed yesterday that this separation from high concentrations would lead to better aroma retention. I see Professor Thijssen agrees.

Chairman: It's also possible that flashing the feed results in entrainment of juice in the vapour.

Mannheim: I was considering entrainment losses to be negligible.

D. T. Shore: (A.P.V.): With a flash of the feed a large number of droplets are formed, giving a high inter-facial area. This will result in the sudden strip of aroma components. Every falling film evaporator, for example, has to have a flash at the inlet in order to distribute liquid, and I do agree with Professor Mannheim that this flashing at an inlet is a factor which adds to the loss of aroma.

G. D. Saravacos (National Technical University, Greece): I agree that good flavour recovery can be achieved through flash stripping. It is our experience that with apple juice, whose flavour compounds are very volatile, flavour is recovered by flashing. On the other hand, the flavours of grape juice cannot be stripped by flashing, but only by using a long residence time in the evaporator. This is because the flavours in grape have a lower volatility. Any component having a volatility of below 3–5 will not be removed by flashing. Thus there are likely to be some components of most fruits which will not be removed by flashing alone, and more stages will be needed to get them out.

F. Bramsnaes: I think you said that it's worthless to add back aroma to jam because you lose the flavour during storage. Is this your own experience and does this apply to berry jams?

Mannheim: As far as jam is concerned, I didn't try any of your other exotic Scandinavian berries; my experience is with peach, apricot, cherry and strawberry. With these four flavour was added back. Tasting after one or two days showed a difference when comparing jam with add back to jam with no add back. However, after one month at 15°C there was no difference in a triangular test.

D. T. Shore: I believe, Professor, you were a little unkind to the climbing film evaporator. When talking of the climbing film evaporator you mentioned between 5 and 30 minutes as the residence time. Now I would not have classified an evaporator with that residence time as really acting as a climbing film evaporator. The latter has a very short heating zone. You in fact described the zones, and generally, by feeding at boiling point or with flash, one avoids that long residence time zone or turbulent flow that exists before the onset of the true film. In this way, even with very long tubes, short residence times, comparable to those of the falling film tubular evaporators, can be obtained. This is in fact shown in your Table III for the single pass climbing film unit.

D. E. Blenford (RHM Research Limited): First, I have a comment to make. During the concentration process viscosity goes up enormously, particularly if the concentration ratio is large. Therefore a mixed combination of evaporation systems is often used very successfully, and this leads to a very much more economic application and less damage.

Mannheim: I agree with your comment. I think when I spoke about the agitated film evaporator, that I said this unit could be used with advantage for the final stage of tomato concentration.

Blenford: I would like also to ask a question. In general, the higher the concentration of the extract being dealt with the more stable it might be considered to be in terms of normal qualities and aromas. If various concentrations of an extract having the same qualities are stored in similar conditions, the more concentrated extracts have a greater stability. If this is so, why is there not a higher utilisation of reverse flow practice in industry?

Mannheim: I do not entirely agree because in my experience some deleterious effects, such as browning, do occur in tomato juice concentrates at high temperatures.

Chairman: I've had the same thought as Mr. Blenford many times; the most concentrated product is almost always given the lowest temperature treatment. Apparently the ambient temperature storage stability does not correlate with the high temperature rate of degradation.

M. Karel: (M.I.T.): I will try to answer Mr. Blenford's question. I would differentiate between two major effects. First, in general, the rates of reaction increase with concentration—just mass action. This effect would, for many reactions, increase the rate of deterioration of the concentrate. On the other hand, there is the effect of mobility. As you increase the concentration, viscosity and other related effects decrease the mobility of the reactants, and this is important, for example, in aroma loss which is diffusion controlled. It is also important in some enzyme reactions which are diffusion controlled.

But in the case of browning, I would say the concentration is the important factor.

Chairman: There is one aspect of evaporation which we haven't really covered, Professor.

Mannheim: There must be more than one.

Chairman: An impression has been given that in using multiple-effect evaporators there are no problems. For most of the food products that we wish to handle, it is an unfortunate fact that each effect has to be approximately the same size in that the quantity of vapour that is re-used has to be the same in each effect. If a high concentration ratio is being used, then in the last effect of the evaporator there may not be enough liquid to wet the wall. Therefore the need for recirculation arises as you approach the concentrated end of the evaporator. So what happens is that the number of effects is reduced and a separate finishing evaporator is added. Normally the finishing evaporator is dealing with a viscous material and it can use live steam giving a high temperature difference.

And that really brings me to the point that was mentioned earlier about the stability of the product becoming better as the concentration goes up. It is an interesting fact that all milk powder is sold in the U.S.A. and I think generally around the world, on the native (or undenatured) whey protein figure. Work that was done by Bell and Webb as far back as the late 1930s showed that although the temperature of pre-heating had an enormous effect on the native whey protein figures in milk, the actual temperatures of the evaporator or even the drier that followed, didn't. Thus the conditions in the later stages of evaporation do not matter so far as protein denaturation is concerned, though these will be important when considering other factors such as browning.

Mannheim: This shows that you really have to examine each product evaporation system separately, because you cannot make a generalised statement just as you cannot even make a generalised statement about deterioration and storage. In fact, as I mentioned in the paper, the new APV Plate is intended to overcome the multiple-effect problem. Also perhaps the importance of thermal economy is decreasing compared with the value of the product.

B. L. Patience (Stork-Amsterdam): You mentioned briefly during your lecture that steam injection tends to reduce the degree of fouling in the evaporator. Why is this so? Is it perhaps because of the role of protein in the fouling?

Mannheim: Yes, I believe this is so. Protein denaturation before evaporation appears to reduce the fouling problem.

Chairman: Another explanation is that steam injection pre-heating ensures that the heat transferred within the evaporator tube is all boiling heat transfer. This is a separate effect and would reduce fouling quite independently of the denaturation effect in the pre-heater itself.

Varsel: You mentioned the Vincent waste heat evaporator. Is this being used for anything apart from citrus molasses concentration?

Mannheim: Vincent does not claim the unit to be suitable for anything else; if you know the system do you believe it would be applicable to citrus juice concentration?

Varsel: It would depend on the quality that is wanted.

Mannheim: Well, suppose at least a good taste is required.

Varsel: Then I don't think it would be usable.

Mannheim: The unit has a long residence time. No, I gave it just as another example of economy. In other words, I'm not belittling economy, but I would not like to overemphasise economy at the expense of quality. There is always a struggle between the technologists who want quality and the management who want economy.

A New Concept for the Freeze-concentration Process

W. E. L. SPIESS, W. WOLF, W. BUTTMI and G. JUNG

Bundesforschungsanstalt, Karlsruhe, W. Germany

Freeze-drying is known as the drying process that best retains food quality. For drying liquids or pastes the advantages of the process lie in the complete transfer of the unbound liquid water into a frozen state and the rapid transition of this frozen water and the adjacent bound water into vapour during the drying process. A further advantage is the low processing temperature. The rapid transition of the freezable water from liquid into the frozen state and from the frozen state into vapour prevents loss of water-soluble substances and minimises the extent of degradation that occurs during conventional drying, such as non-enzymatic and enzymatic browning reactions and protein denaturation. The low processing temperatures aid these quality improvements and help to reduce flavour loss.

Despite these high expectations, the freeze-drying process is commercially applied to only a few products, mainly because of its high processing costs. It therefore seems desirable to develop a process where the advantages of freeze-drying, namely freezing the liquid water prior to separation, rapid separation of ice, and low processing temperatures, could be combined with an economical operation.

Efforts to overcome the unfavourable economics of the process in the past have been directed towards the improvement of machinery and equipment necessary for the freeze-drying operation.

Little attention has yet been given to the possibilities provided by the product itself, to improve the process economics. Liquids and semi-liquids are the most promising and challenging product types from this viewpoint. In this product category the conservation of a given shape during a dehydration process is of low importance, whereas the preservation of constituents has high priority.

In the freeze-drying process, modifications to the product are possible which allow the sublimation of ice under less restrictive conditions. The aim of such modifications would be the complete suppression of any type of transport resistance in the product and an improvement of the heat supply to the sublimation zone.

Since the solid matter in liquids and pastes as well as in solid food is capable of binding only a certain amount of water against freezing (Riedel 1959) (approx. 0.2–0.4 g water per 1 g of solid substance), roughly 90% of the water present is freezable as almost pure ice. The mechanical separation of ice seemed feasible if the frozen product could be transformed into a state where the integrated structure of the frozen material might be broken and ice mechanically separated out of the mixture by a selective process.

A successful introduction of the proposed process step should result in a starting material for the freeze-drying process with an increased dry matter content compared to the untreated material, thus improving the economic efficiency of the process as a whole.

As a first approach to the problem the possibilities of a structural disintegration of the frozen liquids and frozen pastes were analysed. It was thought that the difference in brittleness and density between the ice and gel fractions (solid matter plus bound water) could be used as a means of accomplishing the mechanical separation. Experiments were therefore carried through to study both factors using gelatin, gelatinised potato starch, milk curds, apple pulp and coffee.

In general frozen liquids or semi-liquid products can be regarded as two-phase systems consisting of the ice phase and the gel phase containing the solid matter plus the bound water. Size and shape of the ice crystals depend on the type of constituents present in the raw material and the rate of crystallisation (Luyet 1968, Némethy 1968).

The enthalpy–concentration diagrams compiled by Riedel (1960) show that this phenomenon, namely the freezing out of water and the subsequent formation of ice crystals within a food product exposed to temperatures below 0°C, is a function of the temperature and the solids concentration.

Tammann (1925), furthermore, has shown that also the size of these ice crystals can be controlled: slow freezing results in large crystals and quick freezing in small ones.

If we now try to break down a frozen liquid into particles smaller than the original ice crystals we can expect three types of particles:

(a) pure ice particles,
(b) pure gel particles,
(c) mixed (i.e. ice–gel) particles.

The mixed particles are comparable to the original product—we still have a unit which contains ice and solid material. The important difference, however, is that the ice is now found at the surface—if the particle is small enough—so that surface sublimation should be possible.

If it is possible now to separate the pure ice particles from the other particles we have already a highly concentrated ground material.

EXPERIMENTAL METHODS

Disintegration

For a study of the disintegration process the raw material was frozen immediately after preparation on metallic trays in layers of 10 mm depth. The freezing temperatures applied were −3°C down to −20°C, and the subsequent storage temperature was −40°C for at least 10 h.

The frozen material was then ground in different types of grinding apparatus (disc attrition mill, ball mill, double-roll crusher) at a temperature of −40°C. The ground material was then sieved in a testing sieve shaker with sieves according to DIN 4188 (sieve openings in mm: 0.09; 0.125; 0.2; 0.3; 0.4; 0.5; 1.0; 1.6; 2.0). The material on each sieve was weighed and examined under a cold room microscope. In addition the moisture content of each sieve fraction was analysed to find out if any enrichment of ice or gel (dry matter and bound water) occurred in the various fractions.

Density

For the density measurements a standard 50 ml pycnometer with hexane as test liquid was used. Starch and gelatine samples were equilibrated at different moisture levels ranging from 0 to 30% H_2O on a wet basis. The temperature range for the experiments was −31° to +30°C.

EXPERIMENTAL RESULTS AND DISCUSSION

Material and disintegration

The results of the experiments are shown in the following particle-size-distribution curves.

On the abscissa of Fig. 1 the particle size in μm is plotted. On the ordinate the cumulative amount larger than stated size, in percent, is plotted so that the measured data provided curves as shown. These can be interpreted as follows.

If one considers a special particle size one can read on the ordinate the total amount of the ground material with a larger particle size than this special size. In addition the measured moisture content of each sieve fraction is represented in this graph. The shape of the water-content curve shows the enrichment of the ice particles towards the smaller particle sizes.

If one calculates now from this particle-size-distribution curve and moisture-control curve for the whole ground material, the curves of cumulative amounts larger than stated size separately for both phases, the ice (or the water-phase) and the dry matter phase, one obtains curves as shown in Fig. 2. In addition the cumulative amount of water in the fractions larger than stated size is calculated and plotted in this figure.

The shapes for the two curves indicate very clearly that enrichment of dry matter occurs in the range of larger particle sizes and that enrichment

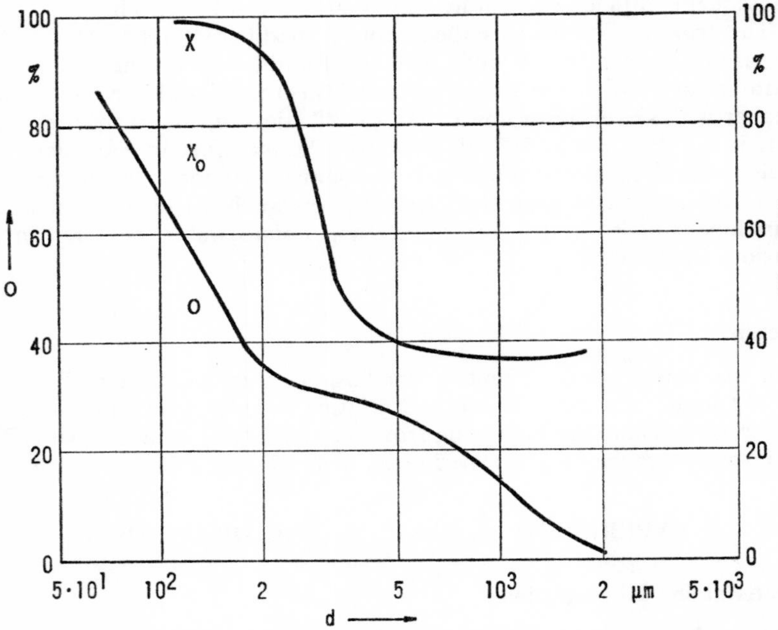

FIG. 1. Cumulative oversize distribution O and water content X as a function of the particle size d of distintegrated gelatine; freezing temperature $-5°C$; grinding temperature $-50°C$; initial water content $X_0 = 80\%$ (w.b.); grinding device, disc attrition mill.

of ice (or water) occurs in the range of smaller particles. This means a concentration of the original material is achieved only by means of a mechanical selective process.

For example, if the disintegrated material shows these particle-size-distribution curves, and one takes a sieve with an opening of, for instance, 0.3 mm, the material retained on the sieve will consist of 90% dry matter and only 10% of the original amount of water. Furthermore, from Fig. 2 the degree of separation of ice from dry matter as a function of the particle size can be calculated since the perpendicular distance between both curves is a measure for the degree of separation. This function is shown in Fig. 3, in which the particle size is plotted on the abscissa and the degree of separation is plotted on the ordinate.

The results of the analysis of the ground material show that with the equipment used particle sizes between 0.02 and 1.00 mm were obtained; in most cases, however, 70–80% of the particles had a size of 0.3 mm or

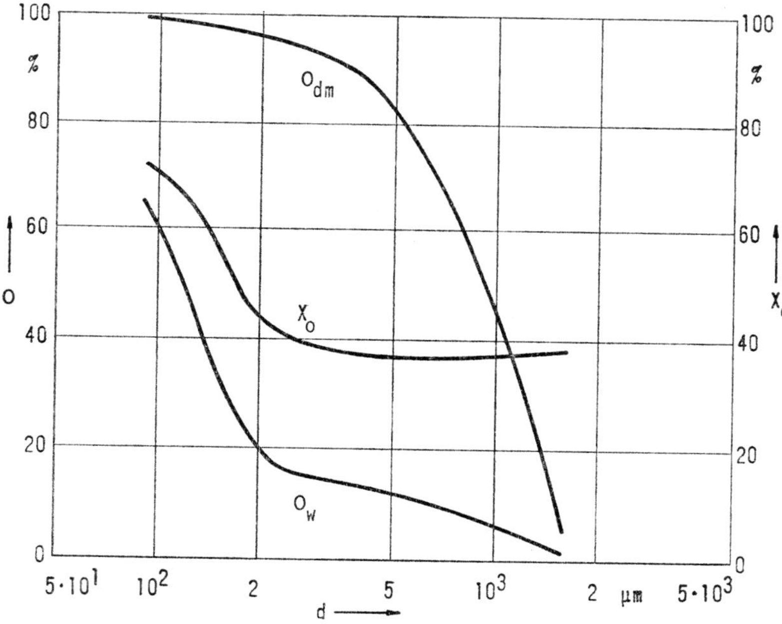

FIG. 2. Water content of the cumulative oversize X_o and the cumulative oversize distribution of ice O_w and dry matter O_{dm} as a function of the particle size d of disintegrated gelatine. Experimental conditions, see Fig. 1.

smaller. Microscopic examination of the ground material indicated a more or less complete disintegration of the ice and gel fractions of all products used in the experiments with the exception of coffee.

This phenomenon is explained by the fact that all products showed very clear cut boundary lines between the ice and the gel fraction, giving the system sufficient possibilities for a break between the two phases. The degree of disintegration of the two phases and the degree of size reduction of ice and gel varied from product to product and with the different grinding equipment used.

Most important, however, were differences in the brittleness between the ice and the gel fractions, a fact which resulted in a different breaking behavior of the gel materials compared to the ice, so that in the various particle size fractions obtained by sieving, different moisture contents were found. The separation effect was most pronounced in the case of gelatine and gelatinised starch so that the use of a mechanical ice separation before the thermal dehydration process is a possibility in these cases.

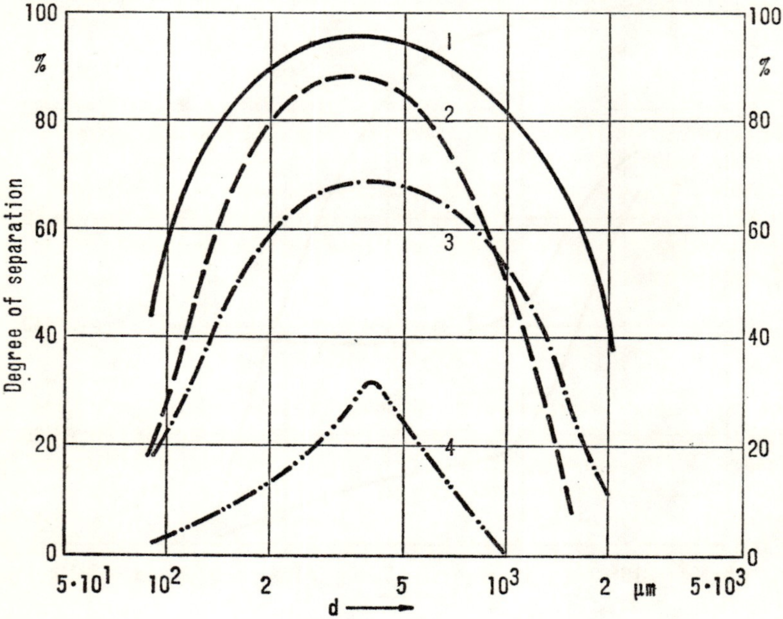

FIG. 3. Degree of separation as a function of the particle size *d* for various products. Curve 1: ideal separation of gelatine ($X_0 = 90\%$ w.b.); curve 2: disc attrition mill; gelatine ($X_0 = 90\%$ w.b.); curve 3: double roll crusher, gelatine ($X_0 = 90\%$ w.b.); curve 4: disc attrition mill; milk curd.

In the case of coffee, practically no differences in the water content were achieved. Increases in water content, however, were observed also in the small particle fractions of milk curds and apple pulp. Furthermore, it was observed that in the different particle size fractions of all products pure ice crystals were present as well as pure gel particles and in some cases mixed particles. In this case a separation by density seems necessary as a second refining step. The general tendency was that in the size range of 0.5–0.6 mm either pure ice or pure gel particles were present; the amount of mixed particles (ice and gel) was only 5–10% (relative to the number of ice particles). Apple pulp showed approx. 25% mixed particles in these size fractions. At sizes below 0.2 mm almost no mixed particles were observed. The observations for coffee, however, showed in most cases no sharp boundary lines between the ice and the solid-matter regions; pure ice crystals as well as crystals with different colours varying from a light yellow to a dark brown appeared. The amount of pure ice crystals in the size range of 0.5–0.6 mm for coffee was approx. 20%; it increased in the size ranges below 0.3 mm to roughly 40%.

Density

The results of the density measurements show that pure ice crystals and gel particles have different densities which is most pronounced at lower temperatures. The density of starch with a water content of 30% (wet basis) at −30°C is approx. 1.4 g/ml, whereas the density of gelatine under the same conditions is 1.3 g/ml. The ice density at a temperature of −30°C is below 0.998 g/ml.

CONCLUSION

The experiments described have shown that it is possible to disintegrate certain frozen foods in such a way that pure ice and pure gel fractions are present. Preliminary separation experiments, based on brittleness and density difference between the ice and the gel fractions, have shown that a partial separation of the two fractions is possible and thus an improvement of the economics of the freeze-drying process can be achieved. To utilise this phenomenon on a commercial scale, further large scale experiment work is necessary.

REFERENCES

Luyet, B. J. (1968). 'The formation of ice and the physical behaviour of the ice phase in aqueous solutions and in biological systems', in *Low Temperature Biology of Foodstuffs* (Ed. J. Hawthorn and E. J. Rolfe), Pergamon Press, Oxford.

Némethy, G. (1968). 'The structure of water in aqueous solutions', in *Low Temperature Biology of Foodstuffs* (Ed. J. Hawthorn and E. J. Rolfe), Pergamon Press, Oxford.

Riedel, L. (1959). *Kältetechnik*, **11** (2), 41–43.

Riedel, L. (1960). *DKV-Arbeitsblatt, Kältetechnik*, 12 (12).

Tammann, G. (1925). *The States of Aggregation*, Princeton University Press, Princeton.

DISCUSSION

M. Karel (M.I.T.): First, a comment. It would seem that the differences you are observing between gelatine and starch on the one hand and coffee on the other, are due to the amount of soluble solids present which would depress the temperature of solidification. At the temperatures you are using a softer gel face is obtained and the frozen material is not brittle enough to fracture at the ice/gel interface.

The question I want to ask you is, have you done any experiments on the freeze-drying of the different fractions?

W. Wolf: No, that will be the second step. We are building a freeze-drying tower, about 6 m in height and 10 cm in diameter, which can be heated to 600° C, which would realise our surface sublimation process.

P. J. A. M. Kerkhof (Eindhoven University): I have some comments on your experiments. My colleague, Mr. Wilkinson, has performed several experiments which are similar to yours. He ground frozen malt extract solutions and made sieve fractions for the purpose of freeze-drying experiments and also for holding experiments to determine aroma loss.

On analysing the sieve fractions to find the dissolved solid content he found that all fractions—and they were all in the same size range as yours—had the same dissolved solid contents, though with dextrin a separation effect was obtained. The behaviour of malt extract seems to be similar to that of coffee extract, and maybe this is due to the fact that there are more smaller molecules present. I am not sure about one thing; you said when you observed the granulated frozen coffee extracts, you saw particles of different colours. In our experience you cannot see the distinction between the ice phase and the coffee phase.

Wolf: Under the microscope pure ice particles and pure coffee, which is dark brown, can be seen and then other particles of the various colours, but we didn't see any sharp boundary between the ice phase and the coffee phase.

C. J. King (University of California): I suspect that lack of a sharp boundary is another sign of Professor Karel's collapse that he suggests occurs upon attempting to grind the material at too high a temperature.

Karel: Our early experiments on coffee showed that even down to $-50°C$ some fouling effects occurred in the grinding and sieving operation.

A. S. Michaels (Alza Research): Is there any chance that the coffee at $-50°C$ is actually plastic and somewhat tacky, so that it adheres to the ice granules or is spread out on the ice granule surface in the course of the grinding process? This would, of course, be one good reason why separation could never be achieved.

Karel: It may be that at lower temperatures separation could be achieved, but I think the brittleness of the coffee constituents is similar to the brittleness of the ice, because as we saw from the microscope pure ice and pure coffee particles exist as separate entities—but both particles have the same particle size, and they couldn't be separated by sieving. The density separation process could be used in this case.

S. Barnett (General Foods): I would like to support Dr. Michaels' theory, having worked extensively with freezing and freeze-drying of coffee extract. At temperatures down to $-85°F$ it's still flexible and it won't shatter.

Sucrose Dehydration by the Heat of Crystallisation

W. M. NICOL

Tate and Lyle Ltd.,
Reading, Berks, U.K.

Crystallisation of sucrose is a unit process which when conducted in the traditional way effects a tenfold diminution of impurities over the mother liquor analysis with a crystal yield of some 50%. However, on occasions, dehydration of the sucrose syrup instead of purification is both adequate and satisfactory. Sucrose solutions, because of their ability to sustain substantial supersaturation for considerable periods without nucleation, are difficult to spray-dry. But this characteristic can be exploited by a more thermodynamically efficient method of dehydration. It is possible with sucrose to evaporate an aqueous solution until it is supersaturated yet unnucleated, and of such a concentration that when all of the sucrose crystallises there is sufficient heat liberated to evaporate the remaining water. The only other requirement is that crystallisation, and therefore heat liberation, is fast enough to prevent heat losses detracting from the water evaporation.

As can be appreciated the thermal efficiency of the process is high because once nucleation has been initiated the heat for evaporation is self-generated in contrast, for instance, to spray-drying where the low coefficient of heat transfer between air and solid is a disadvantage. This phenomenon, called transformation, when applied to sucrose is not new—indeed, it was, and still is, the method whereby the more primitive sugar technologists produce solid sugar from syrups. In the hands of modern technology the process has many interesting potentialities. There are many patents, particularly from the U.S.A., which exploit this effect, but the crucial requirement for efficient transformation is keeping adequate control over the rate of crystallisation to be able to produce the product of the form required. The selection and design of plant is very important in this respect, since this determines the physical characteristics of the sugar produced. The slides show a range of particulate matter possible.

Several interesting features of this form of sugar are:

(1) The ability according to method of granulation to vary the bulk density.

(2) Its ability to act as a carrier medium for flavours, etc., which can be added to the magma during crystallisation at temperatures which need not be above 100°C.
(3) Brown sugar can be produced in a free flowing form.
(4) Faster solubility than comparable normally crystallised sugar.
(5) Higher chemical reactivity in non-food industrial application.

THEORETICAL

The heat of crystallisation of sucrose, ΔH, is positive and increases with temperature. It is equal in magnitude to the heat of solution of crystals in the saturated solution. In this way Jackson (unpublished, Ph.D. thesis, University of Bristol, 1950) has calculated values from the best available data (which are sparse, not only in the sucrose field).

Temperature (°C)	25	60	90
ΔH, J/g	30.3	57.0	107.6

Since ln ΔH has practically a linear relationship with temperature, the data may be extrapolated to higher temperatures:

Temperature (°C)	100	110	120	130
ΔH, J/g	138	180	234	310

The latent heat of vaporisation of water does not vary appreciably from 2.26 kJ/g in the region of 90–130°C. From this data, a table may be constructed balancing heat required to vaporise water with heat available from crystallisation for 100 g of solution (Table I).

From Table I the solution temperature can be matched with concentration. The degree of freedom is further reduced on considering the method of effecting concentration. If the solution is concentrated by boiling, at atmospheric pressure (which is convenient), the minimum temperature for complete dehydration is 123°C. In practice, when temperatures below 125°C are used some auxiliary drying is necessary. Characteristically, however, higher temperatures are employed to effect better drying and to accelerate the rate of production. Unfortunately, under these circumstances crystallisation control is poor.

The effect of associated dissolved impurities is first to require higher temperatures for satisfactory transformation, typically about 130°C, and when the impurity level becomes too high, say over 15%, transformation is incomplete due to the difficulty of crystallisation and association of water.

TABLE I

Sucrose (g per 100 g *solution*)	*Heat required: latent heat of vaporisation* (kJ)	*Heat available from crystallisation* ΔH *at* 110°C (kJ)	ΔH *at* 120°C (kJ)	ΔH *at* 130°C (kJ)
86	31.6			
87	29.4			26.9
88	27.1			27.3
89	24.9		20.9	27.6
90	22.6	16.2	21.1	27.9
91	20.3	16.4	21.3	
92	18.1	16.6	21.6	
93	15.8	16.7	21.8	

DISCUSSION

E. Seltzer (Rutgers University): I take it that brownulated sugar is made in this way, is it? Is that the trade name of the American sugar?

W. M. Nicol: Yes, it is made in a similar way. They evaporate to about 93% by weight sucrose. It is then nucleated by vigorous agitation and the transformation is effected. The brownulated sugar process then demands that the sugar be milled very finely and reagglomerated into particles of a given size. But, of course, there is no need to carry out the milling and reagglomeration. With good control one ought to be able to produce the agglomerates of the right shape and size in one fell swoop.

Seltzer: How low bulk density could you get?

Nicol: It's quite easy to halve the bulk density of normal crystalline sugar in this way, and we have reached 30% in particular cases.

F. Bramsnaes (Danmarks Tekniske Højskole): If you didn't mill the transformed sugar finely and agglomerate, would you still obtain a good final white colour?

Nicol: Yes, this is correct; indeed, one can have a sucrose solution of a much higher colour than is possible in normal crystallisation which, because of the small particles created in transformation, still results in a product of greater whiteness. The larger crystal gives a sparkle whereas the transformed sugar gives a whiteness.

Seltzer: I wondered where the sugar industry has been all this time not using that available heat? Where was all this heat, what did you do, throw it away? Did you have chemical engineers in the industry who know how to use it?

Nicol: There seems to be certain reluctance in the market for this kind of product. It is true that in the normal crystallisation of sugar there is heat evolved, and this merely helps to keep the solution boiling, thereby reducing the steam load in the refinery; but then the heat is just lost.

Chairman: Well, you actually take it out with cooling water, don't you?

Nicol: The heat is extracted, yes, in the condensors and poured into the river.

Session III

MEMBRANE CONCENTRATION

Introduction

DANIEL I. C. WANG

Department of Nutrition and Food Science,
Massachusetts Institute of Technology,
Cambridge, Massachusetts, U.S.A.

The use of membranes as a means of concentrating and/or separating dissolved solutes has advanced from a laboratory curiosity to full-scale industrial practice in the past ten years. The use of membranes in the laboratory can now be found as frequently as some of the other conventional laboratory tools. Manufacturers of membranes and associated equipment in the U.S. and Europe can now supply in a routine fashion a variety of equipment to process liquid volumes ranging from several millilitres up to several litres in quantity. Large-scale industrial operation using membrane permeation has also reached full-scale reality. In addition, there are presently 10–20 industrial organisations which are able to provide membranes and associated equipment for a variety of operations. Lastly, reported applications using membrane permeation in food processes include: concentration and purification of macrosolutes such as proteins and enzymes, concentration of beverages and fruit juices, desalting of protein solutions, concentration of emulsions, treatment of processing wastes, and membrane-coupled operations.

The basic underlying principles that govern membrane permeation will be shown to be relatively straightforward. On the other hand, deviations from the ideal situations can be anticipated. Considering the diverse nature of food systems, how does the food technologist or a food engineer address the pertinent selection criteria as to the feasibility of membrane permeation as applied to their systems? Some of these basic questions envisioned by this chairman include the criteria in membrane selection, criteria in membrane system selection, cost analysis, and potential and existing applications. The specific questions associated with each of these areas will be briefly outlined. It is hoped that speakers in this session will address themselves, at least in part, to some of these questions.

When one considers the criteria in membrane selection, some of the obvious questions that are raised concern solvent flux and solute rejection. There exist today many membrane manufacturers that produce a variety of membranes. Each membrane will be reported to possess a nominal permeation flux that corresponds to a certain nominal rejection or retention capability. In general, the flux is usually reported for a pure solvent such as water, and the rejection characteristic is often reported for a single component such as a specific protein or a specific ion. On the other hand, when the use of membrane for food systems is considered, one can rarely assume that these systems can be simulated as a simple and well-defined mixture. Furthermore, it is quite possible in food systems that certain components in extremely low concentrations and difficult to quantify could contribute in a major fashion with respect to the favourable characteristics as a food product. One is, therefore, faced with the dilemma in its membrane selection criteria as to the optimisation of parameters in membrane permeation along with pertinent properties of a food product.

Other considerations in membrane selection for food systems include the physical and chemical constraints which might be necessary for maintaining during permeation. These include factors such as resistance to pH variation, effect of temperature variations during processing and cleaning, influence of operating pressures, resistance to standard sanitation chemicals, etc. The answers to some of these questions from speakers in this session would then allow some *a priori* judgements to be made without resorting to time-consuming test programmes.

The second general area one might question is the selection of appropriate membrane systems. In particular, the fact that concentration polarisation can be anticipated during membrane processing is well recognised. Vast amounts of research and development have been devoted to minimising this effect. For example, pilot and plant scale membrane systems have been developed which employ both laminar flow and turbulent flow concepts. The advantages of these concepts with respect to permeation rates will be very enlightening. On the other hand, discussions as to the merits of the various types of systems in light of food processing criteria cannot be overlooked. Some of these criteria include sanitation, cleaning operations, inspection, microbiological considerations, maintaining the functionality of the product, reliability, and lastly scaleability. It is hoped that speakers in this session will also address the useful criteria which one can employ in the selection of a system for its specific food process used.

The last two general areas in considering membranes for processing of food stuffs are interrelated. These are what specific products which membrane permeation can be applied to; and, more importantly, what is the cost associated with such processes. With respect to applications, one might consider first the technical rationales in the selection of membrane permeation over existing conventional methods which can achieve the same goal. Furthermore, it would be enlightening to consider the application of membrane systems first as a potential replacement and/or complement to known and existing processes and second as a projection to the future as to new and novel systems. Lastly, the ultimate judgement factor with respect to economic feasibility is cost. Specifically, what is the capital investment and operating cost for membrane processing with respect to the types of food systems and scale of operation? Also, what are the major uncertainties such as membrane life and associated system life when one contemplates the use of a membrane as a routine food process operation?

Tailored Membranes

ALAN S. MICHAELS

President, Alza Research and Vice-President, Alza Corporation, Palo Alto, California, U.S.A.

INTRODUCTION

The past decade has seen the rapid emergence of industrial interest in, and successful large-scale utilisation of, pressure-activated molecular separation processes employing permselective membranes for the treatment of aqueous solutions of importance to the food, pharmaceutical, chemical, petroleum, and metallurgical industries. These processes (which involve such operations as dehydration or concentration, demineralisation, impurity-removal, or fractionation of multicomponent mixtures) have become technically and economically feasible principally because of the discovery and successful commercialisation of techniques for the fabrication, from a variety of synthetic and modified natural polymeric materials, of specially structured films or membranes which display unusually high permeability to water, while possessing the ability to prevent the passage of certain solute molecules. The objective of this paper is to review—with a brevity almost certain to omit acknowledgement of many important contributions to the field—the technology of preparation of these 'tailored membranes', and the state of our scientific knowledge about how they are formed, and how they function.

At the outset, it is useful to identify two 'classes' of pressure-driven membrane separation processes performed on aqueous solutions, which are now well-established in the literature; these are (1) 'reverse osmosis' (or 'hyperfiltration'), and (2) 'ultrafiltration'. The former, which is dealt with in great detail in the monograph edited by Merten (1966), involves the selective removal from solution of micromolecular solutes (usually, but not exclusively, simple electrolytes) whose molecular dimensions are of the same order as those of the water molecule. The latter, which has been described in some depth by this writer in an earlier publication (Michaels, 1968) involves the separation from solution of relatively much larger solute molecules or colloids. Operationally, the two processes are very similar, in that both involve the contact of a membrane surface with a solution under pressure, with selective passage through the membrane of solvent (and certain solutes) devoid of solutes retained by the membrane. In reverse osmosis, the solutes retained by the membrane usually have sufficient thermodynamic activity in solution to necessitate application of quite high pressures to overcome their osmotic pressure, and achieve significant water-throughput and/or solute removal. In ultrafiltration, on the other hand, the solutes to be retained by the membrane are of such high molecular weight that their osmotic pressures are frequently negligible; hence, ultrafiltration normally requires far lower operating pressures to achieve useful transmembrane flows than reverse osmosis.

The principal difference between reverse osmosis and ultrafiltration is to be found in the structure and function of the membranes tailored for the two processes (Michaels, 1968). Reverse osmosis membranes, it is now generally accepted, pass water by a process of dissolution and molecular diffusion through the (essentially homogeneous) polymer phase comprising the membrane barrier; solute species retained or rejected by the membrane are found usually to have substantially reduced solubility (relative to water) in the membrane, and/or to diffuse much more slowly than water through the membrane substance. Thermodynamic and molecular kinetic treatment of this solvent/solute solution-diffusion process leads to a prediction of the permeation and solute-rejection properties of such a membrane which is in excellent agreement with experimental observations—namely, that the water flux through the membrane is a linear (increasing) function of the applied pressure, while the solute rejection coefficient (the fraction of solute present in the original solution which is retained by the membrane) rises with applied pressure at low pressures, approaching an asymptotic limit at higher pressures. The parameters which determine the specific hydraulic permeability and solute rejection capacity of a reverse osmosis membrane are to be found in those characteristics of the membrane substance which influence the solubility and diffusivity within the polymer of water and solute molecules (Kedem, 1972; Merten, 1966). Evidence is persuasive that it is the *chemical constitution* and *molecular architecture* of the polymer which critically determine its reverse osmosis performance.

Ultrafiltration membranes, on the other hand, are capable of retaining only relatively large molecules; permeability of these membranes to water is far too high to be explained by any rational solution-diffusion mechanism. It has been concluded—and largely substantiated by electron microscopic evidence—that such membranes are demonstrably microporous, containing voids or pores whose dimensions are comparable with or smaller than those of the molecules the membrane can retain. In such microporous structures, water (with accompanying microsolutes) flows through the membrane under applied pressure by essentially laminar, viscous flow; indeed, measurement of the hydraulic permeability and porosity of such membranes is sufficient to allow calculation of a reasonable estimate of mean pore size, which is usually in good agreement with that observed microscopically. With ultrafiltration membranes, therefore, the chemical composition or molecular configuration of the polymer comprising the membrane is of relatively minor importance in determining its water permeability or solute-rejection characteristics; rather, it is the *processing* method for creating the microporous structure which is the critical parameter. This is not to imply that polymer properties are unimportant in the fabrication of tailored ultrafiltration membranes; as will be shown below, there are some very important property requirements which must be satisfied to allow successful membrane production. However, from a single suitable polymer it is usually possible, by adjustment of fabrication conditions, to prepare a variety of membranes covering a range of pore-sizes, water permeabilities, and solute-rejection characteristics.

Because of these rather important differences in structure and function of reverse osmosis and ultrafiltration membranes, it is appropriate to describe their fabrication and properties separately; however, as will become evident, many of the fabrication procedures, and process parameters affecting membrane properties, are common to both classes, as are many of the physicochemical phenomena involved in membrane formation.

REVERSE OSMOSIS MEMBRANES

Modern reverse osmosis technology can be said to have had its origin with the observation by Reid and Breton (1959) that homogeneous, thin films of commercial cellulose 'diacetate', when contacted with sodium chloride solution at pressures of *c.* 50–100 atm, yielded a membrane permeate containing only *c.* 5% of the salt present in the pressurised solution. Evaluation of a large number of other film-forming polymers indicated that, while many showed varying degrees of salt-rejection, cellulose diacetate was far superior in this respect. However, for the thinnest film (*c.* 50–100 μm thick) which could be fabricated free of pin-holes and withstand the applied pressure without rupture when supported by the then-available porous substrates, the water permeation rate through these membranes at pressures of *c.* 100 atm was only about 0.01–0.02 gal/ft^2 day.

In the late 1950s, Loeb and Sourirajan (1962) made the important discovery that, by casting films of cellulose acetate from what at that time were regarded as esoteric mixed solvent solutions, and drying and water-washing these castings under rigidly controlled conditions, they were able to prepare membranes which, when submitted to reverse osmosis with saline solutions under conditions comparable with those used by Reid and Breton, yielded aqueous permeates nearly as well depleted in salt as found by Reid but displayed water permeation rates of the order of 1–5 gal/ft^2 day—over 100-fold higher than achieved with conventional films of the same polymer. They also found that these membranes had another anomalous characteristic: they were salt retentive only if the side of the membrane exposed to the air in the casting process was in contact with the pressurised solution.

Subsequent studies (by Lonsdale *et al.* 1965; Kesting *et al.* 1965; and others) of the structure of this membrane established it to be both 'anisotropic' or 'asymmetric', and ultramicroporous; that is, the membrane is essentially a laminate comprising a nearly defect-free, ultrathin (submicron) film of dense polymer overlying a substantially thicker layer of microporous polymer. This is shown schematically in Fig. 1; an electron

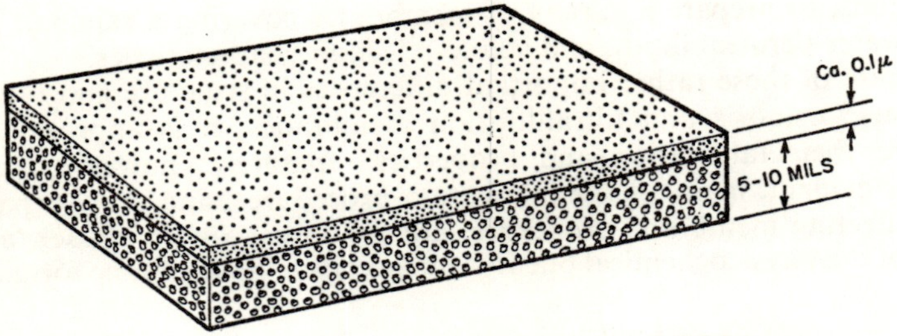

FIG. 1. Schematic diagram of the cross section of an asymmetric reverse osmosis membrane.

micrograph of the cross section of a typical Loeb/Sourirajan cellulose acetate membrane is shown in Fig. 2.

Since the hydraulic permeability of the ultrathin 'skin' layer is substantially lower than that of the porous substructure, and since the substructure displays virtually no salt-rejection capacity, this membrane behaves in reverse osmosis essentially as an extremely thin film of homogeneous cellulose acetate, sufficiently well-supported by the underlying microporous material to prevent its rupture under the quite high stresses to which it is subjected in use.

The incentive for reverse osmosis membrane research since the Loeb/Sourirajan development has been (1) to understand and explain the inherent water-transport and salt-rejection properties of cellulose acetate, (2) to understand and explain the processes leading to the formation of asymmetric membranes of this type, (3) to find or synthesise other polymers with inherent water- and salt-transport properties comparable or superior to those of cellulose acetate, and (4) to improve or modify membrane fabrication processes to yield higher water permeability, higher salt rejection, more durable and reproducible membranes, which can be manufactured in quantity both economically and with high reliability. It is undoubtedly a fortuitous quirk of nature that cellulose diacetate proves to be a

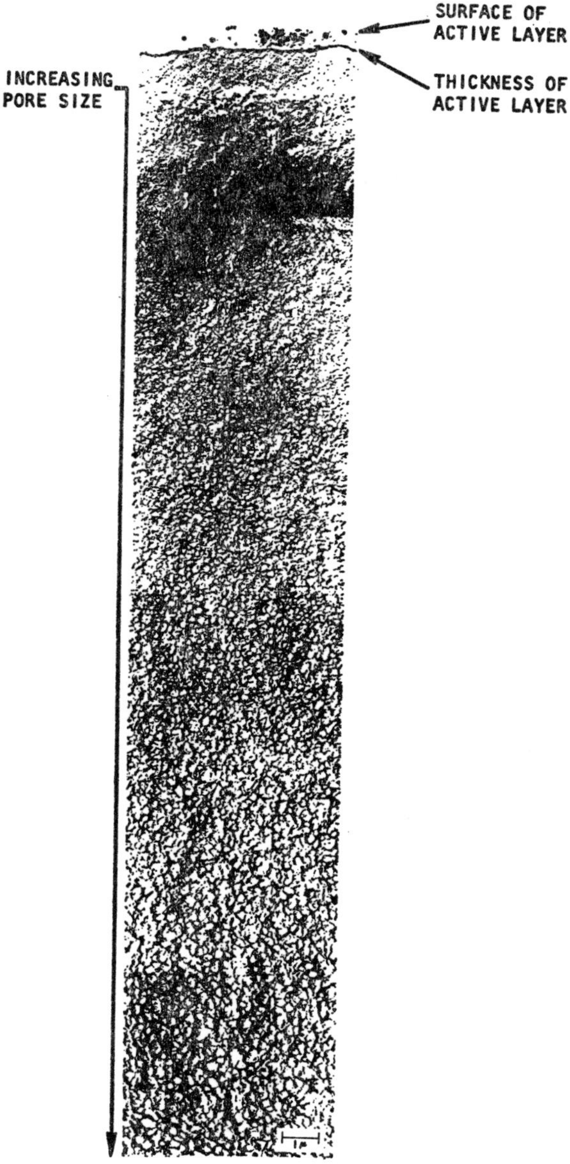

FIG. 2. Electron micrograph of the cross section of an asymmetric cellulose acetate reverse osmosis membrane. (Courtesy of R. L. Riley, Gulf Environmental Systems, Inc.)

polymer which (1) posssesses chemical constitutional and configurational features that render it reasonably highly water permeable yet quite highly impermeable to electrolytes, (2) is amenable, by virtue of its mechanical, viscoelastic, and solution-properties, to fabrication as an asymmetric membrane (for reasons to be discussed below), and (3) possesses thermo-mechanical, and chemical, stability sufficient for it to withstand the rather significant stresses (both physical and chemical) to which it is necessarily subjected in reverse osmosis operations. It is thus no surprise to find that, through many years of intensive research, cellulose diacetate remained alone as the polymer of choice for water desalination applications. Only in the relatively recent past—consequent to the realisation that there are only two *fundamental* polymer requirements for the production of useful reverse osmosis membranes, namely, a high *inherent* permselectivity to water relative to salt, and the capacity to produce ultrathin films of such polymers with low defect-density—have a significant number of new polymers (and new 'ultrathin' membrane fabrication methods) emerged as superior candidates for the production of commercially useful osmosis membranes.

The art and technology of 'ultrathin' membrane fabrication owes its practical success to a fundamental principle of membrane transport physics: the flux of any substance across a membrane, under a given set of upstream and downstream boundary conditions, is in virtually all known instances *inversely proportional* to the membrane thickness—at least down to thicknesses approaching molecular dimensions. Thus, *any* polymer which possesses adequately large inherent discrimination between two permeating species will be useful as a permselective membrane if it can be made thin enough to yield useful flux. On the other hand, a 'defect" (e.g. pinhole) in a membrane which can pass solution without solvent/solute discrimination will operate to reduce the apparent selectivity of the membrane, in proportion to the fraction of the total solution flux through the membrane which passes through the defect. So long as the *defect density* (or size) in a membrane does not increase more rapidly with decreasing membrane thickness than the gross permeability of the membrane itself, the consequence of reduced membrane thickness will be increased flux without sacrifice (or possibly with improvement) in effective selectivity. Based on experimental data obtained with a variety of 'ultrathin' membranes in the thickness range of 0.05–1.0 μm, this generalisation appears to be applicable.

POLYMERS SUITABLE FOR REVERSE OSMOSIS MEMBRANES

The origins of the inherent water-to-salt permselectivity of cellulose acetate have been the subject of study and speculation for many years (Kedem, 1972; Kesting, 1971; Merten, 1966) and there continue to be nearly as many explanations as researchers. The commercial diacetate (containing approximately 2.5 acetate groups per glucopyranose unit), which has been most exhaustively studied, is unusual in that it absorbs relatively small amounts of water (about 15% by weight), yet exhibits a diffusivity to water but little lower than the self-diffusion coefficient of water, and an activation energy for water diffusion only slightly higher than the viscous activation (or self-diffusion) energy for bulk water (Michaels, *et al.*, 1965). In contrast, the partition coefficient of simple electrolytes into the polymer from aqueous solutions is quite small (of the order of 0.02–0.05), their diffusion coefficients are quite low (*c.* 10^{-8} cm^2/s), and their diffusion activation energies much larger than those observed in aqueous solution (Merten, 1966; Michaels *et al.*, 1965). As the degree of acetylation of the polymer is altered over the range from the mono- to the triacetate, water-sorption and diffusion coefficient decrease significantly, but sorption and diffusivity of electrolytes decrease much more markedly. Hence, while the triacetate is a significantly better 'salt-barrier', its water permeability was, for a long time, considered too low to make it a useful membrane candidate; the 2.5 acetate has the 'balance' of adequate salt retention and water permeability to have rendered it the most satisfactory membrane material.

The unusually high water permeability of cellulosic esters despite their water sorptivity is probably most rationally explained by (1) the relatively high 'population density' of moderately polar atoms (primarily carbonyl- and ester-oxygens and hydroxyl groups) with which individual water molecules can but weakly associate, and (2) the relatively rigid backbone structure of the cellulose chain, which hinders chain-packing and minimises chain–chain interactions at the expense of water molecule–chain interactions. These characteristics favour reasonably homogeneous distribution of water molecules throughout the polymer without their substantially altering chain-packing, and permit relatively free movement of water molecules between binding sites. The anomalously low sorptivity for and diffusivity to simple ions, on the other hand, is probably due (1) to the quite low (relative to water) dielectric constant of the moist polymer (which, as is true with simple organic liquids (Kedem, 1972), requires the thermodynamically unfavourable dehydration of the ion for its transfer into the polymer), and (2) the significantly larger dimensions (relative to water) of the partially hydrated ions dissolved in the polymer.

In recent years, considerable progress has been made in the development of a number of new polymeric materials which display water- and salt-permeability characteristics equivalent or superior to those of cellulose acetate, while possessing other properties (mainly mechanical, thermal, and chemical stability) in which cellulose 2.5 acetate is deficient. These include mixed esters (e.g. acetate/propionate/butyrate) of cellulose, cellulose triacetate, network polymers prepared from partial cellulosic esters of methacrylic and/or acrylic acid, polyacrylonitrile and its copolymers, polyamides, and polyurethanes (Applegate and Antonson, 1972; Chen *et al.*, 1972; Credali *et al.*, 1972; Dresner and Johnson, in press). It has been shown by Lonsdale (1973) and Eirich *et al.* (1973) that if all polymers displaying useful properties for reverse osmosis desalination, irrespective of their chemical constitution, are compared with respect to their intrinsic water- and salt-permeabilities, there is a consistent and monotonic trend toward lower salt permeability with declining water permeability. There therefore appears to be a practical upper limit to salt rejection (which, fortunately, is in the range of 99.9+% at operating pressures of *c.* 100 atm) above which water permeability becomes too low to permit attainment of attractive water fluxes with even the thinnest low-defect membranes which can be fabricated. Among the most promising new polymeric materials for desalination applications are an expanding family of aromatic polyamides, polyimides, and other nitrogen-containing aromatic-backbone polymers (Applegate and Antonson, 1972; McKinney, 1972). These materials—which incidentally conform to the molecular/structural criteria proposed above—are destined to dominate the scene for reverse osmosis desalination applications, because of their unusually high salt-rejection characteristics (99.9% sodium chloride rejection from concentrated salt solutions at pressures only moderately above osmotic), adequate water permeabilities, ability to be reproducibly fabricated in ultrathin membrane form, extraordinarily high mechanical strength, creep-resistance, and thermal and chemical stability.

POLYMERS SUITABLE FOR ULTRAFILTRATION MEMBRANES

Virtually all commercially available ultrafiltration membranes are asymmetric microporous structures, the 'barrier layers' of which appear to contain ultrafine pores of diameter ranging from *c.* 10 to 200 Å. A schematic diagram of the cross section of an idealised asymmetric ultrafiltration membrane is shown in Fig. 3; an electron micrograph of the cross section of such a membrane is shown in Fig. 4. In contrast to reverse-osmosis membranes, ultrafiltration membranes possess extremely high hydraulic permeabilities (water flux values of 50–500 gal/ft^2 day at driving pressures of 5–50 psi), and are seldom if ever subjected to pressures exceeding 100 psi.

The essential requirement of a polymer useful for ultrafiltration membrane fabrication is that it is capable of retaining its ultramicroporous structure during the fabrication process, and under the conditions of thermal, mechanical, and chemical stress to which it is exposed in use (Michaels, 1971a). This requirement is satisfied by polymers which (1) are rigid or glassy, and display little to no creep or cold-flow at their use-temperatures, (2) are not detectably softened or plasticised by water, and (3) are relatively insensitive to hydrolytic and/or oxidative degradation in aqueous environments. As will be elaborated upon below, polymers of high molecular weight, high glass-transition temperatures, and/or tendency to develop paracrystalline structure will tend to satisfy the rigidity requirements; and polymers of low polarity relative to water are usually unplasticised by and sorb very little water.

Representative polymers from which useful ultrafiltration membranes can be prepared include many of the high-modulus thermoplasts which find application as engineering materials or synthetic textile fibres, e.g.

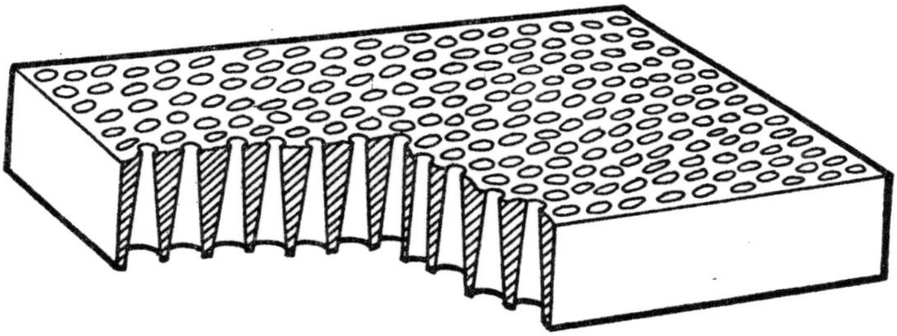

FIG. 3. Schematic diagram of the cross section of an asymmetric ultrafiltration membrane.

poly(methyl methacrylate), poly(vinyl chloride), polystyrene, polyacrylonitrile, the nylons, the rigid cellulosic esters (Michaels, 1971b). Of particular interest and utility are thermoplasts developed for high-temperature service in corrosive environments, which are usually employed in film or fibre form; these include the aromatic polyamides or polyimides, aromatic polycarbonates, aromatic polysulphones and aromatic polyethers. Membranes produced from these polymers are characterised by high resistance to collapse or consolidation under pressure, resistance to change in properties on exposure to elevated temperatures (whether wet or dry), high mechanical strength and abrasion-resistance, and excellent durability in acidic, alkaline, and oxidative environments. While the cellulose esters (particularly the acetates and nitrates) were among the first polymers successfully utilised for ultrafiltration membrane fabrication, such membranes have a propensity to 'creep-consolidate' (and lose water permeability) in normal use, have a quite low (40–60°C) upper use-temperature, are prone to hydrolytic degradation in strongly acidic and even mildly alkaline media and are susceptible to microbial and enzymatic attack.

It is important to recognise that a polymer converted into the form of

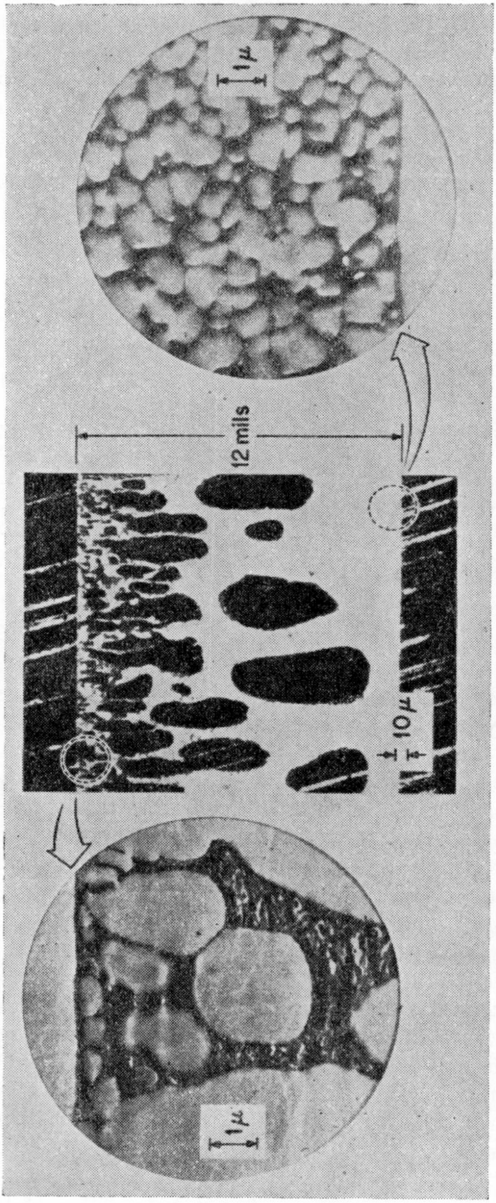

FIG. 4. Electron micrograph of the cross section of a typical asymmetric ultrafiltration membrane; the 'skin layer' is approximately 0.2 μm thick, and contains pores about 40–50 Å in diameter. The pores in the substructure are of the order of 5–20 μm in diameter. (Courtesy Amicon Corporation.)

an ultrafiltration membrane, because of its porosity and fine pore-texture, presents an enormous surface area per unit mass to its gaseous or liquid surroundings. Because of this, the susceptibility of the membrane-matrix to chemical attack is usually far greater than that exhibited by the polymer in bulk. Consequently, many polymers which might be expected, on a basis of their reported chemical stability in bulk form, to yield useful and stable membranes prove to be quite vulnerable to environmental breakdown when fabricated as membranes. It is for this reason that the 'super polymers' designed for very high temperature and corrosive environment service have proved to be among the best materials for ultrafiltration membranes.

PREPARATION AND CHARACTERISTICS OF ASYMMETRIC REVERSE OSMOSIS AND ULTRAFILTRATION MEMBRANES

The process utilised in the fabrication of the Loeb/Sourirajan asymmetric cellulose acetate membrane, and variants upon that process which have been developed for preparation of other reverse osmosis membranes, are also utilised in the manufacture of asymmetric, microporous ultrafiltration membranes (Michaels, 1971a, 1971b). For brevity, a general description of the process and an explanation of the influence of process variables on membrane structure and properties will be provided here, and should be understood to apply to both types of membrane. Theories of the mechanism of formation of microporous, asymmetric membrane structures have been elaborated by many researchers over the past 10–15 years, and the reader is directed to contributions by Kesting (1971), Eirich *et al.* (1973), Strathmann *et al.* (1971), Klein and Smith (1972), and Frommer and Lancet (1972), and Michaels (1968), among others, for details and differences of opinion. What follows here is an hypothesis which this writer has found useful to guide the preparation of functional membranes, and helpful in explaining the influence of polymer properties and processing variables on membrane structure and properties.

As a rule, if a solution of virtually any (linear) polymer of suitable high molecular weight and sufficiently high polymer concentration in a reasonably good solvent is cast as a film upon a smooth support, and if the surface of the film is exposed to a liquid (e.g. water) which is a non-solvent for the polymer but is miscible with the solvent, a solid, opalescent or translucent, film will be formed which can be shown to be (1) microporous or ultramicroporous, and (2) structurally non-uniform in cross section normal to the film-plane, displaying increasing porosity and/or pore-size with distance from the non-solvent-contacting film surface. The detailed fine structure of such an asymmetric film, its porosity, the rate-of-change of its microstructure with distance from the film surface, and its stability to structural change with time—all of which are essential in determining its utility as a reverse osmosis or ultrafiltration membrane—are highly dependent upon such parameters as polymer structure and composition, solvent composition, casting solution solids concentration, and precipitating bath composition and temperature.

In the original Loeb/Sourirajan process, a solution of cellulose acetate in a mixed solvent comprising water, acetone, and magnesium perchlorate is cast as a film on a glass plate, exposed to air at a low temperature for a short time, then immersed in ice water; after immersion for sufficient time to permit virtually complete extraction of acetone and salt, the film is then briefly 'annealed' in hot water, and stripped from the plate. By appropriate variations in polymer solution composition, it has been found (Frommer and Lancet, 1972; Klein and Smith, 1972) that satisfactory membranes can be fabricated via this process without the necessity for the initial drying or terminal 'annealing' step, and avoiding the use of low gelling-bath temperatures; such a simplified procedure has proved greatly to facilitate large-scale membrane manufacture.

An understanding of the physicochemical processes which take place in a polymer solution under the conditions of precipitation by non-solvent as occur in such an operation is essential not only to account for the unique asymmetric structure of these membranes, but also to the development of a rationale for predicting and controlling the flow and solute retention properties of such membranes. The phenomena governing asymmetric membrane formation are, fortunately, rather simple in concept, and few in number: they are principally (1) the thermodynamics of polymer solutions; (2) the kinetics of unsteady state diffusion; (3) the viscoelasticity of polymers and polymer solutions; (4) the kinetics of phase transformations; and (5) capillary forces in porous solids.

Polymer solution properties

An inherent property of (linear) polymers of moderate to high molecular weight is their tendency to dissolve only in solvents whose 'cohesive energy density' (which reflects intermolecular force-field intensity) is very similar to that of the polymer itself (Klein and Smith, 1972). In such 'good' solvents—those whose C.E.D. or solubility parameter closely matches that of the polymer—the polymer swells and dissolves virtually athermally to yield solutions comprising individual polymer molecules interspersed with solvent molecules, and both polymer and solvent can be regarded as miscible in all proportions. Such solutions are usually optically clear and homogeneous, and viscous; viscosity is strongly dependent on polymer concentration and molecular weight. In liquids whose solubility parameters are even moderately disparate from those of the polymer, solution does not take place; rather, the polymer sorbs a limited amount of the solvent, yielding a two-phase system comprising solvent virtually devoid of polymer, and a polymer phase containing a limited amount of dissolved solvent. With increasing polymer/solvent solubility parameter disparity, the fraction of solvent sorbed by the polymer precipitously decreases. The lower is the concentration of solvent in the polymer phase, and the poorer is its solvency power, the more viscous or rigid is that phase.

Even within the range of solubility parameters where a solvent is able truly to 'dissolve' a given polymer, the 'state of solution' of the polymer is very sensitive to the 'match' of solubility parameter. Only if the 'match' is virtually perfect will the polymer molecules be truly independent of one another and fully extended in solution; with slight 'mismatch', the polymer molecules tend either to coil upon themselves or associate with one another in solution. In *dilute* solutions, this manifests itself by reduced viscosity (relative to solutions in good solvents) and opalescence; in *concentrated* solutions, by *increased* viscosity or gelation, and opalescence. The allowable 'mismatch' of solubility parameter between solvent and polymer which will permit compatibility of the two components is frequently very small, particularly if the molecular weight of the polymer is up to 50 000 or higher.

If, to a solution of a polymer in a 'good' solvent, is incrementally added a 'poor' solvent miscible with the good solvent, a critical concentration of the 'poor' solvent will ultimately be reached at which the polymer can no longer remain in solution. This critical concentration will, of course, depend upon the difference in solubility parameter between the two solvent components, and will decrease as the solubility parameter mismatch increases.

At that critical concentration of poor solvent, phase-separation will take place, yielding a virtually polymer-free liquid phase comprising the two solvent components, and a polymer-rich phase also containing both solvents. However, the *solvent content* of the polymer-rich phase will strongly depend upon the 'goodness' of the good solvent and the 'poorness' of the poor solvent: if the solubility parameter gap between 'good' solvent and polymer is small, and that between 'good' and 'poor' solvent is large, competition of polymer molecules versus 'poor' solvent molecules for 'good' solvent molecules will favour the polymer, and the polymer-rich phase will contain a relatively large amount of 'good' solvent, and thus be relatively polymer-dilute and 'poor'-solvent-deficient. On the other hand, if the 'good' solvent is only marginally adequate to dissolve the polymer, and if the 'poor' solvent has a solubility parameter close to that of the 'good' solvent, then the polymer-rich phase which will be formed will necessarily be more concentrated in polymer and contain little of either solvent component. These differences can be more clearly appreciated by examining a ternary diagram of the types illustrated in Fig. 5, wherein the 'phase boundary lines' for two particular polymer/solvent/

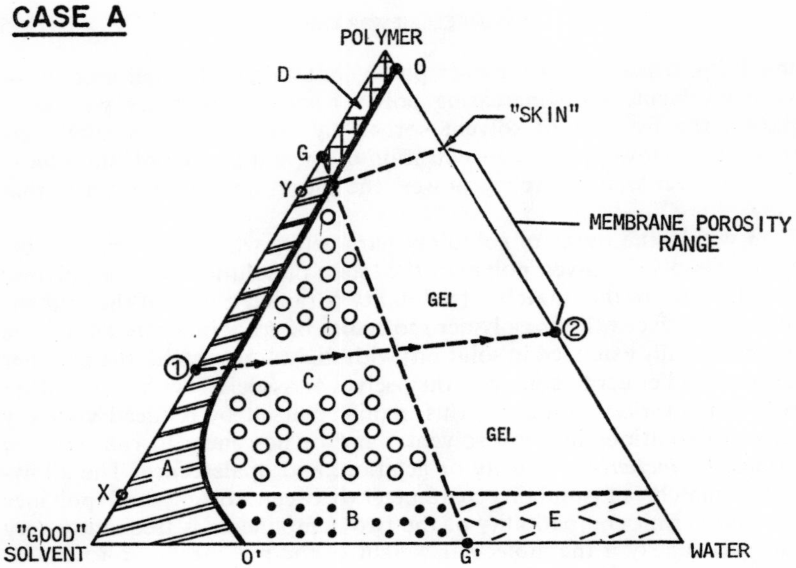

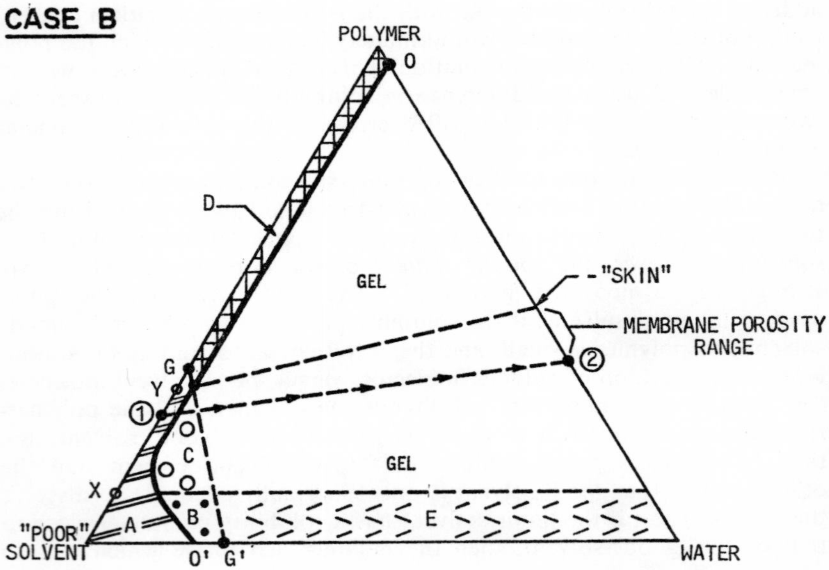

FIG. 5. Equilibrium ternary composition diagrams for two polymer/solvent/non-solvent (water) systems. Case A: A 'good' solvent for the polymer. Case B: A 'poor' solvent for the polymer. O–O′ is the two phase boundary. In region A, a homogeneous polymer solution exists; in D, a homogeneous, solvent-containing, solid

non-solvent/systems are shown. Not only will the polymer phase precipitated from the poorer solvent be richer in polymer—the 'tolerance' of the solution for non-solvent will be lower, meaning that phase separation will occur at lower non-solvent concentrations or less solution-dilution.

The 'solubility parameter' of a solvent for a particular polymer cannot only be altered by changing the chemical constitution of the solvent; it can also be controlled by mixing two or more (miscible) liquids whose solubility parameters 'bracket' those of the polymer. Thus, binary mixtures of a 'good' and 'mediocre' solvent will permit adjustment of the solvency power to values intermediate between those of either component alone, and frequently, binary mixtures of (miscible) solvents which independently are 'poor' polymer solvents will behave as 'good' solvents.

As a general rule, if the solubility parameters of *both* components of a binary solvent for a polymer lie *between* that of the polymer and that of the precipitating liquid, the polymer phase precipitated on addition of non-solvent will contain little solvent, whereas if one component of the binary solvent has a lower (or higher) solubility parameter than either the polymer or non-solvent, the precipitating polymer phase will be relatively rich in that component. For example, polystyrene is soluble in acetone, dimethylformamide, and mixtures thereof; the solubility parameter of polystyrene is closer to that of acetone than DMF. Addition of water to acetone/polystyrene solutions precipitates a polymer phase rich in acetone and poor in water, while that precipitated from DMF solution is poor in either DMF or water. Addition of water to polystyrene solutions in mixtures of acetone and DMF precipitates polymer progressively poorer in acetone and DMF as the ratio of DMF to acetone increases. For asymmetric membrane production, it is usually necessary that the polymer phase formed by precipitation be poor in solvent, for reasons to be explained below.

Kinetics of solvent/non-solvent interchange during membrane formation

When a thin layer of polymer solution is exposed to a non-solvent such as water (miscible with the solvent), rapid counter-diffusion of solvent and

---

polymer. To the left of GG′, the polymer-rich phase is liquid; to the right, it is solid. In region C, the polymer phase only is continuous; in regions B and E, the solvent phase only is continuous. The region labelled 'GEL' is a pseudo-stable region wherein both the polymer and solvent phases are continuous, and a microporous gel exists. The line 1 → 2 represents the change in composition of a thin film of polymer solution following immersion in water. The final membrane composition lies on the polymer–water boundary, within the brackets labelled 'Membrane Porosity Range'. Polymer solutions of compositions lying between X and Y can be expected to yield useful asymmetric membranes.

non-solvent takes place across the contact interface. Close to this interface, the polymer solution composition changes with extraordinary rapidity, causing virtually instantaneous precipitation of polymer and depletion of solvent. At increasing distances within the polymer solution layer removed from the precipitating interface, however, the *rate* of arrival of non-solvent (and removal of solvent) becomes progressively slower—the rate of composition-change varying no less than inversely with the square of the distance from the film/liquid interface. Moreover, if the polymer phase precipitated in the process is continuous (whether it be porous or void-free), the polymer itself will offer significant additional resistance to transport of either solvent or non-solvent, thereby further retarding changes in liquid composition in underlying portions of the film. Consequently, the precipitative processes occurring upon exposure of the polymer/solvent film to non-solvent take place with explosive rapidity within a short distance (of the order of 1 μm or less) of the exposure-surface, but with rapidly declining rates further into the film.

Membrane 'asymmetry' is virtually exclusively attributable to this process.

The role of phase transformation kinetics

In polymer solutions as in any multicomponent system, establishment of conditions for phase-separation requires a 'critical supersaturation'; the more rapidly this supersaturation is generated, the faster does the phase-separation take place, and most importantly, the *smaller the scale of size* of the phase elements which are formed. In systems containing macromolecules, the morphology of the phases produced by the separation process is governed largely by the rapidity with which those macromolecules can move with respect to one another and the solvent molecules with which they are associated.

If a polymer solution is relatively dilute, if the polymer is of relatively low molecular weight, and/or if conditions for precipitation are established relatively slowly, the process of phase transformation will involve the formation of isolated particles or droplets of solvent-rich polymer, which agglomerate or coalesce into a bulk, separate phase. If, however, the polymer concentration is suitably high, the polymer of high molecular weight, and/or conditions for precipitation established rapidly, the polymer-rich phase will tend to separate in the form of quite small, frequently fibrillar elements which tend to associate strongly with one another, yielding a three-dimensional network whose interstices are occupied by polymer-free liquid. Under appropriate conditions, this gel comprises *two interpenetrating continuous phases*—a polymer phase and a liquid phase; this structure is the *sine qua non* of an asymmetric membrane. Under less favourable conditions, the polymer-rich phase may spontaneously consolidate or coalesce, with consequent coalescence and/or expulsion of the liquid phase; this leads to loss of continuity of the liquid phase, and either formation of dense, void-free polymer, or formation of a 'closed cell' foam-structure containing isolated liquid-filled cavities. The formation of void-free, dense polymer appears to be an essential feature of the 'skin' barrier layer of asymmetric reverse osmosis membranes.

The role of polymer, and polymer solution, viscoelasticity

The phase-separation processes described above obviously require molecular movements of solvent and non-solvent molecules with respect to one another, of solvent and non-solvent molecules with respect to polymer molecules, and of polymer molecules with respect to one another. The rapidity with which these movements can take place, relative to rapidity with which solution composition is changing during the precipitation process, ultimately governs the microstructure of the membrane.

Of all these molecular movements, the slowest by far—and therefore the rate-limiting element in polymeric phase transitions—are polymer–polymer movements. In solid polymers, constraints on chain-movement due to interchain molecular forces and interchain entanglements render these materials resistant to plastic deformation; their tendency to creep or flow under stress decreases very rapidly with such structural parameters as increasing chain-length (molecular weight), increasing interchain cohesion (polarity), and increasing chain stiffness. As a rule, linear polymers with weak interchain forces and highly flexible backbones (e.g. polyisobutylene) behave as soft, elastomeric and plastic substances at all but very low temperatures. In contrast, polymers with strong interchain forces and/or rigid backbones (e.g. polyvinyl chloride, polystryrene, polyacrylonitrile) are rigid, glassy solids at normal temperatures.

These bulk viscoelastic properties of polymers also manifest themselves (albeit to a smaller degree) in their solutions. Solution viscosity is profoundly dependent on polymer molecular weight and polymer concentration; it is also dependent upon chain flexibility, and (particularly in concentrated solutions) on *interchain attractions*. If these attractions are relatively strong (as is the case in mediocre solvents), chain-contacts will be established throughout the solution yielding a three-dimensional polymer network; under these conditions the 'solution' ceases to behave as a viscous liquid, but becomes an elastomeric gel. The resistance of this gel to plastic deformation is sharply dependent on polymer concentration, and 'goodness' of solvent. This gives rise to a frequently sharp discontinuity in solution viscoelasticity—the so-called 'sol–gel transformation'—with increasing polymer concentration or decreasing solvency-power of the solvent. Once a polymer solution enters the gel state, the relative positions of polymer molecules within the gel become essentially frozen, and further changes in the network structure which would be expected to accompany changes in solvent content environment are greatly impeded. It is thus possible to form expanded gel structures which are thermodynamically unstable, but which are 'pseudo-stable' because of the extremely slow rate of polymer chain rearrangement or relaxation. Indeed, all microporous membrane structures are inherently unstable, but persist solely because of the slowness of the process of creep or cold-flow of the polymeric matrix.

The conditions under which a polymer solution will convert, upon solvent/non-solvent exchange, into a coherent, stable, microporous gel with essentially complete continuity of void-space is largely determined by the structural characteristics of the polymer. Polymers of high molecular weight, high interchain cohesion or propensity for crystalline ordering, and high chain rigidity (the latter two criteria being reflected into a high glass-transition temperature) tend to yield such gels readily. The higher the molecular weight of the polymer, the lower the polymer concentration in solution which will yield a stable gel; as a rule, polymer solutions containing between 15 and 35% polymer by volume will yield useful membrane structures, although with polymers of molecular weight of one million or more it is possible to fabricate useful membranes from (naturally quite viscous) solutions containing as little as 5% polymer by volume.

The consequences of contacting a polymer solution with a solvent-miscible non-solvent are thus obvious: at the contacting interface the polymer solution is rather rapidly brought to a condition requiring phase separation. If the polymer-rich phase which forms is solvent-rich and of relatively low viscosity, and if that phase can grow and consolidate during the necessarily short time that solvent remains in that layer, the film which initially forms will be quite dense and relatively void-free, and what voids do exist within it will be of relatively large dimensions. If, on the other hand, the initial polymer solution, when diluted with non-solvent, tends to yield a polymer-rich phase of low solvent content, then the solution will gel before significant phase-separation can occur; as solvent is replaced by non-solvent, the resulting film will be a 'frozen' microporous network of high porosity, and fine pore-texture. The higher the polymer concentration in the original solution, and/or the more marginal the solvency power of the solvent, the more rapid will be the onset of gelation, and the finer the pore-texture.

Further down into the polymer solution from the contacting interface, the diffusion processes of non-solvent penetration and solvent extraction occur progressively more slowly. The more slowly conditions suitable for phase-separation are reached, the greater the opportunity for flow and consolidation of both the polymer-rich and polymer-free phases, and thus the coarser becomes the texture of the microporous structure. In some instances, the polymer-rich phase which initially separates is still quite dilute in polymer, but is rapidly gelled by the continuing exchange of solvent for non-solvent. As a result, the terminal structure will be found to consist of a continuous matrix of porous polymer of quite fine pore-size, containing quite large cavities dispersed within that matrix.

Obviously, any change in precipitation conditions which tends either to slow down the rate of solvent/non-solvent interchange in the film, or to increase the fluidity of the polymer-rich phase that precipitates, will tend to reduce the porosity of the resulting film and increase the pore size. Raising the temperature of the process, while it increases diffusion rates, always increases polymer fluidity to a higher degree; this yields membranes with denser 'skin' layers, and coarser porous substructures. The addition of a solute (e.g. salt) to the precipitating liquid (e.g. water) reduces the activity of the precipitant, and slows down the precipitation process; this, too, yields denser (less porous) skin layers, and more coarsely porous substructures.

The role of capillary forces in asymmetric membrane formation and structure

The foregoing analysis is helpful in explaining the development of porosity and structural asymmetry in membranes found by the precipitative process, and the variations in void-fraction and pore size resulting from variations in solvent composition, polymer structure, and precipitation environment. There is, however, one further phenomenon which must be considered in accounting for the *final* structure of asymmetric, microporous membranes: this is the effect of *capillarity* or *interfacial tension* which always exists in two-phase systems.

In any microporous medium, there are surface or interfacial tensile forces acting at every interface defining a pore/matrix boundary. These forces give rise to the development of hydrostatic pressure differences across these phase boundaries which place the phases in tension or compression. These hydrostatic pressures are directly proportional to the interfacial tension between the two phases, and essentially inversely proportional to the pore radius. In the case of a microporous polymer solid whose voids are liquid (e.g. water) filled and which is immersed in liquid, the solid matrix is under tensile stresses which operate to deform the solid in a manner tending to collapse the finer pores at the expense of larger ones, if not to expel the liquid phase entirely. If the pore size is initially large enough, the stresses so generated will be insufficiently large to cause plastic flow and deformation of the solid matrix, whereupon the microporous structure will be 'pseudo-stable'. If, on the other hand, the initial pore size is very small, these stresses can become much larger than the yield stress of the polymer, and collapse of the pore-space takes place rapidly; this is accompanied by a marked decrease in porosity, and either the growth of much larger pores or complete elimination of pores. This leads to a peculiar paradox: as conditions of precipitation are adjusted to yield increasingly finer pore size in a membrane (e.g. more rapid gelation, higher polymer concentration in solution), the pore size monotonically decreases until a critical pore size is reached, whereupon spontaneous capillary collapse of the pore structure ensues, with consequent complete loss of porosity, or generation of a rather coarsely porous structure. This critical condition is, of course, most likely to occur at the 'skin layer' of an asymmetric membrane, since it is in this region where precipitation and gelation occur most rapidly, and thus the pore size is smallest. It is precisely this 'capillary collapse' phenomenon which appears to give rise to the formation of 'dense' skin layers of asymmetric reverse osmosis membranes. Obviously, any environmental variable which reduces the yield stress of the polymer will facilitate the collapse process; this explains why thermal annealing of an asymmetric membrane densifies the skin layer. That brief annealing selectively densifies the skin without causing collapse of the porous substructure is due to the fact that the skin layer has the finest pore-texture, and thus the greatest capillary stress. Exposure of the membrane to yet higher temperatures, or to moderate temperatures for extended times, will permit capillary consolidation of the porous substructure, yielding in the limit, a dense, void-free film.

The principles of capillarity also predict that, if the interfacial tension between the polymer and void-liquid can be reduced, the consolidation stresses attributable to capillary forces will be mitigated, and the propensity to spontaneous collapse of finely porous membrane structures reduced. This is confirmed by the experimental observation that, if surface active agents are added either to the polymer solution or the precipitating liquid (usually water), the mean pore size (both in the skin layer and substructure) of the resulting asymmetric membrane is significantly reduced, and the retention of ultrafine porosity in the skin layer becomes possible.

Capillary forces leading to collapse and consolidation of microporous, asymmetric membranes are substantially greater if such water-saturated membranes are allowed to dry by evaporation. Under these conditions, the collapse-stress imposed upon the polymer matrix is equal to the negative pressure imposed in the void-liquid by the curvature of the liquid–air menisci, as given by the Kelvin equation:

$$\frac{\Delta p = 2\gamma \cos \theta}{r}$$

where (γ) is the surface tension of the liquid, (θ) is the angle of contact between the liquid and the pore-wall, and (r) is the pore radius. In finely porous membranes, these negative pressures can easily reach values of several hundreds of atmospheres—well in excess of the yield stress of most polymers. The result is that most asymmetric reverse osmosis membranes, and many ultrafiltration membranes, will undergo virtually complete collapse (and thus almost complete loss in useful permeability) on drying. Only in the case of membranes comprised of polymers whose wettability by water is poor (i.e. where (θ) is in the vicinity of 90° or more), can drying be carried out without major structural alteration. Since the storage and handling of water-wet membranes is both inconvenient and costly, techniques for allowing the dessication of (water-wettable) asymmetric membranes have had to be developed; two methods have been found to be suitable. One is to equilibrate the wet membrane with aqueous surfactant solution; this allows drying to take place at relatively low consolidation stress, due to the low surface tension of the pore liquid. The other is to equilibrate the membrane with an aqueous solution of a non-volatile, water-miscible organic substance such as glycerol or polyethylene glycol. In this case, drying of the membrane leaves behind, to occupy the pore spaces, the non-volatile liquid, which is readily extracted when the membrane is recontacted with water.

If one determines the porosity of a typical asymmetric membrane formed by the precipitative process and compares the *volume fraction* of polymer in the membrane with the volume-fraction of polymer in the original casting solution, it is invariably found that the former is greater than the latter (Marcinkowsky *et al.*, 1966). This reflects the fact that capillary consolidation processes are at work throughout the precipitation stages of membrane formation. Another consequence of this shrinkage process is obviously a convective movement of liquid (i.e. solvent and water) out of the membrane into the precipitating bath; this convective liquid traffic is in a direction opposite to the diffusion of precipitant (water), and therefore retards the transport of water into the membrane substructure. As a result, the rapidity of precipitation or gelation below the membrane surface is slower than one would expect for the process of diffusion alone, and thus the rate-of-change of pore texture with depth is accentuated. It is not unlikely that the extreme thinness of the 'skin layer' of both ultrafiltration and reverse osmosis asymmetric membranes (typically between 0.1 and 1.0 μm) is in part attributable to this convective phenomenon.

The skin-structure of the asymmetric membrane

It is, of course, the microstructure of the skin layer of an asymmetric membrane, as well as its thickness, which determines its transport and rejection characteristics. Unfortunately, because of the extraordinary thinness of this layer and the great rapidity with which it forms, it has not yet been possible adequately to characterise its structure and thereby to correlate precisely its flow/solute-retention behaviour with membrane fabrication variables.

With reverse osmosis membranes, the skin layer must be virtually void-free; if such membranes are fabricated under conditions where voids are created and remain within this layer, the trans-membrane water fluxes are inordinately high, and the salt or solute-retention significantly reduced. That this elevation in flux and diminution in retentivity is due to bulk solution-leakage through pore-defects in the skin layer is supported by the observation that the presence of trace-amounts of colloidal or macro-molecular material in the feed-solution delivered to such a membrane will usually cause a marked reduction in flux and an *increase* in solute retentivity.

With ultrafiltration membranes, the skin layer is demonstrably microporous, as evidenced by surface-electron microscopy, gas- and liquid-permeation measurements, and 'critical bubble pressure' measurements, wherein water-wet membranes are subjected to increasing gas pressure until gas flow commences by displacement of the water. The *porosity* (i.e. void fraction) of the skin layer is always much lower than that of the macroporous substructure; it is seldom greater than 50%, and frequently as low as 2–5%. Skin porosity (which governs hydraulic permeability but not solute retentivity) is determined by the polymer glass-transition temperature and molecular weight (increasing with both), and usually decreases with increasing polymer concentration in the casting solution. Precipitation variables which favour early gelation (e.g. high concentrations of polar 'mediocre solvents' in the casting solution) or reduce capillary consolidation forces (e.g. surfactants in the precipitation bath) also favour high skin porosity and thus high water flux.

Skin-layer pore size is also dependent on the same parameters which control porosity. Polymer concentration in the casting solution is a critical variable (pore size decreasing rapidly with increasing polymer concentration); and with the precipitation variables mentioned above which favour high skin porosity, also tends to favour finer pore texture.

What is particularly astonishing about the skin-layer structure of ultrafiltration membranes is their *extraordinarily narrow pore-size distribution*. This is most clearly reflected by examination of their 'solute rejection spectra'—the variation in their retentivity for solutes of differing molecular size. Representative data for several different membranes, covering a broad range of retentivities, are shown in Fig. 6. Solute retentions varying from 10% to nearly 100% over a solute molecular weight range of only twofold is not unusual. Since (for spherical macromolecules) molecular diameter varies roughly as the cube root of the molecular weight, the pore-size variability in such a membrane cannot be much greater than ±10% from the mean value. The most appealing explanation for this extraordinary pore-uniformity is that the extreme rapidity of skin formation necessarily requires that every pore be generated at the same instant, under virtually identical conditions.

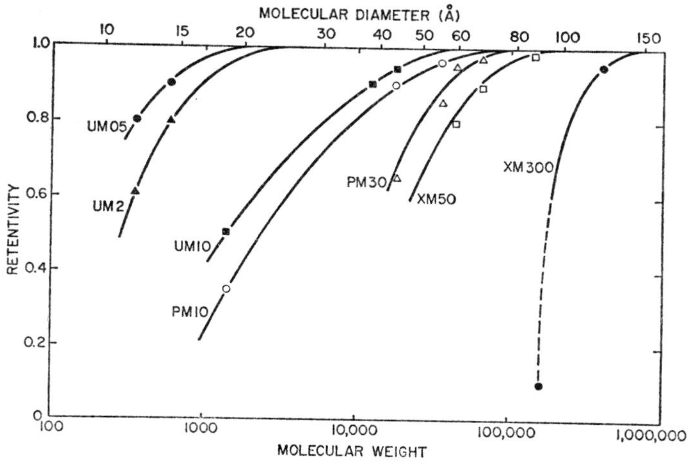

FIG. 6. Typical 'solute retentivity spectra' and estimated pore-size distributions for a series of typical ultrafiltration membranes. (Courtesy of Amicon Corporation.)

'DIRECT' OR 'DRY' PROCESSES OF ASYMMETRIC MEMBRANE FORMATION

The foregoing discussion dealt with the formation of asymmetric membrane structures by exposure of polymer solutions to solvent-miscible non-solvents such as water. As has been shown by Kesting (1971), it is also possible (and considerably more convenient) to produce such structures by casting films of solutions of polymers in solvent/non-solvent mixtures, and removing the solvents by evaporation into air under controlled conditions. The principal requirement for development of asymmetric, microporous structures via this process is that, during the evaporation process, the solvent composition in the film must change in the direction of decreasing polymer compatibility. This is, of course, accomplished by selecting solvent mixtures in which the 'good' solvent component has the higher volatility. (It is a curious fact that the protective and decorative coatings literature is rich in information relating to the importance of mixed solvent relative volatility in determining the integrity of paint- and lacquer films; indeed, the 'undesirable' solvent systems and drying conditions which give rise to 'bloom' or opalescence in resinous coatings are precisely those which yield asymmetric membranes.)

The gelation, precipitation, and void-formation phenomena which give rise to asymmetry in these 'dry' membrane fabrication processes are essentially the same as those involved in the 'wet' process, with the added complication of unsteady-state heat transfer and surface cooling accompanying solvent evaporation. The former process is, however, somewhat less versatile than the latter, in that solvents and non-solvents must be selected from compounds which not only have the proper 'polymer compatibility' balance, but also the proper relative volatility. For large scale, continuous membrane manufacture, however, the dry process has many practical advantages, since precipitation and washing baths are eliminated, terminal drying avoided, and solvent-recovery facilitated. To the best of this writer's knowledge, the 'dry' process is presently used only in the preparation of reverse osmosis membranes.

ULTRATHIN COMPOSITE MEMBRANES

Recognition that the unique high flux/high solute retention characteristics of asymmetric reverse osmosis membranes is due to the thin homogeneous polymer skin on these structures led to the realisation that, were it possible to prepare defect-free, ultrathin films of high-salt-rejection polymers, and to laminate these films upon a porous substrate of high hydraulic permeability, high porosity, and sufficiently small pore size to provide adequate support of the ultrathin film against high hydrostatic pressure, a superior membrane might result. One of the most important benefits to be gained from the development of such a composite membrane is the complete separation of the requirements of the salt-rejecting membrane material from the (totally unrelated) requirements for the porous support. Yet another benefit is the possibility of producing ultrathin films substantially thinner than those achievable by the asymmetric membrane process.

The research of Rozelle *et al.* (1970), and of Riley and coworkers (1971, 1973) over the past few years, has indeed established that such composite ultrathin reverse osmosis membranes can not only be prepared, but can be produced by techniques amenable to large scale membrane manufacture. Both groups have confirmed that it is feasible to prepare reverse osmosis membranes comprising an ultrathin, defect-free barrier layer of cellulose acetate of thickness between 250 and 1000 Å, supported on a microporous, high-pressure-resistant, mechanically strong support-film, capable of sustained operating service at pressures of 100 atm, whose salt-rejection properties very closely approach the limit expected for the permselectivity of the barrier film. To put this accomplishment in proper perspective, it should be recognised that a polymer film of this thinness is a 'black' or 'iridescent' film—it is thinner than the wavelength of visible light, and is only of the order of 25–100 polymer chains in thickness (assuming that the polymer molecules have their long axes parallel to the film plane).

The Rozelle group actually prepared their ultrathin (*c.* 200 Å) membrane by spreading a solution of cellulose acetate in a suitable water-immiscible solvent (e.g. cyclohexane) upon water, and allowing the solution to evaporate. The resulting film was then transferred to the surface of a microporous support membrane (e.g. a Millipore filter membrane, or an asymmetric ultrafiltration membrane fabricated from a polysulphone polymer). Subsequently, they found that, by prewetting the microporous support with an aqueous solution of water soluble polymer (e.g. polyacrylamide), and contacting the wet support with the solution of cellulose acetate in organic solvent under conditions of gravity drainage and evaporation, they were able to achieve comparable results.

The Riley group has employed a somewhat different (and possibly simpler) procedure, from which have emerged composite structures of still thinner yet reproducible barrier-layer membranes (Riley *et al.*, 1971, 1973) Utilising an asymmetric ultrafiltration membrane support-structure prepared from a mixture of cellulose acetate and cellulose nitrate, they coat the 'skin side' of this membrane with an aqueous solution (dispersion?) of polyacrylic acid, and dry the coating. The polyacrylic acid serves to plug the pores in the substrate skin, and also 'fills in' irregularities on the skin surface, thereby providing a virtually optically flat surface to receive the ultrathin membrane barrier. Since the polyacrylic acid is soluble or redispersible in water, it is readily and rapidly extracted from the membrane when it is exposed to water or salt solution in use.

The ultrathin barrier film of cellulose acetate is applied to the (pretreated) support membrane by simple dip-coating of the top-surface of the support in a water-immiscible solvent (e.g. chloroform) solution of the polymer, followed by gravity-drainage and solvent-evaporation. Barrier-layer thickness is controlled by (1) solution viscosity (i.e. polymer concentration), and (2) rate of (vertical) travel of the film out of the coating bath. A schematic diagram of the cross section of this type of membrane is shown in Fig. 7.

It has been found that barrier-films as thin as 250 Å can be prepared reproducibly by this technique, and that composite membranes so fabricated perform excellently under reverse osmosis desalination conditions at pressures upwards of 1500 psi. Riley *et al.* (1973) report that the pore size in the 'skin layer' of the asymmetric support film is of the order of 500 Å and less, indicating that so long as the pore openings across which the ultrathin membrane is deployed are of the same scale of dimension as the film thickness, adequate support against rupture at high pressure is provided.

In principle, the hydraulic permeability of a composite membrane with a 250 Å ultrathin cellulose acetate film should be about 10 times higher than that of an asymmetric cellulose acetate membrane with a 'skin' thickness of 0.25 μm, so long as the hydraulic resistance offered by the porous substructure in either case is small compared to that of the barrier layer. In practice, it is found that, if the thickness of the ultrathin layer of a composite membrane is in a range substantially greater than the pore size of an immediately-adjacent support-film, the flux varies inversely with the

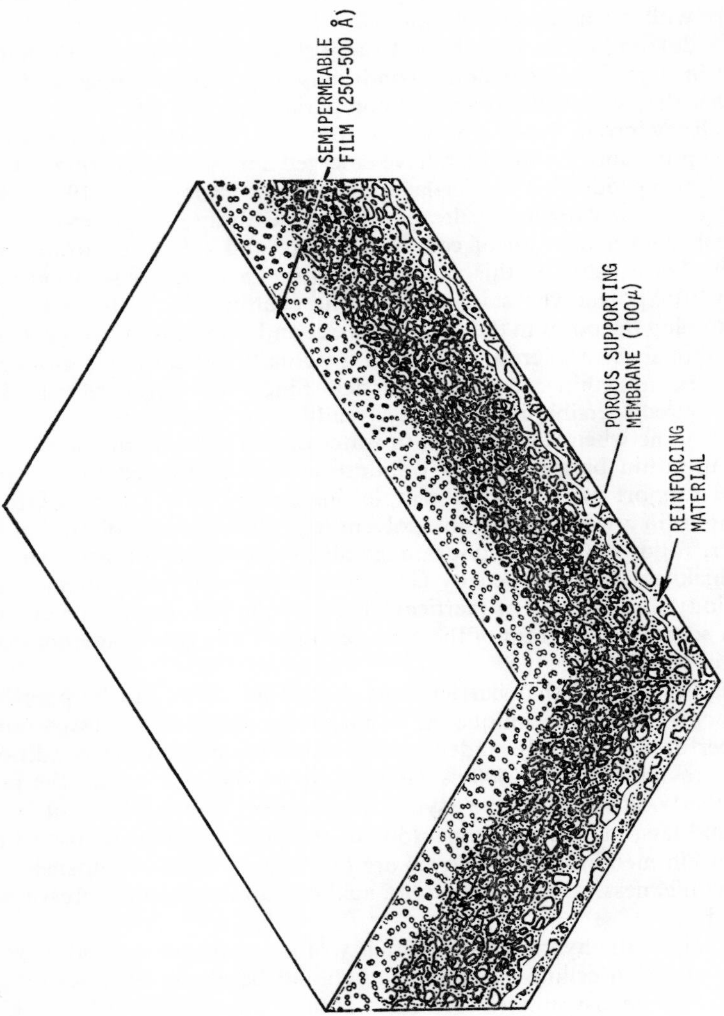

FIG. 7. Schematic diagram of an ultrathin-film composite reverse osmosis membrane. (Courtesy of R. L. Riley, Gulf Environmental Systems, Inc.)

layer thickness; however, when the ultrathin film thickness approaches the underlying pore dimensions, the flux increases more slowly than inversely with thickness. The reason for this is quite straightforward: part of the supporting substructure contacting the film is necessarily a solid, virtually water-impermeable polymer; the 'permeation active' areas of the ultrathin film are therefore restricted to those regions which overlie substrate pores. Hence the 'limiting permeability' of such a composite membrane will be the product of the permeability of the ultrathin film in the unsupported state and the porosity of the skin layer of the asymmetric support. Riley reports that surface porosities of his support membranes lie between 15 and 30%, meaning that flux-constraints imposed by the support are small relative to the increase in flux achievable with barrier film thicknesses in the 250–500 Å range.

The laboratory and field performance of composite reverse osmosis membranes of this type has been gratifying. For sea water demineralisation at pressures of 100 atm, sustained fluxes of close to 25 gal/ft^2 day have been achieved, with salt-rejections in excess of 99.3%—close to the theoretical value for cellulose 2.5 acetate. Also of particular note have been (1) the reproducibility of membrane performance, (2) the durability of the membrane in extended sea water service, and (3) the resistance of the membrane to pressure-consolidation and flux-decay over long time-periods in high pressure uses. This latter benefit stems, of course, from the fact that the asymmetric microporous support membrane can be fabricated from mechanically strong, stable, creep-resistant polymers which themselves need have no special permselectivity.

While most of the development work on composite membranes has been directed toward cellulose acetate barriers and water desalination, it is obvious that the technique and methods are equally applicable to all other film forming polymers useful for reverse osmosis applications. Moreover, the technique is also useful for the preparation of 'ultrathin' membranes for such processes as gas mixture separations, organic liquid mixture separations, and the like. In this writer's opinion, the composite ultrathin membrane development is of engineering significance at least as great as that of asymmetric membrane development, and is certain to extend the utility of membrane separation processes into large-scale industrial applications previously considered unachievable or uneconomic.

HOLLOW FIBRE MEMBRANES

Preoccupation with planar membrane geometry automatically leads one to develop techniques to reduce membrane thickness as much as possible to maximise throughput per unit area. If, however, one removes the constraint of planarity, it becomes obvious that there are certain membrane geometries which make it possible to achieve high permeation throughput per unit volume of membrane system without resorting to extreme membrane thinness. By far the most appealing is the small-diameter hollow cylinder, the wall of which comprises the permselective membrane. Since the area-to-volume ratio of a cylinder is inversely proportional to its diameter, and since (for a given OD/ID ratio) the wall thickness of a cylinder is directly proportional to the diameter, the permeation flux per unit volume of a hollow cylindrical membrane increases inversely as the square of the diameter. For reverse osmosis or ultrafiltration applications, it can easily be shown that a reasonably-sized bundle of small diameter cylinders of 'hollow membrane fibres' of outside diameter of *c.* 250 μm and wall thickness of *c.* 50 μm should provide transmembrane flows equivalent to those obtained with many hundreds of square feet of ultra-thin planar membranes.

Another important benefit which accrues from the use of hollow fibre membrane geometry for pressure driven separations relates to the problem of membrane support. While most membrane polymers have low inherent tensile strengths, when fabricated into hollow fibres of sufficiently small dimensions, the stresses to which the material is subjected under a large *hydrostatic* pressure difference across the fibre wall can be made much smaller than the yield or failure stress. If high pressures are generated inside the fibre lumen, the wall is essentially in tension; the rupture pressure is proportional to the tube wall thickness, and inversely proportional to the diameter. If pressure is applied to the exterior of the fibre, the wall is in compression, and the pressure that can be withstood without wall-collapse is far higher; moreover, when collapse occurs, the fibre merely flattens, and flow through the lumen is interrupted. In neither 'inside' nor 'outside' pressurisation is it necessary to provide free-draining mechanical support for the membrane, since in both cases the membrane adequately serves as its own support.

Techniques for continuous spinning or extrusion of hollow fibres of small diameter date back to the early days of the synthetic fibres industry. By appropriate spinneret design (usually involving an annular extrusion orifice, coupled with a means for continuously feeding gas or liquid into the hollow core during spinning), it has been found possible to produce uniform-bore, uniform diameter hollow monofilaments as fine as 150–200 μm OD and 50–100 μm ID Melt-, dry- (i.e. dope) and wet- (i.e. coagulation bath) spinning techniques can be utilised.

Early efforts to develop hollow fibre membrane systems for reverse osmosis and related uses were undertaken by McLain and Mahon (1969) of Dow Chemical Co. with cellulose triacetate, and by Hoehn and Pye (1970) of the DuPont Co. with nylon. In both cases, the fibre walls were comprised of essentially dense, void-free polymer, and displayed water- and salt-permeability values characteristic of those materials in bulk. The fibres so produced are assembled into multifibre bundles, 'potted' with resin at the bundle-ends to isolate the fibre-lumen channels from the external fibre spaces, and inserted into a tubular shell. Pressurised fluid is delivered to the shell-side of the fibre bundle, permeate or filtrate passing through the fibre walls and out of the lumen. While transmembrane fluxes in reverse osmosis (at *c*. 600 psi) are only in the range of 0.1–0.5 gal/ft^2 day with such systems, the membrane area per unit vessel volume is enormous —of the order of 500–1000 ft^2/ft^3. Thus, water throughputs of the order of 50–500 gal/(day) (ft^3) are achievable. Systems of this type have now been in commercial use for several years, and their performance (particularly for brackish water demineralisation) has been quite satisfactory. More recently, DuPont has introduced a hollow fibre system utilising an aromatic polyamide membrane polymer in lieu of the earlier aliphatic polyamide; this material has both higher water permeability and salt-rejection.

The application of techniques for producing asymmetric membranes in hollow-fibre configuration has also been successful, using both 'wet' and 'dry' spinning methods (Cohen *et al*., 1972; Orofini, 1971); this yields a fibre whose wall is an asymmetric microporous membrane, the skin of which lies on the external fibre surface. By this means, water fluxes through the fibre wall (externally pressurised fluid) have been increased 5 to 10-fold over 'dense membrane' fibres of comparable wall-thickness, without significant sacrifice in solute retention. This flux increase, of course, substantially reduces the area requirements for a given throughput capacity, and thereby allows the use of smaller fibre bundles.

Efforts to elevate the wall permeability of hollow fibre membranes as a means of reducing membrane unit size and cost are, however, frustrated by a hydrodynamic limitation: namely, as wall permeability increases, the traffic of filtrate or permeate which must pass through the fibre-lumen also increases. The end result of this is an increase in the hydrostatic pressure drop along the fibre axis driving the permeate out the fibre-end. This lumen pressure, if it becomes great enough, will not only limit water-flow through the fibre wall, but will operate to reduce the solute-rejection efficiency. It is this writer's belief that, for fibres of the order of 250 μm in outside diameter, and of any reasonable length (e.g. one foot or more), transmembrane fluxes of *c*. 10 gal/ft^2 day may already be in a range where lumen pressures can adversely affect flow and solute rejection.

Another limitation of 'shell-side-feed' hollow fibre systems is the problem of fouling of the interfibre spaces by particulate or colloidal debris present in the feed fluid, and the difficulty of removing such debris from the fibre bundle after it accumulates. As a rule, thorough prefiltration or chemical pretreatment of feed streams is recommended and employed in an effort to minimise this problem in the field.

The use of 'shell-side-feed' hollow fibre membrane systems for *ultrafiltration* applications has been virtually impossible, because of the problems of membrane solute polarisation and fouling. However, the fact that most ultrafiltration operations are carried out at very *low* pressures (5–50 psi) provides an opportunity to use hollow fibre membranes by supplying pressurised feed to the fibre-lumen, and allowing permeate to pass out through the fibre wall. Under such flow conditions solute rejected at the membrane surface can both readily diffuse and be convected into the flowing feed-stream without adversely affecting transmembrane flux.

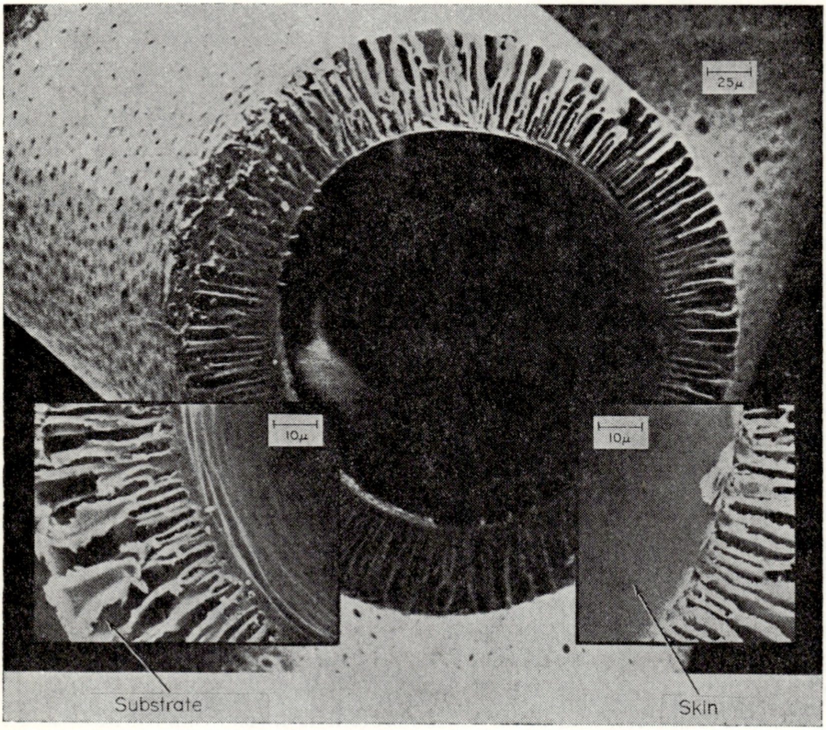

FIG. 8. Electron micrograph of the cross section of an asymmetric hollow-fibre ultrafiltration membrane. External diameter *c.* 200 μm; internal diameter *c.* 100 μm. (Courtesy of Amicon Corporation.)

Clearly, hollow-fibre, asymmetric ultrafiltration membranes suitable for lumen-feed must be structured such that the 'skin' layer is *inside* the fibre. This has necessitated the development of a rather novel fibre-spinning and gelation/precipitation technique, with which a family of 'controlled solute cutoff' skin-inside membrane fibres are now being commercially produced. An electron micrograph of the cross section of such a fibre is shown in Fig. 8; a laboratory-scale, 12 000-fibre module produced from such fibres is shown in Fig. 9. The water permeabilities and solute rejection properties of such hollow-fibre membranes are found to match quite closely those of the same membranes prepared in sheet form. Since, however, a 12 000-fibre bundle (measuring about 8 in long by 4 in diameter) contains about 15 ft^2 of membrane area, the transmembrane flow achievable with a unit of this size is in the neighbourhood of 200–300 gal/day.

FIG. 9. A 12 000-fibre ultrafiltration module of membrane surface area 1.6 m^2; the unit is about 4 in in diameter and 8 in long. (Courtesy of Amicon Corporation.)

The principal limitation of these 'inside flow' hollow fibre ultrafiltration systems is obstruction of the fibres by particulate solids in the feed. Evidence suggests that, so long as particles of dimensions greater than 10% of the fibre-lumen diameter (which is typically about 150–200 μm) are absent, fibre-plugging seldom occurs. Hence, prefiltration of feed streams with quite coarse (10–20 μm) filters is usually adequate to avoid or minimise this problem.

From both economic and engineering design considerations, it would appear that inside-flow hollow fibre ultrafiltration systems are destined to become of increasing importance for large scale ultrafiltration applications, and may well dominate the field within the next decade.

'DYNAMICALLY FORMED' MEMBRANES

In 1965, research at Oak Ridge National Laboratories (Marcinkowsky *et al.*, 1966) led to the important observation that, if saline solute containing a low (millimolar) concentration of tetravalent thorium salt were pressure-filtered through a suitably finely (submicron) porous medium, the filtrate so produced was significantly depleted in salt, and filtration rates even at modest driving pressures were substantially higher than those realised with the then-available reverse osmosis membranes. Since the porous filter medium itself is totally devoid of salt-retention capability, it was deduced that colloidal hydrogel particles of thorium hydroxide (the normal hydrolysis product of thorium (IV) salts in aqueous solution at near-neutral pH) were entering and lodging within the pores of the filter medium, and that these hydrogel particles effectively formed a 'secondary membrane' which displayed salt-retention properties. It was also found that only by continuously supplying small amounts of thorium as a feed additive would the salt-rejection properties of the membrane be continuously maintained; this indicated that the substance constituting the secondary membrane was undergoing constant erosion or loss by leaching or fluid flow, and therefore required constant replenishment. Hence the appropriateness of the term 'dynamic' to describe composite membranes of this type.

Subsequently, Kraus, Johnson and their co-workers at ORNL (Marcinkowsky *et al.*, 1966) were able to show that not only thoria, but a large number of polyvalent metal salts which yield colloidal hydrous oxides with electrolyte character (e.g. Fe(III), Zr (IV), Sn (IV, and U (VI)) are able to form high-flux dynamic membranes with measurable salt rejection capacity. They also found that a variety of water-soluble or colloidally dispersible organic polyelectrolytes were as effective or more effective than the inorganic materials.

In virtually all instances where salt rejection has been observed with dynamic membranes, the hydrogel substance forming the secondary membrane has significant ion-exchange capacity, and it appears that salt rejection is primarily a consequence of Donnan co-ion-exclusion by the electrostatically charged hydrogel matrix. This is evidenced by (1) the marked decline in salt rejection with increasing ionic strength (electrolyte concentration) in the feed solution; (2) the reduction in salt rejection as the valency of ions in the feed solution opposite in sign-of-charge to that of the hydrogel is increased; and (3) the frequently marked dependence of salt rejection on feed solution pH. As a rule, anionic polyelectrolytes appear to be less pH and ion-valency sensitive than cationic substances like the hydrous metal oxides; indeed, it has been found that a 'two-step' treatment involving initial exposure of the porous medium to a cationic gel-former followed by exposure to an anionic polyelectrolyte (e.g. polyacrylic acid) yields a more durable and more efficient salt-rejecting membrane.

A variety of microporous support-media have been used successfully for the preparation of dynamic membranes: porous metals and ceramics, porous carbon, porous plastics prepared by sintering and/or leaching methods, microporous filters of the Millipore or Gelman type, and asymmetric ultrafiltration membranes. The principal requirements are (1) an effective pore size below 1 or 2 μm, and (2) relatively high hydraulic permeability. One of the most appealing features of these dynamic membrane systems is that the porous support-media can be fabricated in whatever configuration is desired (tubular shapes are preferred for obvious reasons) and assembled into finished modules; the secondary membrane is applied when the module is put into service. Continuous or intermittent introduction with the feed of the appropriate secondary-membrane-forming materials provides continuity of membrane performance, and dismantling of the module for cleaning or rejuvenation is seldom necessary.

The major limitation of dynamic membrane systems for reverse osmosis applications is their relatively low electrolyte-rejection efficiency (seldom over 90%), and their ability to operate effectively with only relatively dilute feed solutions. However, the water flux rates realised with these membranes are impressive (upwards of 100 gal/ft^2/day at pressures of 700–900 psi); in fact, at these transmembrane fluxes, the salt rejection is almost certainly polarisation-limited, so that 'true rejection' by the membrane is undoubtedly much higher than the reported values. For water-renovation applications, and for the purification or moderate concentration of dilute aqueous process streams containing simple electrolytes, dynamic membrane systems merit serious consideration.

The dynamic membrane concept also has a potentially important applicability in ultrafiltration, but efforts to exploit it in this context have yet to be reported. Studies of solute-polarisation processes in ultrafiltration (Blatt *et al.*, 1970) have shown that, if micro- and macrosolutes are simultaneously present in aqueous solutions subjected to ultrafiltration, the ultrafiltrate is frequently found to be depleted in solutes which (in the absence of macrosolute) would freely pass the membrane. In such instances, it is almost certain that the retained macrosolute forms a 'secondary membrane' on the surface of the primary ultrafiltration membrane, whose solute permeability is significantly lower than that of the primary membrane. While this has usually been cited as an inherent limitation of ultrafiltration for solute fractionation applications, it can be turned into an asset. For example, a given water-soluble polymer will form a secondary membrane on an ultrafiltration membrane whose degree of hydration and gel-microstructure will depend greatly upon the polymer chemical constitution and degree of electrolyte-character. This will be reflected in both the hydraulic permeability of that secondary membrane, and its solute-rejection spectrum. It should, therefore, be possible to develop a family of water-soluble polymers which, when added in trace-quantities to a given solution of mixed micro- or macrosolutes, will each allow the formation of a 'composite' membrane of specified (but differing) solute-rejection characteristics on the same ultrafiltration membrane support. By this means, it might be possible to separate an aqueous mixture of, say, oligosaccharides with amino acids or short-chain polypeptides into two streams, one containing only the saccharides, and the other, only the amino acids and polypeptides; and in subsequent ultrafiltration with other added polymers, separate each of the two mixtures into their individual components. In the writer's opinion, this represents an intriguing area for research which might broaden substantially the applications for pressure-activated membrane separation processes.

PRESENT STATUS AND FUTURE PROSPECTS

In summary, progress over merely the past fifteen years in membrane research and development has been both remarkable and prodigious, and has brought the combined skills of the polymer chemist and physicist, physical chemist, chemical engineer, mechanical engineer, and plastics technologist to a high level of achievement in a field of primary importance to the chemical process industries. It is probably safe to say that there are very few separation problems involving aqueous solutions for which there does not now exist a membrane inherently capable of effecting the desired separation under conditions and rates which are both achievable and economical on a large scale. Of perhaps even greater importance is the fact that such membranes can today be fabricated in a variety of configurations by continuous, low-cost manufacturing procedures, with a high degree of uniformity, reproducibility, and reliability.

The step, however, from a membrane with attractive and useful permeability and permselectivity to a reliable, efficient, economical, and easily-maintained *membrane separation process* for large scale industrial use is a major one indeed; and it is perhaps no surprise to find that, at present, we are living through the painful phase of systems and process development, with much yet to be learned about membrane management and utilisation. The status of membrane process development is a subject to be dealt with by other participants in this conference; suffice to say here that much headway has been made in this area over the past few years, and it is almost certain that, by 1980, membrane processes will occupy as important a place in our unit operations arsenal as such more common separation processes as distillation, absorption and extraction, evaporative concentration and ion exchange.

REFERENCES

Applegate, L. E. and Antonson, C. R. (1972). 'The Phenomenological Characterization of DPI Membranes', in *Reverse Osmosis Membrane Research* (Ed. H. K. Lonsdale and H. E. Podall), Plenum Press, New York.

Blatt, W. E., Dravid, A., Nelson, L. and Michaels, A. S. (1970). 'Solute Polarization and Cake Formation in Membrane Ultrafiltration', in *Membrane Science and Technology* (Ed. J. Flinn), Plenum Press, New York.

Chen, C. T., Eaton, R. F., Chang, Y. J. and Tobolsky, A. V. (1972). *J. Appl. Polymer Sci.*, **16**, 2105.

Cohen, M. E., Grable, M. A. and Riggleman, B. M. 'Hollow Fiber Reverse Osmosis Membranes', in *Reverse Osmosis Membrane Research* (Ed. H. K. Lonsdale and H. E. Podall), Plenum Press, New York, 1972.

Credali, L., Chiolli, A. and Parrini, P. (1972). *Polymer*, **13**, 503.

Dresner, L. and Johnson, J. S. Jr. (to be published). 'Hyperfiltration (Reverse Osmosis)', in *Principles of Desalination*, 2nd edn. (Ed. K. S. Spiegler and A. D. K. Laird), Academic Press, New York.

Eirich, F. R., Su, M. T., Strathmann, H. and Baker, R. W. (1973). *Polymer Letters*, **2**, 201–205.

Frommer, M. A. and Lancet, D. (1972). 'Mechanism of Membrane Formation', in *Reverse Osmosis Membrane Research* (Ed. H. K. Lonsdale and H. E. Podall), Plenum Press, New York.

Hoehn, H. H. and Pye, D. G. (1970). U.S. Patent 3,497,451 (to the DuPont Co.), February 24 (1970).

Johnson, J. S., Jr. (1972). 'Dynamic Membranes', in *Reverse Osmosis Membrane Research* (Ed. H. K. Lonsdale and H. E. Podall), Plenum Press, New York.

Kedem, O. (1972). 'Water and Salt Transport in Hyperfiltration' in *Reverse Osmosis Membrane Research* (Ed. H. K. Lonsdale and H. E. Podall), Plenum Press, New York.

Kesting, R., Barsh, M. and Vincent, A. (1965). *J. Appl. Polymer Sci.*, **9**, 1873.

Kesting, R. (1971). *Synthetic Polymeric Membranes*, McGraw-Hill, New York.

Klein, E. and Smith, J. K. (1972). 'Solubility Parameters and Asymmetric Membranes', in *Reverse Osmosis Membrane Research* (Ed. H. K. Lonsdale and H. E. Podall), Plenum Press, New York.

Loeb, S. and Sourirajan, S. (1962). *Adv. Chem. Ser.*, **38**, 117.

Lonsdale, H. K., Merten, U. and Riley, R. L. (1965). *J. Appl. Polymer Sci.*, **9**, 1341.

Lonsdale, H. K. (1973). 'Reverse Osmosis Membrane Systems', Conference on Recent Progress in Membranes, Case Western Reserve University, Cleveland, Ohio, May 8.

Marcinkowsky, A. E., Kraus, K. A., Phillips, H. O., Johnson, J. S. and Shor, A. J. (1966). *J. Am. Chem. Soc.*, **88**, 5744

Mattson, R. J. and Tomsic, V. J. (1969). *Chem. Eng. Progress*, **65**, 62.

McKinney, R., Jr. (1972). 'Properties of Aromatic Polyamide and Polyamide–Hydrazide Membranes', in *Reverse Osmosis Membrane Research* (Ed. H. K. Lonsdale and H. E. Podall) Plenum Press, New York.

McLain, E. A. and Mahon, H. I. (1969). U.S. Patent 3,423,491 (to Dow Chemical Co.), January 21.

Merten, U. (Ed). (1966). *Desalination by Reverse Osmosis*, M.I.T. Press, Cambridge, Mass.

Michaels, A. S. (1968). 'Ultrafiltration', in *Progress in Separation and Purification* (Ed. E. Perry), Interscience, New York.

Michaels, A. S., Bixler, H. J. and Hodges, R. M. (1965). *J. Coll. Sci.*, **20**, 1034.

Michaels, A. S. (1971a). 'High Flow Membrane', U.S. Patent 3,615,024, October 26.

Michaels, A. S. (1971b). 'Preparation and Properties of Ultrafiltration Membranes', Conference on Reverse Osmosis and Ultrafiltration, Chemical Centre, Lund University, Lund, Sweden, May 17.

Orofino, T. A. (1971). 'Development of Hollow Fibre Filament Technology for Reverse Osmosis Desalination Systems', Office of Saline Water Research and Development Progress Report No. 549, U.S. Govt. Printing Office, Washington, D.C.

Reid, C. E. and Breton, E. J. (1959). *J. Appl. Polymer Sci.*, **1**, 133.

Riley, R. L., Lonsdale, H. K. and Lyons, C. R. (1971). *J. Appl. Polymer Sci.*, **15**, 1267.

Riley, R. L., Hightower, G. R., Lyons, C. R. and Togami, M. (1973). 'Thin Film Composite Membranes for Single Stage Seawater Desalination by Reverse Osmosis', Gulf-En-A12293 (Rev); presented at 4th Int. Symposium on Fresh Water for the Sea, Heidelberg, Germany, 9–14 Sept.

Rozelle, L. T., Cadotte, J. E. and McClure, D. J. (1970). *J. Appl. Polymer Sci.*, Applied Polymer Symposium No. 13, 61; see also *Reverse Osmosis Membrane Research* (Ed. H. K. Lonsdale and H. E. Podall), Plenum Press, New York, 1972, pp. 419–435.

Strathmann, H., Scheible, P. and Baker, R. W. (1971). *J. Appl. Polymer Sci.*, **15**, 811–828.

DISCUSSION

M. Karel (M.I.T.): You stressed the need for low water-absorption in the ultrafiltration membrane. Is there a limit, in the sense that you still must have a positive spreading coefficient?

A. S. Michaels: The wettability of the polymer is important to ensure that you do get air displacement from the membranes, if you start with a dry membrane. As a practical matter, the more hydrophobic the membrane is, the more durable it tends to be in storage and in service.

Karel: In view of what you call an accumulation of junk at a microporous layer, why not put in composite membranes—a skin on both sides to avoid the accumulation of junk in the membrane itself?

Michaels: But you don't need it, really. If you get the skin on one side and it is retentive, then nothing comes into the permeate.

Karel: Referring to the recent work by Kesting, in addition to the solvent precipitation method which you mentioned, there are several others including track etching. What do you think about them?

Michaels: The limitation with track etching is pore size, because you can't get down to much below a tenth of a micron by track etching techniques. The maximum porosity that you can generate in such a membrane is necessarily low—no more than 15–20% of the total membrane area. The other thing is that the thickness of the membrane is still in the range of 1 mil, so you're putting cylindrical holes of 1 mil long, and the resistance to those holes is pretty high. I'd rather look at them as being competitors to traditional filters, where you are talking about particulate removal in the range of a fraction of a micron rather than as competitors to U.F. membranes.

Chairman: You said that the chemistry of membrane manufacture is way ahead in the whole game. How would you, as a membrane manufacturer, know what type of membrane to use which will be retentive for certain components when you don't even know exactly the nature of what you're holding back?

Michaels: It would be nice to have at least some type of chemical identification of the compounds that you are trying to retain or permeate before you begin to do anything about membrane design. You can make a rational selection of polymeric materials on a basis of some knowledge about the chemical constitution of the components you want to retain. The major guiding rule which I have found to be useful is that, if you want impermeability, find a polymer in which the particular components that you want to retain are very insoluble.

S. A. Goldblith (M.I.T.): At the beginning you named a number of materials in which membranes could be cast. Did you investigate whether or not some of the constituents of these materials passed into the foodstuff? Did you have to get FDA approval to use them? Did you do any testing to find out whether or not there was a breakdown of components?

Michaels: The answer is no, we didn't have to go to the FDA for approval—but that day will be coming I'm sure.

With respect to extractables from these membranes, it turns out that the very process for making these polymer films involves solvent casting and exhaustive washing as part of the manufacturing step.

D. E. Blenford (RHM): You mentioned the advances that have been made—not all of them dramatic and I would like you to indicate where you think the breakthrough is likely to be in the food indutry in the next five or ten years.

Michaels: Important breakthroughs are required in order for us to see the penetration in the food industry which we all seek—I frankly feel that the dairy applications are the ones which hold the greatest promise; because those are the ones that can make best use of the separative capabilities of ultrafiltration. Whether this is removing lactose from whey, or whether it's concentration of skimmed milk with the exclusion of the electrolytes which are usually undesirable components, I'll leave others who are more skilled than me to comment. But that certainly is one area. The area of protein recovery from meat packing, dairy plant or cannery waste-streams not only as a pollution abatement measure, but as a means of recovering useful and saleable protein or other nutrients—I think is an extremely important potential application.

F. Bramsnaes (Copenhagen): I am concerned with hygiene problems and your statement about the chemical stability was very promising. Does that mean, for instance, that you could use nitric acid or caustic soda as they use in a dairy for cleaning?

Michaels: The membranes will take caustic up to pH 12 and the one resistant to acids down to pH 1. I think that covers a broad enough range in aqueous operating conditions to cover those requirements. In fact the sulphone-type membrane has been used in an application in contact with 60% KOH for extended time periods. So if it can stand that, I would hope that it would be able to take foods!

Bramsnaes: Two years ago you predicted that this could be done, and you are still predicting it. What is keeping the membrane from being used industrially?

Michaels: It's a complicated story. Most of the companies who are in the membrane business right now are not making membranes—they are developing and making systems. You'll find very few membrane-manufacturing organisations that are willing to sell membrane stock. The result is that those organisations which are ahead in membrane development are undertaking to develop better systems in which to use those membranes but they're not releasing membranes for industrial application until they have the systems.

Membrane Concentration

R. F. MADSEN

A/S De Danske Sukkerfabrikker, Driftteknisk Laboratorium, Nakskov, Denmark

The processes named Ultrafiltration, Hyperfiltration and Reverse Osmosis are characterised by being filtration of a solution where some of the ions or molecules in the solution are able to pass the filter (membrane), and some are rejected. When the membrane rejects ions and molecules of a molecular weight less than 500, the process is normally called hyperfiltration (reverse osmosis), because the filtration in most cases has to act against an osmotic pressure of importance to the process. The process is called ultrafiltration when the rejection of molecules of a molecular weight less than 500 is small, and the osmotic pressures are then of less importance. In the following a short description of the theory behind these processes, and the experience of how the theory generally works in practice is given.

HYPERFILTRATION

According to normally accepted laws the water flux (filtration rate) through a membrane is governed by:

$$J_a = K_a \cdot S(\Delta P - \Delta\Pi_m) \tag{1}$$

where J_a = permeate flux (litre/m^{2}h); K_a is a membrane constant; S is the membrane area; ΔP is the differential pressure over the membrane; $\Delta\Pi_m$ is the difference in osmotic pressure of the two sides of the membrane.

When the process takes place the concentration of rejected material on the membrane surface increases and will always be higher than in the bulk liquid on the concentrate side. It is therefore common to write formula (1) as:

$$J_a = K_a \cdot S(\Delta P - (\Pi_c \cdot M - \Pi_p)) \tag{1a}$$

where Π_c is osmotic pressure in the bulk concentrate; M is a 'polarisation concentration factor'; Π_p is the osmotic pressure on the permeate side.

The art of building a high capacity hyperfiltration plant is first of all to increase K as much as possible, i.e. making high capacity membranes, and to decrease M as much as possible, i.e. to construct a system where the membranes are flushed as effectively as possible with the lowest possible power consumption.

All membranes normally allow a certain passage of the ions and molecules they are to reject. The term 'permeability' (B) normally means:

$$B = \frac{\text{concentration in permeate}}{\text{concentration in bulk concentrate}} \times 100 \tag{2}$$

The term 'rejection' (D) normally means:

$$D = 100 - B \tag{3}$$

The permeability for a certain ion or molecule and a certain membrane depends on membrane characteristics, pressure difference, molecular size and shape, electrical loading, and whether the component is a solvent for the membrane material.

The flux of salts and other rejected material through the membrane is described by the formula:

$$J_s = K_s S(C_s^c - C_s^p) \tag{4}$$

where J_s is the flow of salt through the membrane; K_s is a membrane constant; S is the membrane area; C_s^c is the solute concentration on the membrane surface; C_s^p is the solute concentration in the permeate.

This formula states that the solute (salt) flow through the membrane is pressure independent, i.e. regarded as a diffusion process. This formula gives rather satisfying results when used on salts and hyperfiltration membranes with rather high rejection, giving the formula for permeability:

$$B = \frac{K_s S(C_s^c - C_s^p)}{C_s^c(K_a S(\Delta P - (\Pi_c \cdot M - \Pi_p)) + K_s \cdot S(C_s^c - C_s^p))} \tag{5}$$

and for cases where $C_s^c \gg C_s^p$ and $K_a \gg K_s$, therefore $\Pi_c \gg \Pi_p$:

$$B = \frac{K_s}{K_a(\Delta P - \Pi_c \cdot M)} \tag{6}$$

Results of experiments on sea water and salt solutions on our 985–999 membranes agree with this formula, i.e. qualitatively, but not quantitatively, giving the result that high pressure is an advantage in order to obtain low permeability, but the effect is lower than described by the formula.

Polarisation concentration

The polarisation concentration factor M (the ratio between concentration on the membrane surface and in the bulk concentrate) will destroy both flux and rejection, if it increases much above one.

The M factor depends first of all on the conditions of flow over the membrane, but also on viscosity, membrane flux and membrane rejection, increasing with growing viscosity and membrane flux, decreasing with falling rejection.

As is obvious from eqns (1a) and (6), it is important that M is kept low in order to get high flux and good rejection figures, but it is in many cases also important to keep it low in order to prevent scaling and crystallisation on the membrane surface, as increased concentration often means that the liquid on the membrane surface can be supersaturated, and therefore start to crystallise. If the liquid contains gel-forming compounds (pectins, proteins, etc.), a high M can mean gel formation on the membrane surface, causing formation of a secondary membrane on the membrane surface which can change the membrane characteristics completely.

In practice two types of flow are used to decrease M:

(1) Pure turbulent flow. Tubular reverse osmosis systems are the most typical example of this design.
(2) Thin channel systems, where the removal of boundary layers is based on liquids flowing in a very thin layer over the membrane over a short distance. The Reynolds' number belongs within the laminary region. Our plate and frame system, the spirally wound modules, the capillary modules, and some tubular modules with internal rods are examples of this type of design.

In the literature you will find discussions (Sourirajan, 1971; Murkes and Bohman, 1972; Srinivasan *et al.*, 1967; Gill *et al.*, 1969) as to whether one or the other type of design is best, seen from a theoretical point of view. The difficulty in solving this problem from this viewpoint is that whereas the turbulent tubular design can be treated rather satisfactorily in this way (Sourirajan, 1971; Murkes and Bohman, 1972), at least for liquids where the viscosity remains unchanged with changing concentration, the thin channel designs that are actually used are very difficult or impossible to treat from a theoretical, mathematical point of view. The thin channel flow between two parallel membranes, however, has been subject to some mathematical treatment, showing that this type of flow should be of great advantage. On the other hand Murkes and Bohman (1972) doubt the advantage of laminar flow.

Practical experiments by Bixler and Cross (1969) clearly show the advantage of the thin channel design.

Pressure drop/percent permeate produced is the most important figure for a hyperfiltration plant in this respect, i.e. how much the pressure in the system drops every time 1% of the feed is transformed into permeate. This figure is called ΔP_p or pressure drop/percent water recovery:

$$\Delta P_p = \frac{F \cdot \Delta P_c}{Q} \cdot 100$$

where F is total permeate flux; Q is total feed; ΔP_c is pressure drop above plant on concentrate side.

The reason why this figure is so important, is that it determines the energy needed to flush the membranes and at the same time the maximum water recovery possible without use of booster pumps. In both cases it has a major effect both on investment costs and operating costs.

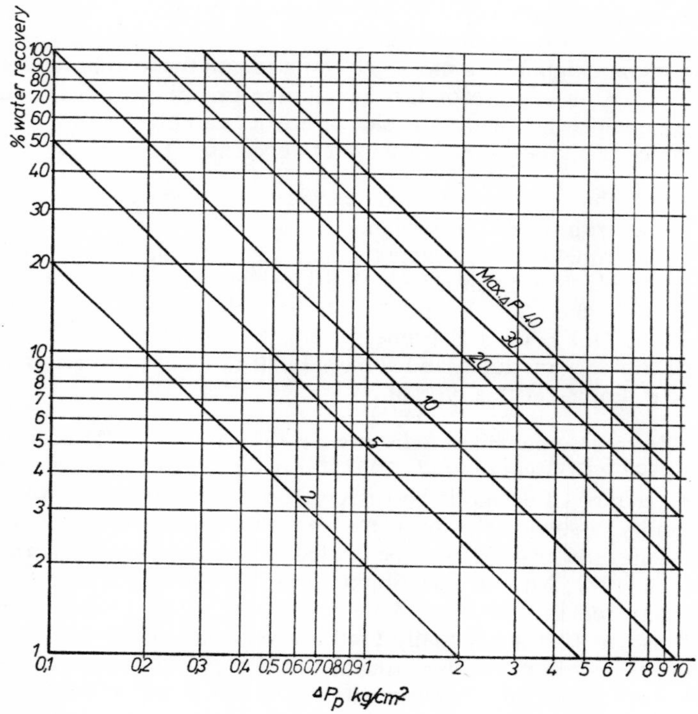

FIG. 1. Possible water recovery for definite ΔP_p and max. allowed differential pressures above the plant.

Figure 1 gives the possible water recovery for definite ΔP_p and the maximum allowed differential pressures above the plant, illustrating the importance of low ΔP_p figures very well.

Practical use of hyperfiltration systems shows that an irrecoverable decline of flux occurs if the membrane flushing is below a certain critical value (Sheppard and Thomas, 1970, 1972), showing that the boundary layer removal has to be above a certain critical value to prevent heavy scaling.

Sheppard and Thomas (1972) report a critical speed of 9 ft/s for untreated river water. The starting membrane flux is 30 gfd, decreasing to 15 gfd within a month for 0.52 inch I.D. and 2 ft long tubes. At a membrane flux of 10 gfd the critical velocity is 4 ft/s.

Calculating on the first figure the flow per tube is 22.5 litre/min, and the pressure drop is approx. 1.3 kg/m^2 per m^2 filtering area, corresponding to a pressure drop of 0.7–0.35 kg/cm/percent water recovery, if we assume that the tubes are arranged in series with no increased pressure drop from interconnecting pipes.

We have no direct comparison of this figure with thin channel plants, but as an illustrative example we have results from a 28 m^2 DDS (De Danske Sukkerfabrikker) module that treats water from a municipal biological sewage plant. With a feed volume of 45 litre/min, and a production of 14–16 litre/min permeate, corresponding to 720–820 litre/m^2/24 h (17.5–20 gfd), on a 95% NaCl rejection membrane, and an inlet pressure of 42 atm, and an outlet pressure of 33 atm, we have kept a constant capacity. It has been necessary to clean with detergent once a month, as the capacity drops from 820 to 720 litre/m^2/24 h during a month, regaining full capacity after cleaning. The pressure drop is in this case 0.25–0.29 kg/cm^2/percent water recovery, which indicates that this figure is of the same dimensions as for the tubular plant.

On whey hyperfiltration plants we have carefully examined the size of flow which it is necessary to have over the membranes in order to prevent a rapid and drastic decrease in capacity.

With one daily cleaning at 10°C we can keep a capacity of 310 litre/m^2/day ~ 7.5 gfd for a 2:1 concentration of a 6% TS (total solids) whey with a pressure loss/percent water recovery of 0.7–0.75 kg/cm^2; this is to be regarded as an average figure.

For a 1:4 concentration this figure increases to 1.1–1.2 kg/cm^2 due to increased viscosity.

Practical experience has shown that if this figure is lowered to 0.5 kg/cm^2/percent water recovery on a 1:2 concentration of whey, or 0.7 kg/cm^2/percent water recovery on a 1:4 concentration it will result in an unacceptable decline in flux within a few hours.

Another example from our own experience is concentration of egg white on DDS 40-modules. The feed is 30 litre/min, the permeate production 2.9 litre/min, the inlet pressure 30 kg/cm^2, the outlet pressure 0 kg/cm^2. ΔP_p is 3.1 kg/cm^2/percent water recovery. 3 × 90 membranes are arranged in series. The membranes in series have increasing water flux in order to keep the permeate flux on egg white constant with decreasing pressure in the module. A change to a membrane with a higher initial flux corresponding to $\Delta P_p = 5$ kg/cm^2/percent water recovery gives an irrecoverable flux drop due to protein layers.

All these examples show that for each application in a specific plant a specific ΔP_p is necessary in order to prevent a long term membrane fouling effect. Today we usually determine this minimum ΔP_p by long term experiments on the different fluids we are to treat. It would be a great advantage if it was possible to show the flow rates necessary to keep membranes clean on the basis of theoretical calculations, and, if needed, brief experiments.

In the literature on this subject (Sourirajan, 1971; Murkes and Bohman, 1972; Srinivasan *et al.*, 1967; Gill *et al.*, 1969; Bixler and Cross, 1969; Sheppard and Thomas, 1970, 1972), the theoretically most economic M factors have been calculated. The flow rates stated in these papers are in practice only valid if clean solutions with no scale- or gel-forming material are used. In order to reduce scaling and gel formation to a minimum it is, according to our experience, necessary in most practical cases to work with higher flows over the membranes, in theory giving M factors very close to one for the main solute to be removed.

The rather sharp limit for increased membrane fouling found in practice can be explained theoretically as follows.

When colloidal gel particles come to the surface, they will immediately work as a secondary membrane, which means that part of the total differential pressure between feed and permeate is now above the gel particles, pressing them down towards the membrane surface. As the gel particle is rather large-sized, and the force pressing the particles down towards the membrane can be strong, diffusion forces have no possibility of removing the particle from the boundary layer, and so the only possibility of removing the gel particle from the membrane surface is if it slides along the wall. In order to start the motion of such a particle, the forces deriving from the pressure drop must be stronger than the friction between the particle and the wall; if not, the particle will remain stationary.

If this theory is correct, it means that the parameters of importance for the prevention of this type of fouling are water permeability for membrane and gel particle, pressure drop above the membrane, pressure drop along the membrane, and the coefficient of friction. The important factor in system design is to get the highest possible pressure drop along the membrane with the lowest possible power consumption, and to get as short a flow along the membrane as possible. This is best obtained with the thin channel design.

ULTRAFILTRATION

What previously has been said about hyperfiltration plant performance is valid for ultrafiltration plants, too, except for the osmotic pressure being so low that it is of no practical importance for the process.

Flow pressure dependence

Figure 2 shows a typical flow pressure performance for an ultrafiltration plant. It is seen that when the pressure increases, flux increases proportionally at first, but very rapidly reaches a constant level. The figure also shows that this level not only depends on the membrane, but first of all on the flow along the membrane. The reason for this is that the gel formation on the membrane forms a secondary membrane on the surface which increases with the pressure (and increasing flux), but decreases with increasing flow over the membrane.

The thin channel flow has, according to the experience of Michaels (Flinn, 1970; Bixler and Cross 1969) and our own experience, very great

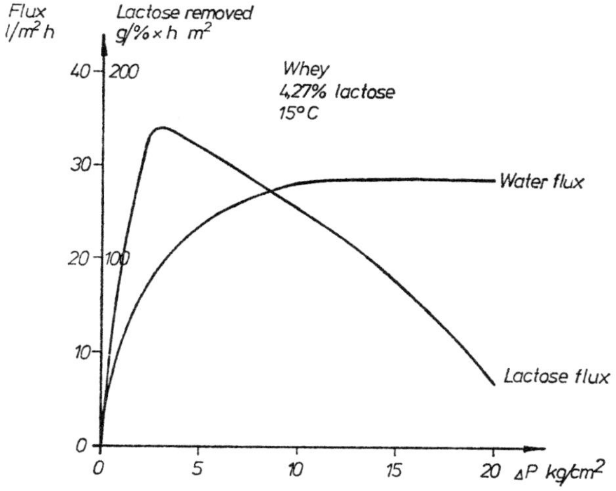

FIG. 2. Typical flux-pressure figures for whey ultrafiltration. The lactose flux is in gram percent lactose in the whey, per hour, and per m^2 filtering area.

advantages in connection with the removal of this secondary membrane, especially in high viscosity fluids.

Rejection figures in ultrafiltration are normally required, varying from a rejection of 100% for some components to a rejection of 0% for other components. Timmins (Symposium, 1973) Michaels (Flinn, 1970) and others give curves for membrane rejection of components with different molecular weights for different membranes. These papers as well as our own experience show that each membrane changes rejection from 0 to 100 over a rather wide range of molecular weights, and also the electrical loading and the shape of a molecule is of importance for the rejection of different molecules.

The case is further complicated, as in addition to this the membrane changes characteristics when pressure and other operating characteristics are changed.

DDS HYPERFILTRATION AND ULTRAFILTRATION

I have now given a brief introduction into some of the theory behind the processes, and will now describe our membranes and equipment, and give some practical examples.

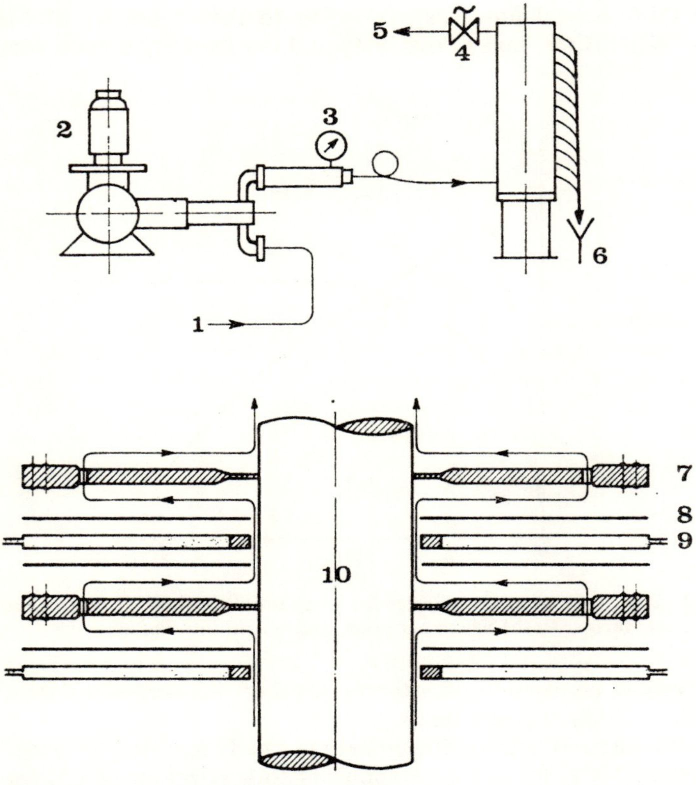

FIG. 3. Construction of the DDS system. 1, inlet; 2, pump; 3, manometer; 4, pressure relief valve; 5, concentrate outlet; 6, permeate outlet; 7, membrane spacer; 8, membrane; 9, membrane support plate; 10, centre bolt.

The DDS system is constructed as a plate-and-frame system (Fig. 3) where the liquid to be concentrated flows in a narrow channel (0.3–0.5 mm) over the membranes radially and only in contact with the same membrane in a rather short pass (~150 mm). The liquid normally has an average velocity of about 0.5 m/s, and the flow is laminar.

In the systems based on turbulent flow the boundary layer has to be removed by a liquid motion back into the bulk liquid. In the short pass laminar flow system the boundary layer is moved parallel to the membrane and is removed over the end of the membrane. For low viscosity liquids the two systems of boundary layer removal both function satisfactorily, and there is little difference in the performance. For higher viscosities the laminar flow short pass system has very great advantages, as it still functions in the normal way, whereas viscosity increase very quickly destroys the turbulent flow systems, as real turbulence is almost impossible to achieve with high viscosity.

In medical and food industry applications this is of great importance, because, as will be seen later, high concentrations in the end products are very important in order to get the correct results. Our experience shows that reasonable performance is possible with the laminar flow short pass system; a viscosity of up to 20 cP and treatment up to 80 cP is possible.

Equipment

The filters we use are shown in Figs. 4, 5 and 6.

We produce a range of laboratory and pilot equipment with modules 20 cm in diameter, covering a hydraulic easy-assembly laboratory module with 0.36 or 0.72 m^2 filtering area and pilot modules with 1.8 and 5.4 m^2 filtering area. In addition we produce a range of modules for technical applications 40 cm in diameter and with filtering areas from 4 to 28 m^2 per module. The latter types of module are best suited for most applications within the food industry; the module can be delivered in a fully sanitary construction.

At the Achema exhibition we introduced a new 30 cm module with $\sim$20 m^2 filtering area. This module is especially intended for industrial high pressure applications, first of all for water purification and then for applications with suspended matter in the liquid to be treated; for most ultrafiltration applications it has not as good a performance as the 40 cm module.

The liquid volume on the concentrate side is a very important figure, because it determines:

(1) The minimum time of treatment.
(2) The necessary volumes of flushing water and cleaning solutions.

In Table I we have compared the volumes and times for the DDS system, a 10 mm$^{\varnothing}$, and a 30 mm$^{\varnothing}$ tubular system, and some practical examples are calculated.

When skim-milk is ultrafiltered for soft cheese production, a concentration ratio of 1:6 is normally necessary. An average flux of 7–10 litre/m^2/h is normal. In a continuous system this means an average treatment time in the DDS system of 30 min, from when the skim-milk enters the plant until the same skim-milk on average leaves the plant, whereas a 30 mm$^{\varnothing}$ tubular system has an average hold-up time of approx. 170 minutes. This is important for the bacteriological quality of the product.

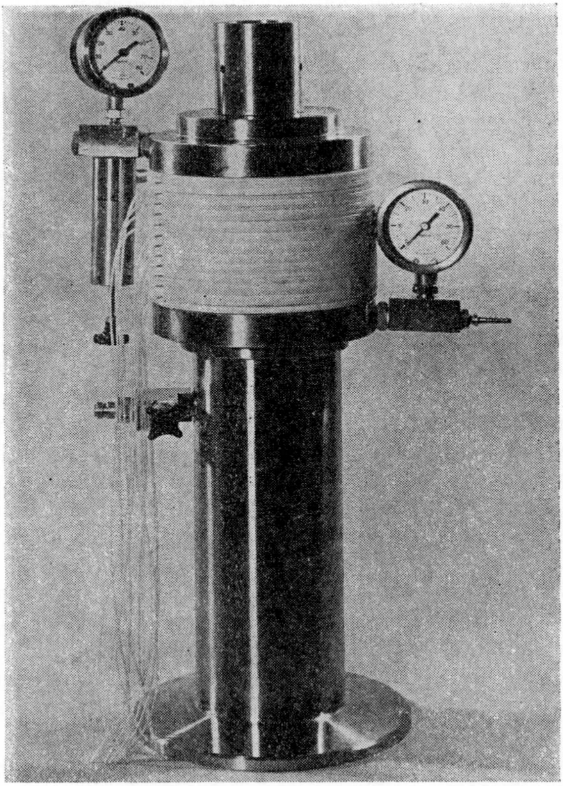

FIG. 4. An 0.30 m^2 LAB-module with hydraulic assembly.

The DDS system treats for this example a volume of approx. 5% of the produced concentrate in 20 h. When a system is stopped for cleaning, this volume is of extreme importance, because it is proportional to the amount of flushing water needed, the losses under equal conditions, and the amount of cleaning solution needed. Under the same conditions the 30 mm$^{\varnothing}$ tubular system treats a volume of 28% of the produced concentrate in 20 h.

Table I gives similar figures for a whey ultrafiltration 1:20.

Further, all our modules have the following characteristics.

(1) Each pair of membranes can be individually controlled, and in case the permeate is unsatisfactory, it can be returned or stopped.
(2) Membrane exchange is possible on site giving a very reasonable membrane exchange price ($\sim$150 D.kr./m^2 for the 40 cm modules). Membrane exchange can be accomplished in one working day.

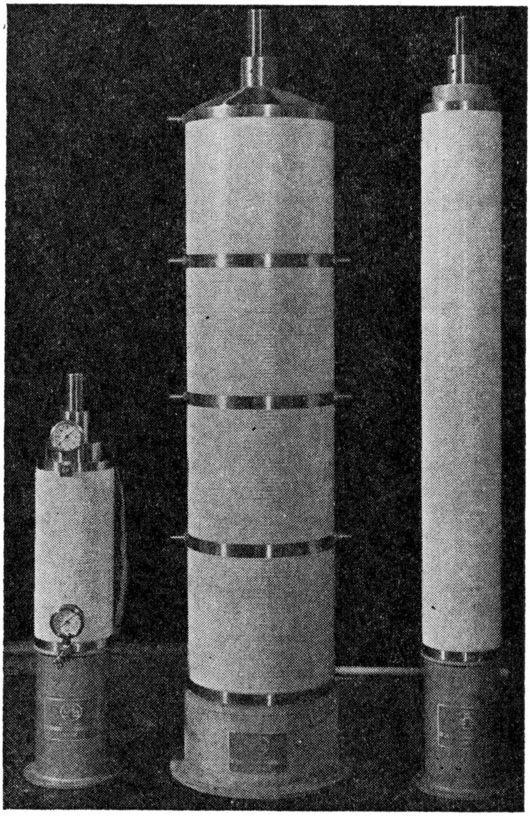

FIG. 5. 3 DDS modules, 7 m^2, 28 m^2, and 5.4 m^2 columns.

(3) All our modules can be modified in such a way that we can determine how many membranes can be run in series with the same liquid. This is important, because it allows us to use the best possible compromise for the pumps between the amount of liquid pumped and the pressure loss in the system (pump pressure).

Membranes

In Table II are shown the more important types of membrane produced by DDS. Types 985, 995, and 999 are hyperfiltration membranes, and 800, 600 and 500 are ultrafiltration membranes. All these membranes are made

FIG. 6. A 30 cm module for operation pressures up to 80 kg/cm^2.

of cellulose acetate. We have several years of experience with these membranes and are able to give performance guarantees for a number of applications with the membranes. An important fact is that all the membranes are truly anisotropic with an extremely thin active layer preventing internal clogging in the membranes.

AR8 is a new ultrafiltration membrane, it is also of cellulosic origin and truly anisotropic. It has the great advantage of being alkali resistant up to

1% NaOH solutions, whereas the other membranes are only resistant up to pH 7–8. The other characteristics of this membrane are almost the same as for the membrane type 800. For some applications on protein solutions the new membrane type AR8 has great advantages.

TABLE I

	DDS *Module 40*	*Tubular* *10* mm$^{\varnothing}$	*30* mm$^{\varnothing}$
Theoretical concentrate volume			
litre/m^2	0.4	2.5	7.5
U.S. gal/ft^2	0.01	0.063	0.19
Practical concentrate volume including piping			
litre/m^2	1.5	3.6[a]	8.6[a]
U.S. gal/ft^2	0.04	0.091	0.22
Average treatment time for continuous plant at concentration ratio 1:6 flux 8 litre/m^2/h ~4.7 U.S. gal/ft^2/24 h	0.49 h	1.2 h	2.8 h (~skim-milk)
Concentrate volume % produced concentrate 20 h	4.9%	11.6%	28%
Average treatment time for continuous plant at concentration ratio 1:20 flux 20 litre/m^2/h ~12 U.S. gal/ft^2/24 h	0.31 h	0.77 h	1.8 h (~whey)
Concentrate volume % produced concentrate 20 h	7.2%	17.2%	41%

[a] Figures adding the same volume for 'pipe volume' as for DDS system. In practice these figures are probably higher.

Design of plants

Plants for ultrafiltration as well as hyperfiltration can be built both as batch and continuous plants.

Figure 7 shows the normal layout of a batch plant, and Fig. 8 shows the layout of a continuous plant.

The flux (permeate flow) depends very much on the concentration of the liquid. In order to get the same average capacity with a continuous plant as with a batch plant, it is necessary to have several steps in the continuous plant. In the continuous plant the feeding pump provides the back pressure for the modules. The recirculation pumps, which give normally a much

TABLE II

SALT REJECTION AND OPERATING CHARACTERISTICS FOR DIFFERENT DDS MEMBRANES

Membrane type	*Na-rejection* %	*Sucrose-rejection* %	100% *rejection of*	*Max. operating pressure* (kg/cm^2)	*Capacity of UF/HF permeate* ($litres/m^2/24$ h)
AR 600[e]	0	5	MW 20 000, rennin, trypsin	20	5 800[c]
AR 800[e]	5	15	MW 6 000, insulin, pepsin	20	2 000[c]
500	0	0	MW 64 500, haemoglobin	10	6 000[d]
600	0	5	MW 20 000, rennin, trypsin	20	5 500[c]
800	5	15	MW 6 000, insulin, pepsin	20	1 800[c]
870	65	90	MW 1 000, peptides	50	1 900[b]
975	78	99.5	MW 200, saccharose	50	2 200[b]
985	87	100	MW 200, saccharose	50	2 000[b]
990	92	100	MW 200, saccharose	50[a]	1 400[b]
995	97	100	MW 200, saccharose	50[a]	1 000[b]
999	99	100	MW 200, saccharose	80[a]	350[b]

pH operating range: 3–7.

[a] 100 kg/cm^2 allowable with reduced lifetime.

[b] Measured at 42 kg/cm^2.

[c] Measured at 10 kg/cm^2.

[d] Measured at 8 kg/cm^2.

[e] pH operating range: 3–14.

The above data were measured after 24 h operation in a 20 cm DDS UF/HF module, equipped with 30 membranes with a total membrane area of 0.54 m^2. All results are measured at 25°C. The influent was tap water with a hardness of 22.7°dH (German Degrees of Hardness) and a conductivity of 1500 μmho/cm.

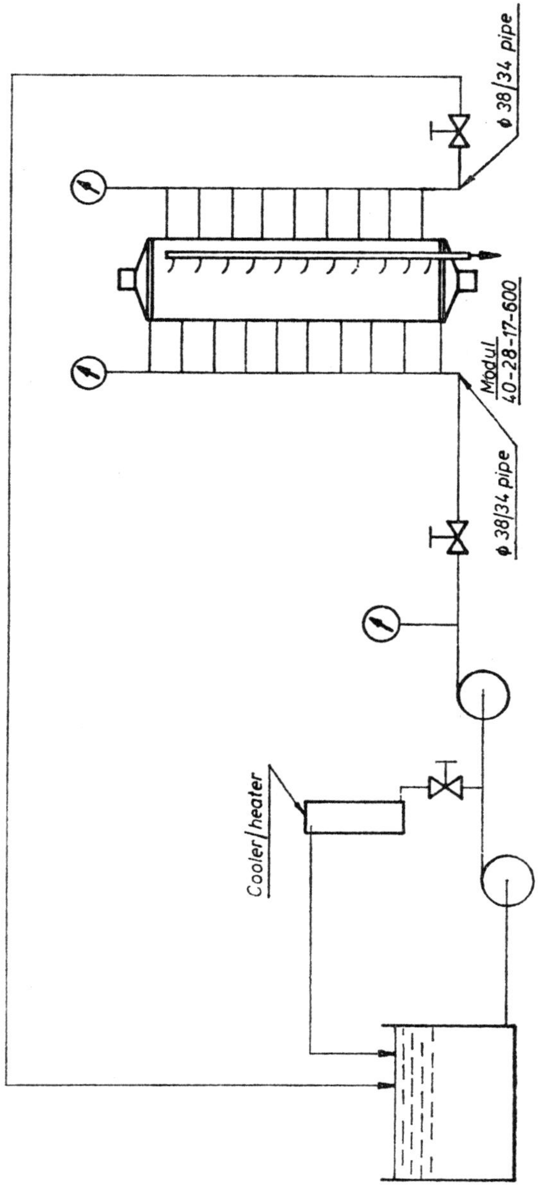

FIG. 7. The normal lay-out of a batch plant.

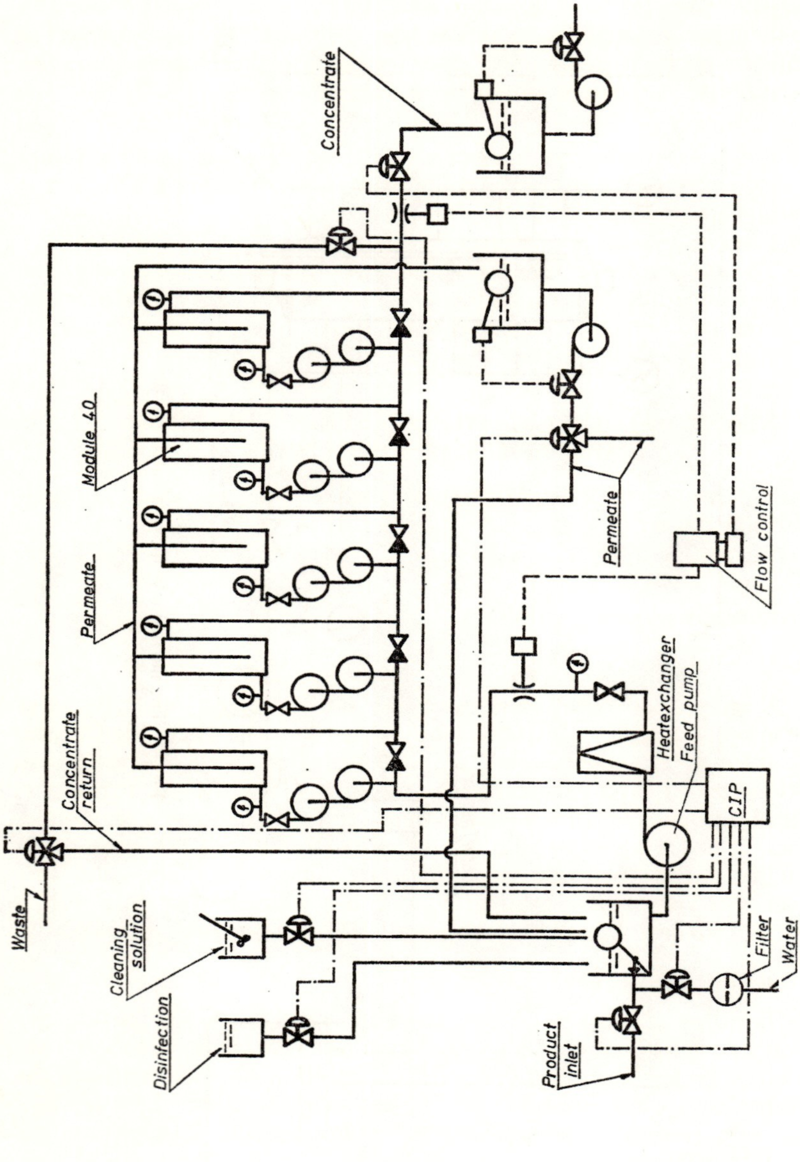

FIG. 8. The lay-out of a continuous plant.

higher flow than the feeding pump, have only to give pressure enough to press the liquid through the modules. Thus the electricity consumption for continuous plants is much lower than for batch plants. Furthermore the continuous plants have the following advantages:

(1) The hold-up time in the plant is very short, thus treatment at a little higher temperature than in batch plants is often allowed. This means higher capacity.
(2) The concentrate is produced continuously.
(3) The concentration ratio can easily be controlled automatically.
(4) For hyperfiltration plants the use of a relatively small high pressure pump and booster pumps means a lower investment cost.

INDUSTRIAL APPLICATIONS WITHIN THE FOOD AND MEDICAL INDUSTRIES

In the following I shall give a few examples of processes which are used industrially today.

Hyperfiltration (reverse osmosis)

WHEY

After some years of pilot plant operation the first large industrial plant came into operation in May, 1972, at the dairy Val d'Or at St. Aignan-des-Gues in France. Figure 9 is a picture of this plant. The plant concentrates 80–90 t whey/day at a concentration ratio of 1:4. The average capacity is 220 litre permeate/m^2/20 h, and experience so far has shown a membrane lifetime of at least 8 months at this capacity. Experience has shown that the whey has to be centrifuged in self-cleaning centrifuges and low pasteurised. The plant is a batch plant and is cleaned after each ~20 h operation. Figure 10 gives operating figures for the first year of operation.

THE USE OF HYPERFILTRATION ON WHEY

Hyperfiltration of whey is an alternative to evaporation. The advantages are low operation costs, simple installation and low labour consumption; also steam boilers are not necessary. This makes it possible to install the equipment at normal cheese dairies. The disadvantages are that the capacity varies with concentration, and the maximum possible concentration is around 25% TS.

Examples of cases where hyperfiltration of whey is justified are as follows.

Preconcentration before transport

If whey is preconcentrated at the dairy before transport to a drying plant or a lactose plant, the transport costs, which are often D.kr. 20–40/1000 litre ~ \$12–25/1000 USG removed permeate, are saved and the same evaporation is saved at the drying or lactose plant.

Preconcentration before delivery back to farmers

Transport costs are saved. The pigs' consumption of preconcentrated whey may be bigger than that of ordinary whey, as the consumption of this is limited by the large volume of liquid, and thereby the use can be increased.

FIG. 9. Industrial plant at the dairy Val d'Or at St. Aignan des Gues in France. Hyperfiltration of 100 t whey/day 4:1.

Concentration to skim-milk TS content and mixing with skim-milk before delivery to the farmers is often an excellent solution.

Direct use of whey concentrate instead of whey powder

Whey concentrate from hyperfiltration can be used instead of whey powder for many products; for instance ice cream, saving drying costs.

Preconcentration before electrodialysis

To increase electrodialysis capacity it is an advantage to preconcentrate the whey.

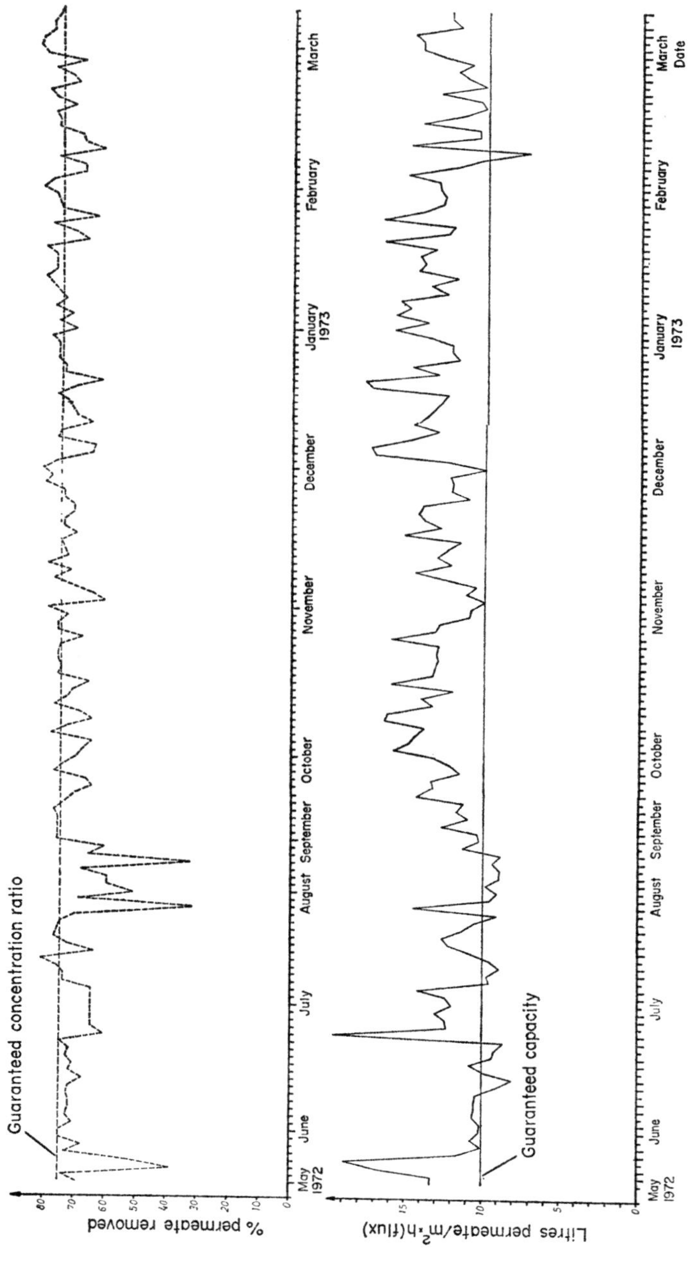

FIG. 10. Operating figures for the first year of operation of the plant in France.

COMPARISON BETWEEN HF AND EVAPORATION OF WHEY

Subject	*Hyperfiltration*	*Evaporation*
Steam consumption	0 (only for pasteurisation)	250–550 kg per 1 000 litre evaporated water
Electricity consumption	Continuous plants: approx. 10 kWh per 1 000 litre removed water. Batch plants: up to 20 kWh per 1 000 litre removed water	Varies. Often approx. 5 kWh per 1 000 litre removed water
Labour	4 h per day	Normally two operators during the whole operation. (Boiler house and evaporator)
Cooling water consumption	Continuous plants: 0–7 000 kcal per 1 000 litre removed water. Batch plants: 0–14 000 kcal per 100 litre removed water	125 000–290 000 kcal per 1 000 litre removed water
Waste and wash water. BOD and heat pollution	Sweet whey: 0.2–0.8% of the whey BOD in total losses. Acid whey: 1–3% of the whey BOD in total losses. Efficient security against high BOD by mistakes. No heat contamination	BOD varies with evaporator drop catchers and cleaning procedure. Overboiling gives extreme high BOD. Hot waste water
Economical plant size	From 6 000 litre per day and upwards with no limit	From 80 000 to 100 000 litre per day
Concentration in final product	Max. 25% TS. Capacity varies with concentration	Up to 60% TS

MEMBRANE CHOICE, WASTE WATER BOD

The normal membrane used for hyperfiltration of whey is membrane type 985. BOD (BiO_5) in the permeate from this membrane is during a concentration on sweet whey 200–500 mg O_2/litre corresponding to 0.2–0.8%

of the BOD in the whey ending in the permeate and waste water. The lactose losses are less than 0.1% of the lactose content.

With membrane 990 these figures can be reduced about 25%, but this is normally not justified, as the capacity is decreased 25%. On acid whey the BOD in waste water is only decreased to 1–3% of BOD in original whey, because the membranes have low rejection of acids.

Hyperfiltration 4:1 reduces acidity in acid whey by 15–25%.

AUTOMATICS

DDS delivers pressure relief valves, automatics and automatic CIP (Cleaning in Place) cleaning of our own design for the plants, reducing labour consumption to 1–4 h/day, even for a big plant.

PLANT DESIGN

Hyperfiltration plants for whey can in principle be built by two different methods.

Batch plants

Batch plants have the advantage that such small plants, from 1 to 4 modules, 40–28.8–2–985 treating for 6–15 h per day, are making it possible, in an economical way, to preconcentrate whey, even at very small dairies.

Figure 11 shows a diagram, with dimensions of a small batch plant with 2–4 modules.

The whey is recycled through the modules back to the storage tank. The feed for each module has to be 54 litre/min. No special automatics is required and the plant can operate overnight without an operator.

The operating costs are somewhat higher than for the continuous plants.

Continuous plants

In continuous plants the whey is only pumped once under high pressure and then in a number of steps recycled with centrifugal type booster pumps over a number of modules, see Figs. 12 and 13. The advantages are low installation and operating costs, continuous production and low hold-up time allowing operating temperatures up to 25°C.

The number of steps needed depends on the concentration ratio. With a low number of steps, the modules will work on too high a concentration, giving too low a permeate flux.

The following table gives the guaranteed capacity in percent for different numbers of steps and different concentration ratios compared with capacity for batch plants operating at 12°C = 100%. The temperature at continuous operation is from 12°C to the maximum of 25°C, and the heating is the sole product of pump energy.

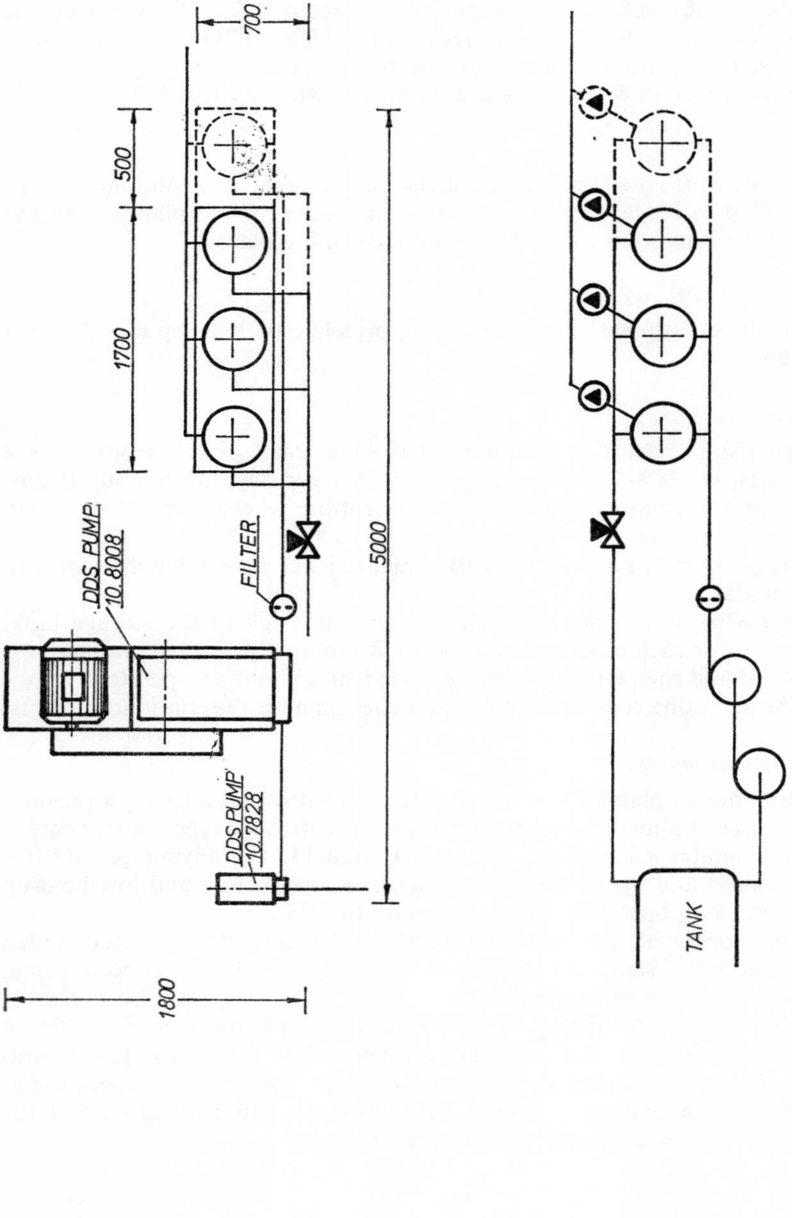

FIG. 11. Dimensions and diagram for a small batch plant with 2–4 modules.

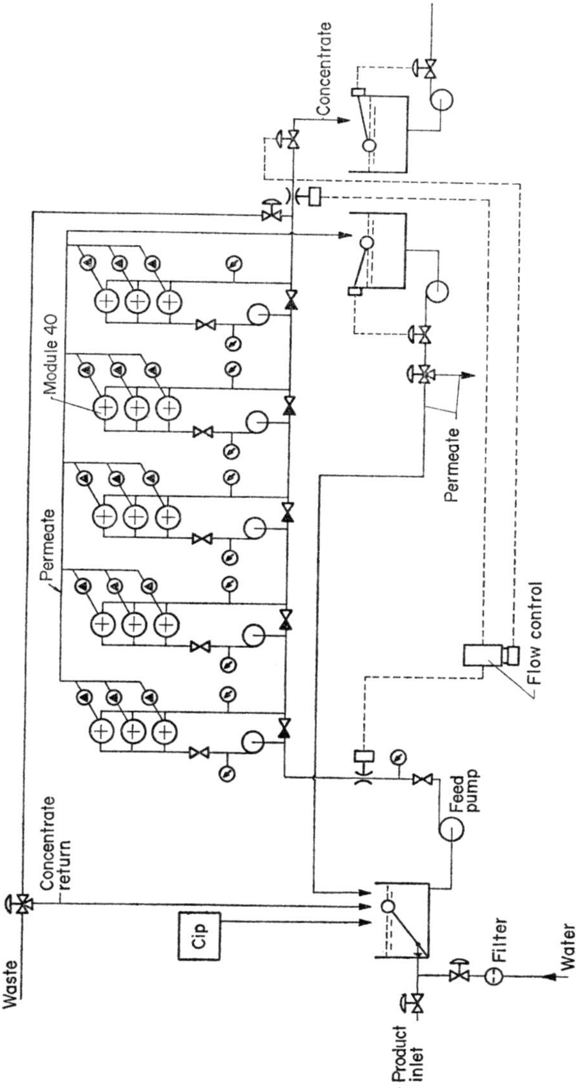

FIG. 12. Continuous UF-system with CIP cleaning.

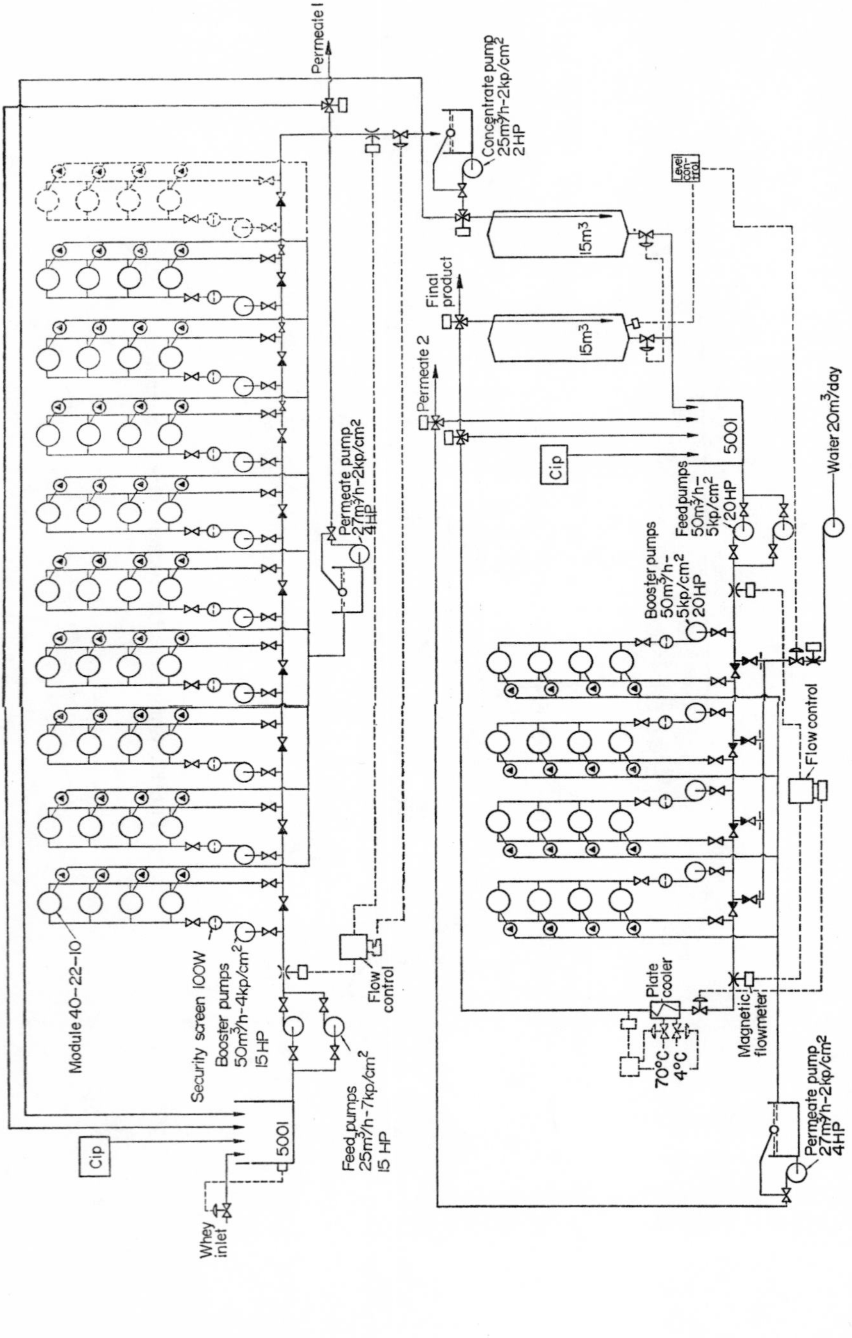

FIG. 13. UF plant with wash for production of 70% whey proteins from 4–500 t whey/day.

No. of steps	Concentration ratio 2:1	3:1	4:1
1	80	45	30
2	95	75	70
3	100	85	80
4	100	100	90
5	100	105	100
6	100	105	100

The table shows that 2 steps are reasonable at 2:1 concentration, 4 steps at 3:1 concentration, and 4–5 steps at 4:1 concentration.

The continuous plants can, with a ratio control between inlet and outlet flow, easily be controlled to give a constant product.

GUARANTEED CAPACITIES FOR HF OF WHEY

Guaranteed capacity in 1000 litre treated whey per day equal to 20 h operation, based upon sweet pasteurised (72°C, 15 s) whey with pH >6.2, fat content <0.05%, and membrane 985 is as follows.

% TS in start	Concentration ratio 2:1	2.5:1	3:1	3.5:1	4:1
6.5	14.0	10.2	8.2	6.7	
6.0	14.7	11.1	9.2	7.5	6.2
5.5	15.4	11.8	10.2	8.4	7.0
5.0	16.4	12.8	11.2	9.5	8.2
4.5	17.4	13.6	11.8	10.1	9.8

Capacity per module 40–28–3.

When acid whey with pH >4.5 is treated on membrane 985 the guaranteed capacities are to be reduced by 25%.

For membrane 990 and sweet whey with pH >6.2 the guaranteed capacities are to be reduced by 25%; with acid whey and pH >4.5 the reduction is 35%.

Electricity consumption

CONTINUOUS PLANTS

In the continuous plants the whey is pumped by a feed pump to 25–30 kp/cm^2 and then with the high pressure DDS booster pump No. 10.7880 recycled over up to 4 modules 40–28–3 for each booster pump. The electricity consumption in kWh per day and per 1000 litre permeate will, including cleaning, be:

Treated l whey per module	*Concentration ratio*		
	2:1	*3:1*	*4:1*
6 000			16
8 000		14	12.5
10 000		11.5	
12 000	13	10	
14 000	11.5		
16 000	10.5		

BATCH PLANTS

In the batch plants the electricity consumption will vary with configuration. Small batch plants (< 5 modules) have approximately twice the electricity consumption of continuous plants, bigger plants have decreasing electricity consumption to 1.5 times the consumption of continuous plants.

Cooling water consumption

CONCENTRATION AT CONSTANT TEMPERATURE

For both batch and continuous operation where constant temperature is required, cooling, normally with ice water, is necessary according to the heat introduced by pumping. This will be for continuous plants in kcal per day and per 1000 litre permeate.

Treated l whey per module	*Concentration ratio*		
	2:1	*3:1*	*4:1*
6 000			11 000
8 000		9 700	8 600
10 000		8 000	
12 000	9 000	6 900	
14 000	8 000		
16 000	7 300		

For batch plants the cooling water consumption will be 1.5–2 times as high. To this must be added the possible cooling before the product reaches the plant to give the necessary storage temperature.

CONCENTRATION STARTING AT 12°C ALLOWING THE CONCENTRATE TO LEAVE A CONTINUOUS PLANT AT MAX. 25°C

Concentration in this way often prevents internal cooling. The following table gives approximate concentrate temperature, internal cooling in kcal per 1000 litre permeate, and external cooling per 1000 litre permeate to cool the permeate to 12°C.

	Concentration ratio								
	2:1			*3:1*			*4:1*		
Treated l whey per module	*conc. temp.*	kcal *int.*	kcal *ext.*	*conc. temp.*	kcal *int.*	kcal *ext.*	*conc. temp.*	kcal *int.*	kcal *ext.*
6 000				25	500	6 500	25	4 100	4 300
8 000				24	0	6 000	25	0	4 300
10 000				22	0	5 000	23	0	3 700
12 000	18	0	6 000	20	0	4 000			
14 000	17	0	5 000						
16 000	16	0	4 000						

Cleaning solutions

Each module in a reasonably sized plant has, including piping, pumps, etc., an internal volume of approximately 30 litres.

With existing prices the costs for cleaning solutions are approximately D.kr. 7.50 per day and per module and D.kr. 2.20 per day and per module for sterilising solutions, a total of D.kr. 9.70 per day and per module = US $1.55. All plants have in this respect to be calculated with one module more than is installed to allow for tank volume, etc. Water consumption for flushing is 1–1.5 m³ per day and per module.

Hyperfiltration and evaporation, economical comparison

Operating costs for continuous hyperfiltration plants are a total of 6.80–7.40 D.kr. per 1000 litre permeate. 20% interest + depreciation gives a total of 9–10 D.kr. per 1000 litre permeate.

The difficulty in making a fair comparison between hyperfiltration and evaporation is that whereas the hyperfiltration plant is almost a self-contained unit, evaporation involves also steam production, cooling water supplies, etc.

An evaporator vaporising 7.5 t/h for 20 h per day has together with the boiler house approximately the following operating costs:

2 shift workers in 24 h/day at D.kr. 25/h (boiler house and evaporator)	=	D.kr. 1 200	
Fuel oil 3.0 t at D.kr. 220	=	660	
Cleaning agents	=	50	
Cooling approx. 30 000 000 ~ 1 500 m^3 cooling water at D.kr. 0.30	=	450	
+Ice water, if a cold product is needed			
Electricity consumption about 800 kWh at D.kr. 0.125	=	100	D.kr. 2 460
Maintenance			100
		Total	D.kr. 2 560

equal to more than D.kr. 17.10 per 1 000 litre evaporated.

Approximate investment:

Evaporator	D.kr. 700 000
Boiler	250 000
Cooling water system	100 000
Pipe installations, etc.	200 000
	D.kr. 1 250 000

20% interest + depreciation per year = D.kr. 250 000 will be equal to D.kr. 6.65 per 1000 litre produced condensate. Compared with evaporation the results for hyperfiltration are per year with 250 operation days:

	Hyperfiltration	*Evaporation*
Operating costs	D.kr. 277 000	703 000
Interest + depreciation	336 000	250 000
	D.kr. 613 000	953 000

or D.kr. 340 000 in favour of hyperfiltration.

The difference in investment is:

D.kr. 1 680 000 − 1 250 000 = D.kr. 430 000

SKIM-MILK

Several small industrial plants are now in operation for the hyperfiltration of skim-milk. Although you may think the main application would be preconcentration before drying, the main application so far has in fact been preconcentration before production of ice-cream.

Normally skim-milk powder is used to give an increase of the fat-free dry substance in ice-cream. Instead of this a hyperfiltration of skim-milk is possible. The advantages are:

(1) Better flavour of the ice-cream, all 'powder' taste is removed.

(2) Better texture, allowing reduced fat content to obtain the same quality.
(3) Great economical savings.

Each 28 m^2 module can treat 9000 litre skim-milk producing ~3900 litre permeate/day. In normal Danish ice-cream mix this means that skim-milk powder corresponding to 4000 litre skim-milk/day can be substituted by skim-milk. According to Danish conditions this is a saving of approximately D.kr. 900 per day and per module. The operating costs are approx. D.kr. 100/day. The pay-back time for the plant is about 150 days.

For dessert mix the savings are equivalent.

EGG WHITE

Several industrial plants for the concentration of egg white are installed with our modules used by the Danish company SANOVO. The capacity is approximately 7 t egg white concentrated to half volume in 20 h.

A special membrane configuration and special care with the pumping has been necessary in order to prevent destruction of the whipping properties.

Egg white powder produced from concentrate has a higher bulk density and better whipping properties than direct spray dried egg whites.

Ultrafiltration

Although the above hyperfiltration applications are important today the most important applications within the food and medical industries are probably in the ultrafiltration field.

DAIRY INDUSTRY

Today ultrafiltration of skim-milk and whey is used for several different applications.

Skim-milk

This is ultrafiltrated with our membrane type 600. A concentration ratio of 1:3 is possible at 2–4°C, 1:6 or 7 is possible if the last part of the ultrafiltration takes place at 50°C. The capacity of the plant is 6.0–7.5 litre permeate/m^2/h.

Our 40 cm modules with 17 intermediate flanges are used for this application. Normal inlet pressure is 5–6 kg/cm^2.

Many pilot plants and semi-technical plants have been in operation for more than one year, and the first technical continuous plant was started in February, 1973, in Germany.

Ultrafiltrated skim-milk concentrate is used for:

(1) spray- or freeze-dried protein powder for dietetic products;
(2) junket, quarg, or cheese manufacture according to the INRA/ANVAR (Maubois and Mocquot, 1971) method, where whey proteins remain in the product increasing the protein output by 15–20%.

FIG. 14. A commercial plant for UF of skim-milk.

For this ultrafiltration we use our 40 cm diameter modules with 17 intermediate flanges giving 18 parallel runs of 1 m^2 filtering surface each, and the membrane we use is our 600 membrane (cellulose acetate). Figure 14 shows a commercial plant for the ultrafiltration of skim-milk.

The pump we use is a centrifugal pump with a pressure up to 5 kg/cm^2. The flow of liquid through the module is laminar, and it is dependent on the pressure drop, i.e. viscosity.

ULTRAFILTRATION FOR CHEESE PRODUCTION

Camembert cheese

The main reason for interest in skim-milk ultrafiltration is the new invention by Maubois and Mocquot (1971) of a cheese manufacturing method where skim-milk is ultrafiltrated before the curd is produced.

Ultrafiltration

Skim-milk is concentrated 6–7 times volumetrically until the required protein/dry matter content has been obtained.

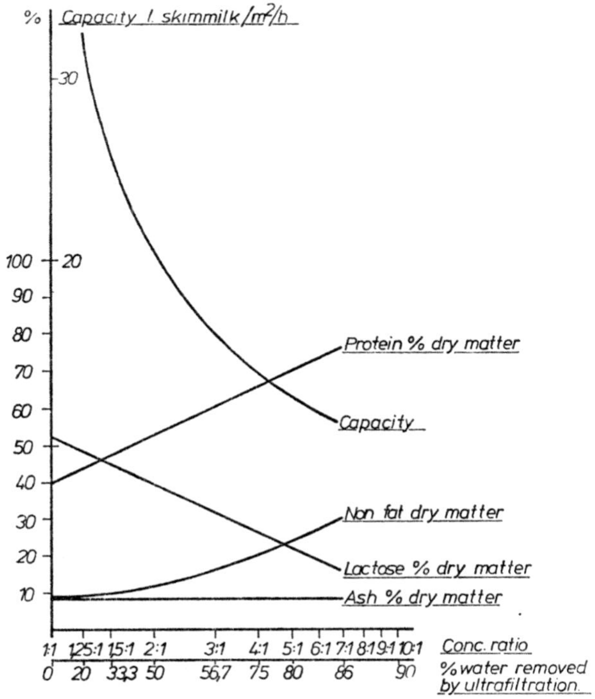

FIG. 15. The composition of protein-enriched concentrate as a function of the concentration ratio.

Figure 15 shows the composition of the protein-enriched concentrate as a function of the concentration ratio. The dry matter content and the capacity at the various concentration ratios are also shown in this figure.

From the figure it appears that by removal of 86% water a concentrate with a protein content of about 75% can be obtained without any washing.

CHEESE PRODUCTION

The ultrafiltrate concentrate is mixed with cream, starter, and renin enzyme, whereupon the mixture is put directly into moulds, where the cheese at once assumes its final form.

By using this method the whey proteins remain in the cheese, which gives about 20% greater yield, and the discharge of whey is reduced to zero.

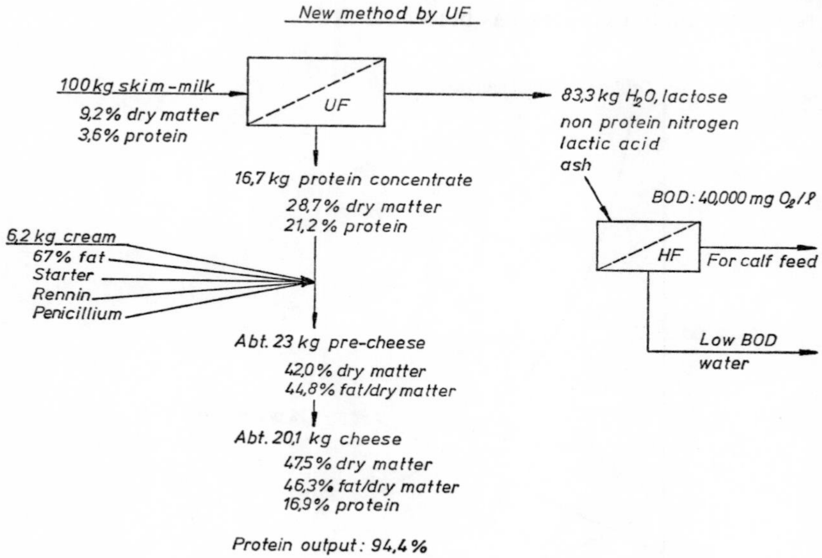

FIG. 16 (a and b). Mass balance for production of Camembert cheese.

Figures 16a and 16b show the mass balance for production of Camembert cheese.

Furthermore, it appears from the figures that the total pollution degree is reduced by nearly a quarter due to the reduction in BOD and to the reduction in the amount of the produced whey.

Operating experience—skim-milk ultrafiltration

The two biggest and most important problems in ultrafiltration of skim-milk are:

(1) to obtain an efficient cleaning procedure;
(2) to remove sufficient lactose with the permeate during the ultrafiltration.

Cleaning of the system

Within the dairy industry the efficiency of the cleaning agent is very important and decisive for the use of the process and for the equipment as well.

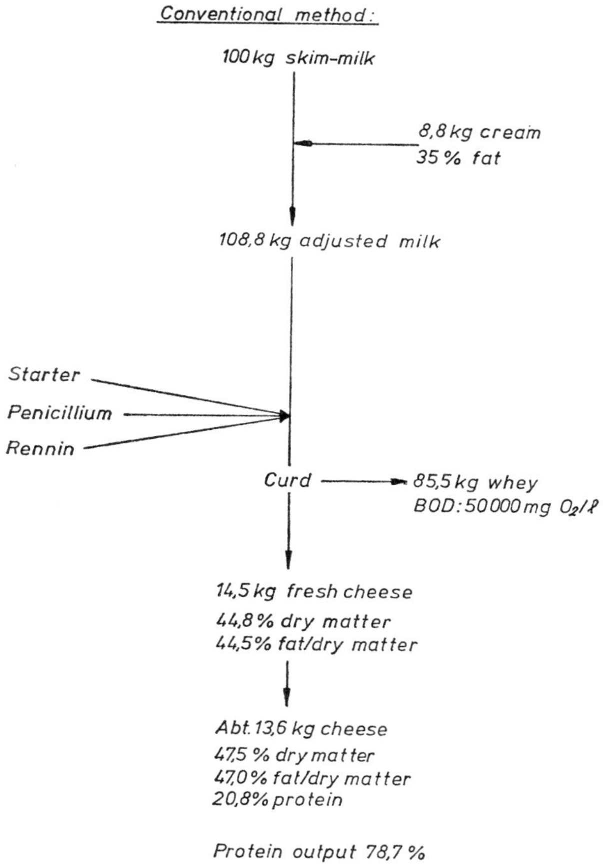

FIG. 16 (b).

The inexpensive cleaning agents normally used, such as caustic soda and nitric acid, cannot be used on cellulose acetate membranes. This gives a limitation on both the pH and the temperature.

The knowledge of the character of the coatings (protein/fat/various minerals) has made it necessary to use comparatively expensive cleaning agents such as proteinase enzymes and detergents.

As a disinfectant hypochlorite, chloramine, iodophor or formalin can be used, all dependent on what is permitted in the country concerned.

The new alkaline resistant membranes can be cleaned with caustic soda.

Lactose removal

One of the main problems in skim-milk ultrafiltration is the fact that in order to get the correct products it is essential that the lactose concentration

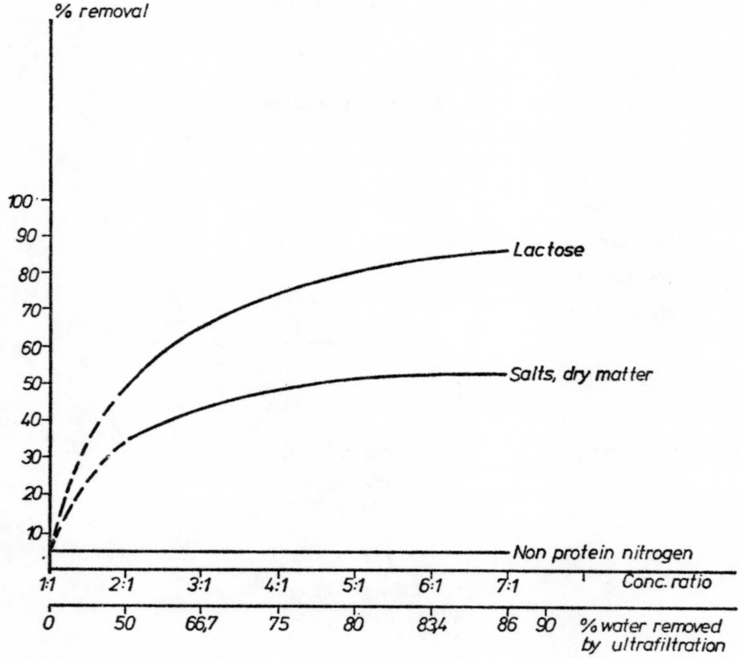

FIG. 17. Reduction of different components in the skim-milk as a function of the volumetric concentration ratio.

in the permeate is almost the same as in the concentrate. With a decreased lactose removal acid production during ripening of the cheese increases, the fresh cheese curd gets a sweet taste, and the ripening cheese gets too soft.

Figure 17 illustrates the reduction of different components in the skim-milk as a function of the volumetric concentration ratio. The removal in percent is calculated as kg material leaving the system as ultrafiltration permeate divided by the kg material at the start.

From the figure it appears that there is no lactose rejection since the lactose reduction at a removal of for example 50% liquid exactly equals 50%. On removal of 66.7% liquid the lactose reduction is 66.7% and so on.

Removal of minerals

Another important point in cheese production is the salt concentration and the dispersion of the salts in the concentrate.

Here Ca^{2+} is of special interest as increasing quantities of Ca^{2+} have the same effect as increasing quantities of H^+, that is, the cheese has a tendency to get too sour and too soft. Figure 17 also shows that there is a rejection of minerals.

By mass balance it is apparent that only one quarter to one third of the total amount of Ca and P_2O_5 is leaving the concentrate during the ultrafiltration. The total amount of Mg is diminished by one half.

However, only a tenth of the K–Na-quantity is left. The main reason for this is the protein layer; K and Na normally pass the membranes without any rejection.

Proteins

All the proteins remain in the concentrate. Only two thirds of the NPN (non-protein nitrogen) is leaving the concentrate with the permeate.

Temperature, capacity

In a batch process we operate at 2–4°C to a 3-fold concentration to obtain the best bacteriological conditions. At this temperature the capacity is limited. After 3-fold, it is necessary—due to the viscosity of the concentrate—to raise the temperature to 50°C.

Regarding the quality of the milk concentrate, it must only be kept at this temperature for a limited time. At 50°C you also obtain reasonable bacteriological conditions.

Figure 18 shows the relationship between temperature, viscosity, and the flux rate as a function of the concentration ratio.

Besides the temperature the capacity varies somewhat with the milk quality. Good bacteriological conditions and efficient centrifugation to 0.1% fat content, depending on the concentration ratio, give the best results.

In a continuous plant it is possible to operate at 50°C during the whole concentration time due to the retention time being reduced to 5–30 min depending on the sides of the plant. This gives around 50% larger capacity.

Figure 8 shows a five-stage continuous plant. The principle of the continuous plant is that some of the milk is recycled in order to get the right flow over the membranes.

Process control on the figure is shown as flow control. This gives a constant concentration ratio.

Another method is control by means of a refractometer, which reacts on the outlet valve. This gives a constant dry matter in the concentrate.

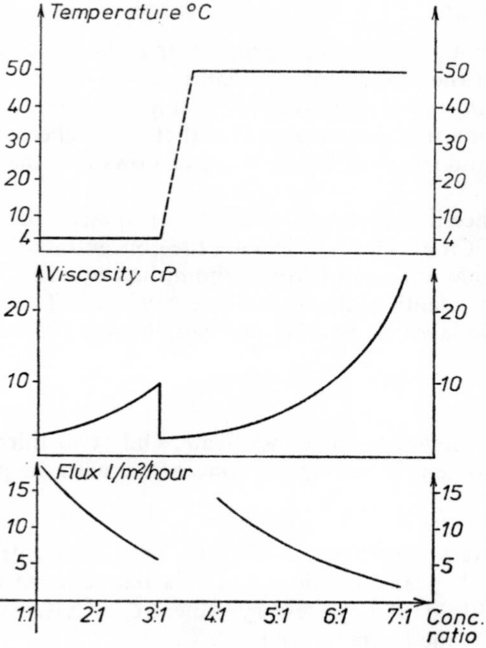

FIG. 18. The relationship between temperature, viscosity, and the flux rate as a function of the concentration ratio for skim-milk.

The plant

The technical plant for cheese production in the way mentioned is not yet available, as the final machinery is not yet ready. Today, the method is commercially used for production of high protein skim-milk powder.

However, at the end of 1974, a continuous working plant is going to be put into operation for ultrafiltration of skim-milk for production of Camembert cheese in France.

ICE-CREAM PRODUCTION

The main reason for interest in this product is that this is one of the applications of the hyperfiltration system. By this process all the milk solids are concentrated at a rather low temperature of 4°C to about 15% dry matter.

We have produced ice-cream in collaboration with a Danish dairy.

Figures 19(1) and 19(2) show the mass balance for production of ice-cream. Figure 19(2) shows one of the traditional production methods with addition of skim-milk powder.

By using hyperfiltration instead of skim-milk powder, you obtain considerable economic advantages, and the ice-cream quality is as good as normal.

At a production of only 6500 kg of ice-cream mix there is a surplus of about 500 D.kr. a day due to the high intervention price for skim-milk powder (5 D.kr. per kg) on the EEC market.

Whey is ultrafiltrated with our membranes types 600, 800, or AR8; the best results are obtained with membranes 800 or AR8.

Several companies have started small-scale production of 40–75% protein powders. These powders and the concentrate can be produced in very good bacteriological condition and are fully soluble. The powder has an excellent nutritional value.

THE USE OF ULTRAFILTRATION ON WHEY

The products which can be produced by ultrafiltration of whey have qualities varying with the type of whey, with the pretreatment, the method of ultrafiltration, and the after-treatment.

Sweet whey

The whey must be low pasteurised (72°, 15 s) and centrifuged directly after the cheese making.

The characteristics of foodstuffs produced from sweet whey are that they have almost no denaturation, the solubility is 99.9% if high temperature is avoided, and they are almost tasteless. Powders with a protein content of 25–75% on fat free TS or even higher can be produced.

The nutritional value of the whey proteins is excellent. The Committee on Amino Acids of the Food and Nutrition Board of the National Research Council (U.S.A.) has pointed out that 17.4 g of whole egg protein or 28.4 g of cow's milk protein will furnish the average daily adult requirements of amino acids for a 70 kg man. Only 14.5 g of whey protein would be needed to provide these same quantities of amino acids.

The materials are very good for the production of dietetic and baby-food, and the chocolate industry and bakeries show great interest in them. Later in this paper information about these products and their production will be given.

Protein concentrate made from sweet whey is also used directly for the following applications. In the cheese manufacturing the concentrate is returned to the cheese milk after a heat coagulation at 75° increasing the

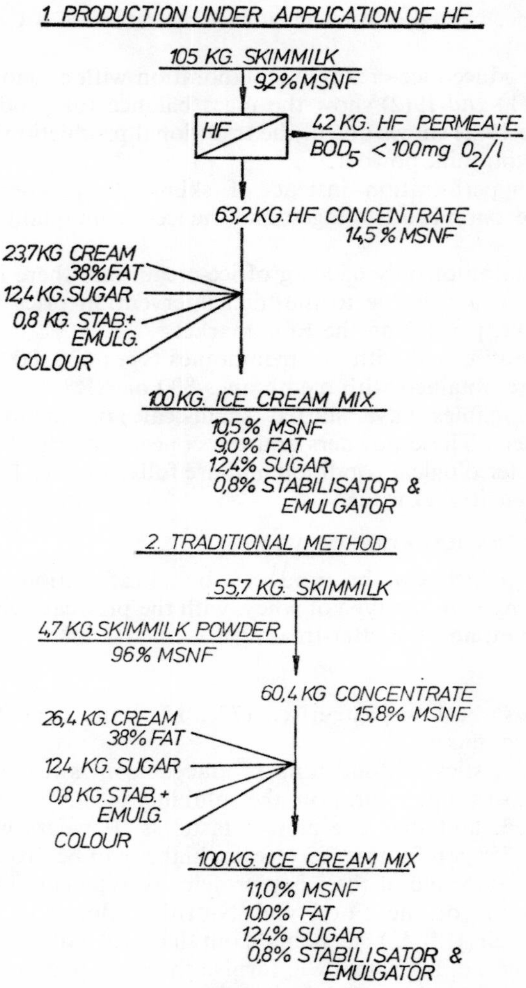

FIG. 19 (1 and 2). Mass balance for production of ice cream.

cheese output with 15% on skim-milk. In ice-cream production the concentrate can substitute a part of the skim-milk powder in the ice-cream mix.

Acid whey

Due to the high content of acid and minerals in acid whey from casein and cottage cheese production it is not suitable for direct use for pigs' and calves' feeding in larger quantities.

Ultrafiltration of acid whey removes to a large extent the lactic acid and the minerals and gives a concentrate which can easily be spray dried without neutralisation. The lactic acid taste can, however, not be removed completely, and although the protein powder from acid whey is highly soluble, the same high solubility as for sweet whey protein powder cannot be achieved.

Whereas acid whey as such has almost no value, this method can transform the whey proteins to valuable products.

An ultrafiltration process combined with electrodialysis can produce excellent products with a high protein content.

Pretreatment and storage of whey before ultrafiltration

To ensure the performance of the ultrafiltration plant and the quality of the final product all insoluble material as casein fines must be removed from the whey, and the fat content should normally be reduced to below 0.1%.

An initial heat treatment of the whey is also needed, even where the incoming cheese milk has been pasteurised. High-temperature (72–73°C) and short-time (15–20 s) pasteurisation is recommended to ensure good bacteriological quality of the final product and to prevent capacity reduction on sweet whey. Bacteria and yeast growth reduce the plant capacity, and the pasteurisation prevents this. The plant capacity varies with the pH of the whey as shown in Fig. 20.

It is important that the centrifugation takes place after the pasteurisation in order to remove coagulates, and a self-cleaning type of cream separator is recommended for this combined purpose.

A proper working temperature, normally below 10–12°C, is often chosen according to the bacteriological claims for the final product, and an initial cooling of the whey should take place immediately after completed pretreatment. Treatment at higher temperature may be chosen for continuous plants, when bacteriological conditions are good.

The pre-treatment must take place immediately after the whey is produced. When the whey has to be used for valuable products it is necessary that it is not regarded as a waste product in the first part of its life. It is important to ensure that no unnecessary contamination takes place at or under the cheese-making machine, and that also the first part of the system is properly cleaned up to hygienic standards.

EQUIPMENT

Modules

Sanitary modules, type 40 with 10 intermediate flanges are used for all whey ultrafiltrations. They each have a filtering surface of 24.2 m^2. With membrane 800 they are named 40–24.2–10–800.

PLANT DESIGN

Ultrafiltration plants for whey can in principle be built by two different methods.

Batch plants

Batch plants have the advantage that small plants, from 1 to 4 modules 40–24.2–10, are making it possible, in an economical way, to ultrafiltrate

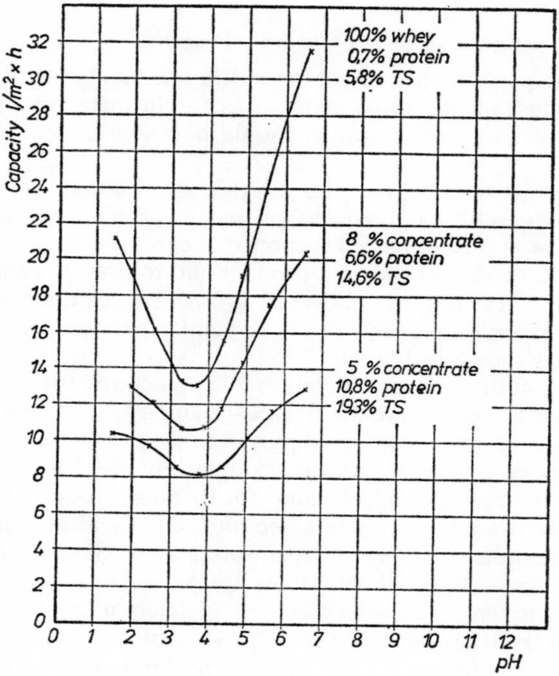

FIG. 20. The variation of ultrafiltration capacity with pH (UF of whey at 10°C).

whey even at very small dairies, for instance for returning the concentrate to the cheese milk.

The whey is recycled through the modules back to the storage tank. The feed for each module has to be 170 litre/min. No special automatics is required and the plant can operate overnight without any operator. The operating costs are somewhat higher than for the continuous plant.

Continuous plants

In continuous plants the whey is only sent once to the plant and then in a number of steps recycled with centrifugal type booster pumps over a number of modules. The advantages are low installation and operating costs, and continuous production allowing operating temperatures up to 30°C, for membrane 600 up to 50°C.

The number of steps needed depends on the protein content in the whey and in the final product. With a small number of steps, the modules will work on too high a concentration, giving a lower capacity.

WASHING

Depending on the whey quality a simple ultrafiltration can be used to produce products with up to 45–60% protein. If a higher protein content is desired, it is necessary to add water and continue the ultrafiltration in order to remove more lactose from the product. This washing can be arranged in three different ways depending on the plant size.

Batch plants

When 6.5% protein is reached in the concentrate, water is added to the batch in the same quantity as the produced permeate. When sufficient lactose is removed, the batch is concentrated to 12–13% protein.

Continuous plants

In a middle sized continuous plant, where the concentration to 6.5% TS is made continuously, the section for the continuous washing can be too small. For that reason the final washing and concentration is made batch-wise in 2–5 h.

Big plants can be arranged to operate continuously in all three steps giving a frictionless operation.

Automatic control of continuous plants to give constant products is easy, because concentration ratios and addition of water to the concentrate can be kept constant by automatic ratio flow control.

CONCENTRATE AND POWDER FROM SWEET WHEY

As a guideline to show which products can be obtained by ultrafiltration of whey, here are quoted some analyses of concentrate and powder produced from sweet whey at AKAFA in Denmark.

The whey is here pasteurised at 72°C for 15 s, centrifuged and cooled to 10°C. The batch process takes place at 12–13°C, and the ultrafiltration is continued until the protein content is about 60% of TS. Depending on the quality of the whey the concentration ratio is between 20:1 and 25:1.

A typical analysis of the whey is:

Total protein (N × 6.37)	0.70%
Lactose	4.30%
Ash	0.50%
Fat	0.05%
Total dry solids	5.55%
pH	6.5
Acidity	0.09% (as lactic acid)

Typical analyses of protein powders compared with ordinary whey are given in Table III.

TABLE III

	Ordinary whey powder	*50% whey protein powder*	*70% whey protein powder*	*Claims for use in baby-food*
% Humidity	4.0	4.66	5.77	max. 3
% Ash	8.0	4.04	3.61	max. 6
% Lactose	71.0	33.7	15.8	approx. 45
% Protein	13.6	49.4	68.3	40–50
% Fat	1.4	5.00	5.99	max. 5
% Lactic acid		0.09	0.12	
% Not identified compounds	2.0	2.11	0.41	
Na (% of ash)	8.0	7.5		
K (% of ash)	29.2	24.7		
Ca (% of ash)	3.3	6.5		
Mg (% of ash)	1.6	2.2		
P_2O_5 (% of ash)	7.3	8.1		
pH	6.1	6.8	6.7	6.0–6.8
Starch	+	+		max. 30% denatures P
Solubility	0.1–0.2	0.1	0.1	max. 0.5
Density, g/cm³	0.67	0.31	0.39	
Sedimentation	A	A	A	min. B
Total count/g	2 000	2 200	2 000	max. 5 000
Coliform/g	0	0	0	max. 5
E. Coli/g	0	0	0	max. 0.1
Yeast/mould/g	0	< 100	< 100	
Haem. bact./g	0	0	0	max. 10
Feed units per kg	125	138	148	

Amino acids

The composition of the amino acids in protein enriched whey powder compared with ordinary whey powder is given in Table IV together with the N-content of the amino acids. Except for the rather low content of methionine, the whey proteins contain a very useful amount of amino acids with a high nutritive value. The compositions seem to show that the normal factor of 6.37 for multiplying the nitrogen content to get the protein content in reality is about 6.9.

TABLE IV

	N-content of amino acid %	*Protein enriched whey powder*				*Ordinary whey powder*	
		1	*2*	*3*	*4*	*a*	*b*
Tyrosine	8.6	2.9	3.1	1.3	1.2	3.2	4.4
Phenylalanine	9.5	3.2	4.0	3.2	3.5	3.8	4.4
Methionine	10.7	2.1	2.1	2.0	2.1	2.3	1.2
Glutamic acid	10.8	17.8	16.9	20.3	18.1	18.3	18.3
Aspartic acid	12.2	10.6	10.1	10.8	10.7	10.5	10.7
Isoleucine	12.4	5.1	5.9	5.8	5.9	5.1	6.3
Leucine	12.4	10.1	9.7	9.6	10.0	11.5	10.0
Cystine	12.6	1.9	2.6	2.0	2.4	2.3	1.9
Treonine	13.8	7.3	6.8	6.6	6.9	5.1	6.2
Valine	14.1	4.6	5.9	5.7	5.9	5.2	6.3
Proline	14.4	6.4	5.5	6.1	6.0	5.6	6.4
Tryptophan	15.0	2.2	1.8	2.0	2.3	1.3	1.3
Serine	16.1	5.6	5.3	4.9	5.0	4.8	4.5
Alanine	19.7	5.0	5.0	4.6	4.8	4.7	4.4
Lysine	21.8	9.1	8.9	8.6	9.2	9.4	6.7
Glycine	24.5	2.6	2.4	1.8	1.9	2.1	2.0
Histidine	30.6	1.3	1.7	1.8	1.7	2.2	2.4
Arginine	35.8	2.2	2.3	2.7	2.4	2.6	2.6
N in % of protein		14.47	14.45	14.43	14.60	14.50	14.35
Calculated factor for Kjeldahl		6.91	6.92	6.93	6.85	6.90	6.97
Calculated factor in % of 6.37		108.5	108.6	108.8	107.5	108.3	109.4

The amino acids are enumerated in increasing N-content, which is corrected for 1 mol H_2O. The analyses were carried out at Valio, Finland (1 and a), INRA, France (2) and Landøkonomisk Forsøgslaboratorium, Denmark (3, 4 and b).

Ash

The composition of the ash is improved by ultrafiltration. Analysis of the ash in 60% protein powder compared with the ash in ordinary whey powder, both from the same whey, illustrates this.

Percent of minerals	*60% protein powder*	*Ordinary whey powder*
Na	7.5	8.0
K	24.7	29.2
Ca	6.5	3.3
Mg	2.2	1.6
P_2O_5	8.1	7.3

Degree of denaturation

A comparison of the degree of denaturation of the whey concentrate determined at its isoelectric point, pasteurised at 65°C for 15 s and in the powder spray dried at an inlet temperature of 150°C is quoted:

	Concentrate before pasteurisation	*Concentrate after pasteurisation*	*Powder*
Denaturated protein	1.4%	2.1%	4.6%
Undenaturated protein	93%	91%	89%
NPN	5.6%	6.9%	6.4%

Guaranteed capacity

The guaranteed capacity per module 40–24.2–10–800/600/AR6 in 1 sweet whey at 10–13°C in 21 h per day is as follows:

% protein of f.f. TS in the whey	*0.7% protein (total) in the product % protein of f.f. TS in the product* 20	*30*	*40*	*50*	*60*
8	14 000	11 000	9 500	9 000	8 500
10	15 500	11 500	10 000	9 000	8 500
12	18 000	12 000	10 000	9 000	8 500
14	20 500	12 500	10 000	9 500	9 000

Capacities, however, vary with protein content, dry matter content, temperature, and pH. Time does not allow me to go into details with this problem.

ENZYMES

For the concentration and purification of a large number of different enzymes a great number of our plants are in operation.

Today there are technical DDS-plants in operation for the following products: Bacterioproteinase, Amyloglycosidase, Rennilase, Pectolase, and Trypsin plus further products, the names of which we still do not even know.

The working and cleaning parameters for enzyme concentration plants in general are given in the next sections.

Membranes

Membrane type 600 is used almost exclusively, it has a molecular weight cutoff about 20 000. A few customers use, however, type 800, which has a molecular weight cutoff of 6000–8000.

Modules

As ultrafiltration is involved, low pressures should be applied. The individual sections in the modules should not be too large, as a large section gives a high pressure loss, this means a high inlet pressure in order to preserve a sufficiently high flow of liquid along the membrane surface. A typical module for enzymes is one with 11 individual parallel coupled sections, DDS module type 40–22–10.

In special cases, when the final product has a viscosity higher than, for example, 5 cP, a module with 18 sections is recommended, DDS module type 40–18–17. The figures 10 and 17 in the specifications refer to the number of inserted intermediate flanges or separation flanges between the individual sections.

Modules and equipment are always used hygienically. An example of such a plant is given in Fig. 21.

As there is a variety of enzymes in different products, it is impossible to state specific capacities. Permeate capacities between 5 litres/m^2/h and 60 litres/m^2/h may be expected, and figures of about 20–30 litres/m^2/h are common.

Pre-treatment of the product

The product should be cleaned as much as possible. The costs for the cleaning process are paid for by an essentially longer membrane lifetime, higher plant capacity, and easier cleaning and disinfection procedure. Filtration on filters with diatomite or asbestos is widely used in connection with ultrafiltration. Sometimes a sterile filtration is made before the ultrafiltration, this is not done primarily for the ultrafiltration, but because the viscosity of the final product is so high that a sterile filtration is impossible to carry out.

In certain cases it is necessary to adjust pH of the product to within the range 3–8.

Temperature

Normally it is desirable for the bacteriological quality of the product to work at low temperatures, 4–6°C. It should only be realised that the

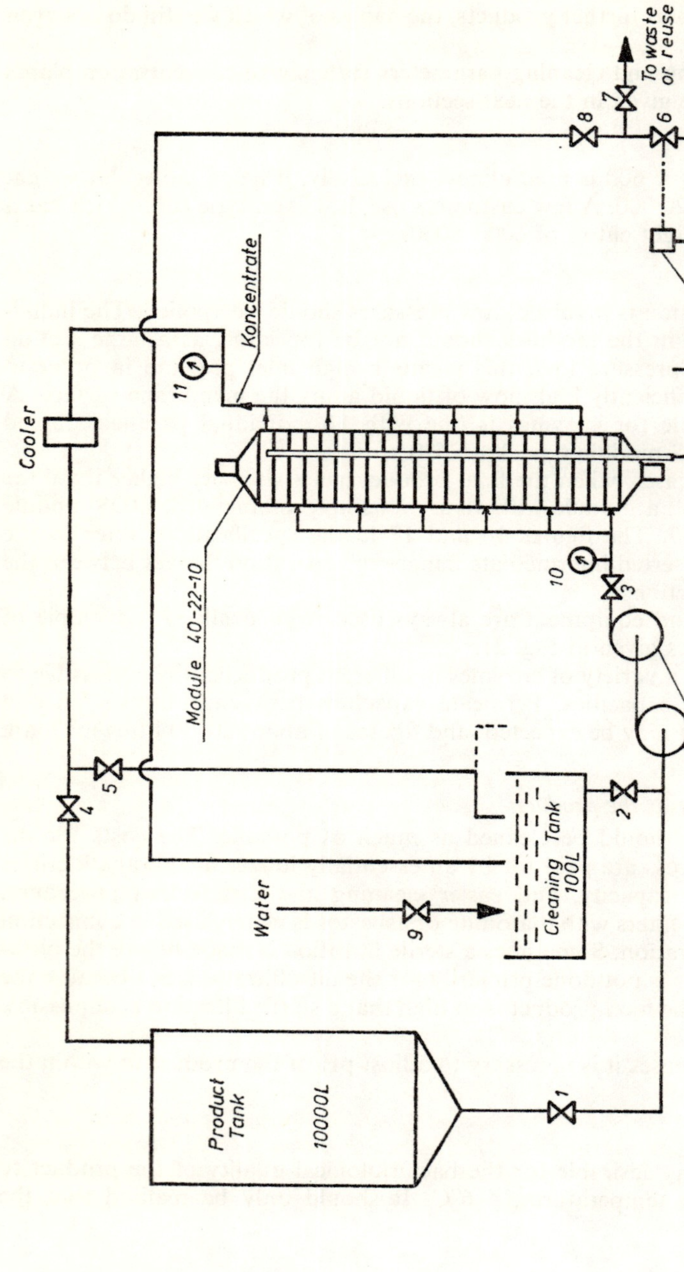

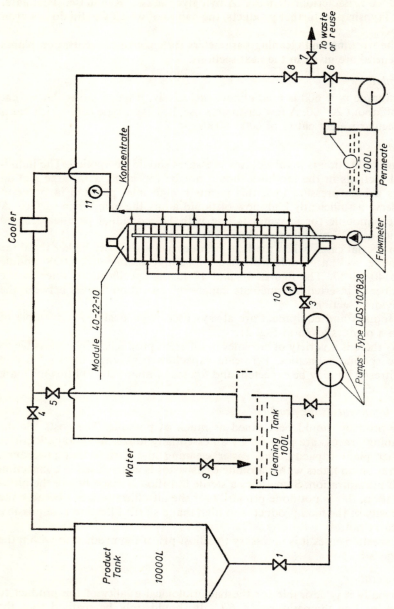

FIG. 21. Diagram of a sanitary batch UF plant for skim-milk.

capacity of the plant decreases 1–3% for each degree the temperature is lowered, as well as that the viscosity of the product increases at decreasing temperature. The plant and the membranes work excellently right up to 30°C. Consequently the operating temperature is fixed in cooperation with the customer in accordance with the claim of the product.

Pressure

Low pressures, right down to 3–4 kp/cm^2, are applied. Generally no capacity increase is obtained at higher pressures. The result is only that the membranes are compressed and blocked.

High pressures are here about 20 kp/cm^2. If general rules for operating pressures are to be laid down, it would be that low-viscosity solutions with a small amount of dissolved dry substance should be processed at 4–8 kp/cm^2, while heavier solutions with a large amount of dissolved dry substance can be processed at 6–15 kp/cm^2.

Detention or loss of activity

In general there is no loss of activity through faultless membranes. It happens, however, that a certain loss of activity is seen, and no activity is found in the water passing through the membranes, if analysed. The mistake most commonly made is to ignore the concentrate remaining in the module. If, however, this has been taken into consideration, or if the volume is so small compared to the total concentrate that it could be missed, the matter can be more complicated. Some reasons can be suggested, although it must be admitted that there is no ready solution to the problem.

On the membrane surface there is a higher concentration of enzyme and ballast substances than in the actual solution, consequently one can imagine that a partial decomposition of the enzyme is taking place in this area owing to a changed ballast substance concentration or a high enzyme concentration.

For enzymes which are activated by ions naturally present, like Ca^{2+}, it happens during the concentration that Ca^{2+} passes through the membrane, consequently the enzyme solution receives a smaller amount of Ca^{2+}, which in some cases results in apparent decreases in activity, but if only one compensates for the lost activator, the enzyme loss is reduced to very small values, 1% or less.

In certain cases, however rare, it is not possible to find any actual reason for the loss of enzyme. It will then be necessary to study the problems more thoroughly before plants of industrial size are planned. For this work the DDS laboratory module is excellently suited, only it should be clear that if a mechanically sensitive high molecular enzyme is to be treated, one can get better results with the laboratory module, which is equipped with a flow-speed piston pump, than with industrial plants equipped with centrifugal pumps or mono pumps, owing to the different influence of the pumps on the product. Membranes 800 are primarily used when losses have to be absolutely prevented. Membranes 600 are used when a purification is wanted.

INSULIN

An ultrafiltration plant has been installed in a new insulin factory in England (Moreau, 1972).

For animal products the main problem has been the fact that an efficient fat separation is necessary. When this has been achieved, the plants will run very efficiently.

The operating pressure for this application is normally between 5 and 12 kg/cm^2.

CONCLUSION

I have now given a survey of some of the main applications for the processes in use today. One application I have not mentioned, although it is probably the most important in future: the production of clean water and the purification of waste waters. We are working intensively within this field and have delivered a number of plants for such applications.

Until now only a small number of the very great possibilities for the processes has been developed. Some of the large industries, where we are just beginning today, are the cellulose and the sugar industry. On the other hand we are now at the stage where the processes are technical realities in a number of applications.

REFERENCES

Bixler, H. J. and Cross, R. A. (1969). 'Final report on control of concentration polarization in reverse osmosis desalination of water', Office of Saline Water, Res. Develop. Progr. Rept. No. 469.

Bundgaard, A. G., Olsen, O. J. and Madsen, R. F. (1972). *Dairy Industry*, **37**, 539–546.

Flinn, James E. (1970). *Membrane Science and Technology*, Plenum Press, New York.

Gill, W. N. *et al.* (1969). 'Mass transfer in laminar and turbulent hyperfiltration systems', Office of Saline Water, Res. Develop. Progr. Rept. No. 403.

Maubois, J. L. and Mocquot, G. (1971). *Le Lait*, **51** (508), 495–533.

Moreau, D. (1972). *Manufacturing Chemist & Aerosol News*, **43** (8), 42 (1972).

Murkes, J. and Bohman, H., *Desalination* (1972). **11**, 269–301.

Sheppard, J. D. and Thomas, D. G. (1970). *Desalination*, **8**, 1–12.

Sheppard, J. D. and Thomas, D. G. (1972). *Desalination*, **11**, 385–398.

Sourirajan, S. (1971). *Reverse Osmosis*, Logos Press, London.

Srinivasan, S. *et al.*, 'Simultaneous development of velocity and concentration

profiles in reverse osmosis', Office of Saline Water, Res. Develop. Progr. Rept. No. 243, 1967.
Symposium on Freeze Drying, Lucerne 1973 (in press).

DISCUSSION

H. R. Lovell (Blackpool College of Technology and Art, U.K.): I think a fundamental problem which has not been looked at in great depth is the microbiological aspect. The very nature of the process concentrates bacteria.

In the work that I did on pasteurised skimmed milk using tubular membranes and with recycling, the operating temperature was 35°C. With this system, after four hours I had a continuous fermenter! Therefore my question is what sort of microbiological quality of feedstock do you use: and, with regard to the whey, is it going to animal feed, or is it going for human consumption and, again, what sort of microbiological quality in the end product are you getting?

R. F. Madsen: With whey products I showed you the products we have been able to produce with microbiological analyses on whey powder. Normally when we make an ultrafiltration of skimmed milk starting off at 4°C and concentrating up to three times, the viscosity becomes so high that one can no longer pump the liquid. We then raise the temperature to 50°C. In a continuous system the total time at 50°C is around fifteen minutes. My experience has been that this plant can run without microbiological problems for twenty hours. In France to produce Camembert cheese, one can only begin with non-pasteurised skimmed milk, which is not the best starting material. With this plant in a batch process, sixteen hours at 2–4°C, followed by four hours at 50°C can be achieved. But we don't like to use intermediate temperatures, because then you get fermentation. To get the best cheese by this process you have to remove more calcium from the skimmed milk than is lost by ultrafiltration. We first concentrate threefold by ultrafiltration at 2–4°C and then acidify down to pH around 6.0, before a short ultrafiltration stage up to around sixfold concentration.

B. S. Horton (Abcor Inc., U.S.A.): Residence time is a function of flux, and the comparison made between thin channel systems and the tubular systems does not take into account the possibilities that you have different flux rates. In fact, it is possible to build a commercial plant with a 25 mm diameter tube which would show a residence time of one hour or less. And, from data which have been published, if you're processing skimmed milk, you'll find at 48–50°C, depending on the particular organisms in the milk, you will get a net 'kill' over the first hour.

I believe that the cleaning business will prove to be at the end of the list of the cost elements, depreciation, labour, membrane replacement, and so on, involved in the overall process.

Madsen: Most of us operate at both high and low temperatures, but as far as I know the results of all our systems are almost the same if we have the same operating temperature. But we have different opinions about temperature. The reason is perhaps that the acid whey is no problem at high temperatures and you can make a good product: but when you make a first class product for human consumption from skimmed milk or sweet whey, you can run into troubles at 50°C because you have thermophyllic bacteria in the product and these can clog up the whole system.

A. S. Michaels: Polarisation or gel accumulation problem in the membrane really is a two-step process. It is initiated by accumulation by convective transfer of solute to the membrane surface and produces a layer which offers a hydraulic resistance to flow in addition to that offered by the membrane: that's step number one. This process is counterbalanced by a tendency for the material to diffuse back into the flowing stream; but this is a rate limiting process limited by counter-diffusion.

The straight membrane in the absence of any solute at all behaves as a porous medium: the flux is directly proportional to the pressure. But if you get solute present in the solution and you flow liquid at constant velocity tangential to the membrane and measure the flux as a function of applied pressure, invariably what you find is a curve where the flux is no longer directly proportional to pressure. As you decrease the fluid velocity, the curves tend to plateau off at low values, because a dynamically steady state condition is reached where the rate of convective transfer of solute to the membrane is counter-balanced by the rate of diffusive transport back the other way. When you raise the pressure a little bit, you tend to thicken this layer and increase its hydraulic resistance in exactly compensating proportions to the increase in pressure. The net result is that the overall resistance of the membrane plus the layer goes up proportionately to the applied pressure, and for this reason you end up with a constant flux.

If the pressure is increased, even though the flux doesn't change, the fraction of the pressure drop which occurs across the 'cake' of the gel layer increases. If the flux drops to half its value in the absence of any solute, it is obvious that half the ΔP is carried by the overlying layer—the other half by the membrane. That pressure tends to cause a slow consolidation of the cake of the gel layer, which results in a further reduction in its hydraulic permeability.

If you make measurements over hours or days, for a fixed fluid velocity and a fixed operating pressure, you find that the flux declines with time at a rate that is dependent upon the applied pressure. This is because you're consolidating the cake more and, as a result, the hydraulic permeability is decreasing. The natural thing you'd think to do is to raise the pressure. 'I should get a higher flow at a higher pressure, all things being equal.' Well indeed you do, but not for long. The flow will rise for a short period of time, and then will rapidly decline to a lower value; and it will be a lower value than you would have accomplished if you had merely let the system operate at the same constant pressure for a longer period of time. The correct thing to do to increase the flux in this situation is to reduce the pressure. If it turns out that the fluid velocity conditions and the shear stress of the wall are sufficient to disengage the accumulated gel in due course of time, reducing the pressure will indeed result in a slow increase in the flux. So things work upside-down in this type of situation.

Let me just finish by saying that the best conditions for operating an ultrafiltration system, therefore, are obviously at the lowest possible ΔP initially, which provides conditions for minimum possible cake compaction. This has another advantage in that the cake that is formed in these circumstances is relatively loose and relatively easily disengaged, so the fluid velocity of the shear stresses of the wall that are necessary to remove it or to keep it moving is low. Once you've allowed the pressure to rise and the film to consolidate, the forces very often required to remove it are greater than you can supply by conventional fluid mechanical movement over the surface of the membrane and mechanical cleaning is required.

Madsen: It is correct—you could, in practice, get the same with very low operating pressure; but the problem is that, if you have a low operating pressure, you must increase the volume you pump. The piston pump cost is, in fact, the major part of the total plant.

D. E. Blenford (R.H.M.): I understand that you have approached the government in terms of putting your plants into operation. The only comment you made concerning their reaction to your application was in terms of hygiene. Was there any comment made about the use of your membranes in conjunction with dairy produce?

Madsen: I can answer that question for our plant because the materials used in our membranes and in the rest of the systems are approved now in the U.S., Germany and Denmark. We have told these countries, in confidence, exactly how we have made our membranes and what materials are used in our equipment; and then they gave their approval.

S. A. Goldblith: It is physically cleaning the material out in a membrane process that seems to be important to its capacity over a practical, economic period of time. These things are clear to us in the States and have, in fact, blocked the potential use of the process in milk processing. Whey is a waste material in the United States generally: milk is wholly food, and there is tremendous resistance on the part of our industry to take the big step in processing milk without the stamp on the equipment which says '3A'. Now we could go through and get a standard written round our design, but it is a four-year effort, normally, for a new 3A standard on a process or a type of equipment to emerge.

D. T. Shore (A.P.V. U.K.): Could you say something about the further processing of the material from the ultrafiltration process, which is still extremely prone to infection because it is still at low concentration, and how you converted it to the powder?

Madsen: This processing is done on a sweet whey which is first pasteurised at 72°C for 15 seconds: the operation here was a batch ultrafiltration at about 10°C, and the count in the concentrate is normally a little lower than what you would expect after allowing for the concentration effect.

We then pasteurise the concentrate at 60° for 20 minutes—that is a long time low temperature pasteurisation because we could not get hold of a normal pasteuriser in which to do it. This concentrate, with 21% solids content, and 60% of the solids proteins, can be stored at 4–6°C for up to three days before spray drying without any problems.

(The Chairman announced the adjournment of the meeting).

Studies of Membrane Deposits in Ultrafiltration

PETR DEJMEK and BENGT HALLSTRÖM

Department of Food Engineering, Lund University, Sweden

At the Lund University in Sweden the departments of Chemical Engineering and Food Engineering have together worked on pressure driven membrane processes since 1969. We have mainly devoted our resources to engineering problems and studied very little the membrane as such and its manufacture. It is the aim of this paper to review some of the work we have done in this field.

In these modern membrane separation processes with high flux membranes the flux decline caused by the accumulation of non-permeable solutes at the membrane surface (concentration polarisation and fouling) is a most serious problem. These problems are notable especially in food industry applications. In these applications ultrafiltration is the most interesting process and for these reasons we have come to study membrane deposits in ultrafiltration. Also, very little is reported in the literature about these problems.

Our activities are following two lines; one line is the running of a test station for ultrafiltration and reverse osmosis, open to industry and scientists; the other line is our research work.

The test station was founded and is financed by the Swedish Board for Technical Development and the purpose is to promote membrane technology in Sweden. In industry there is a great interest in testing membrane processes in different applications but at the same time there is a certain hesitation against investing in expensive test plants. One purpose of the test plant is to overcome these difficulties. The plant was started up in May this year. It includes:

1 pilot plant RO, 1.8–4.6 m^2;
1 pilot plant UF, 1.6–2 m^2;
1 laboratory plant RO, 0.6 dm^2;
2 laboratory plants UF, 1.4 dm^2.

The equipment includes in addition to the usual components (modules, pumps, piping, tanks) also rather advanced instrumentation for measuring and control (pressure, pressure drop, temperature, flow, conductivity, pH, ultrafiltration, etc.). As the plants are supposed to work mainly on food problems we have tried to make the equipment as hygienic as possible. We have several industrial contracts and commissions and we are just now involved in separation problems concerning, e.g. whey, potato waste water, waste water from vegetable processing and sugar solutions.

In our research work on deposits in ultrafiltration we are both trying to study and measure the build-up and to find techniques for minimising the influence of the deposits on flux.

MEASUREMENT OF MEMBRANE DEPOSITS

The method developed involves the use of radioactively labelled macromolecules. In the experiments we used casein. The build-up of the concentration polarisation layer in a thin ultrafiltration channel was then observed by means of a radiation meter. In this way the build-up could also be followed during running of the unit. The experiments have been described elsewhere (Dejmek *et al.*, 1973).

THE INFLUENCE OF PRESSURE ON FLUX

It is a well-known fact that increasing the pressure in ultrafiltration of macromolecular solutions often results in little or no increase in permeate flux. This is in contrast with the almost linear increase of flux with pressure in reverse osmosis of microsolutes, e.g. salts. We have conducted a series of experiments which suggest that compression of the macromolecular deposit under pressure is responsible for an appreciable part of the observed behaviour.

In the first experiment two ultrafiltration membranes with deposits from the running of a casein solution were put together with the active sides facing each other. The pressure drop was measured and compared to that of a single membrane with deposit. The twin membrane pressure drop was significantly higher than twice that of a single membrane under similar conditions. We suppose this results from compression of the protein layer, as in the twin membrane test the compression pressure over the layer is higher than that in the single membrane test. In a similar experiment a membrane with deposits from casein solution was compressed by means of a mechanical spring. Increasing the spring pressure increased the pressure drop over the membrane plus the layer thus confirming the earlier results.

Both these experiments further prove the advantage of using open membranes and low pressures in ultrafiltration.

FURTHER STUDIES ON PERMEABILITY OF DEPOSITS

In these experiments we use a small laboratory cell, originally equipped with a stirrer. A known amount of a macromolecular solution is transferred to the cell. A thin porous disc is placed in the cell, level with the solution, to prevent mixing of the original solution with buffer that is added subsequently. The solutions are ultrafiltered without stirring.

When the amount of permeate corresponding to the volume of the original solution has been collected, most of the original solute is deposited on the membrane. Thus the pressure drop across the deposit may be correlated to the amount of deposited material and other variables. We hope to be soon in a position to publish the results.

MEMBRANE-BOUND ENZYME

When ultrafiltration and reverse osmosis are applied to protein-rich liquids, frequent cleaning is necessary. A preferred method in the dairy industry has been the use of different proteases in the cleaning solutions.

We have investigated whether proteases could be bound directly to an ultrafiltration membrane and if any changes in the fouling of the membrane could be noticed. Related work was reported by Envirogenics. The procedure was in brief the following: an ultrafiltration membrane was cast after the van Oss recipe (25 g cellulose acetate E 383–3,75 ml acetone, 50 ml formamide). The membrane was mounted in an ultrafiltration cell, activated in 8 min by cyanogen bromide solution at pH 11, washed thoroughly and an enzyme solution (trypsin or papain) was added. The coupling was allowed to proceed for 8 h at 4°C. The enzyme not covalently bound to the membrane was removed by thorough washing with buffer solutions of different pH's and different ionic strengths.

The activities of the membranes of 25 cm^2 were determined to be 0.5–1 mol BAEE/min (after Takanaka) or when using casein as substrate 0.02–0.04 units/min (after Kunitz). The permeability to water, which was 0.002 cm/s, before the enzyme coupling, decreased to about one half. The membranes with bound trypsin and papain were used to concentrate haemoglobin and casein solutions of different concentrations in a stirred cell. The permeate flux and fouling tendency were the same as with untreated membranes. No difference in the spontaneous recovery of flux between individual experiments could be found.

The observed activity of the membranes exceeds the activity to be expected from a monomolecular layer of trypsin upon the nominal surface of the membrane. Thus we guess the enzyme has become attached inside the membrane structure, thus blocking some of the larger pores. The achieved activity is apparently not sufficient to reduce fouling.

TURBULENCE PROMOTERS IN ULTRAFILTRATION OF WHEY

Turbulence promoters of various types have been proposed for use in ultrafiltration and reverse osmosis equipment and some manufacturers have incorporated such devices in their modules.

We have investigated the effect of a turbulence promoter of the 'static mixer' type. Similar tests have been conducted for reverse osmosis (Pitera and Middleman, 1973). Two strips of metal are twisted (right-hand and left-hand twist respectively) and cut into 180-deg-twisted sections. The sections are joined together with right- and left-hand twists juxtaposed, and with the trailing edge of one section at right angles to the leading edge of the next. An assembly of 52 sections was inserted into a $\frac{1}{2}$-inch membrane tube cut from Havens 215 VDR ultrafiltration module. The tube was tested with a whey concentrate containing 6.5% protein and 11% TS. At a pressure of 7 bar, the usual decrease of flux with time was almost eliminated already at a flow rate of 1.5 litre/min (0.2 m/s). With a flow rate of 2 litre/min, a flux of 0.28 m^3/m^2 day (7 gfd) was achieved. The pressure drop of a 1.8 m^2 module under same conditions was estimated to be 3.5 bar.

Comparing these results with other investigations on whey and whey concentrate it seems that these type of turbulence promoters would give slightly better performance than other devices or simply increasing the velocity when the comparison is based on the same pressure drop. Another advantage of the system would be the low feed rate necessary. The results will be published (Dejmek *et al.*, 1974).

PERIODIC BACK PRESSURE WASH

The research team of the Paper and Pulp Manufacturers League has investigated in detail the effects of the so-called osmotic wash on fouling of reverse osmosis membranes and found this method indispensable in treatment of the difficult wastes encountered in the said industries (Amerlaan and Wiley, 1969). We have followed the idea of osmotic wash further to include the use of positive pressure on the permeate side of the membrane.

A Universal Water laboratory flow cell was modified to suit the experiments (Fig. 1). In the modified apparatus, a circular membrane, 1, of 5 cm diameter is supported in the usual way on a porous stainless steel disc, 2. Opposite the membrane at about 0.03 cm distance, a porous disc, 3, in plate, 4, provides the other wall of the feed channel. Feed liquid approaches the membrane through a 0.03 cm slit round the edge of the plate, 4, flows centripetally past the membrane and leaves through the porous disc, 3.

When pressure is applied to the permeate outlet, the membrane is lifted off its support, 2, but its movement is checked by the porous plate, 3. At the same time, permeate is flowing backwards through the membrane, and an eventual fouling layer is removed. It is obvious that the present apparatus is very rough on the membranes, particularly because the slit between plate 4 and the body of the apparatus must be abridged. We have not observed any increase in pure water flux, or evidence of gross structural damage. However, rejection decreases. Of the membranes tested, the Iris membrane, supplied kindly by Mr. Richard of Rhone-Poulenc, has been the most successful with a decrease of rejection from 99+ to 97% on Blue

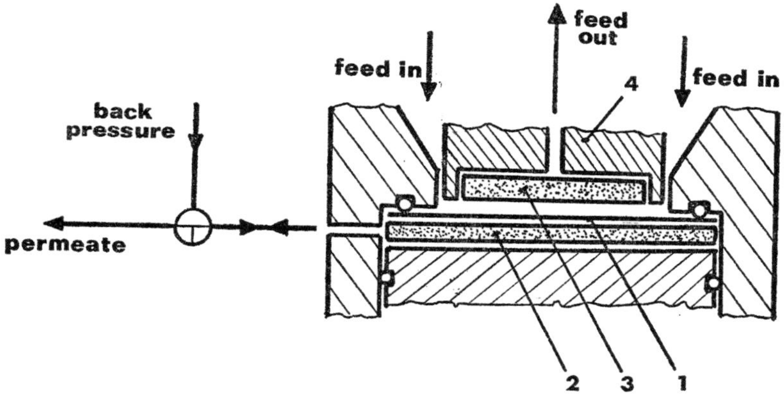

FIG. 1.

Dextran, molecular weight 2×10^6. It is an asymmetric membrane supported by a monofilament cloth.

On the point of diminishing the effect of concentration polarisation and fouling, the idea of frequent back pressure application appears to be promising. No optimisation has yet been done. Using a back pressure burst of 4 s every minute, no trace of deposit was found on a membrane after 24 h of ultrafiltering a dilute solution of Blue Dextran and starch hydrolysate. The flux was constant at 2.4 m^3/m^2 d (60 gfd). This is more than double the capacity of the apparatus under the same conditions without backwashing. In comparison a commercial thin channel apparatus at a circulation rate of 2 litre/min started at 2.2 m^3/m^2 d and dropped to 1.4 m^3/m^2 d during the first 15 min of the experiment.

REFERENCES

Amerlaan, A. C. F. and Wiley, A. J. (1969). *Tappi*, **52**, 118.

Dejmek, P., Hallström, B., Klima, A. and Winge, L. (1973). *Lebensm.-Wiss. u. Technol.*, **6**, 26.

Dejmek, P., Funeteg, B., Hallström, B. and Winge, L. (1974). *J. Food Sci.*, in press.

Envirogenics Co., 'New technology for treatment of wastewater by RO', report to EPA.
Pitera, E. W. and Middleman, S. (1973). *IECh Proc. Des. Develop.*, **12**, 52.

DISCUSSION

A. S. Michaels (Alza Research, U.S.A.): What was the concentration of Blue Dextran in that experiment, and what was the frequency of the backwash?
Hallström: The frequency was one back pressure wash every minute, and we washed for four seconds. For the concentration, I'm sorry I don't have the details here, but I think it was a rather low concentration.

Industrial-scale Ultrafiltration and Reverse Osmosis Plants in the Food Industries

BERNARD S. HORTON

Abcor Inc. Brussels, Belgium

INTRODUCTION

Ultrafiltration and reverse osmosis have been in commercial use in two major industrial fields (the electrocoat painting of automobiles and appliances, and the treatment of whey) for nearly three years.

A detailed presentation of the advantages and disadvantages of various designs of equipment has been given elsewhere (Horton 1972). Here, it is sufficient to say that no one design has *all* the virtues one would like when processing food solutions.

This paper therefore gives a brief review of the status of one design approach which has proven to be an excellent compromise in practice.

The design referred to in this review involves, for ultrafiltration, tubular membranes, 25 mm in internal diameter and 3 m long. Average pressures of operation are generally 2.0–2.5 atm. Temperature of operation for whey is usually 50°C. Flow rates are 80–120 litre/min per tube. The individual membrane tubes are connected in combinations of series and parallel passes and are housed in stainless steel cabinets to both collect permeate and provide in-place cleaning and sanitation of the permeate side of the system. Details of this design are described in Horton (1972) and Selitzer (1972).

For reverse osmosis, tubular membranes are 12 mm in internal diameter and 1.5 m long. Average pressures of operation are 25–60 atm, depending on the specific membrane selected for maximum retention of mineral salts or for partial demineralisation during concentration. Temperature of operation is 35–40°C. Flow rates are typically 1.5 litre/min per tube. Twenty individual tubes are connected in series/parallel passes by stainless steel tube sheets and headers. Plastic balls are employed in the tubes as turbulence promoters.

INDUSTRIAL PLANTS

Three large industrial-scale cheese whey ultrafiltration plants are now in operation in New Zealand, the United States and France.

The world's largest reverse osmosis plant in a dairy is now in its second year of operation in the United States as a concentrator of ultrafiltration permeate (lactose).

New Zealand Co-operative Dairy Company

This plant processes lactic acid casein whey and has a nominal capacity of 200 000 litre/day. In fact, the design proved to be very conservative and the plant can process considerably more whey each day. The plant has now finished two seasons of operation.

Dr. Kenneth Kirkpatrick of the NZDRI reported (Horton and Kirkpatrick, 1972):

'Design capacity of the plant has been readily achieved and equipment performance and reliability have been satisfactory.

'Restoration of plant flux daily by detergent cleaning, and sanitation by standard procedures have been effectively achieved in practice. Where circumstances led to a deterioration in the sanitary status of the plant, it proved possible to restore the plant to a completely satisfactory level of operation.

'Some progressive loss of flux has been noted, but a one year operational life for membranes appears entirely feasible.'

Crowley Foods, Inc.

This dairy and food products company is one of the largest producers of cottage cheese in the United States, with two plants for this product in northern New York State. The company and its products, as well as the ultrafiltration and reverse osmosis plant, are described in considerable detail in Selitzer (1972).

The design capacity of the new whey processing plant was 140 000 litre/day, to be processed in 20 h. The ultrafiltration section (270 m^2) was to concentrate the protein 12-fold to yield products with 50% protein on a dry basis. (On occasion Crowley Foods have produced 30-fold concentrates—75% protein on a dry basis.) Two 10-hour batches were scheduled.

The reverse osmosis section (475 m^2) was designed to concentrate the lactose solution, in once-through continuous operation, from 6% solids to 24% solids (75% lactose on a dry basis).

The plant has been in operation since June 1972.

EXPERIENCE WITH ULTRAFILTRATION

(1) Actual capacity of the ultrafiltration equipment is 30% above design capacity. Two batches were not justified and the utilisation of the plant has not been optimum. It has been operated for 2500 h instead of 6000.

(2) The ability to clean the tubular membranes physically and in-place

with spongeballs has proved to be invaluable in restoring plant performance when the introduction of poor quality whey causes fouling of the membrane surfaces. The last four months of operation show no spongeball flushing is necessary in normal operation.

(3) The ultrafiltration plant is able to turn out consistently protein concentrates with total plate counts of less than 50 000/ml, and usually less than 10 000/ml. Coli counts are less than 1/ml. The quality of the concentrate is a function of the quality of the whey because the ultrafiltration equipment does not contribute to quality of the protein concentrate.

(4) Cleaning and sanitation may be carried out using the reverse osmosis permeate (effluent) which is purer than the local well water, thereby helping the total dairy plant effluent situation.

(5) Membrane lifetimes of at least one year can be expected, though prolonged unnecessary storage with sanitisers accelerated membrane ageing.

(6) Membrane tube failures due to defects in manufacture are running at a few percent. The ability to change single tubes (0.2 m^2) has proven of advantage in maintaining the best overall plant operation.

(7) Recovery of concentrate without dilution by the water used to push it out of the ultrafiltration system has proven entirely successful since the water acts in plug flow. The hold-up volume of the tubular system is of no real consequence in the operation.

(8) Operating costs are now generally in agreement with predicted values.

EXPERIENCE WITH REVERSE OSMOSIS

(1) At installation the reverse osmosis section yielded a 28% solids concentrate. Modules were removed to reduce this level in order to avoid risk of lactose crystallisation should the temperature of operation not be up to normal.

(2) Plant capacity is now up to 10% above design capacity.

(3) Approximately 10% of the original modules (1.3 m^2 each), the first ones in series in each of the six parallel stainless steel cabinets, were changed due to chemical attack of the membrane by cleaners and sanitisers. These were used at levels above recommended levels for reverse osmosis membranes in order to restore plant performance.

(4) While histories of TPC's, yeasts, moulds and coli show that lactose concentrates with excellent quality can be produced routinely, cleaning of reverse osmosis equipment has been more difficult than cleaning of ultrafiltration equipment. We ascribe this primarily to the smaller passages in the reverse osmosis tubes.

(5) Cleaning and sanitising of the permeate side of the reverse osmosis tubes is important for good general plant operation and more so if the permeate is to be re-used, as at Crowley Foods.

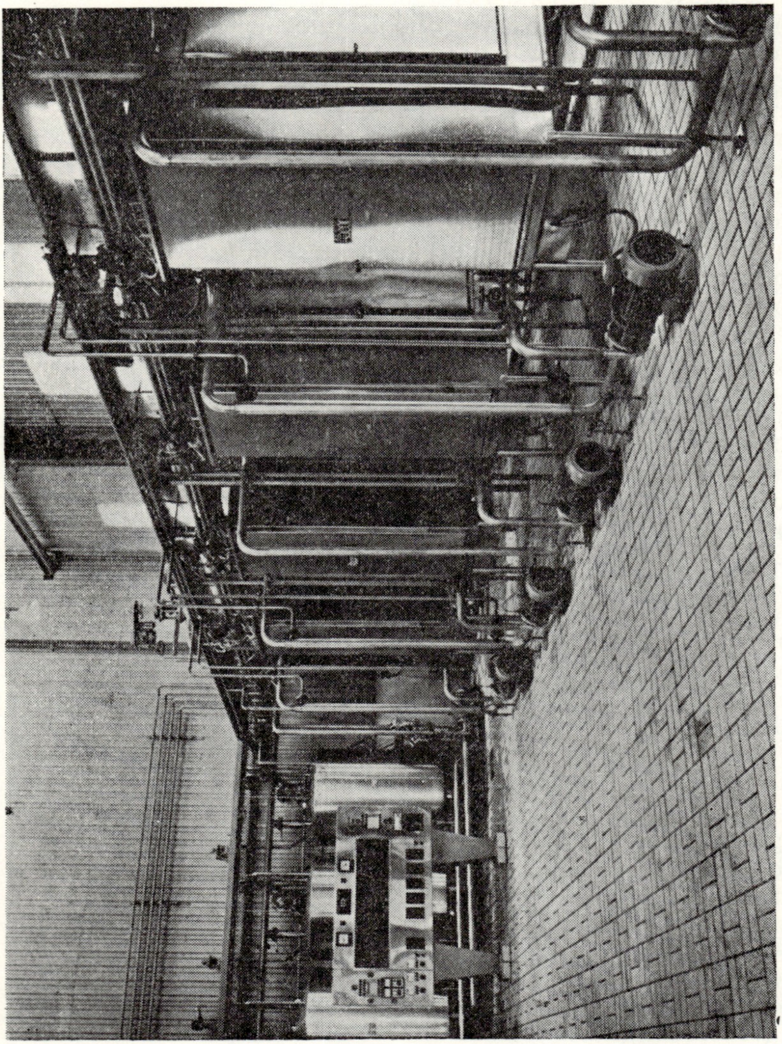

FIG. 1. View of continuous 250,000 litre/day whey ultrafiltration plant at La Prospérité Fermière, St. Pol-sur-Ternoise, France (September 1973).

(6) Operating costs have been running higher than we would usually predict because of the distribution of labour costs between ultrafiltration and reverse osmosis compared to our preference for loading all labour on the ultrafiltration operation (because of the accepted higher value of the protein product.)

La Prospérité Fermière

The first large ultrafiltration plant of continuous (stages-in-series) design has just gone into operation in St. Pol-sur-Ternoise, France, at a new whey processing facility of La Prospérité Fermière. This plant processes approximately 250 000 litres/day of different sweet wheys. The plant incorporates five ultrafiltration stages, each having 45 m^2 of membrane area. A photograph of this plant is shown in Fig. 1. Each stage may be switched off separately for cleaning, permitting 24-hour operation of 4 stages.

SUMMARY

(1) Ultrafiltration can be operated at 50°C with membrane lifetimes of one year or more.

(2) The quality of the local water used for cleaning, sanitising and storage must be controlled for successful operation. Good quality water can even lead to significant savings in cleaning costs.

(3) The rapidly rising cost of fuel oil in the U.S.A., as compared to low costs of electricity, makes reverse osmosis look better for concentration of whole whey, prior to hauling, or for preconcentration to expand existing evaporator capacity.

REFERENCES

Horton, B. S. (1972). 'Plant Design of RO and UF for Food Applications', presented at the meeting, Applicazione Dei Processi per Membrana nell'Industria Latterio-Casearia, of SIPIA, Milan, Italy; 18 May.

Horton, B. S. and Kirkpatrick, K. J. (1972). 'Whey Processing Progress in Other Countries', presented at Whey Products Conference/1972; Chicago, Ill.; 14–15 June, 1972. Published as pp. 100–140 in the proceedings, ERRL Publ. No. 3779, Eastern Regional Research Laboratory, U.S. Dept. of Agric., 600 E. Mermaid Lane, Philadelphia, Pa. 19118, U.S.A.

Selitzer, R. (1972). 'Crowley Begins Membrane Processing of Cottage Cheese Whey', *Dairy and Ice Cream Field*, June.

DISCUSSION

R. F. Madsen (Danish Sugar Company, Denmark): Where do you use all that labour? We normally do not have more than 5%.

Horton: The question is 'where are you putting the plant?' If you have no other processes going on around it; you're stuck with one man. You need other things that the man can do, as in the case of the large reverse osmosis pilot plant I showed you where he ran the evaporators. We have an alarm in the system and he just checked the control panel every two hours.

The labour requirement is in the cleaning time. However, our three commercial plants tell us that our 5% projection might not be what is charged to this process. What we think should be done may not be done in reality.

S. A. Goldblith (M.I.T., U.S.A.): I don't think I would feel complacent about the 10 000/g bacteriological count—for several reasons.

In the first place, what are the organisms involved, secondly, what is the material going to be used for and, thirdly, what is the treatment that is going to be given to it afterwards? I should want to know these before I sat back and did not worry about a 10 000/g count.

Horton: At Crowley they are using some of the liquid concentrates directly in their own products, including possibly yoghurt, where I believe the bulgaricus count is probably rather high and we might be diluting the product—I don't know. On the other hand, the additional processing steps of drying for baby food are going to knock the count way down. We are told that 10 000 g is a fairly happy situation.

Breaking down the total plate count is obviously something that varies from plant to plant.

D. T. Shore (APV Ltd., Crawley, U.K.): When selling whey concentrate or powder, there is a specification for the lactic acid figure a big change in which indicates bacterial contamination. Do you have any figure for that?

Horton: Sure—the first two commercial plants are running on acid whey, and the acidity is high. New Zealand probably neutralise a good deal of their product for some of their applications. The counts at the French plant—it is one month old and it's too soon to tell—but as far as we know there is no pH change; there are no indications that any fermentation has taken place.

A. S. Michaels (Alza Research, U.S.A.): The desire of some people to build commercial systems is pulling things along way ahead of some of the basic knowledge we would like to have. Particularly, let's say in modelling and following the bacterial growth in a staged ultrafiltration plant operating on skimmed milk at 30°C (at an ideal pH for growing bugs). People won't wait for that, and what is happening is that everything is very empirical.

We would like to put a tracer on some protein molecules and follow them through an ultrafiltration system and see what the real distribution is during a couple of months of average running in a typical dairy or food plant.

I'm afraid we'll never get the chance to be that scientific or have that good engineering base for producing products.

Session IV

SPRAY DRYING

Introduction

EDWARD SELTZER

Food Process Engineering, Department of Food Science,
Rutgers University, New Brunswick, N.J., U.S.A.

It is useful for the chairman to establish some perspective within which an all-day symposium can have some boundaries. Though now in an academic institution, this chairman has the sense that spray dryers are intended predominantly for manufacturing (in this case, foods), and we must reconcile an economic basis with our scholarly interest as well as the engineering information needed to improve such an operation.

The manufacture and sale of spray drying equipment is highly cyclical but its use in industry is fairly even because of some major volume items such as non-fat milk powder, instant coffee and tea, fixed flavours and corn syrup solids. In the food industry, unlike old soldiers who fade away, spray dryers usually remain in use for at least a couple of decades. Consequently any one given year's estimate of expenditures for new spray drying equipment is not really a reliable guide to current capacity in a large country such as the U.S.A. Present expenditure in the U.S.A. for food spray dryers is about 15 million dollars annually. This total includes those spray dryers designed and fabricated within the food industries as well as those purchased from spray dryer manufacturers. This volume is about 12–14% of total new dryers annually purchased or installed for foods (including shelf, tunnel, belt conveyer, drum, fluidised bed, flash, double cone, kiln, freeze and other types). A more impressive basis for the importance of spray dryers to the food industry is to estimate the value of foods that have been spray dried. This chairman's guess is that the magnitude in the U.S.A. is of the order of one billion dollars per year.

During my early career, spray drying was an obscure art and not a very reputable one. Design was often by guess work and there were enough charlatans among the bona fide engineers to give the field a bad name. Today, with 30 years of progress, scores of publications, monographs and some few books, spray drying is a well-accepted unit operation and process. By trial and error and much engineering ingenuity and creativity, spray drying has broadened to a wide range of applications and very great flexibility. In fact, we have reached the stage where 'what is new in spray drying often is the improving of the old'.

It would be an imposition for the chairman to intrude upon the subject matter of the lectures in this stage of the programme, but some subjects may be mentioned that have some novelty or current importance in spray drying:

(1) Design of Spray Dryers for Improved Thermal Economy and Size Economy. An example is the use of the multi-tiered centrifugal atomiser wheels having high capacity in one unit; one feed line serves several levels of atomiser outlets therefore creating multiple 'umbrellas'. One can also atomise several materials simultaneously and independently (e.g. vitamins).
(2) Recovery of Waste Solids. Whey is an example of recovery of solids such as by evaporating, crystallising and spray drying.
(3) Conducting Reactions in the Spray Dryer. Carbonates are said to be formed from hydroxides by exposing a solution to a carbon dioxide atmosphere. An alternative or reciprocal of this is to scrub SO_2 out of industrial exhaust gases (as drying medium) by spraying a sodium carbonate solution, thereby forming sodium sulphate. This has a higher conversion rate than the scrubber method, about 100% removal of sulphite versus 88% by water scrubbing. There are said to be many other reactions carried out in spray dryers, but most information is proprietary.
(4) Enrobing, Coating and Encapsulating. In the pharmaceutical field, an active ingredient is coated with a water soluble coating, e.g. gum arabic. The coating and the coated materials must be compatible in order to give continuous coating. Some materials are protective-coated, as, for example, metals with resins, to prevent oxidation. Coating may be done by spray chilling or congealing a slurry in molten wax. Ambient or refrigerated air may be blown during the spraying. Dissolved sugar is sometimes used for spray drying in order to coat another simultaneously sprayed material. Or a dusting powder may be 'fogged in' as fine crystals in order to coat an interior active material.
(5) Spray Agglomeration Systems. There are numerous such techniques. In many cases the spray dried beads are directly clustered within the dryer or the beads may first be pulverised and then run through a second type of spray tower for clustering into agglomerates.
(6) Closed Cycle Drying. With gas-tight tower construction, inert gases such as N_2 can be used successfully in the food industry, an example being modified or pregelatinised corn starch which is pyrophoric, thereby preventing combustion. The N_2 is scrubbed free of moisture and returned for reheating and reintroduction to the tower. Modified

starch is a particularly appropriate case where spray drying has merit above filtration followed by tunnel or belt drying.

(7) Special New Food Uses. Food flavours as well as perfumes are now very commonly spray dried. Currently soybean protein of high purity can be spray dried following solvent extraction. Single cell protein (yeast) from petroleum is now said to be commercially spray dried.

Some operating considerations that are presently of note are:

(1) Use of fuel oils for direct firing that do not contaminate food.
(2) Spray evaporation.
(3) The use of clean-in-place (CIP) concepts for spray dryers; portable or fixed equipment may be operated at very high pressures, up to 2000 psi, for fast clean-up of spray dryers.
(4) OSHA standards in U.S.A. (Occupational Safety Hazard Act) now must be adopted for safe construction of dryers. The user is responsible for factors such as safe stairways, explosion panels, access to operating parts, elevators in place of man-ways, etc.

Here are some challenges: There is a trend toward non-aqueous solvents for which different thermodynamic tables need to be constructed as against the conventional steam tables and psychrometric charts for water–air mixtures. As this kind of operation expands or becomes needed, the properties of the solvent need to be available, such as specific heat, latent heat and the solvent content of the product as it relates to the relative humidity and the temperature of the outlet air.

A colleague of mine is evaluating *mass transfer potential* from knowledge of sorption isotherms of foods. This approach should have some importance to spray drying technology. Such isotherm data also make possible estimating of monomolecular layer moisture content, a practical criterion for drying end-point. Water activities at outlet air humidities as they relate to product moisture constitute important practical and theoretical tools for intelligent spray drying. This is the kind of thinking we expect of the new generation of practitioners.

Effects of the Latest Developments on Design and Practice of Spray Drying*

O. G. KJAERGAARD

Niro Atomizer Ltd., Copenhagen, Denmark

INTRODUCTION

Three decades of research and development have made spray drying a highly competitive means of drying, whether the situation is a delicate operation under sterile conditions or a high tonnage production of chemical products. The range of chemical products suitable for spray drying continues to expand. Unit sizes and capacities have increased substantially through advancements in technical know-how and an appreciation of spray-dryer mechanisms, especially in the atomisation field.

The recent expansion in the application of spray dryers to industrial uses has necessitated great attention being paid to design and techniques, in order to meet the stringent demands which are being made on plant performance. Figure 1 shows a photograph of a typical spray dryer installation.

This paper not only covers advances related to food processing, but to other industries as well. As the subject is spray drying, the developments in, for example, re-wet agglomeration are not mentioned.

Basic principles of spray drying

GENERAL LAYOUT

The spray drying operation is commonly known, but it is opportune at this stage of the lecture to summarise the fundamentals of this drying process.

Spray drying is the transformation of a feed from the fluid state to the dried form in one continuous operation. Feeds in solution, suspension, and paste form can be dried. The feed is sprayed into a hot drying medium, and depending upon the physical and chemical properties of the feed and

* All illustrations are by courtesy of A/S Niro Atomizer.

the dryer design and operation, a dried product conforming to powders, granules, or agglomerates is produced.

Spray drying consists of four process stages: (1) atomisation of feed into a spray, (2) contact of atomised spray with drying air, (3) evaporation of

FIG. 1. Typical spray dryer installation showing air heater, chamber and cyclone battery.

moisture from the spray, (4) separation of dried product from the exhausted air.

Figure 2 shows the principal flow sheet of a spray dryer. The most important components are: the *atomiser* (No. 5), through which the feed is atomised into the hot drying air; the *air disperser* (No. 12), for dispersion of the drying air, when entering the chamber, and the *drying chamber* (No. 16), in which the atomisation and drying take place.

Further equipment comprises: The *air heater* (No. 10) for heating of the drying air. The *fans* (Nos. 8 and 31) for transport of air. A *recovery system* for fines entrained in the outlet air including the *main cyclone* (No. 28). *Pneumatic transport system* for collection of the finished product.

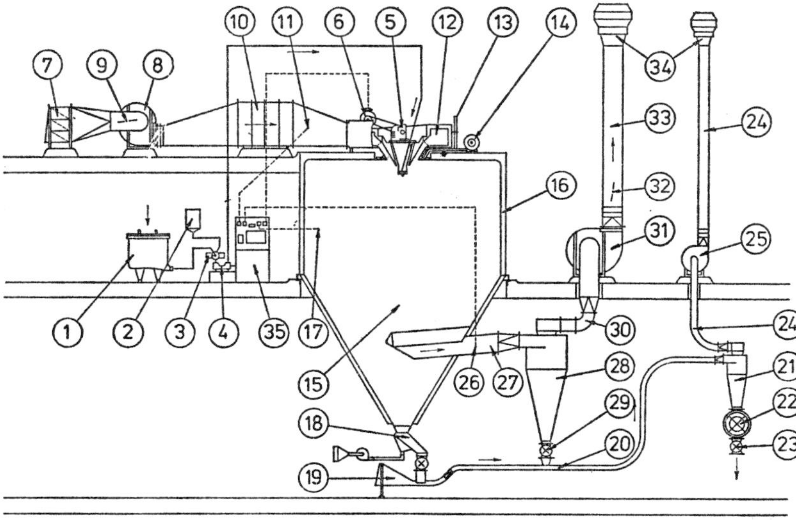

FIG. 2. Diagram of spray dryer.

MECHANISMS OF ATOMISATION

The selection and operation of the atomiser are of supreme importance. The prime function of atomisation is the production of:

(1) a high surface-to-mass ratio resulting in high evaporation rates;
(2) a finely divided product of the desired particle shape and density.

To meet these requirements, many atomisation techniques have been used in spray dryers, the most common being atomisation by:

(1) Pressure energy as in Centrifugal Pressure Nozzles;
(2) kinetic energy of air as in Two-Fluid Nozzles;
(3) centrifugal energy as in Rotary Atomisers (Wheels).

The mechanism of atomisation has been studied by many workers and the subject is still highly controversial, in spite of much published data.

The mechanism now accepted as applicable to commercial atomisers has resulted from work by pioneers on the stability and collapse of simple liquid jet (ligament) systems. Simple relationships were then projected, and these were extended to the more complicated conditions which actually occur during the atomisation process.

The progress in the development of mechanisms applicable to commercial atomisation can easily be traced. As in so many research studies, one man

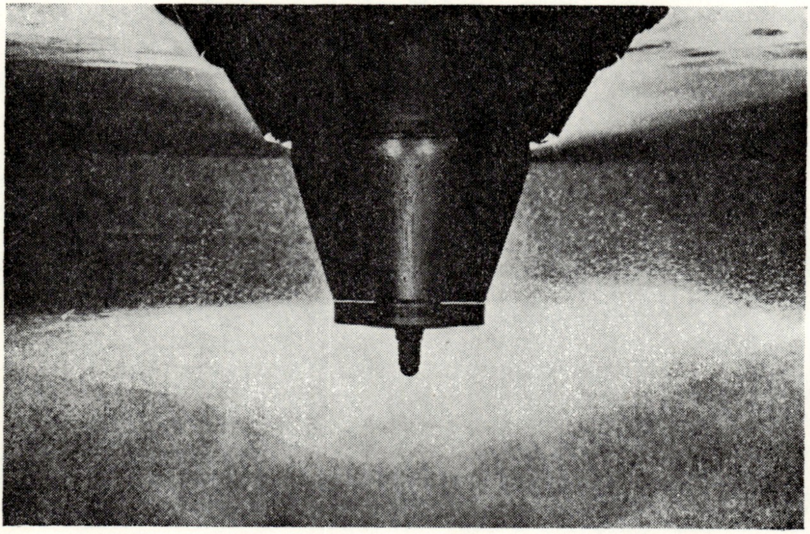

FIG. 3. Working rotary atomiser.

is credited with their innovation, in this case Lord Rayleigh who in 1878 published his work on the mathematical break-up of non-viscous liquid jets under laminar flow conditions.

In this connection it is of at least historic interest to note that the first patent on spray drying was granted to S. R. Percy, 1872.

Figure 3 shows a rotary atomiser (Niro Atomizer design) in operation.

SPRAY–AIR CONTACT AND FLOW PROFILE

The way in which spray is contacted with air is determined by the position of the atomiser in relation to the drying-air inlet. The actual flow profile is dependent upon the design of the air disperser incorporated into the dryer. Air flow within spray drying chambers is a major consideration in obtaining optimum product quality at maximum evaporation capacities. The movement of air predetermines the rate and degree of evaporation by influencing the passage of spray drops through the drying zone, the concentration of product in the region of the chamber wall, and the extent to which semi-dried particles re-enter hot spots around the disperser (through eddy-flow effects).

Air flow is classified as cocurrent, countercurrent, or mixed flow (a combination of the former two), as related to the passage of product from the atomiser to the drying chamber outlet.

EVAPORATION FROM DROPLETS

When droplets, formed during atomisation of the feed, contact the hot air, evaporation of moisture takes place. The initial evaporation from the droplet leads to the formation of a crust on the surface of the particle. Evaporation is completed when this dried layer extends down to the centre of the particle. Permeability and strength of this crust are determining factors for the appearance of the end-product.

The rapid evaporation keeps the particles cool. This is essential when handling products that are heat sensitive. Due to the evaporation/cooling effect the product thus seldom reaches the temperature of the air at the chamber outlet.

POWDER RECOVERY

Product separation is closely connected to chamber design, and pneumatic transport systems, cyclones, bag filters, and scrubbers are used. The topic of dried product handling is a subject in its own right.

FINAL REMARKS

The spray drying is characterised by the following advantages:

(1) The product will dry without getting into touch with hot metal surfaces.
(2) Product temperature is low—also when the inlet drying air is of relatively high temperature.
(3) As the evaporation takes place from a large surface, the time of the drying operation is a matter of a few seconds.
(4) The finished product is a stable powder, easy to handle and transport.

Spray drying has been accepted for the solution of many drying problems because the operation has proved not only efficient but economic. Overall cost analysis shows how competitive spray drying is in comparison with other forms of drying, especially at high capacities.

For further information reference is made to the comprehensive work of Masters (1972), which also includes up-to-date references.

TRENDS IN DEVELOPMENT OF SPRAY DRYER DESIGN AND PRACTICE

General

The basic designs of the standard types of spray dryers have not changed for many years, but a number of individual designs for specific products and drying processes have been developed. These special designs will be dealt with below.

Some of the components of the spray drying plant have been subject to an intense development work. An interesting example is the new designs of rotary atomisers as mentioned below.

Generally speaking, the efforts towards improvement of the general design have concentrated on increasing plant capacity. This involves at least one of the following two means.

Qualitative

(1) Increasing the effective temperature drop across the dryer, which is equivalent to increasing the inlet temperature.

Quantitative

(2) Increasing the amount of drying gas through the drying chamber.

By increasing the inlet temperature, two aspects must be considered: the heat resistance of common materials of construction, and the problem of avoiding deposits in the hotter zones (i.e. the ceiling) of the chamber when drying heat sensitive products.

Today the maximum inlet temperature is 600–800°C for non-sensitive products, and 400–500°C for heat sensitive products.

Increase in amount of drying air of course means larger plant components. The limit for the time being is the air dispersers: approximately 300 ton/h of air through one plant corresponding to approximately 30 ton/h of water evaporation at, for example, 350°C to 105°C at the inlet and outlet of the dryer respectively. This actually involves both possible types of air disperser at the same time, the so-called ‘sandwich-drying-system’, as half of the air is introduced through the ceiling above the atomiser, and the other half from below through a central air disperser of the chimney type.

An example of a recent change in design of the drying chamber is shown in Fig. 4. The W-chamber is a more compact and space-saving design. Approximately 25% reduction in building height is obtained by having the cone tip built up into the dryer bottom cone. The product recovered at the bottom of the chamber is collected by a rotating arm and a screw conveyor.

FIG. 4. Bottom of W-type chamber.

NEW ATOMISER DESIGNS

General

As mentioned previously, the atomiser is a most important part of the drying assembly.

The rotary atomiser is selected whenever possible, as it features the following advantages over pressure nozzle atomisers.

(1) Droplet size can be varied by adjusting the design and speed of the wheel independent of the feed rate, which could be changed from zero to maximum.
(2) A full spectrum of sizes from a few kilograms per hour to 30–50 tons per hour is available.

The two-fluid atomiser, however, also has the characteristic of point (1) as the atomisation can be varied by adjustment of the flow of atomising air independent of the feed rate, but all nozzle atomisers are limited to maximum capacities of approx. 2 ton/h per nozzle. Thus a large number of nozzles must be used in parallel to make up for one rotary atomiser.

The advantages of the nozzles have been the possibilities of:

(1) enabling production of large particles from big drops, and

(2) making wear-resistant atomisers simply by constructing the orifice plate of wear-resistant material, e.g. tungsten carbide.

The prilling atomiser

Typical average particle size for a rotary atomiser is between 30 μm and 150 μm depending on the physical properties of the feed and the atomiser.

With high speed centrifugal atomisation, the liquid feed is distributed centrally in the wheel and accelerates to the wheel edge, usually guided by vanes to prevent slippage, so that on discharge from the wheel the liquid has attained the peripheral speed of the wheel. The vanes also concentrate the liquid at the wheel edge, producing a liquid sheet similar to that considered in pressure nozzle techniques.

The formation of drops on disintegration of this liquid sheet is dependent on a combination of the viscosity and surface tension of the liquid feed, the inertia of the liquid feed at the wheel periphery, and the frictional effects at the liquid–air interface.

With speeds generally used in commercial atomisation, inertia and air friction are the main contributing factors to the mechanism of drop formation.

A new design of rotary atomiser has been made to make it possible to increase the particle size when atomising feeds based on melted products for spray congealing.

A new design of rotary atomiser has been made to make it possible to increase the particle size, for instance when atomising feeds based on melted products for spray congealing. This process normally involves the desire for production of very large particles (typically 0.5–1 mm diameter) with a narrow size distribution.

In this new type of atomiser the warm molten feed enters through a heated feed pipe and is atomised into large droplets when passing a number of holes in the outer wall of the wheel. The speed of the wheel is much lower than normal. The droplet size is governed by the size of the hole and the contraction factor. The assembly is pre-heated before start-up and kept warm by circulation of a heating medium around the feed pipe and in the wheel.

An interesting application of spray cooling is found in the production of wash-active enzymes for detergents. The enzymes are manufactured by fermentation followed by spray drying of the concentrated enzyme solution. A slurry of the dried enzyme product is made in melted wax. Titanium dioxide and sodium sulphate can be applied—the TiO_2 to give the final product the same colour as the detergent—the Na_2SO_4 to adjust the product to the final activity.

The wax used for this coating will be a water-soluble wax or so-called non-ionic, i.e. a surface active material already used in the detergent. The melting point is around 55°C.

The enzyme/wax slurry is spray cooled in a plant similar in construction to a spray dryer, but cold air is used to congeal the particles.

The requirement to prilled enzyme preparation calls for a particle size with about 99% greater than 100 μm. Average particle size is normally within 200–500 μm.

The resulting free-flowing product is now ready to be incorporated into detergents and will in this form exhibit all the properties wanted by the end user, i.e. stability and controlled release rate.

The wear-resistant atomiser

While, as previously mentioned, it has been easy to design a wear-resistant pressure nozzle by manufacturing the orifice plate of, for example, tungsten carbide as a sintered press-body, a wear-resistant atomiser wheel presented more of a problem.

The atomiser wheel design (Moeller and Straarup, 1969) for handling abrasive and corrosive mineral concentrates is illustrated in Fig. 5. The atomiser wheel features wear-resistant sintered bushings so placed to protrude within the inner chamber of the wheel. Such arrangement produces a protective layer of solids to be formed right from the start of the spray drying process. Solids separate out from the feed as a layer along the outer wall of the inner chamber. The thickness of the layer corresponds to the projection distance of the bushings. The protecting layer of solids, together with the use of wear-resistant bushings and a wear-resistant baseplate enables the wheel body to be fabricated in stainless steel and yet possess a long operating life. The atomiser wheel possesses the most effective and cheapest built-in abrasion control, since the abrasive solids in fact provide their own protective layer.

Use of wear-resistant bushings and baseplate exceeds by several hundred times the life of conventional wheels, when operating with abrasive and corrosive feeds. However, even with wear-resistant bushings, wear does occur slowly over the surface of the bushings in direct contact with feed concentrate. Bushing life is quadrupled by repositioning the bushings to equalise wear over the entire surface. This is accomplished by rotating the bushings 90° at a time.

This design has proved entirely satisfactory during long periods of operation.

An interesting application of this atomiser is the plant for drying cement raw meal slurry shown on the flow sheet of Fig. 6 (Gude and Lund, 1972; Damgaard-Iversen and Kruse, 1973; Hansen *et al.*, 1973).

A spray dryer (of the W-design) is mounted directly on the outlet from the rotary kiln. The inlet temperature to the spray dryer is 600–900°C. The spray dried raw meal collected from the drying chamber and the cyclones is fed to the rotary kiln.

Experience has shown that where a spray dryer is linked to a kiln, many advantages are offered.

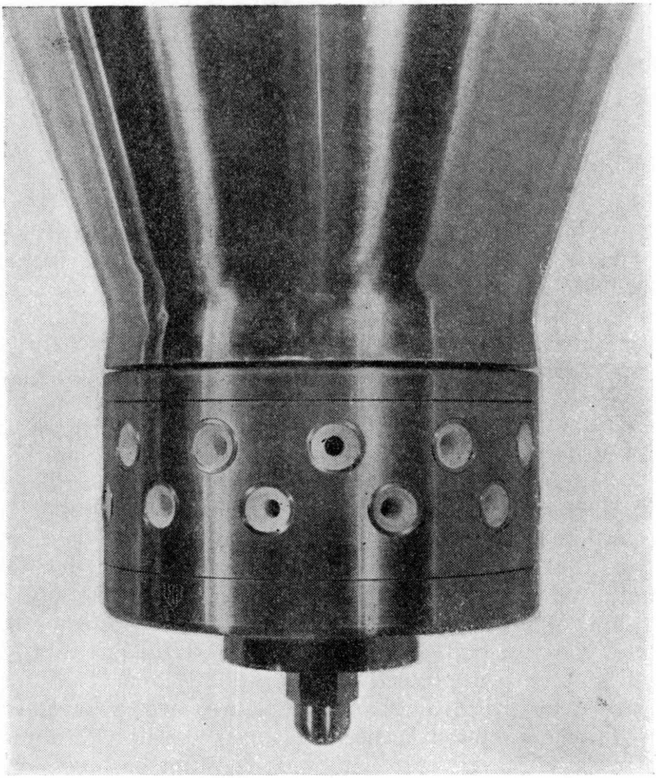

FIG. 5. Wear-resistant atomiser wheel.

(1) The kiln operation is greatly simplified by complete elimination of crosses and chains, and the evaporative load is transferred to the spray dryer.
(2) Production is increased as the whole kiln is utilised for heat treatment.
(3) Savings in fuel consumption are achieved by keeping the outlet gas temperature from the dryer as low as 130–150°C.
(4) The short residence time characteristic of the spray dryer enables plant start-up to be completed quickly, whereas during operation,

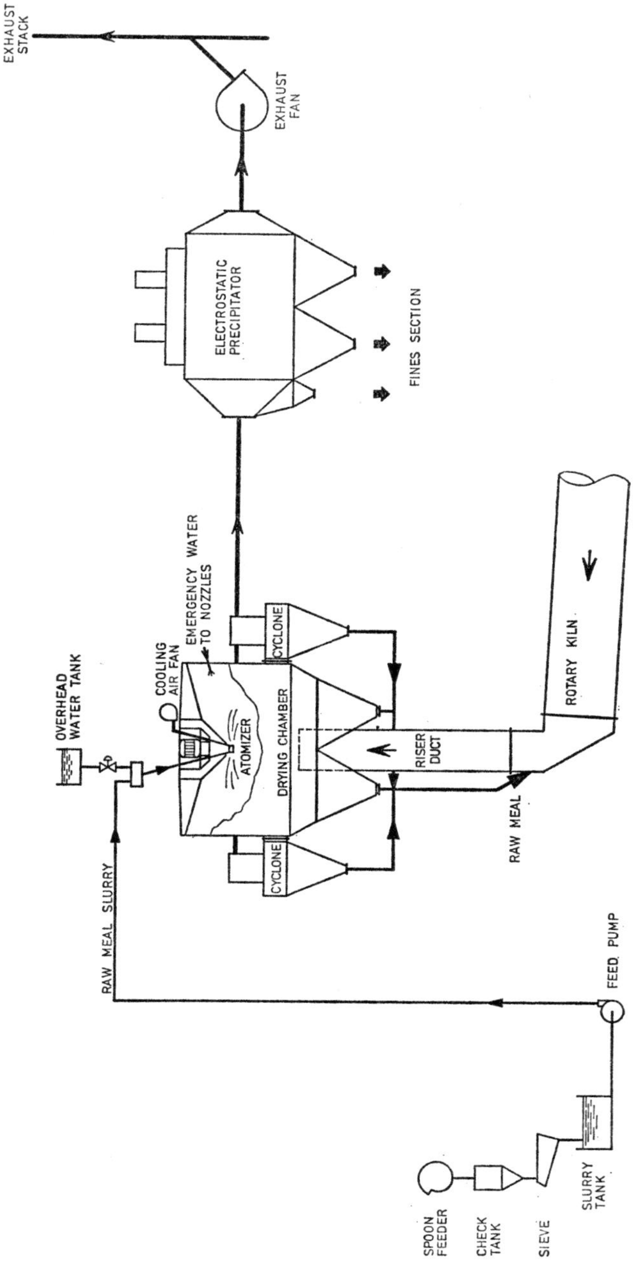

FIG. 6. Flow diagram for drying of cement raw meal.

processing conditions of the combined dryer/kiln system remain much steadier than if a conventional wet kiln operated alone. This may in turn lead to a higher run factor.

(5) Furthermore, a slurry can be processed that is considerably heavier than that normally handled in the chain section of a kiln.

The first two plants of this type in operation have shown that the capacity of a given kiln could be increased by 50–70% at the maximum. Such economic and operational advantages make application of spray drying a sound proposition in the cement industry.

NEW DESIGNS AND PRACTICES OF SPRAY DRYING EQUIPMENT

General

The latest developments can be divided into two groups, as follows.

(1) Developments concerned with meeting desired product requirements, and
(2) new types of plants to perform special drying operations.

Developments to meet various product specifications

Meeting product requirements is closely linked with equipment design and operation.

Spray drying is a procedure which by various adjustments is capable of meeting many industries' end-product specifications for subsequent processing or direct consumer usage. Desired product requirements can be listed under the properties of:

(1) particle size distribution;
(2) bulk density and particle density;
(3) moisture content;
(4) flowability;
(5) friability;
(6) wettability, penetrability, sinkability, dispersibility and solubility;
(7) retention of colour, flavour, activity and aroma.

This will be illustrated by the following examples:

(1) Drying of enzymes illustrating (1), (4) and (7).
(2) Drying of skim milk illustrating (1), (2) and (6).
(3) Drying of whey illustrating (4), (5) and (6).
(4) Drying of tomato pulp illustrating (3), (5) and (7).

(5) Lecithination of whole milk powder illustrating (1) and (6).

SPRAY DRYING OF WASH-ACTIVE ENZYMES

The use of enzymes in detergents can be traced back to the beginning of this century, but large scale industrial production of proteolytic enzymes for washing purposes did not take place until after 1960.

Wash-active enzymes were soon accepted and in 1969 it was estimated that more than 50% of the detergents used in Europe contained enzymes.

About the same time attention was focused on the potential health hazards resulting from inhalation of enzyme dust. This problem was thoroughly investigated by The Federal Trade Commission in the U.S.A. from 1969 to 1971. The conclusion of this extensive work clearly pointed out that detergents containing enzymes were not injurious to health and that the use of enzymes might even have a positive effect on water pollution.

In the meantime work was carried out to change the physical appearance of the products, and today most enzymes are produced as free-flowing and stable powders.

Spray drying is used to convert the purified enzyme solution or suspension into a dry, stable and dust-free product with 3–5% residual moisture. In order to dry the enzyme preparation, a certain amount of Na_2SO_4 has to be added. Generally speaking, the end product will contain 50% or more of Na_2SO_4. Without addition of Na_2SO_4 the enzyme product is sticky and difficult to dry.

By applying spray drying alone, an activity loss of up to 30% (depending on type of enzyme) is often observed. Spray drying combined with fluid-bed drying can reduce this activity loss to 1–5%. The principle of this system is shown in the flow sheet in Fig. 7 (Mahler and Markussen, 1972).

The drying conditions are selected in such a way that the powder leaves the spray dryer with a moisture content of 14–18%. The moist powder falls into the fluid bed, and final drying at low product temperature continues through the first fluid bed. In the last fluid bed ambient air is used to cool the product down before bagging-off. By dividing the drying into two steps, maximum activity can be maintained.

Small particles which are collected in the cyclones are returned pneumatically to the atomisation zone for agglomeration with the evaporating spray.

The air velocity in the fluid cooling bed is high enough to separate off the fines. The powder leaving the fluid cooling bed is dust free, free-flowing and has retained more than 95% of the original activity.

SPRAY DRYING OF SKIM-MILK

Bulk density of milk powder

Bulk density influences the packaging and transport costs of milk powders and is influenced basically by the type of atomisation.

Such variables as the degree of concentration, viscosity of the concentrated milk, drying air temperature, particle size distribution, the degree of particle agglomeration, and the handling of powder also affect the packing density.

As powder specifications are becoming harder to meet with respect to solubility, wettability, flowability, etc., there is a trend to make a compromise between all these requirements hence the packing density to a certain extent is not viewed with such a critical eye as it has been in the past.

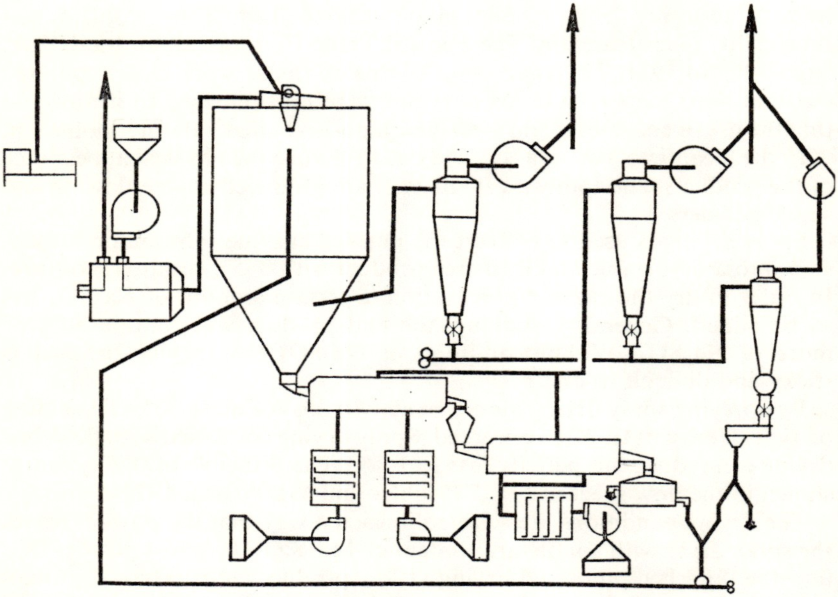

FIG. 7. Flow diagram of spray dryer with fluid beds attached.

As a large fraction of the production of skim-milk powder is not used directly for consumption, but serves as a raw material for a number of industries, there is a strong interest in the manufacture of high density skim-milk powder.

The ideal structure of such a product is:

Macro structure

(1) Single un-agglomerated particles with a certain wide size distribution to allow smaller particles to fill the space between larger ones, thus a minimum interstitial air;

Micro structure

(2) The highest possible density of the solid part of the individual particles and the lowest possible amount of occluded or entrapped air. The occluded air also affects the shelf life of the powder.

It has been relatively easy to establish the optimum values for concentration, temperature and viscosity of the feed and the drying conditions:

(1) maximum solids content of feed;
(2) minimum viscosity of feed to facilitate atomisation;
(3) low drying temperatures, i.e. slow drying, to avoid evaporation inside the droplets.

Much work has recently been done to improve the atomisation in order to give the best particle size distribution and minimum occluded air.

Some years ago a special atomiser wheel was developed featuring special S-shaped channels (Nyrop, 1958). The increased length results in a thinner liquid film and improved atomisation. Due to the shape, less air is whipped into the feed, and due to the extra centrifugal and shear forces some of the air in the feed is removed just prior to atomisation.

Normally the bulk density of spray dried skim milk is approx. 0.55 g/cm^3 (bagging density, the measuring device is tapped 100 times) and the amount of occluded air is around 40 cm^3/100 g.

With the S-vaned wheel bulk density was improved by 10% to just above 0.6 g/cm^3 as the amount of occluded air decreased to approx. 20 cm^3/100 g.

The latest development (Písecký and Soerensen, 1972) is based on the removal of air from the feed by letting the feed boil in the channels of a specially designed atomiser wheel. The heating medium is steam injected into the feed in the atomiser wheel itself or in the feed pipe.

The resulting powder has a density of just below 0.8 g/cm^3 (33% increase!) and the amount of occluded air is as low as 6 cm^3/100 g.

This system has so far also been tried on an emulsion of vitamin A and whey in gelatine. The bulk density was only improved from 0.50 to 0.55 g/cm^3 (tapped 100 times), but the amount of occluded air was decreased from 40 to 20 cm^3/100 g, which means improved stability of the vitamin.

Meeting 'instant' properties in spray drying

Closely connected with product specification is the demand nowadays for 'instant' products. Such products are free-flowing in form and are easily wetted and dispersed when forming a solution.

Spray drying equipment can be designed to meet 'instant' powder requirements. Powder can show improved instant properties by incorporating a fluid cooling bed at the base of the chamber. Powder of superior instant properties can be produced in a so-called straight-through instantiser, a special development.

The flow diagram of Fig. 7 also illustrates how instant skim-milk powder results from agglomeration in two separate stages. The initial agglomeration of fines is carried out in the atomisation zone amidst the evaporating spray. Further powder agglomeration occurs within the drying chamber. The spray dryer is operated to maintain a powder moisture content of 6–8% at the dryer outlet.

Under these conditions the powder possesses thermoplastic properties which readily promote agglomeration through self-adhesion of particles. The moist warm powder passes out of the chamber conical base into vibrated fluid beds (Vibro-Fluidizers)* for completion of drying and subsequent cooling. The agglomerated product is finally screened and any fines are returned to the agglomeration zone of the chamber.

The product consists of large porous agglomerates. This structure is characteristic of instant products with excellent dispersibility and wettability. The product is also free-flowing and non-dusty, due to the absence of fine particles.

With some slight modification this type of plant is also used for the production of non-caking whey powder, or products with high fat content.

PRODUCTION OF NON-CAKING WHEY

The high nutritive value of whey was known well before dairying became an industry and yet whey is still the subject of research to discover new biological uses and market potential. In spite of its high nutritive value, whey has until recently been considered a troublesome by-product with a great proportion of its production being discharged into rivers.

Spray drying opened up the way for whey utilisation by producing whey powder for feeding purposes. After this first step the benefits of using dried whey for feeding were very soon recognised and then this stimulated an effort to remove the drawbacks of ordinary whey powder, namely its caking and dustiness tendencies. Whey powder has become a constituent in many types of human foods, for example bread and other bakery goods, ice-cream, sherbet, soup powder, melted cheese, and even baby food.

The main reason for caking, hygroscopicity, stickiness and all the difficulties of drying whey is amorphous lactose, which constitutes approx. 75% of the solids content of an ordinary non-crystallised whey powder. Transforming amorphous lactose into alpha-lactose monohydrate is one of the methods which leads to a caking resistant powder. However, there are other conditions which have to be fulfilled, as the hygroscopicity of whey powders is caused also by proteins and salts.

A 100% crystallisation cannot be achieved during the manufacturing

* Trade-mark of Niro Atomizer.

process, but in practice a maximum of 90–95% is obtained. Throughout the whole manufacturing procedure conditions must be maintained to keep the crystallisation process continuing.

The principles of the plant for production of non-caking whey are seen in Fig. 7 also.

The whey concentrate after pre-heating and concentrating to about 50% solids is pre-crystallised in special tanks. Some 70–85% of the lactose is transferred into the desired alpha-form.

The spray drying proceeds at an outlet temperature which retains 6–7% of moisture in the product ensuring good agglomeration. The product enters the vibrated fluid beds directly from the drying chamber for final drying and subsequent cooling. The product made by this process is one of relatively small agglomerates, good flowability and relatively high bulk density.

The very latest design incorporates an after-crystallisation stage on a special belt conveyor between the drying chamber and the fluid beds. The spray dryer is operated at an outlet temperature of about 55°C. Under these conditions the powder leaving the chamber has a moisture content of 10–14%, as this is essential to achieve an after-crystallisation.

The final product consists of very coarse agglomerates having a relatively low bulk density. This is the most preferred process being able to handle the more difficult types of whey and giving very good instant non-caking properties. The successive crystallisation ensures that the final product has up to 95% of total lactose in the crystalline form. If higher bulk density and smaller agglomerates are desired the product may be treated in a hammer mill.

DRYING OF TOMATO PULP

Tomato pulp is a typical example of a product which is very difficult to dry as the powder is sticky and caking due to its hygroscopic and thermoplastic behaviour. It has however, through careful analysis of the problems, been possible to design a spray drying plant capable of producing a rich, red free-flowing product that reconstitutes to a wet product and compares favourably with the colour, flavour, grade, and serum separation characteristics of tomato pastes used for remanufacture.

The solids content of a fresh tomato is around 5%, that of commercial tomato paste is 30–32%, and that of a spray dried powder 96.5–97%.

Tomato solids in powder form have many advantages:

(1) shipping powder instead of paste can reduce transport costs by 50% or more, as a packing container filled with powder contains more tomato solids than the same container filled with paste;
(2) powder packing eliminates the drum clingage losses associated with paste;

(3) powder directly meets the demand for tomato solids in dry mix use; and
(4) tomato solid handling is much easier with solids in powder form.

The processing of ripe tomatoes, fresh from the field, involves three stages to produce tomato paste ready for spray drying: washing and sorting, pulping (breaking) and pulp concentration. Normally this preparation takes place according to the so-called 'hot break' method: rapid heating causes destruction of enzymes which prevents decomposition of the pectin in the tomatoes. It also liberates the gummy materials which surround the tomato seeds, and which—together with the pectin—help to give 'body' to the finished paste.

By the 'cold break' method the tomatoes are crushed at room temperature. The enzymes (liberated during crushing) will act as a catalyst for the decomposition of the pectin, resulting in a paste which is easier to spray dry than a paste obtained by the 'hot break' method.

The powder achieved from 'cold break' paste has less desirable characteristics on reconstitution, as the solids will settle after approx. 60 s. The reconstitution of 'hot break' powder does not form this settling, but remains in homogeneous suspension.

'Hot break' pastes are spray dried at concentrations ranging between 26 and 32% solids. 'Cold break' pastes are concentrated up to 38% or more prior to spray drying.

The spray dryer is specially designed to cope with the properties of tomato powder. A non-standard drying chamber construction is used. If conventional chamber designs were employed, the thermoplastic and hygroscopic properties of the powder would create such problems of operation that batch operation would have to be adopted. The spray dryer layout is shown in Fig. 8. It features a cocurrent drying chamber having a jacketed wall for air cooling and a conical base. Ambient air is drawn through the jacket prior to entering the chamber via the air heater. Cooling air intake is controlled to enable close maintenance of a wall temperature, which in the range 38–50°C maintains continuous operation. The control panel contains instruments for indication of temperatures over the cylindrical and conical portions of the drying chamber.

The paste is sprayed into the drying air entering the chamber at a temperature of 138–150°C (280–300°F). The drying air to the heater is supplied from the cooling air wall jacket supplemented by atmospheric air intake.

The product settles out of the air flow on to the chamber wall, building up to loose layers (3–5 cm thickness) before breaking away and falling as nodules to the base of the chamber. This build-up is important for completion of evaporation. Increased drying temperatures cannot be used as heat degradation of the product will result.

The majority of the product falls from the chamber base into an enclosed band conveyor. Cool dehumidified air flows countercurrently slowly over the band surface. The product nodules are cooled from approx. 50°C down to 24–30°C. This enables the nodules to become brittle, and readily shatter into powder, as they pass a specially designed vibro-fluidiser for disintegration, drying, cooling and sifting.

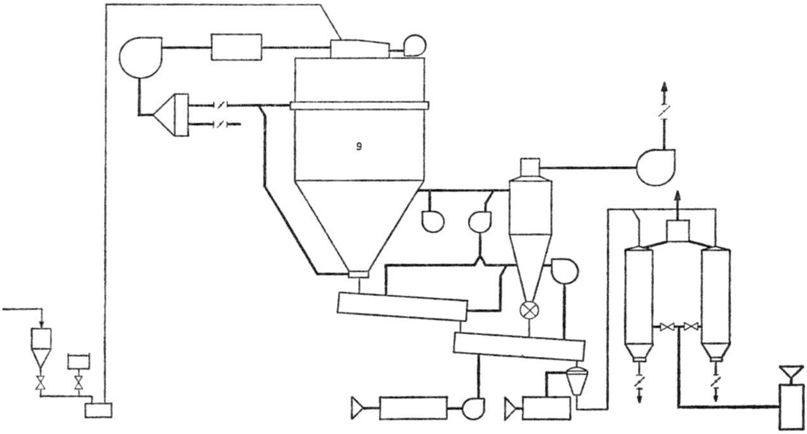

FIG. 8. Flow diagram of spray dryer for tomato pulp.

Some 15–20% of the throughput does not settle on the wall, but passes out of the chamber with the exhausted drying air. All the entrained product is recovered in a cyclone and is fed back to the fluidiser. The resulting free-flowing powder is conveyed in dehumidified air to the storage hoppers for packing.

LECITHINATION

The demand for fat-containing dairy products, such as whole milk powder, is rapidly increasing. For fat-containing milk powders, however, agglomeration alone is not sufficient to obtain instant properties because the wettability is poor. These powders contain free fat which forms a coating on the surface of the particles making them water-repellant.

One modern process is based on the recognition that it is especially the part of the free surface fat of the powder present in the solid state which has an impairing effect on the wettability. A fractionation of butterfat is therefore carried out prior to drying of the milk concentrate. The high-melting fraction of the fat is recycled to the milk concentrate before it is spray dried while the low-melting fraction is applied as a coating on the dried powder. The water-repellent effect of the solid part of the free fat is thereby suppressed.

Recent developments (Písecký and Westergaard, 1971, 1972) have made it possible to simplify this process. Whole milk powder with excellent reconstitution properties in cold water may be obtained if certain measures are taken as regards the content of free fat, particularly that part of the fat that is liquid at room temperature, and the quantity of an edible wetting agent such as lecithin in the surface of the product. Once again the flow sheet of Fig. 7 can be used for a brief explanation of the process.

The plant is adjusted for production of agglomerated whole milk powder according to the straight-through method.

Between the outlet of the first and the inlet of the second vibro-fluidiser the hot agglomerates are treated with lecithin. A warm mixture of lecithin and butter oil is sprayed onto the falling powder by a two-fluid nozzle. Final lecithin concentration should be 15–20% of the remaining liquid portion of the surface fat after crystallisation.

During the fluidisation in the second vibro-fluidiser the lecithin content is equalised over all the agglomerates. At the same time, the last of the fines are removed and transported to a cyclone, where they are separated and bagged off. The final powder is then fed directly to a filling machine, where it is gas-packed.

After the lecithination process such a whole milk powder will dissolve even in ice cold water within a few seconds.

New special designs

In the last few years spray drying has become more than a process of water removal yielding powders consisting of small particles or agglomerates.

Today a great many processes involve solvents other than water or involve drying under aseptic conditions, and sometimes also special particle size requirements must be met.

This has called for new special designs, which will be presented below.

(1) The Spray-Fluidiser;
(2) the Aseptic and Sanitary Design;
(3) the Closed Cycle Drying System, and finally
(4) the Self-Inertising System.

THE SPRAY-FLUIDISER

Where very large ball-shaped particles of the size 0.5–2 mm are required, these may be obtained directly during the spray drying depending on the properties of the product. A new type of spray dryer, shown in Fig. 9, has been developed for this purpose.

The plant is in fact a hybrid between a counter-current spray dryer with nozzle atomisation and a fluid-bed dryer.

The chamber is equipped with a perforated plate in the conical part for the distribution of the drying air which is introduced to the cone.

When a suitable amount of product in a powder form is put into the chamber and the amount of drying air is adjusted, a fluid bed is formed on the plate.

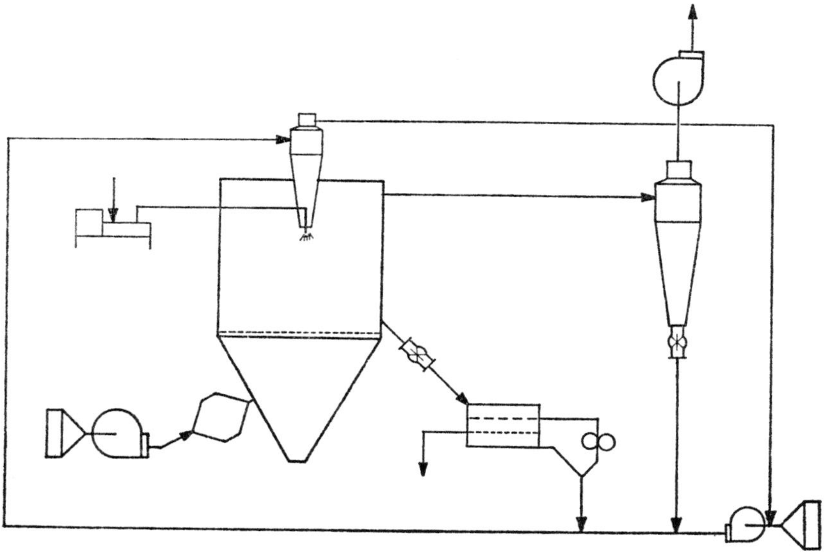

FIG. 9. Flow diagram of spray fluidiser.

The feed is atomised downwards onto the particles of the bed and drying takes place from layers of feed on the particles in the bed. The particle size is increasing and by adjusting the sieving of raw product, the possible milling of oversize and recycling of entrained fines, it is possible to maintain steady conditions at the desired particle size. An exact balance of number of individual particles within the system must be maintained. All the fines are cycled back by a closed pneumatic transport to a cyclone. The fines are introduced around the spray of the nozzle.

The process is limited to products which exhibit a certain physical behaviour related to binding strength. So far, excellent results have only been obtained on inorganic products such as sodium sulphate and ceramic clay (Kjaergaard and Storm, 1972).

The advantages of this plant are:

(1) direct production of large-size particles,
(2) a compact design as the evaporative capacity of a fluid bed layer per volume is very high.

ASEPTIC SPRAY DRYING

Companies manufacturing pharmaceuticals and biochemicals have shown an increasing interest in the application of spray drying.

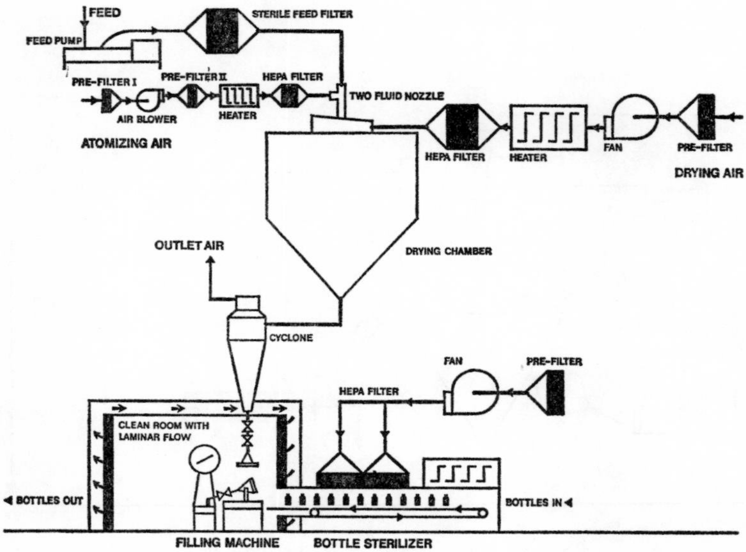

FIG. 10. Flow diagram of aseptic spray dryer.

The development of High Efficiency Particulate Air Filters (HEPA Filters) has made it possible to perform drying under extremely clean conditions. A non-contaminated product cannot, however, be expected by just installing HEPA filters for the drying air on a normal spray dryer.

Contamination of the product does not only come from the air, but also from any sliding or rotating devices, e.g. rotary valves and rotary atomisers. Therefore after sterile filtration of the liquid, the product must not be contacted or handled by such devices.

To prevent contamination, special dryer designs, materials, surface finish, packing materials, and maintenance are extremely important for aseptic operations. The plant design for a typical aseptic spray dryer for serum hydrolysate is shown in Fig. 10.

Two-fluid nozzle atomisation is used and both drying air and air for atomisation are filtered through HEPA filters before entering the drying chamber.

The chamber is kept under slight overpressure to prevent any leakage into the system. The discharge of powder from the cyclone takes place in a laminar-flow cabinet or 'clean room' through a special double valve ensuring a dust-free discharge. The powder container is interchangeable with the one on the filling machine. The filling machine has no rotating parts. The dust-free filling of bottles and ampoules is by automatic vibration weighing.

Bottles and ampoules pass through a sterilisation tunnel. All other items used in the clean room enter through a double-doored autoclave.

The plant is easy to operate and clean. Sterilisation before operation is conducted with 250°C air.

CLOSED CYCLE DRYING

Closed cycle operation leads spray dryer application into fields where:

(1) the vapours released during evaporation are poisonous, inflammable or must be recovered;
(2) the drying must take place in an inert atmosphere to prevent undesirable side reactions, e.g. oxidation or powder explosion;
(3) the product is poisonous or, for other reasons, it is undesirable to permit traces of product to be emitted from the plant to atmosphere.

In principle the closed cycle drying system is based on re-conditioning of the drying gas for returning it to the drying zone and recovering of the evaporated solvent. Figure 11 shows a typical flow diagram for such an installation.

The vapour content of the drying gas (normally nitrogen) leaving the dryer has increased and solvent must be removed from the gas to re-establish a certain constant vapour content before the gas is heated and re-used.

This regeneration is performed in a scrubber/condenser where cold solvent is circulated. Part of the solvent is introduced as a countercurrent spray at the bottom to cool the exhaust gas from the outlet temperature to the adiabatic saturation temperature. Another part is flowing over a system of trays in stage-wise countercurrent to the gas ensuring that the gas leaving the top of the condenser is saturated with vapour at a temperature close to the low inlet temperature of the circulating solvent.

The solvent is cooled in a heat exchanger and an amount equal to the evaporation in the dryer is discharged continuously for reuse in the preparation of the feed.

In order to limit the losses of nitrogen and solvent vapours, despite the overpressure maintained to avoid intake of false air, the plant is constructed to be very gas-tight. The fan shafts, for instance, have special labyrinth-seals applying gaseous nitrogen or liquid solvent. The product is recovered either discontinuously in gas-tight containers or continuously from a product hopper as shown on the diagram, depending on capacity.

The oxygen content in the closed cycle is surveyed by a continuous analyser activating the automatic nitrogen supply. The pressure is automatically controlled by a purge from the system.

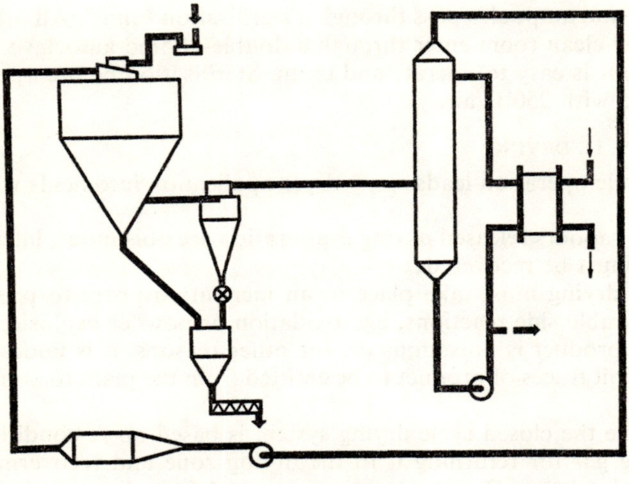

FIG. 11. Flow diagram of closed cycle spray dryer.

Closed cycle plants do require special design considerations. The very nature of the closed cycle system requires detailed process and product know-how to carry through the necessary design calculations. In the open or 'one pass' system the drying air properties are independent of the conditions within the dryer. In the closed cycle plant the drying gas is fed back from the condenser at a temperature and relative humidity depending upon the conditions within the dryer.

The closed cycle dryer is a dynamic system incorporating one or sometimes two feed-back loops. A specialised approach is needed to design such a plant with optimised layout and operating conditions.

The factors governing the economics of a conventional dryer are also relevant to those of a closed cycle dryer. High inlet and low outlet temperatures result in a low gas rate with correspondingly small plant size and low heat consumption.

An additional factor is the effect of the dew point of the recycled gas. The equilibrium conditions between the vapour content in the gas and the volatile content in the product may necessitate a low dew point in order to achieve the residual volatile content desired.

However, a reduction of the recycle temperature will increase the cost of cooling and reheating and will have the adverse effect of reducing the heat capacity of the gas. Consequently, in order to transport the same amount of heat the gas rate must be increased. All these factors must be taken into account when designing a closed cycle dryer.

Finally, there may be a choice of several organic liquids, each presenting different thermodynamic properties.

Typical examples of the adoption of this technique are the manufacturing of:

(1) tungsten carbide (hard metal) presspowder, where wet milling of tungsten carbide has to be carried through in organic solvents such as acetone, ethyl alcohol or others in order to prevent oxidation from water;
(2) pharmaceutical products containing an active substance which is not to be released immediately upon consumption. Here a suspension of the preparation in an organic solvent containing a dissolved coating agent, e.g. a polymer, is dried.

SELF-INERTISING SYSTEM

A modified version of the closed cycle system (Fig. 12) is applicable for drying aqueous feeds when:

(1) the dry product must only be in contact with air of reduced oxygen content to avoid oxidation or powder explosion;
(2) the amount of exhaust gases from the drying system must be limited to the smallest possible amount, e.g. when the exhaust has to be cleaned for air pollution reasons.

The formation of an inert system is accomplished by using a direct fired gas air heater. A controlled amount of combustion air is used and the combustion gases are used as drying medium, maintaining the oxygen content at a low level.

The exhaust gas from the dryer is regenerated in a scrubber/condenser, which removes the majority of the water evaporated from the system. A pressure-controlled purge removes the combustion gases and the rest of the evaporation.

This system quite logically is called 'self-inertising' and has the following advantages compared with the conventional closed cycle system.

(1) There is no demand for N_2 or other inert gas.
(2) There is no demand for an absolute leakproof dryer construction.
(3) Higher drying gas temperatures are possible because of direct heating.

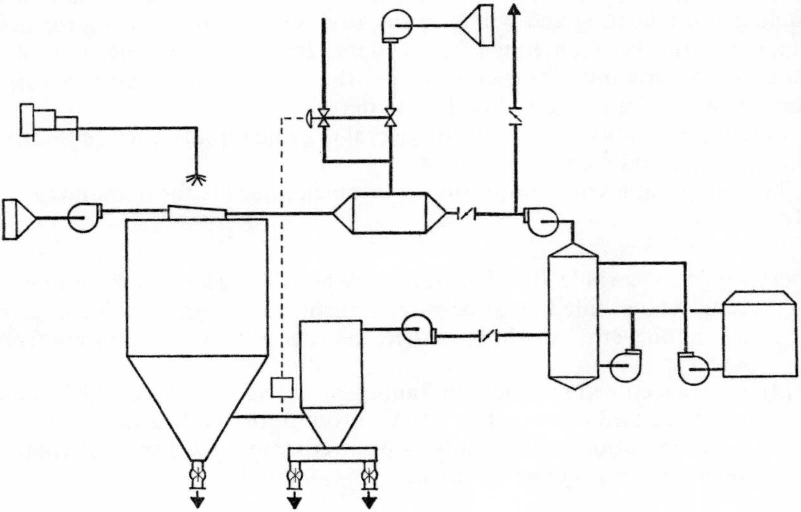

FIG. 12. Flow diagram of self-inertising plant.

REFERENCES

(For patents, reference is made only to British Patents where applicable.)

Damgaard-Iversen, J. and Kruse, F. (1973). (Assignors to Niro Atomizer): British Patent No. 1.316.331.

Gude, K. E. and Lund, B. (1972). (Assignors to Niro Atomizer): British Patent No. 1.283.122.

Hansen, O. *et al.* (1973). (Assignors to Niro Atomizer): German Patent Application No. 2.244.398.

Kjaergaard, O. G. and Storm, J. (1972). (Assignors to Niro Atomizer): British Patent Application No. 57167.

Mahler, J. L. and Markussen, E. K. (1972). (Assignors to Novo Terapeutisk Laboratorium A/S, Copenhagen): German Patent Application No. 2.134.555.

Masters, K. (1972). 'Spray Drying' *Chemical and Process Engineering Series*, Leonard Hill Books, London.

Moeller, Aa. and Straarup, O. (1969). (Assignors to Niro Atomizer): British Patent No. 1.276.000.

Nyrop, J. E. (1958). (Assignor to Niro Atomizer): United States Patent No. 2.850.085.

Percy, S. R. (1872). United States Patent No. 125.406.
Písecký, J. and Soerensen, I. H. (1972). (Assignors to Niro Atomizer): Danish Patent Application No. 4269 (not published before 29/2-1974).
Písecký, J. and Westergaard, V. (1971). (Assignors to Niro Atomizer): British Patent No. 1.301.796.
Písecký, J. and Westergaard, V. (1972). *Dairy Ind.*, **37** (3), 144.
Rayleigh, Lord. (1878). *London Math. Soc. Proc.*, **10**, 4.

DISCUSSION

H. A. C. Thijssen (Eindhoven University of Technology, The Netherlands): Do you get a thermal degradation because you are introducing dry particles in the hottest zone of the spray drier? If a certain fraction of the particles recirculates 3 or 5 times and there is a chance that this occurs again, doesn't that result in an unacceptable quality loss?

Kjaergaard: The answer to the first question would be that the amount of recirculation is very small, only 10 or 15% of the net throughput. In the early models of this plant we did recycle through the ceiling air disperser but now the only hot point in the drier is actually the air disperser itself, since the hot air is coming down into the chamber very close to the body of the atomiser. By introducing the fines from below we avoid their introduction into this very hot zone.

N. Wookey (Tenstar Products Ltd., U.K.): Mr. Kjaergaard hasn't mentioned any specific advances in the general recycling type driers in which something like 70 or 80% of the material emerging from the spray drier is put back into the top of the drier. This is normally done, of course, for crystallisation where you have droplets of syrup impacting on a bed of falling powder. One of the things we have found convenient with this type of drier is that you have the opportunity to obtain specific particle size ranges in your finished product. If you take the main output from the drier and sieve selectively then return to the top of the drier the particle size you don't want, you only obtain the specific particle size required.

Kjaergaard: We have now built the first plant according to this principle and it is operating on a product that would interest you—namely clay!

Wookey: We have had a plant operating at Ashford in Kent recycling about 75% of the powder and in this way you can crystallise a high dextrose syrup.

Kjaergaard: I understand your question now. As far as I know there have not been any improvements in that technique. I have only seen it operated once in our laboratory.

J. D. Mellor (C.S.I.R.O., Australia): You mentioned in your lecture three very important factors for future development: (1) the flexibility of the spray drier: (2) a heater in the feed line nozzle: (3) a highly efficient filtering system and electrostatic precipitator. I feel that these three factors might be important in relation to a development you were interested in about 7 years ago. You did some studies on sub-zero temperature drying with very low humidity air. I am not quite sure whether you got as far as the low temperature spray drying stage or even further to the vacuum spray drier stage—perhaps you could enlighten me on that? In developing a sub-zero spray drier there are some particular problems that present themselves. One is that you have got to evaporate about 35% of the initial moisture to get frozen particles and secondly you need to design a nozzle system so that it does not freeze up.

Kjaergaard: Some years ago, we were interested in doing some drying on a spray drier which would resemble the kind of freeze-drying you are doing on special freeze-drying equipment but this project was laid aside about 3 or 4 years ago. This special atomiser I mentioned is foreseen to be used for spray congealing purposes but of course it would be possible to use it with very low inlet temperatures. The other problem of running a spray drying plant at sub-zero temperatures is filtering—I don't think that the solution to avoid particles sliding around the plant would be a filter since it would need to be a very fine filter which would block up very quickly.

Chairman: Materials are protected by evaporative cooling very effectively in spray drying and if one doesn't have a high air outlet temperature one can get the product out at the bottom, assuming it is not highly hygroscopic at 110°F or lower. During the actual atomisation there is a tremendous driving force with a high rate of heat transfer. In a paper published in the 1930s by Arthur D. Little & Co. in 'Industrial and Engineering Chemistry' a determinate thermodynamic treatment showed that the particles were actually cooled to something like −50°F at that instant. It depends on conditions and on the kind of material and on fine atomisation, but it is actually cooled very substantially below freezing at that instant. At least this is so theoretically. Kleisher, Professor of Mechanical Engineering at Harvard Engineering School, tended to substantiate this view. I don't think that we appreciate the tremendous driving force which can reduce the temperature of the dried particles to so low a temperature instantaneously. Another thing that happens around the zone of atomisation, there is a sudden shrinkage in volume causing turbulence because one goes from an air temperature of 500°F to 220°F very rapidly and you know from the common gas laws the change in volume occurring. Turbulence will keep those particles previously mentioned from entering into the zone but that does not prohibit the possibility of high outlet air temperature causing some damage.

Mellor: I appreciate the fact that theoretically it is possible to get an evaporative cooling effect so that your particles are at −50°F but as far as I am concerned it is only theoretical. It has not been substantiated by experiments and I also know it is very hard to measure.

Chairman: In 'Advances of Food Research' Volume II by Settelmeyer is described how we put probes and thermocouples into the zones of drying and found fairly low temperatures, not that low, but quite low.

S. Barnett (General Foods Inc., U.S.A.): I have not encountered temperatures as low as that myself.

Theoretical Modelling of the Drying Behaviour of Droplets in Spray Dryers

P. J. A. M. KERKHOF and W. J. A. H. SCHOEBER

Department of Chemical Engineering,
Eindhoven University of Technology,
P.O. Box 513, Eindhoven, The Netherlands

NOTATION

a, b, c, d	constants in eqn (53)	(—)
c_d	drag coefficient	(—)
c_p	heat capacity	(J/kg°C)
d	droplet diameter	(m)
d	density of pure substance	(kg/m^3)
$g, \bar{g}$	gravitational acceleration	(m/s^2)
j	mass flux	(kg/m^{2}s)
k	mass transfer coefficient	(m/s)
m	constant in eqn (3)	(—)
m	slope of linearised equilibrium curve (see p. 365)	(—)
n	constant in eqn (4),	(—)
	constant in eqn (47)	(—)
q	heat flux	(J/m^{2}s)
r	radial co-ordinate	(m)
t	time	(s)
$u, \bar{u}$	droplet velocity with respect to fixed co-ordinates	(m/s)
$v, \bar{v}$	air velocity with respect to fixed co-ordinates	(m/s)
v	velocity in mass and heat transfer	(m/s)
$w, \bar{w}$	droplet velocity relative to air	(m/s)
x, y, z	Cartesian co-ordinates	(—)
x	tangential direction in eqns (11), (12)	(—)
x_w	relative amount of water evaporated	(—)
z	transformed distance based on dissolved solids	(kg)
A	activity	(—)

A_d	cross-sectional area of dryer	(m^2)
A_n	coefficient in eqn (47)	(—)
Bi	Biot number	(—)
D	diffusion coefficient	(m^2/s)
$\underset{\cdot}{D}$	multicomponent diffusion coefficient in Stefan–Maxwell equation	(m^2/s)
E	mass or heat transfer efficiency	(—)
F	surface water flux	(kg/m^2s)
Fo	Fourier number, Dt/R^2 or $\lambda t/\rho c_p R^2$	(—)
H	enthalpy	(J/kg)
H_a	aroma Henry coefficient	(m^3/kg)
K_1	constant in eqn (5)	(s^{-1})
K_2	constant in eqn (6)	(s^{-1})
Nu	Nusselt number, $\alpha d/\lambda$	(—)
Pr	Prandtl number, $\mu c_p/\lambda$	(—)
Q	air flow rate	(m^3/s)
R	droplet radius	(m)
R	gas constant	(kcal/kmol K)
R_d	dryer radius	(m)
Re	Reynolds number, $\rho\|w\|d/\mu$	(—)
S_x, S_y, S_z	co-ordinate in the x, y or z direction	(m)
Sc	Schmidt number, $\mu/\rho D$	(—)
Sh	Sherwood number, kd/D	(—)
Su	surface tension parameter, $\sigma d\rho/\mu^2$	(—)
T	temperature	(°C)
T_k	temperature	(K)
U	transformed concentration	(kg/kg)
X	mole fraction	(—)
α	heat transfer coefficient	$(J/m^2s°C)$
γ	defined in eqn (27)	(—)
δ	thickness of film outside the droplet	(m)
θ	angle of deflection from the x direction in horizontal plane	(—)
θ	temperature parameter	(°C)
λ	thermal conductivity	(J/ms°C)
λ_n	coefficient in eqn (47)	(—)
μ	dynamic viscosity	(Ns/m^2)
ν	kinematic viscosity	(m^2/s)
ρ	density, concentration	(kg/m^3)
σ	surface tension	(N/m)
τ	shear stress	(N/m^2)
ψ	concentration parameter, eqn (42)	(kg/m^3)
ω	weight fraction	(—)
∇	gradient	(m^{-1})

Subscripts

a	aroma
cr	critical
e	eddy
eff	effective
j	component j
m	molecular
max	maximum
0	at $t = 0$; at $T = 0$ in eqn (89)
r	in radial direction
s	dissolved solids
w	water
x, y, z	in the x, y or z direction
M	mass transfer
H	heat transfer
θ	tangential direction

Superscripts

$'$	gas phase
$*$	virtual, at tower wall
$\bullet$	steady state
i	interface
s	with respect to dissolved solids
δ	in bulk of gas phase

INTRODUCTION

General

Spray drying is one of the most important processes used in drying liquid foods, including dairy products, fruit juices and tea and coffee extracts. In the chemical and pharmaceutical industry widespread applications are also found (Masters, 1972). In the process of spray drying small droplets are formed by atomisation, which come into contact with hot air. Owing to the high surface area thus created and the high air temperature, short drying times can be achieved. The product is obtained as a powder which can be handled easily. For food liquids several quality demands can be made on the product, such as appropriate bulk density and colour, instant properties, the least possible degradation of original constituents, and a desired flavour pattern. This last requirement can mean either a full retention of originally present aroma components or the removal of undesired flavours. Usually the aroma components are very volatile and are present in very small amounts. The variables influencing these quality aspects are the feed composition, concentration and temperature, the air temperature and humidity, the method of atomisation, the geometry of the dryer, and the direction of air and product flow in the dryer.

Although much knowledge about the influence of process variables on product quality as well as design rules for spray dryers has been obtained empirically, more and more attention is being paid to the fundamental aspects of the drying process. In the last few decades theoretical investigations have been performed in the field of flow pattern in dryers, drying rates and trajectories of droplets, the behaviour of droplets in sprays, loss of aroma components and thermal degradation of heat-sensitive components (Lapple and Shepherd, 1940; Edeling, 1950; Sjenitzer, 1952; Arni, 1959; Thijssen and Rulkens, 1968; Rulkens and Thijssen, 1969; Pita, 1969; Bailey *et al.*, 1970; Hortig, 1970; Adler and Lyn, 1971; Mühle, 1971, 1972; van der Lijn *et al.*, 1972; Stein, 1972; Krüger, 1973; Rulkens, 1973; Schoeber, 1973). Compilation of theoretical and experimental knowledge of spray drying is given by Seltzer and Settlemeyer (1949), Marshall (1954), Dombrowski and Munday (1968) and Masters (1972).

The basic equations of the transfer of heat, mass and momentum have been long known, but the development of larger and more rapid computers, digital as well as analogue, has only recently offered more possibilities of solving complex equations, or solving equations for more complex conditions.

In this paper the equations for the motion of droplets, for the mass and heat transfer in and around droplets, and for thermal degradation will be discussed. With the aid of these equations the solutions to several problems concerned with spray drying will be calculated, viz.:

(1) The motion and evaporation of water droplets for the case of unidirectional motion in still air, and for the case of a swirling air flow in which droplets are injected from a centrifugal pressure nozzle.
(2) The loss of aroma components and the evaporation of water during the internal circulation period after the formation of the droplet.
(3) The drying of rigid droplets of aroma-containing, heat-sensitive food liquids.

Motion and evaporation of water droplets

For the estimation of dryer geometry and dimensions the calculation of particle trajectories in the dryer is necessary. In order to calculate the diameter and the height of a spray dryer, radial and vertical trajectories of the drying droplets must be estimated. Edeling (1950) calculated these trajectories for a spiral flow of non-evaporating particles in air with tangential and axial air flow, using graphical methods; he gave approximating equations for the case that the droplets moved at terminal velocity.

Lapple and Shepherd (1940) present equations for the deceleration in unidirectional flow and for the motion of droplets in rotational air flow. These authors also present a step-by-step method for calculating droplet deceleration. Coulson and Richardson (1968) calculated the trajectory of decelerating droplets under turbulent flow conditions. Kessler (1967), Brauer and Kruger (1969) and Hortig (1970) performed calculations and produced data on combined motion in vertical and horizontal directions. Mühle (1971, 1972) obtained results regarding the motion in three-dimensional rotating air flow of droplets formed by a spinning disc atomiser. Combined calculations of motion and evaporation are given by Sjenitzer (1952), Bailey *et al.* (1970), Stein (1972) and Krüger (1973).

Sjenitzer (1952) described the deceleration and the evaporation of water droplets under simplified conditions. Bailey *et al.* (1970) calculated the two-dimensional simultaneous motion and evaporation of droplets in swirling air flow. The drying of insoluble solid materials in three-dimensional motion in rotating air flow was described by Stein (1972), while Krüger (1973) calculated the drying of droplets likewise in three-dimensional motion, taking into account changes of volume or of density of the particles.

Adler and Lyn (1971) present a continuum-mechanics model of the combined flow of air and droplets during two-dimensional evaporation in swirling flow.

Sjenitzer (1952) clearly showed that the evaporation during the deceleration of the droplets can be considerable, and should therefore be taken into account in the calculation of drying time. From the work of Bailey *et al.* (1970) it follows that the evaporation has also a marked influence on the motion of the drops.

For the calculation of the droplet trajectory knowledge is required of the air flow pattern in the dryer, of the drag forces exerted on the particles and of the initial velocity of the droplet. These aspects will be treated in the next section (p. 355). Furthermore, it is necessary to take into account droplet evaporation, which is in turn influenced by the relative velocity of the droplets with respect to the air. The mass and heat transfer aspects will be discussed in the third section. In the fifth section the method of Sjenitzer (1952) will be compared with numerical calculations of unidirectional motion and evaporation of water drops, and a calculation will be presented for the combined motion and evaporation of water drops in three-dimensional air flow.

The loss of volatile flavour components

One of the primary factors determining the product quality of dried fruit-juices and coffee and tea extracts is the amount of aroma compounds retained. Both the overall loss of aroma components, which flattens the smell and taste of the reconstituted product, and the selective loss of certain aroma compounds, which may strongly affect the flavour pattern, may occur during drying. Experimental investigations (Reineccius and Coulter, 1969; Rulkens and Thijssen, 1972; Thijssen and Rulkens, 1968; Sivetz and Foote, 1963) have shown that in spray drying under optimum conditions the volatile compounds can be retained to a large extent. The explanation of this effect was given by Thijssen (1965), who stated that the transport of water and of volatile aroma components inside the droplet takes place by molecular diffusion, and that the diffusion coefficients of both water and aroma components decrease with decreasing water concentration; the diffusion coefficients of aroma components decrease more sharply than the diffusion coefficient of water. Because of the high evaporation rate of water, soon after the onset of drying a sharp water concentration profile will be built up inside the droplet, and the surface water concentration will drop rapidly to a low value—a 'dry skin' is formed. At low water concentrations the diffusion coefficients of aroma components are so much lower compared to water that the dry skin is virtually impermeable to the aroma components.

The calculation of the drying rate and aroma loss from a droplet in spray drying is very complex, because of the water-concentration and temperature dependence of the diffusion coefficients, the non-linearity of the vapour–liquid equilibrium relationship, the non-isothermal nature of the drying process, and the shrinkage of the droplet due to water loss. Approximate calculations concerning the drying rate of droplets with internal diffusion resistance were performed by Thijssen and Rulkens (1968). Dlouhy and Gauvin (1960) calculated the drying rate by neglecting internal resistance to mass transfer, and found results which agreed well with their experimental work on the spray drying of lignosol solutions.

Van der Lijn *et al.* (1972) first published solutions to the diffusion equation for the water loss from drying droplets of aqueous maltose solutions, with a concentration and temperature dependent diffusion coefficient. Rulkens (1973) extended this work with approximate calculations of aroma transport. Schoeber (1973) included in his calculations a ternary diffusion model for aroma transport.

In the third section the diffusion equations will be treated, while the fourth section will deal with the diffusional properties of liquid foods. Some results of calculations will be given in the fifth section.

After formation, the droplets have a high relative velocity with respect to the air, which causes high shear stresses. Owing to this shear, internal circulation may be induced in the droplet. As long as this circulation continues, the formation of a dry skin is hindered, and volatile aroma components may be lost (Menting, 1969; Thijssen and Rulkens, 1968). The presence of surface active agents may prevent internal circulation. In order to obtain a qualitative idea about the amount of aroma components lost during the internal circulation period, a criterion will be derived, expressing the critical velocity below which no circulation occurs, in relation to the gradient in surface tension and the droplet diameter. The derivation will be given in the next section, and in the third section the water and volatile loss during circulation will be discussed theoretically. In the fifth section some results of the calculations will be presented.

Thermal degradation

The degradation of many heat-sensitive compounds in foods increases with increasing moisture content and temperature (Labuza, 1972; Ball and Olson, 1957). Although spray drying requires only very short drying times, the temperature of the drying droplets may reach dry-bulb temperature before the completion of drying (Trommelen and Crosby, 1970). As the centre of the droplet still has a high moisture content, thermal degradative reactions may occur. Schoeber (1973) included in his calculations the thermal inactivation of phosphatase. In later sections further details will be given.

MOTION OF AIR AND DROPLETS IN SPRAY DRYERS

Air flow pattern

An extensive study of the air flow pattern and configuration of spray dryers was made by Marshall (1954). The large number of possible configurations of spray dryers and of air flow patterns makes it impossible to give a general description of the velocity distribution. With respect to the droplet motion a qualitative description of the air flow direction can be given as cocurrent, countercurrent, and ideally mixed flow. Regarding the motion of the air through the dryer an approximate division can be made between swirling or rotational flow (tangential air inlet) and unidirectional flow (axially directed air inlet).

The flow of air is also influenced by the atomisation process and by the droplet motion. Momentum is transferred to the air in the surroundings of the atomiser, causing a pumping effect (Marshall, 1954). In large spray dryers this effect is only local and does not disturb the air flow in the rest of the dryer considerably (Masters, 1972). Further reviews of air flow patterns and configurations are given by Belcher *et al.* (1963), Masters (1968) and Lyne (1971).

UNIDIRECTIONAL AIR FLOW

The equation for the velocity v_z of air in the case of vertical air flow is evident:

$$v_z = \pm Q/A_d \tag{1}$$

in which Q is the flow rate of drying air and A_d the cross section of the dryer. The positive direction is taken downward.

ROTARY AIR FLOW

The velocity field in a rotary flow pattern in a spray dryer is similar to that in the cylindrical part of a cyclone and has a behaviour intermediate between free vortex flow and forced vortex flow (Marshall, 1954; Bird *et al.*, 1960). No exact solutions to the equations of motion and of continuity for this case have been encountered in the literature.

Edeling (1950) in his calculations of droplet path used the following approximation for the tangential velocity v_θ:

$$\frac{v_\theta}{v_\theta^*} = \left(\frac{r}{R_d}\right)^{0.5} \tag{2}$$

in which v_θ^* is a virtual tangential velocity, extrapolated to the tower wall. He assumed a constant axial velocity and no radial velocity.

Mühle (1971, 1972) proposes the relation

$$\frac{v_\theta}{v_\theta^*} = \left(\frac{r}{R_d}\right)^{-m} \tag{3}$$

with m in the range $0.5 \leq m \leq 0.85$ for the tangential velocity, and for the radial velocity v_r:

$$\frac{v_r}{v_r^*} = \left(\frac{r}{R}\right)^{-n} \tag{4}$$

For $n = 1$ eqn (4) represents a perfect point sink stream.

Bailey *et al.* (1970) propose a velocity distribution field built up of free vortex flow around a 'solid core', i.e. a forced vortex:

$$\begin{aligned} v_\theta &= K_1 \cdot r & r &< R_d/5 \\ v_\theta &= R_d^2 K_1/(25r) & r &> R_d/5 \end{aligned} \tag{5}$$

The constant K_1 is determined from the air inlet velocity. For the radial velocity these authors used

$$v_r = -K_2 \cdot v_\theta \qquad r > R_d/5 \tag{6}$$

with $K_2 = 0.025$. The addition of radial velocity components did not influence very much the calculated trajectories. This was found both for evaporating and non-evaporating droplets.

The same tangential air velocity distribution was used by Stein (1972) for the calculation of the trajectories of drying droplets, combined with a uniform axial air velocity.

Chaloud *et al.* (1957) present tangential velocity profiles for a spray dryer in which the air is fed in tangentially at the bottom of the dryer. The liquid feed is atomised at the top. These authors observe strong dependence of the axial velocity on the radial distance from the tower axis. Probably this is due to the rise of a hot air core in the centre of the tower, possibly combined with a pumping action of the atomiser.

For the calculations in the present paper a uniform downward axial velocity will be assumed, and a tangential velocity profile analogous to the one described by Bailey *et al.* (1970). Radial velocity will be neglected.

The motion of droplets

FORMATION PERIOD

The droplets formed by rotating disc atomisation or by centrifugal pressure nozzle atomisation have high initial velocities, of the order of 20–200 m/s.

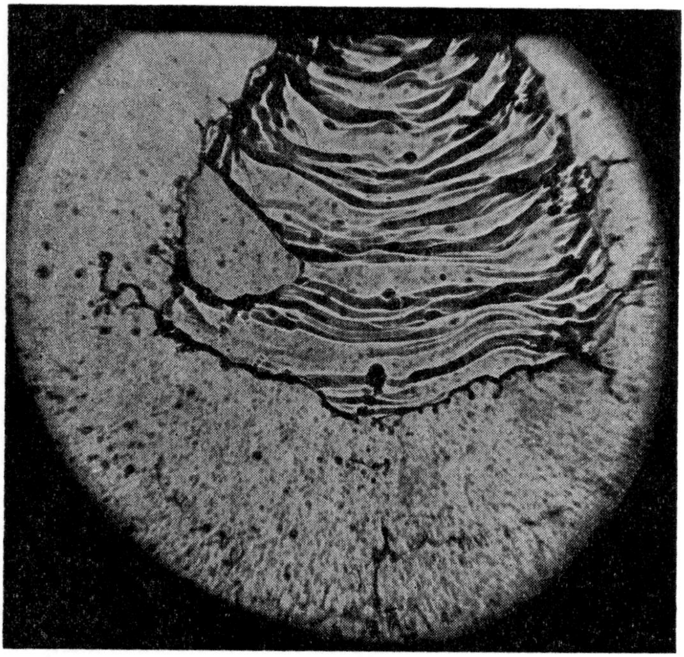

FIG. 1. High-speed photograph of a spray of maltodextrin solution from a centrifugal pressure nozzle. Exposure: 150 ns.

In the case of pressure nozzles a conical sheet is formed first, from which, by several mechanisms including rim disintegration, wave formation, and perforation of the sheet, droplets can be formed (Dombrowski and Fraser, 1953; Dombrowski and Munday, 1968). These mechanisms can be observed in Fig. 1, in which a high-speed photograph is given of the atomisation of an aqueous maltodextrin solution. The sheet is perforated and by contraction threads and drops are formed. More photographs are given by Dombrowski and Munday (1968) and Tate and Marshall (1953). In Fig. 2 a double-flash photograph is shown of the same experiments. It is

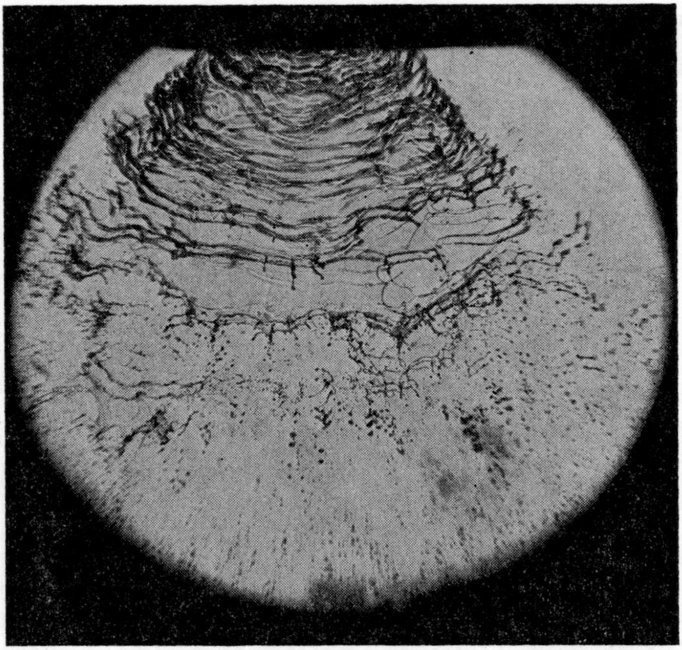

FIG. 2. High-speed double-flash photograph of a spray of maltodextrin solution from a centrifugal pressure nozzle. Exposure: 150 ns; time between flashes: 85 μs.

clear that the velocity at several distances from the orifice remains constant, and even the droplets just formed have approximately the same velocity. This agrees with the findings of Dombrowski *et al.* (1960). The absence of deceleration of the droplets in the first few centimetres is caused by the air flow induced by the atomisation.

As the same pumping effects occur in spinning wheel atomisation, in that case no appreciable deceleration will probably occur in the direct vicinity of the atomiser either.

EQUATION OF MOTION FOR A SINGLE DROPLET

The equation of motion for a spherical particle, moving at velocity $\bar{u}$, with a relative velocity $\bar{w} = \bar{u} - \bar{v}$ with respect to the air velocity $\bar{v}$, reads

$$\frac{d\bar{u}}{dt} = \left(\frac{\rho - \rho'}{\rho}\right)\bar{g} - \frac{3}{8}\frac{c_d\rho'|\bar{w}|\bar{w}}{\rho R} \tag{7}$$

in which ρ and ρ' are the density of the droplets and of the air respectively, $\bar{g}$ is the gravity vector, c_d the droplet radius, and $|\bar{w}|$ the absolute magnitude of the vector $\bar{w}$.

The drag coefficient c_d varies with the Reynolds number,

$$Re = \frac{\rho'|\bar{w}|d}{\mu'} = \nu'|\bar{w}|d$$

in which μ' is the air viscosity and d the droplet diameter. ν' is the kinematic viscosity of the air. The drag coefficient is influenced by droplet deformation and oscillation, and internal circulation. Hughes and Gilliland (1952) graphically presented c_d versus Re, for motion of liquid drops in gases, with a surface tension parameter

$$Su = \frac{\sigma \cdot 2R \cdot \rho'}{(\mu')^2}$$

where σ is the surface tension of the liquid. At a given value of Su the drag coefficient c_d at low Reynolds numbers has the same dependence on Re as holds for rigid spheres, but at higher Re numbers deformation and oscillation may occur, resulting in higher c_d values than for rigid spheres. Inspection of their data reveals that for normal spray drying conditions, $Re < 250$ and $Su > 10^4$, the drag coefficient shows in good approximation the same dependence for liquid drops as for solid spheres.

This behaviour is different from that of liquid–liquid dispersions or gas-bubble-in-liquid systems, in which internal circulation may decrease the drag coefficient by a considerable amount (Elzinga and Banchero, 1961; Klee and Treybal, 1956; Griffith, 1962; Redfield and Houghton, 1965; Peebles and Garber, 1953). That the drag coefficient is not dependent upon the occurrence of internal circulation is due to the high viscosity ratio of liquid and gas (Levich, 1962).

For the relation between c_d and Re various correlations are presented in the literature. Perry (1963) presents the following relations:

$$\begin{array}{rrl} \text{Stokes' law region:} & Re < 0.3; & c_d = 24/Re \\ \text{Transition region:} & 0.3 < Re < 1000; & c_d = 18.5/Re^{0.6} \\ \text{Newton's law region:} & Re > 1000; & c_d = 0.44 \end{array} \tag{8}$$

As this correlation is not continuous at the boundaries of the regions, it is not very suitable for use in calculations for large deviations in Re.

A correlation continuous up to $Re = 2 \times 10^5$ is given by Mühle (1971) after Kaskas (1964):

$$c_d = \frac{24}{Re} + \frac{4}{Re^{0.5}} + 0.4 \tag{9}$$

Starting from theoretical calculations and experimental data, Ihme *et al.* (1972) propose the following correlation:

$$c_d = \frac{24}{Re} + \frac{5.48}{Re^{0.573}} + 0.36 \tag{10}$$

for $Re \leq 10^4$.

The drag coefficients obtained from these relations are valid for steady motion of non-evaporating spheres. Acceleration or deceleration of the droplet will have an increasing effect on the drag coefficient (Mühle, 1971) and evaporation will also influence the drag coefficient (Bailey *et al.*, 1970). No quantitative influence of these phenomena is known at present (Torobin and Gauvin, 1959, 1961). Therefore, and because the deceleration effects take place only over a short distance, the steady-state drag coefficient is seen as a good approximation. The deviation from reality will not have much influence on the trajectories of the droplets, relative to other simplifications such as the exclusion of particle interaction, and particle–air interaction.

Prediction of internal circulation period

In the case of spray drying the occurrence of internal circulation has negligible influence on droplet motion, as shown above. There is, however, a marked influence on mass transfer rate, especially if the dominating resistance for mass transfer lies in the liquid phase, as is the case for volatile aroma components. In the next section this will be treated in more detail. For high *Re* numbers ($Re > 1$) no relations are available (Levich, 1962) to predict whether or not internal circulation will occur. An approximate estimation will be given here, as derived by Kerkhof (1973).

It is known that surface active agents may decrease or even prevent internal circulation due to the build-up of a gradient in surface tension, which exerts shear stress on the droplet in opposite direction to the air friction (Elzinga and Banchero, 1961; Klee and Treybal, 1956; Griffith, 1962; Redfield and Houghton, 1965; Peebles and Garber, 1953).

For the equilibrium between the surface tension gradient and the frictional force exerted by the air the force balance relation holds:

$$\frac{\delta\sigma}{\delta x} = \tau \tag{11}$$

in which x is a direction tangential to the surface. The shear stress τ varies with the angle of the normal with the flow axis, and somewhere on the sphere reaches a maximum value τ_{max}, the place depending on the *Re* number.

It is now assumed that there is a limit to the value of $\delta\sigma/\delta x$ for a given amount of surfactants present, and that this limit can be approximated by

$$\left(\frac{\delta\sigma}{\delta x}\right)_{max} = \frac{(\Delta\sigma)_{eff}}{\pi \cdot R} \tag{12}$$

The criterion for internal circulation can be read from eqns (11) and (12):

$$\frac{(\Delta\sigma)_{eff}}{\pi R} \leq \tau_{max} \tag{13}$$

in which the equality sign stands for the critical shear stress $\tau_{max,cr}$.

From the theoretical results of Ihme *et al.* (1972) a correlation can be derived between τ_{max} and the *Re* number:

$$\frac{\tau_{max}}{\frac{1}{2}\rho' w^2} = \frac{9.4}{Re} + 0.37 \tag{14}$$

Combining (13) and (14), it follows for the critical velocity w_{cr}:

$$w_{cr}^2 + 12.70\,\frac{\mu'}{\rho' R}\,w_{cr} - 1.721\,\frac{(\Delta\sigma)_{eff}}{\rho' R} = 0 \tag{15}$$

The positive root of this equation is

$$w_{cr} = -6.35\,\frac{\mu'}{\rho' R} + \left[40.34\,\frac{\mu'^2}{\rho'^2 R^2} + 1.721\,\frac{(\Delta\sigma)_{eff}}{\rho' R}\right]^{1/2} \tag{16}$$

The order of $(\Delta\sigma)_{eff}$ can be between the surface tension of pure water (0.07 N/m) and zero. As will be shown in the fifth section the initial velocities used in spray drying may well be higher than the critical velocity w_{cr} according to eqn (16). Thus internal circulation will probably occur during a short period, the length of which can be calculated from the equation of motion. In a later section (p. 375) some calculations will be presented of the amount of water and of volatile aroma components evaporated during the internal circulation period.

MASS AND HEAT TRANSFER

General

In the following the well-known 'one-film' model (Bird *et al.*, 1960) will be used for the description of heat and mass transfer in the gas phase. In this model it is assumed that the resistance to mass and heat transfer is entirely located in a thin film at the phase boundary through which the transport of heat and mass can be described as quasi-stationary. At the interface thermodynamic equilibrium between the phases is assumed. The mass and heat transfer coefficients in such a system are given by

$$\frac{1}{k} = \int_{R}^{R+\delta} \frac{\mathrm{d}r}{D'_{\mathrm{e}}} \quad \text{and} \quad \frac{1}{\alpha} = \int_{R}^{R+\delta} \frac{\mathrm{d}r}{\lambda'_{\mathrm{e}}} \tag{17}$$

in which δ is the film thickness, k and α are the mass and heat transfer coefficients for steady-state transport in the absence of net mass transfer, and D'_{e} and λ'_{e} the turbulent diffusion coefficient and heat conductivity of the gas phase respectively. In the case of considerable net mass transfer, corrections to the heat and mass transfer coefficients should be applied. The description of the effect of net mass transfer on heat and mass transfer rates is given by Bird *et al.* (1960).

EFFECT OF NET MASS TRANSFER ON MASS TRANSFER COEFFICIENT

The radial mass flux of water in the binary system air–water with respect to the evaporating surface is given by

$$\begin{aligned} j_{\mathrm{w}}{}^{\mathrm{i}} &= j_{\mathrm{t}}{}^{\mathrm{i}}\omega'_{\mathrm{w}} - D'_{\mathrm{e}}\rho' \frac{\delta\omega'_{\mathrm{w}}}{\delta r} \\ &= (j_{\mathrm{air}}{}^{\mathrm{i}} + j_{\mathrm{w}}{}^{\mathrm{i}})\omega'_{\mathrm{w}} - D'_{\mathrm{e}}\rho' \frac{\delta\omega'_{\mathrm{w}}}{\delta r} \end{aligned} \tag{18}$$

in which j_{w} is the water flux, i stands for interfacial, ω is the water weight fraction, and the prime (′) denotes the gas phase. As the airflow ($j_{\mathrm{air}}{}^{\mathrm{i}}$ through the interface) is zero, it follows that

$$j_{\mathrm{w}}{}^{\mathrm{i}} = \frac{-D'_{\mathrm{e}}\rho' \dfrac{\delta\omega'_{\mathrm{w}}}{\delta r}}{1 - \omega'_{\mathrm{w}}} \tag{19}$$

For a constant flux and density through the film eqn (19) can be integrated to give

$$\ln \frac{1 - \omega'^{\delta}_{\mathrm{w}}}{1 - \omega'^{\mathrm{i}}_{\mathrm{w}}} = \frac{j_{\mathrm{w}}{}^{\mathrm{i}}}{\rho'} \int_{\delta} \frac{\mathrm{d}r}{D'_{\mathrm{e}}} = \frac{j_{\mathrm{w}}{}^{\mathrm{i}}}{\rho'} \cdot \frac{1}{k} \tag{20}$$

or

$$j_{\mathrm{w}}{}^{\mathrm{i}} = k \cdot \rho' \ln \frac{1 - \omega'^{\delta}_{\mathrm{w}}}{1 - \omega'^{\mathrm{i}}_{\mathrm{w}}} \tag{21}$$

which for low concentrations ω'_{w} gives

$$j_{\mathrm{w}}{}^{\mathrm{i}} = k \cdot \rho'(\omega'^{\mathrm{i}}_{\mathrm{w}} - \omega'^{\delta}_{\mathrm{w}}) \qquad \omega'_{\mathrm{w}} \to 0 \tag{22}$$

Let k_{eff} be defined by

$$j_{\mathrm{w}}{}^{\mathrm{i}} \equiv k_{\mathrm{eff}} \cdot \rho'(\omega'^{\mathrm{i}}_{\mathrm{w}} - \omega'^{\delta}_{\mathrm{w}}) \tag{23}$$

then it follows that

$$\frac{k_{\text{eff}}}{k} = \frac{\ln \dfrac{1 - \omega_{\text{w}}'^{\delta}}{1 - \omega_{\text{w}}'^{\text{i}}}}{\omega_{\text{w}}'^{\text{i}} - \omega_{\text{w}}'^{\delta}} = \frac{1}{(1 - \omega_{\text{w}}')_{\ln}} \tag{24}$$

in which $(1 - \omega_{\text{w}}')_{\ln}$ is the logarithmic average of $(1 - \omega_{\text{w}}')$ between the values at the interface and at the place δ. For low values of ω_{w}' it follows that

$$k_{\text{eff}} = k \tag{25}$$

EFFECT OF MASS TRANSFER ON HEAT TRANSFER

The heat flux from the interface q^{i} is

$$q^{\text{i}} = -\lambda_{\text{e}}' \frac{\text{d}T}{\text{d}r} + j_{\text{w}}{}^{\text{i}} \cdot H \tag{26}$$

in which T is the temperature and H the enthalpy of the evaporating water. For the enthalpy H the zero level is taken at the interfacial temperature T^{i}, hence $H = c_{\text{pw}}'(T - T^{\text{i}})$ in which c_{pw}' is the specific heat of the evaporating water.

For constant heat and mass flux through the film eqn (26) can be integrated over the film thickness δ to give

$$q^{\text{i}} = -\frac{\gamma}{\exp(\gamma) - 1}\alpha(T^{\delta} - T^{\text{i}}) \tag{27}$$

with $\gamma = j_{\text{w}}{}^{\text{i}} c_{\text{pw}}'/\alpha$.

Let α_{eff} be defined by

$$q^{\text{i}} \equiv -\alpha_{\text{eff}}(T^{\delta} - T^{\text{i}}) \tag{28}$$

then it follows that

$$\frac{\alpha_{\text{eff}}}{\alpha} = \frac{\gamma}{\exp(\gamma) - 1} \tag{29}$$

EFFECT OF SHRINKAGE ON MASS AND HEAT TRANSFER

Owing to the evaporation of water the droplets shrink. This induces a flow of air towards the droplet, of which the rate is given by

$$j_{\text{air}} = v^{\text{i}}\rho' \tag{30}$$

in which v^{i} is the velocity of the interface with respect to the droplet centre. As this airstream contains water vapour and heat, the mass and heat transfer rates are influenced by this effect. The velocity of the interface follows from an overall mass balance over the droplet, and for a pure water droplet reads

$$v^{\text{i}} \equiv \frac{\text{d}R}{\text{d}t} = j_{\text{w}}{}^{\text{i}}/d_{\text{w}} \tag{31}$$

For the mass transfer rate the following relationship holds:

$$j_{\mathrm{w}} = j_{\mathrm{w}}^{\mathrm{i}} + v^{\mathrm{i}}\rho'\omega_{\mathrm{w}}'^{\delta} = j_{\mathrm{w}}^{\mathrm{i}}(1 - \rho'\omega_{\mathrm{w}}'^{\delta}/d_{\mathrm{w}}) \tag{32}$$

For the heat flux it follows that

$$q = q^{\mathrm{i}} + v^{\mathrm{i}}\rho' c_{\mathrm{p}}'(T^{\delta} - T^{\mathrm{i}}) = q^{\mathrm{i}}\left\{1 + \frac{\rho' c_{\mathrm{p}}'}{d_{\mathrm{w}} c_{\mathrm{pw}}'}(\exp(\gamma) - 1)\right\} \tag{33}$$

in which c_{p}' is the specific heat of the moist air.

MAGNITUDE OF THE CORRECTION FACTORS

For a droplet evaporating in dry air at 200°C, which may be considered an extreme case for spray drying, the driving force on the basis of absolute humidity is equal to $H^{\mathrm{i}} - H^{\delta} = 0.06$ kg/kg dry air at a wet bulb temperature of 44°C. With $\omega^{\mathrm{i}} = 0.06/1.06 = 0.057$ and $\omega^{\delta} = 0$, it follows from eqn (24) that

$$\frac{k_{\mathrm{eff}}}{k} = 1.02 \tag{34}$$

For the correction of heat transfer use will be made of the analogy between heat and mass transfer:

$$Nu = \frac{\alpha d}{\lambda_{\mathrm{m}}'} = Sh = \frac{k \cdot d}{D_{\mathrm{m}}'}$$

in which λ_{m}' and D_{m}' are the molecular heat conductivity and diffusion coefficient respectively. For γ then follows:

$$\gamma = \frac{j_{\mathrm{w}}^{\mathrm{i}} c_{\mathrm{pw}}'}{\alpha} = 1.02\,\frac{D_{\mathrm{m}}'\rho'\Delta\omega_{\mathrm{w}} c_{\mathrm{pw}}'}{\lambda'} \tag{35}$$

Substitution of $D_{\mathrm{mw}}' = 4.12 \times 10^{-5}$ m²/s at 120°C, which is the film temperature and $\rho' = 0.9$ kg/m³, $c_{\mathrm{pw}}' = 1800$ J/kg°C, and $\lambda' = 0.033$ J/m °C gives $\gamma = 0.118$ and

$$\frac{\alpha_{\mathrm{eff}}}{\alpha} = 0.94 \tag{36}$$

Considering the effect of shrinkage the mass transfer rate can be given by

$$j_{\mathrm{w}} = j_{\mathrm{w}}^{\mathrm{i}}(1 - 5.13 \times 10^{-5}) \simeq j_{\mathrm{w}}^{\mathrm{i}} \tag{37}$$

Considering the effect of shrinkage the heat transfer rate is

$$q = q^{\mathrm{i}}(1 + 6.3 \times 10^{-5}) \simeq q^{\mathrm{i}} \tag{38}$$

From these considerations it follows that the effect of net mass transfer should be taken into account in a more accurate calculation, but that the effect of shrinkage is negligible.

COMPARISON OF LIQUID PHASE AND GAS PHASE RESISTANCE TO MASS AND HEAT TRANSFER: THE BIOT NUMBER

The heat flux at the phase boundary of a rigid droplet is

$$-\lambda \left.\frac{\mathrm{d}T}{\mathrm{d}r}\right|_{r=R} = \alpha_{\mathrm{eff}}(T^{\mathrm{i}} - T^{\delta}) \tag{39}$$

in which λ is the heat conductivity of the sphere. By substituting $\theta = T - T^{\delta}$ and $y = r/R$ it follows that

$$-\left.\frac{\mathrm{d}\theta}{\mathrm{d}y}\right|_{y=1} = \frac{\alpha_{\mathrm{eff}}R}{\lambda}\,\theta^{\mathrm{i}} \equiv Bi_{\mathrm{H}}\theta^{\mathrm{i}} \tag{40}$$

The value of the gradient and hence the concentration profile is determined by the value of Bi_{H}. For high values of Bi_{H} a sharp temperature profile is present in the droplet, the resistance to heat transfer lies primarily in the droplet, and θ^{i} approaches zero. A low Bi number causes a flat profile, tending to uniform temperature at decreasing Bi number, and a high value of the dimensionless temperature difference θ^{i}.

The mass transfer of a component j follows analogously:

$$-D_{\mathrm{j}} \left.\frac{\delta\rho_{\mathrm{j}}}{\delta r}\right|_{r=R} = k_{\mathrm{eff,j}}(\rho_{\mathrm{j}}'^{\mathrm{i}} - \rho_{\mathrm{j}}'^{\delta}) \tag{41}$$

in which D_{j} is the liquid phase diffusivity of component j.

Defining $\psi_{\mathrm{j}} = \rho_{\mathrm{j}} - \rho_{\mathrm{j}}^{*}$, in which ρ_{j}^{*} is the equilibrium concentration with $\rho_{\mathrm{j}}'^{\delta}$ and $y = r/R$, eqn (41) becomes

$$-\left.\frac{\mathrm{d}\psi_{\mathrm{j}}}{\mathrm{d}y}\right|_{y=1} = \frac{k_{\mathrm{eff,j}}R}{D_{\mathrm{j}}}\cdot m\psi_{\mathrm{j}}^{\mathrm{i}} = Bi_{\mathrm{Mj}}\psi_{\mathrm{j}}^{\mathrm{i}} \tag{42}$$

in which

$$m = \frac{\rho_{\mathrm{j}}'^{\mathrm{i}} - \rho_{\mathrm{j}}'^{\delta}}{\rho_{\mathrm{j}}^{\mathrm{i}} - \rho_{\mathrm{j}}^{*}} \tag{43}$$

Thus m is the slope of the line between the points $(\rho_{\mathrm{j}}'^{\mathrm{i}}, \rho_{\mathrm{j}}^{\mathrm{i}})$ and $(\rho_{\mathrm{j}}'^{\delta}, \rho_{\mathrm{j}}^{*})$, which both lie on the vapour–liquid equilibrium curve.

High values of Bi_{Mj} lead again to low values of $\psi_{\mathrm{j}}^{\mathrm{i}}$, indicating a dominating resistance to mass transfer in the liquid phase, while in the case of very low Bi_{Mj} numbers flat concentration profiles occur, and the dominating resistance lies in the gas phase.

For the case of internally circulating droplets, too, the Bi_{H} and Bi_{Mj} numbers are indications of dominating resistance.

Formation period

Near the atomiser turbulent flow of liquid and air occurs. No theoretical treatment is known at the moment to describe the mass and heat transfer to the liquid phase in the formation period, and few experimental data are available. Preliminary investigations in our laboratory showed that a small amount of volatile aroma components (up to 10%) is lost from the spray sheet in nozzle atomisation and that the amount lost is governed by the relative volatility rather than by liquid-phase diffusivity. This might be an indication of strong mixing in the liquid phase. As the data obtained were rather inaccurate, no further theoretical relations or correlations have been derived so far.

Internally circulating droplets

GAS PHASE MASS AND HEAT TRANSFER

In extensive reviews Sideman (1966) and Sideman and Shabtai (1964) present correlations for direct contact heat transfer coefficients between liquid drops and a liquid continuous phase. Inspection of the conditions in which these correlations apply teaches us that most of the equations are restricted either to low Reynolds numbers ($Re < 1$) or to high Prandtl numbers,

$$Pr = \frac{\mu' c_p'}{\lambda'} \qquad (Pr > 1 \text{ or } Pr \gg 1)$$

As quoted by Sideman, Bentwich *et al.* (1965) present the relation of Boussinesq (1905):

$$Nu = 1.13(Re \cdot Pr)^{1/2} \tag{44}$$

for the region $2 < (Re \cdot Pr)^{1/2} < 100$.

For mass transfer eqn (44) can be given as

$$Sh = 1.13(Re \cdot Sc)^{1/2} \tag{45}$$

In a recent study Oelrich *et al.* (1973) published numerical calculations of the local mass transfer from internally circulating spheres, in which use was made of the velocity profiles calculated by Haas *et al.* (1972). The values tabulated by Oelrich for $Sc = 1$ can be fitted by the equation

$$Sh = 0.5 + 1.11(Re \cdot Sc)^{1/2} \tag{46}$$

for $Re \cdot Sc > 4$.

Between the *Sh* numbers predicted by eqns (45) and (46) exists only a small difference, which decreases with increasing *Re* number.

In the calculations concerning the internal circulation period eqns (44) and (45) will be used.

MASS AND HEAT TRANSFER IN THE LIQUID PHASE

The equations describing the heat and mass transfer inside an internally circulating droplet can only be solved if velocity profiles in and outside the sphere are known. The velocity pattern was calculated by Hadamard (1911) for very low *Re* numbers. Kronig and Brink (1950) solved the transfer equations with the aid of the stream function of Hadamard for the case of negligible continuous phase resistance. They expressed their solution as a relation between the efficiency *E* and the dimensionless time *Fo* ($Fo = Dt/R^2$ for mass transfer and $Fo = \lambda t/\rho c_p R^2$ for heat transfer). The efficiency *E* is defined as the amount of mass (heat) transferred at dimensionless time *Fo*, and the amount transferred after infinite time:

$$E = 1 - \tfrac{3}{8} \sum A_n^2 \exp(-16\lambda_n Fo) \tag{47}$$

Danckwerts (1951) pointed out that eqn (47) can be used for *Re* numbers up to 2000. The solution of Kronig and Brink was extended by Elzinga and Banchero (1959), who also included external resistance to mass transfer in their equations. They obtained a solution in the same form as eqn (47), in which the coefficients A_n and λ_n are dependent on the *Bi* number. Values of these coefficients for several values of the *Bi* number are given by Elzinga and Banchero and by Sideman and Shabtai (1964).

For the case of no external transfer resistance ($Bi \to \infty$) Calderbank and Korchinski (1956) derived a correlation fitting eqn (47) accurately:

$$E = \{1 - \exp(-2.25\pi^2 Fo)\}^{1/2} \tag{48}$$

OVERALL HEAT AND MASS TRANSFER DURING INTERNAL CIRCULATION

For the Bi_H and Bi_M numbers the following relations can be written:

$$Bi_H = Nu \frac{\lambda'_m}{2\lambda} \tag{49a}$$

$$Bi_M = Sh \frac{mD'_m}{2D} \tag{49b}$$

Calculation of *Nu* from eqn (44) for $Re = 100$ and $Pr = 0.72$ gives $Nu = 9.59$. Substitution of $\lambda' = 0.026$ J/m²s°C and $\lambda = 0.61$ J/m²s°C (Weast, 1970) yields for heat transfer

$$Bi_H = 0.2 \tag{50}$$

From the low value of the Bi_H number it can be concluded that the resistance to heat transfer lies mainly in the gas phase. Thus in good approximation a *uniform temperature in the droplet* may be assumed, and the heat transfer can be directly calculated using eqn (44).

For the transport of water a droplet of an aqueous maltodextrin solution containing 75 wt% water will be taken as an example. From measurements reported by Menting (1969) and Menting *et al.* (1970) the diffusion coefficient of water has been found to be $D_w \simeq 2 \times 10^{-10}$ m²/s. From the sorption isotherm of maltodextrin (Menting, 1969) and the water vapour saturation concentration it follows, for a wet bulb temperature of 40°C and zero bulk humidity in the gas phase, that $m_w \simeq 7 \times 10^{-5}$. With the quantities $D'_w \simeq 2.5 \times 10^{-5}$ and $Sc = 0.60$ then from eqns (45) and (42)

$$Sh_w = 8.75 \quad \text{and} \quad Bi_{Mw} \simeq 40 \tag{51}$$

at $Re = 100$. The intermediate value of the Bi_{Mw} number indicates *resistance to water transport in both phases*. Calculations of the amount of water evaporated should be performed with eqn (47). If the values A_n and λ_n cannot be determined, a rapid estimate can also be computed either by relation (44) or (48). In both cases one of the limitations to water transport is ignored, and so a conservative estimate is obtained.

For the transport of an aroma component the diffusion of acetone in the maltodextrin solution mentioned before will be used. According to Menting (1969) and Menting *et al.* (1970a) the diffusion coefficient of acetone in this liquid is $D_a = 6 \times 10^{-10}$ m²/s. From the relative volatility of acetone with respect to water, $\alpha_{aw} = 46$ (Thijssen and Rulkens, 1968), it follows that at 40°C the equilibrium slope $m = 3.1 \times 10^{-3}$, if at the low concentrations in which aroma components are usually present, the equilibrium curve is a straight line. Substitution of these values and of $D'_a = 1.1 \times 10^{-5}$ m²/s (Perry, 1963) in eqns (45) and (42) gives for $Re = 100$

$$Sh_a = 13.2$$

and

$$Bi_{Ma} = 375 \tag{52}$$

From this high value of the Bi number it follows that *the dominating resistance for aroma transport lies in the liquid phase*. Therefore, the calculation of aroma loss can be carried out with eqn (48).

Rigid spherical droplets

HEAT AND MASS TRANSFER IN THE GAS PHASE

In the same reviews as mentioned in the previous section (Sideman, 1966; Sideman and Shabtai, 1964) a large number of correlations for the continuous phase heat transfer coefficients are given for the case of flow around rigid spheres, of which the majority is of the type

$$Nu = a + bRe^cPr^d \tag{53}$$

Most used at present are values of the constants $a = 2$, $b = 0.6$, $c = 0.5$ and $d = 0.33$ (Ranz and Marshall, 1952) or the Frossling equation with $b = 0.55$ (Frossling, 1938). For mass transfer from a droplet in still air the theoretical relationship

$$Nu = Sh = 2 \tag{54}$$

was confirmed experimentally by Marshall (1955). These correlations are only a few from the literature available; other relations are given by Ihme *et al.* (1972) and reviews are given by Sideman (1966), Sideman and Shabtai (1964), Masters (1972) and Marshall (1954). In the present work the Ranz and Marshall relations will be used for the gas phase mass and heat transfer coefficients:

$$Sh = 2 + 0.6Re^{1/2}Sc^{1/3}$$
$$Nu = 2 + 0.6Re^{1/2}Pr^{1/3} \quad (55)$$

HEAT TRANSFER IN THE LIQUID PHASE

Calculations analogous to those in eqns (49–52), but with the use of eqn (55), give

$$Bi_{\mathrm{H}} = 0.16 \quad \text{for} \quad Re = 100$$
$$Bi_{\mathrm{H}} = 0.04 \quad \text{for} \quad Re = 0$$

As the velocities of the rigid droplets decrease rapidly the assumption of virtually total resistance to heat transfer in the gas phase and, consequently, of a *uniform temperature inside the droplet* is justifiable.

MASS TRANSFER IN THE LIQUID PHASE

In a non-circulating droplet of a liquid food the transport of water, of dissolved solids and of aroma components takes place by molecular diffusion. As a model, the system is regarded as consisting of water, one dissolved solid component, and one aroma component present in a very dilute concentration. The transport of each component is assumed to take place only by molecular diffusion and by convective transport resulting from this diffusion. Further assumptions will be that the diffusion is spherically symmetric, and that no volume contraction occurs upon mixing of water and dissolved solids. It is also assumed that from the beginning of the drying process the droplet exists only in one phase, so there is no crystallisation or bubble formation.

Transport of water

According to Miller (1959) the equation describing the diffusion of water in the ternary system: water, dissolved solid, aroma, with respect to a volume fixed frame of reference, reads

$$j_{\mathrm{w}} = -D_{\mathrm{ww}}\nabla\rho_{\mathrm{w}} - D_{\mathrm{wa}}\nabla\rho_{\mathrm{a}} \quad (56)$$

which for negligible aroma concentration can be written as

$$j_{\mathrm{w}} = -D_{\mathrm{ww}}\nabla\rho_{\mathrm{w}} \quad (57)$$

This binary equation was also derived by Menting (1969), Thijssen and Rulkens (1968) and Chandrasekaran and King (1972a). Rulkens (1973) used the generalised Stefan–Maxwell equation according to Lightfoot *et al.* (1962), which for the water transport reads

$$\rho_w \nabla \ln A_w = \frac{1}{\rho D_{wa}} (\rho_a j_w - \rho_w j_a) + \frac{1}{\rho D_{ws}} (\rho_w j_s - \rho_s j_w) \quad (58)$$

where A_w is the thermodynamic activity of water.

For a volume fixed frame of reference in which no volume contraction occurs the following holds:

$$j_s = -j_w d_s/d_w \quad \text{and} \quad \rho = \rho_w + (1 - \rho_w/d_w)d_s$$

in which j_s is the dissolved solid flux and d_s and d_w are the densities of dissolved solid and pure water respectively. Neglecting the terms $\rho_a j_w$ and $\rho_w j_a$ in eqn (58), some rearrangement produces

$$j_w = -D_{ws} \frac{\delta \ln A_w}{\delta \ln \omega_w} \nabla \rho_w \quad (59)$$

which for $D_{ww} = D_{ws}(\delta \ln A_w/\delta \ln \omega_w)$ is identical with eqn (58).

From a shell balance follows the diffusion equation for water, as

$$\frac{\delta \rho_w}{\delta t} = \frac{1}{r^2} \frac{\delta}{\delta r} (-r^2 j_w) \quad (60)$$

With the assumption of uniform initial concentration profiles the initial condition reads

$$t = 0 \qquad 0 \leq r \leq R_0 \qquad \rho_w = \rho_{w0} \quad (61)$$

The boundary conditions are:

$$t > 0 \qquad r = 0 \qquad \frac{\delta \rho_w}{\delta r} = 0 \quad (62)$$

$$r = R \qquad j_w{}^i = F = j_w'^i \quad (63)$$

The water flux $j_w{}^i$ is equal to the flux of water with respect to the dissolved solids velocity, $j_w{}^s$, at the interface. Analogously to eqn (19) it can be derived for $j_w{}^s$ that

$$j_w{}^s \equiv j_w - v_s \rho_w = \frac{j_w}{1 - \rho_w/d_w} \quad (64)$$

The gas phase water flux at the phase boundary is given by

$$j_w'^i = k_{eff} \rho'(\omega_w'^i - \omega_w'^\delta) \quad (23)$$

$$= k_{eff}(\rho_w'^i - \rho_w'^\delta) \quad (65)$$

The interfacial gas phase water concentration $\rho_w'^i$ can be written as

$$\rho_w'^i = A_w{}^i \cdot \rho_w'^0 \quad (66)$$

in which $\rho_w'^0$ is the water vapour saturation concentration at droplet temperature. A_w varies with $\rho_w{}^i$ as given by the water vapour sorption isotherm.

The overall heat balance for the droplet is

$$q^i = j_w{}^i \Delta H_v + \frac{R}{3} \rho c_p \frac{dT}{dt} \tag{67}$$

in which ΔH_v is the heat of evaporation of water.

The solution of this set of equations and boundary conditions cannot be performed analytically, owing to the strong variation of D_{ww} with water concentration and temperature, the non-linearity of the water vapour sorption isotherm, the temperature dependence of the water vapour pressure and the shrinkage of droplets. Equations similar to (60), but for the drying of slabs, were solved numerically by Menting (1969), Menting *et al.* (1970b), Chandrasekaran and King (1972a), Rulkens and Thijssen (1969) and Kerkhof *et al.* (1972).

Equations (60–67) inclusive were solved numerically first by van der Lijn (1972) and van der Lijn *et al.* (1972) for the drying of droplets of aqueous maltose solutions. In further calculations, Schoeber (1973) and Rulkens (1973) used essentially the same method as van der Lijn. Analogously to the procedure used in the calculations for drying slabs, the concentrations and fluxes were transferred to a new co-ordinate system, based on equal increments of dissolved solid mass or volume. This type of co-ordinate system does not shrink and is invariant with time.

Defining the new distance and concentration variables

$$z = \int_0^r \rho_s r^2 \, dr \tag{68}$$

and

$$U_w = \rho_w / \rho_s \tag{69}$$

eqn (60) can be transformed to

$$\left.\frac{\delta U_w}{\delta t}\right|_z = \frac{\delta}{\delta z}\left(D_{ww} \rho_s{}^2 r^4 \frac{\delta U_w}{\delta z}\right) \tag{70}$$

This equation, with the appropriate boundary conditions

$$t = 0 \qquad 0 \leq z \leq Z \qquad U_w = U_{w0} \tag{71}$$

$$t > 0 \qquad z = 0 \qquad \frac{\delta U_w}{\delta z} = 0 \tag{72}$$

$$z = Z \qquad -D_{ww} \rho_s{}^2 r^2 \frac{\delta U_w}{\delta z} = F = j_w'^i \tag{73}$$

can be solved numerically by a combined semi- and fully-implicit method, analogous to the Crank–Nicholson or the Laasonen finite difference method (Lapidus, 1962; Richtmyer and Morton, 1967). In order to cope with the very sharp concentration profiles without using excessive computer time and storage, it is necessary to use unequal step sizes both in space and time.

Transport of aroma components

Analogously to eqn (56) the flux of an aroma component in a ternary system in a volume-fixed frame of reference is given by

$$j_a = -D_{aw}\nabla\rho_w - D_{aa}\nabla\rho_a \tag{74}$$

This equation was also used by Chandrasekaran and King (1972a, 1972b).

The diffusion coefficient D_{aw} includes a relation between the activity coefficient of the aroma component and the water concentration (Chandrasekaran and King, 1972a). Neglecting the water gradient term, Menting (1969) used a simple binary diffusion equation. Thijssen and Rulkens (1968) and Rulkens and Thijssen (1969) used the ternary Stefan–Maxwell equation, while later Rulkens (1973) and Kerkhof *et al.* (1972) used the generalised Stefan–Maxwell equation (75), in which the driving force of diffusion is the gradient of chemical potential instead of concentration:

$$\rho_a\nabla\ln A_a = \frac{1}{\rho \underset{\bullet}{D}_{aw}}(\rho_a j_w - \rho_w j_a) + \frac{1}{\rho \underset{\bullet}{D}_{as}}(\rho_a j_s - \rho_s j_a) \tag{75}$$

Substitution of Henry's law, viz. $A_a \equiv H_a \cdot \rho_a$, of

$$\frac{1}{D_a} \equiv \frac{\omega_w}{\underset{\bullet}{D}_{aw}} + \frac{1-\omega_w}{\underset{\bullet}{D}_{as}} \tag{76}$$

and of the expressions for j_w and j_s, yields

$$j_a = -D_a\nabla\rho_a - D_a\rho_a\left\{\frac{\delta\ln H_a}{\delta\rho_w} + \frac{1}{\rho}D_{ww}\left(\frac{1}{\underset{\bullet}{D}_{aw}} - \frac{d_s}{d_w\underset{\bullet}{D}_{as}}\right)\right\}\nabla\rho_w \tag{77}$$

As D_{aw} in eqn (74) is directly proportional to ρ_a (Chandrasekaran and King, 1972), the functional description of the aroma flux by eqns (74) and (77) is equivalent. The second term of eqn (77) can have a sign opposite to the direction of the water flux, because of two effects. The first is that the activity coefficient of aroma molecules decreases with increasing water concentration (Chandrasekaran and King, 1971, 1972; Menting, 1969) and thus $(\partial\ln H_a/\partial\rho_w) < 0$.

The second follows from a physical interpretation of the Stefan–Maxwell equation, as given by Rulkens and Thijssen (1969). The term $1/\underset{\bullet}{D}_{aw}$ in eqn (75) represents an interactive force between water and aroma molecules, and $1/\underset{\bullet}{D}_{as}$ represents the interactive force between molecules of the dissolved solid and aroma molecules. At low water concentrations the molecules of the dissolved solid tend to form a more or less structured matrix, with interconnections between the dissolved solid molecules by hydrogen bridges. It is then expected that the interactive force between aroma molecules and the dissolved solids network is much stronger than the interaction between aroma and water molecules. This can be stated mathematically as $(1/\underset{\cdot}{D}_{aw}) \ll (1/\underset{\cdot}{D}_{as})$ which leads to $(1/\underset{\cdot}{D}_{aw}) - (ds/d_w\underset{\cdot}{D}_{as}) < 0$, so that both terms inside the brackets in eqn (77) become negative. The magnitude of the whole water gradient term in eqn (77), which represents the aroma flux caused by the presence of the water concentration gradient, depends upon the value of ρ_w and of $\nabla\rho_w$. At high drying rates sharp water concentration profiles are built up near the evaporating surface, and the water gradient term may become greater than the aroma concentration gradient term, causing an inward flux of aromas near the evaporating surface. At a small distance from the evaporating surface, the water gradient is only very small, and thus the aroma flux is directed outward. This causes a maximum in the aroma concentration (Chandrasekaran and King, 1972; Kerkhof *et al.*, 1972).

The loss of aroma from a drying droplet can be calculated with the diffusion equation, which follows from the shell balance

$$\frac{\delta\rho_a}{\delta t} = \frac{1}{r^2}\frac{\delta}{\delta r}(-r^2 j_a) \tag{78}$$

with the initial and boundary conditions:

$$t = 0 \qquad 0 \leq r \leq R_0 \qquad \rho_a = \rho_{a,0} \tag{79}$$

$$t > 0 \qquad r = 0 \qquad \frac{\delta\rho_a}{\delta r} = 0 \tag{80}$$

$$r = R \qquad \rho_a \simeq 0 \tag{81}$$

This last boundary condition can be so written because of the zero bulk concentration of aroma in the drying air, and the high Biot number Bi_{Ma}, as was also reported by Chandrasekaran and King (1972a) and Menting (1969).

Because the aroma flux is dependent on the water concentration and on the water concentration gradient, solution of the diffusion equation (78) should be performed simultaneously with the solution of the water diffusion equation (60). For the drying of slabs, numerical calculations of aroma transport are given by Menting (1969), Menting *et al.* (1970b), Rulkens and Thijssen (1969), Chandrasekaran and King (1972a) and Kerkhof *et al.* (1972). Calculations of aroma transport in a drying sphere have been reported recently by Schoeber (1973) and Schoeber and Kerkhof (1973). For the solution of the aroma transport the concentration and flux of aroma were also transformed to dissolved-solids based co-ordinates:

$$U_a = \rho_a/\rho_s \tag{82}$$

$$j_a^s = j_a - v_s\rho_a \tag{83}$$

The transformed diffusion equation then reads

$$\left.\frac{\delta U_a}{\delta t}\right|_z = \frac{\delta}{\delta z}(-r^2 j_a^{s}) \tag{84}$$

with the initial and boundary conditions:

$$t = 0 \qquad 0 \leq z \leq Z \qquad U_a = U_{a0} \tag{85}$$

$$t > 0 \qquad z = 0 \qquad \frac{\delta U_a}{\delta z} = 0 \tag{86}$$

$$z = Z \qquad U_a = 0 \tag{87}$$

Thermal degradation of heat-sensitive compounds

In food liquids at elevated temperatures degradative reactions such as browning and enzyme deactivation may occur. Reaction rate does in general not only depend on temperature, but in many cases decreases with decreasing water concentration, as reported in a review by Labuza (1972). In the case of a first-order degradation of a component p, the following equation applies in a volume fixed frame of reference:

$$\frac{\delta \rho_p}{\delta t} = -k_1 \rho_p - \frac{1}{r^2}\frac{\delta}{\delta r}(r^2 j_p) \tag{88}$$

According to Ball and Olson (1957) the effect of temperature on k_1 can be represented by

$$k_1 = k_{1,0} \exp(aT) \tag{89}$$

which proved to fit the experimental results for inactivation of phosphatase in skimmed milk (Verhey, 1972) better than the Arrhenius equation.

TRANSPORT PROPERTIES OF WATER AND AROMA IN FOOD LIQUIDS

Experimental determination of the water diffusion coefficient D_{ww} for aqueous solutions of maltodextrin (Menting, 1969; Menting *et al.*, 1970a), starch gel (Fish, 1958), glucose (Gladden and Dole, 1953), fructose (Chandrasekaran and King, 1972b), sucrose (English and Dole, 1950; Gosting and Morris, 1949; Henrion, 1964), maltose (van der Lijn, 1972), coffee extract (Thijssen, 1960; Thijssen and Rulkens, 1968) and gelatine (Thijssen and Rulkens, 1968) has shown that the diffusion coefficient D_{ww} decreases sharply with decreasing water concentration, especially at water concentrations below about 20 wt%. The magnitude of this variation is of the order of 10^3–10^5. The value of D_{ww} depends not only on the water concentration but also on the temperature. Activation energies have been reported for aqueous solutions of starch gel (Fish, 1958), maltose (van der Lijn, 1972), sucrose (Chandrasekaran and King, 1972b; English and Dole,

1950) and glucose (Gladden and Dole, 1953). The activation energy decreases with increasing water concentrations, and was reported to be of the order of 4–25 kcal/mol.

Water vapour sorption isotherms have been reported for maltodextrin (Menting, 1969), starch gel (Fish, 1958), glucose and fructose (Timmermans, 1960), sucrose (Scratchard *et al.*, 1938; Heiss, 1968), maltose (van der Lijn *et al.*, 1972), coffee extract and gelatine (Nemitz, 1963). Generally the water activity remains close to 1 for water concentrations higher than about 50 wt%, while the water activity falls considerably below about 20 wt%.

The diffusion coefficients D_{aa} and D_{aw}/ρ_a vary only with water concentration and temperature, and not with the aroma concentration itself at the prevailing concentrations in food liquids. Physically this means that at these low concentrations aroma molecules do not interact.

Diffusion data for aroma components in food liquids are given by Chandrasekaran and King (1972b) for ethyl alcohol, ethyl acetate, *n*-butyl and *n*-hexanal in sugar solutions; by Menting (1969) for ethyl acetate, acetone, benzene and carbon tetrachloride in maltodextrin solutions; and by Thijssen and Rulkens (1968) for acetone in coffee extract. A most interesting observation was made by Chandrasekaran and King, who showed that the diffusion coefficient D_{aa} had approximately the same functional relationship to the water concentration for a number of aroma components, only differing from each other by a constant factor $D_{aa}{}^0$, the diffusion coefficient of aroma in pure water. The activation energy for aroma diffusion was given by the same authors for ethyl alcohol in sucrose solutions. Here, too, the activation energy is dependent on the water concentration and is of the order of 4–10 kcal/mol. Comparison of the diffusion coefficients D_{aa} and D_{ww} in relation to the water concentration shows that when the water concentration is low, D_{aa} decreases more sharply with decreasing water concentration than D_{ww}. This leads to a ratio $D_{aa}/D_{ww} < 10^{-2}$, which is indicative of negligible permeability to aroma at low water concentrations.

The aroma activity coefficient in relation to the water concentration is given by Chandrasekaran and King (1971, 1972b) for several aroma compounds in sugar solutions, and for acetone in maltodextrin solutions by Menting (1969). Activity coefficients of aroma increase with decreasing water content and with increasing molecular weight of the aroma component.

CALCULATION OF DRYING PHENOMENA

In this section the solution to some problems concerning the heat, mass and momentum transfer to droplets in spray dryers will be treated. Two main lines can be distinguished, viz.:

(1) The motion and evaporation of pure water droplets.
(2) The drying behaviour of droplets of solutions.

The motion and evaporation of pure water droplets is calculated for two cases.

(a) Evaporation and motion of a water droplet moving vertically in still air. The equations of motion and of mass and heat transfer are solved simultaneously. The results are compared with those obtained by a simplified method proposed by Sjenitzer (1952).
(b) Evaporation and motion of a water droplet in rotating air flow. The droplet is formed by a centrifugal pressure nozzle; motion in three dimensions is calculated simultaneously with evaporation.

The drying behaviour of solutions is considered for two cases.

(a) Evaporation of water and of aroma components during the internal circulation period. The length of this period is calculated, as well as the amounts of water and of an aroma compound evaporated during this period.
(b) The drying behaviour of rigid droplets of liquid foods. The loss of water and of a model aroma component, the droplet temperature and thermal degradation as varying with time are calculated by simultaneous solution of the diffusion and reaction equations and the heat balance. The final aroma retention is also considered as depending on process variables.

Evaporation and motion of pure water droplets

EVAPORATION OF WATER DROPLETS DURING DECELERATION IN UNIDIRECTIONAL VERTICAL MOTION

For the vertical downward motion of a spherical droplet the equation of motion is

$$\frac{\mathrm{d}u}{\mathrm{d}t} = \frac{\rho - \rho'}{\rho} g - \tfrac{3}{4} c_{\mathrm{d}} \frac{\rho' u^2}{\rho d} \tag{90}$$

For the evaporation the following equation holds:

$$\frac{\mathrm{d}m}{\mathrm{d}t} = -k_{\mathrm{eff}} \cdot \pi d^2 \Delta \rho'_{\mathrm{w}} \tag{91}$$

which can be written for the diameter as

$$\frac{\mathrm{d}(d)}{\mathrm{d}t} = -\frac{2k_{\mathrm{eff}} \Delta \rho'_{\mathrm{w}}}{\rho} \tag{92}$$

Equations (90) and (91) or (92) are coupled by the dependence of the drag coefficient and the mass transfer coefficient on the Reynolds number *Re*, and by the dependence of *Re* on *u* and *d*. In the present study eqns (90) and (92) were solved simultaneously on a Burroughs B-6700 digital computer using a third-order Runge–Kutta method (Lapidus, 1962), programmed in BEATHE, a version of Algol-60. The conditions were:

initial droplet velocity:	50 m/s
initial droplet diameters:	100 μm, 200 μm and 400 μm
constant air temperature:	150°C
constant air humidity:	0.01 kg/m^3

Furthermore, it was assumed that the droplets were at wet-bulb temperature from the start of the process, $T = 40°C$. From this follows for the driving force of mass transfer:

$$\Delta\rho'_{w} = 0.035 \text{ kg/m}^3$$
$$\Delta H_{w} = 0.04 \text{ kg/kg dry air}$$

Equations (10) and (55) were used for calculating the drag coefficient and the mass transfer coefficient respectively. For the deceleration period was taken the time during which the droplet velocity was higher than its free-falling velocity based on the *initial diameter*.

Sjenitzer proposed a graphic method of estimating the amount of water evaporated during deceleration. By making some simplifying assumptions, e.g. the ignoring of gravitational forces and of diameter change during deceleration, he obtained an expression for the additional amount of water which evaporated owing to the droplet velocity. With this expression the total amount of water evaporated during deceleration can be written as

$$x_{w} = x_{w}^{\bullet} + 4.42\Delta H_{w}Sc^{-0.67}\int_{Re_2}^{Re_1}\frac{dRe}{c_d Re^{1.5}} \tag{93}$$

in which $x_{w}^{\bullet}$ is the amount of water which evaporates from a non-moving droplet in still air; the second term on the right-hand side represents the additional evaporation described by $Sh = 0.55Re^{0.5}Sc^{0.33}$. x_w is the fractional evaporation, i.e. the amount of water evaporated divided by the amount initially present. Sjenitzer has published graphs for determining the deceleration time and the values of the integral.

The values of the initial and terminal *Re* numbers were determined for the conditions mentioned above, and the deceleration time and fractional evaporation were determined graphically according to the Sjenitzer method. Table I gives the results of both types of calculations.

From this table it can be seen that the simplifications used in the Sjenitzer method may introduce considerable errors, especially in the case of large diameters. The explanation of this is that the effect of gravity on deceleration becomes more important with large diameters, as can be seen from eqn (90). Ignoring the gravity term thus leads to too small deceleration times and too small amounts of water evaporated. This discrepancy between the calculated amounts of water evaporated for the 100 μm droplet, in spite of the nearly equal deceleration time, is due to a difference in deceleration history. As the shrinkage of the droplet would tend to decrease deceleration time, the effect of gravity, on the other hand, increases this time. This, however, does not imply that the histories in terms

TABLE I

COMPARISON BETWEEN THE SJENITZER METHOD AND NUMERICAL CALCULATIONS IN DETERMINING THE EVAPORATION AND DECELERATION OF PURE WATER DROPLETS

d_0 (μm)	u_{term} (m/s)	Re_{init}	Re_{term}	$t_{decel}^{numerical}$ (s)	$t_{decel}^{Sjenitzer}$	$x_w^{numerical}$	$x_w^{Sjenitzer}$
100	0.23	226	1.05	0.087	0.090	0.27	0.21
200	0.69	452	6.2	0.22	0.165	0.21	0.14
400	1.66	904	30	0.54	0.278	0.19	0.093

of mass transfer are equal if deceleration times for both methods of calculation are equal.

As at present more and more use can be made of computer time sharing systems and computer centres, and since algorithms for the numerical solution of this type of equation are readily available in the literature (Lapidus, 1962), it is recommended that the problem of droplet evaporation be solved by computer rather than by simplified methods.

EVAPORATION AND MOTION OF A WATER DROPLET IN ROTATING AIR FLOW

It is assumed that droplets are formed by a pressure nozzle atomiser with a spray angle of 90°C and an initial total velocity of 50 m/s, at the origin of a Cartesian co-ordinate centre $x = 0, y = 0, z = 0$, with initial velocity components in the x and z directions as shown in Fig. 3a. The air velocity consists of a tangential velocity component depending on the radius as given in eqn (5) and in Fig. 3b, and a uniform vertical component v_z. The tower radius was taken as 2 m, the maximum tangential air velocity at the tower wall as 5 m/s. The droplet diameter at the nozzle was taken as 100 μm, 200 μm and 400 μm. The initial droplet temperature was 20°C. The air temperature and humidity were 150°C and 0.01 kg/m^3 respectively. For the drag coefficient and for the mass and heat transfer coefficients eqns (10) and (55), respectively, were used. The equation of motion was written in Cartesian co-ordinates:

$$\frac{\partial u_x}{\partial t} = -\frac{3}{4}\frac{c_d\rho' w_x|\overline{w}|}{\rho d} \tag{94}$$

$$\frac{\partial u_y}{\partial t} = -\frac{3}{4}\frac{c_d\rho' w_y|\overline{w}|}{\rho d} \tag{95}$$

$$\frac{\partial u_z}{\partial t} = \frac{\rho - \rho'}{\rho} g - \frac{3}{4}\frac{c_d\rho' w_z|\overline{w}|}{\rho d} \tag{96}$$

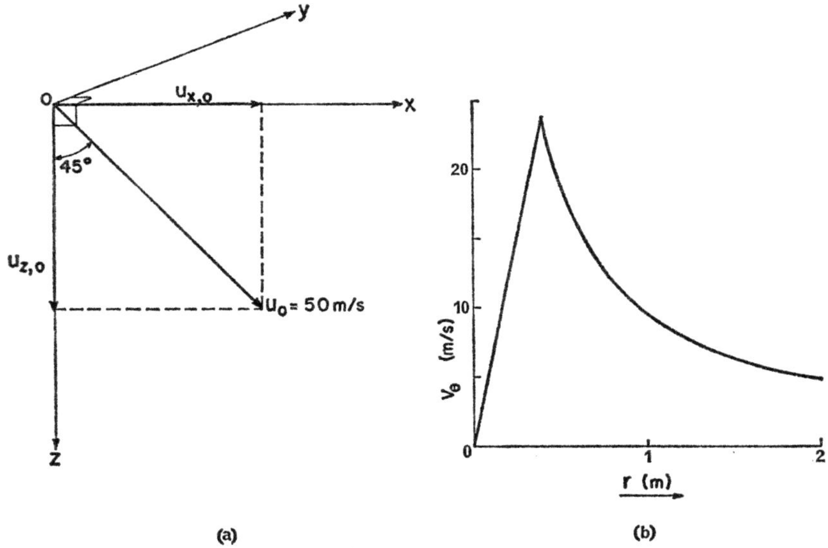

FIG. 3. (a) Co-ordinate system and initial droplet velocity vector used in calculations of motion and evaporation of pure water droplets in swirling air flow. (b) Tangential air velocity profile used in calculations of motion and evaporation of pure water droplets in swirling air flow.

and the tangential air velocity was resolved into x and y components:

$$v_x = -v_\theta \sin \theta \tag{97}$$

$$v_y = v_\theta \cos \theta \tag{98}$$

with

$$\begin{aligned} \theta &= \arctan (S_y/S_x) \quad &\text{for} \quad S_x > 0 \\ &= \arctan (S_y/S_x) + \pi \quad &\text{for} \quad S_x < 0 \end{aligned} \tag{99}$$

with S_x, S_y the x and y co-ordinates respectively. For the droplet temperature eqn (67) was used, while for the droplet radius the following holds:

$$\frac{\mathrm{d}R}{\mathrm{d}t} = -j_{\mathrm{w}}^{\mathrm{i}}/d_{\mathrm{w}} \tag{100}$$

For the solution of these equations, including the determination of the co-ordinates, a BEATHE programme was writted in which the equations

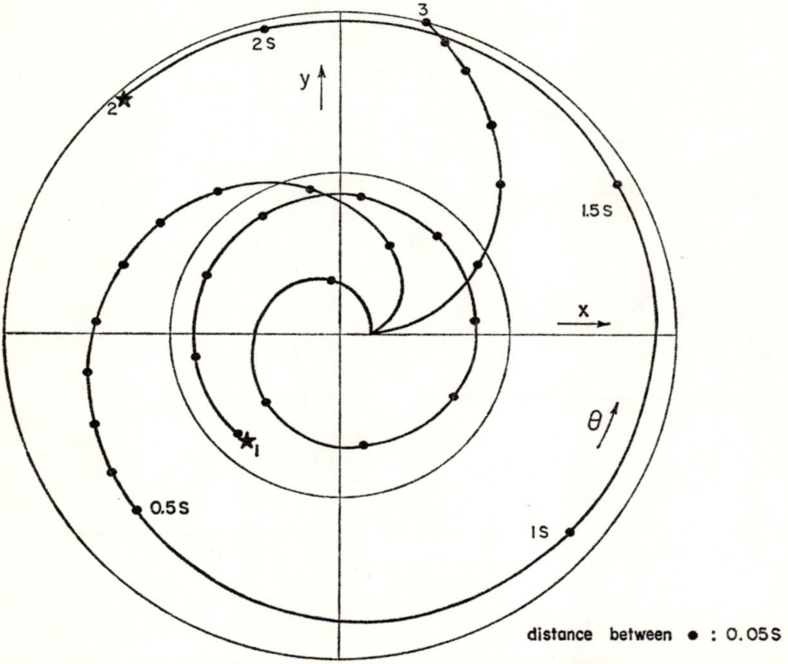

FIG. 4. Horizontal projection of motion of evaporating water droplets in rotational flow. Distance between •: 0.05 s. Droplet diameters: 1: 100 μm; 2:200 μm; 3:400 μm. Downward air velocity: 0.2 m/s.

were integrated simultaneously with a third-order Runge–Kutta method (Lapidus, 1962). A variable time step was used. Calculation times on the Burroughs B-6700 computer of the Computer Centre at Eindhoven University of Technology were 4 min for each run.

In Fig. 4 the path of three droplets is shown in the horizontal plane. The 100 μm diameter droplet is rapidly decelerated and then follows the air flow pattern, with a small outward velocity superimposed. It evaporates within a radius of 1 m from the tower axis. The trajectory of the 200 μm droplet is more extended, but owing to the low tangential velocity near the tower wall, the radial force on the droplet is very small, and the droplet is evaporated before it reaches the tower wall. The 400 μm droplet has the most extended deceleration length in the x-direction, and after 90° rotation hits the tower wall.

Figure 5 shows the vertical trajectories under the same conditions, as well as for a 200 μm droplet in the absence of tangential air flow components. As expected, the vertical trajectory increases with increasing

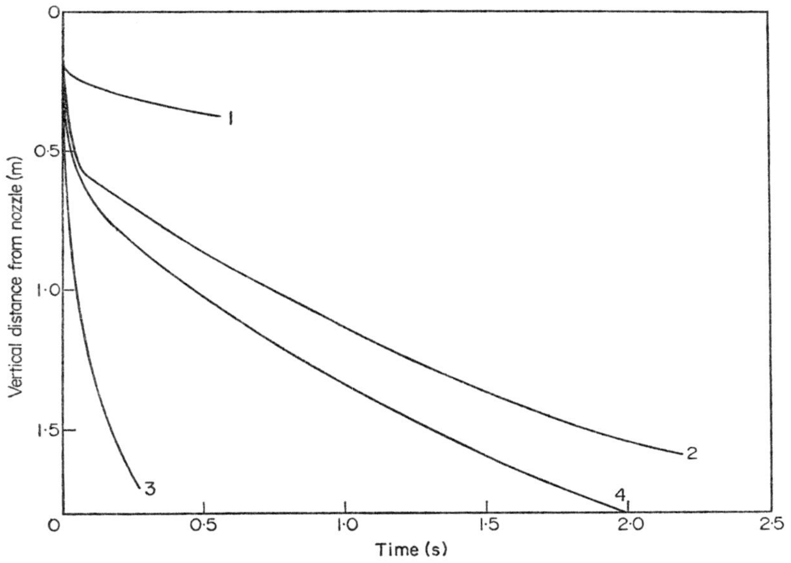

FIG. 5. Calculated vertical trajectories for evaporating water droplets. Droplet diameters: 1: 100 μm; 2, 4: 200 μm; 3: 400 μm; 1, 2 and 3 with tangential air velocity; 4 without. Downward air velocity: 0.2 m/s.

droplet diameter. Although obvious from theoretical considerations the difference between the curves 2 and 4 is interesting. Owing to a higher drag coefficient, the vertical velocity in the case of rotating air flow is smaller than in the case of purely axial flow.

This means that the approximation of using independent equations for vertical and horizontal motion can lead to errors, especially at a high tangential air velocity.

The results calculated here are in good agreement with those presented by Bailey *et al.* (1970), Mühle (1971, 1972) and Stein (1972). The last-mentioned author also included the drying of insoluble solids in his calculations.

Drying behaviour of solutions

LOSS OF WATER AND AROMA COMPONENTS DURING THE INTERNAL CIRCULATION PERIOD

The conditions used in this study were as follows. Droplets were vertically injected with a velocity of 100 m/s in still air of 140°C and a humidity of 0.01 kg/kg. It was assumed that the diffusion coefficients in the liquid phase of aroma and of water were equal to the diffusion coefficients of acetone and of water respectively in a 75 wt% water aqueous maltodextrin solution. From Menting's results (1969) it follows that $D_a = 6 \times 10^{-10}$ m²/s and $D_w = 2 \times 10^{-10}$ m²/s. The transport of water is limited in both phases, the transport of aroma only in the liquid phase. Furthermore, the transport of aroma is considered to be independent of the water transport, which makes possible the use of eqn (48). Equation (45) was used for the water transport in the continuous phase, and eqn (47) for the overall water transport with $Bi_{Mw} = 40$. It was assumed that the droplets were at wet-bulb temperature from the moment of their formation. The critical velocity is given by eqn (16).

The equation of motion and the evaporation equation were solved with a third-order Runge–Kutta method on a Philips P 9200 time-sharing system, programmed in Algol-60. Computing time was 30 s for each run.

Figure 6 shows the droplet velocity, the droplet trajectory and the aroma retention in dependence on time, until the critical velocity is reached. The conditions used for this calculation were

$$d_0 = 50\ \mu\text{m} \quad \text{and} \quad (\Delta\sigma)_{\text{eff}} = 0.01\ \text{N/m}$$

From this figure it can be seen that in 2 ms as much as 20% of the aroma component has evaporated. In Table II are represented the dependence of the critical time, critical velocity and path and the aroma retention on the droplet diameter and on the difference in surface tension, $(\Delta\sigma)_{\text{eff}}$.

Although there is a marked influence of the droplet diameter on the circulation time t_c, there is not much influence on the aroma retention $AR = 1 - E_a$, the latter varying with $Fo_a = D_a t_c / R^2$. More dependence is found for the aroma retention on $(\Delta\sigma)_{\text{eff}}$.

Recalling the approximations made in the derivation, it is evident that there are a number of factors which are unknown. The surface tension gradient has been approximated by a linear gradient over the droplet circumference; it is not known how this gradient varies in reality. It was assumed that the droplet moves through still air, which is certainly not the case in reality. Furthermore, the effects of droplet distortion and oscillation have been neglected, which leads to under-estimation of drag coefficients, and of mass and heat transfer coefficients. The conclusions derived from the calculated results are thus only qualitatively applicable to real drying situations. However, this model provides an indication that internal

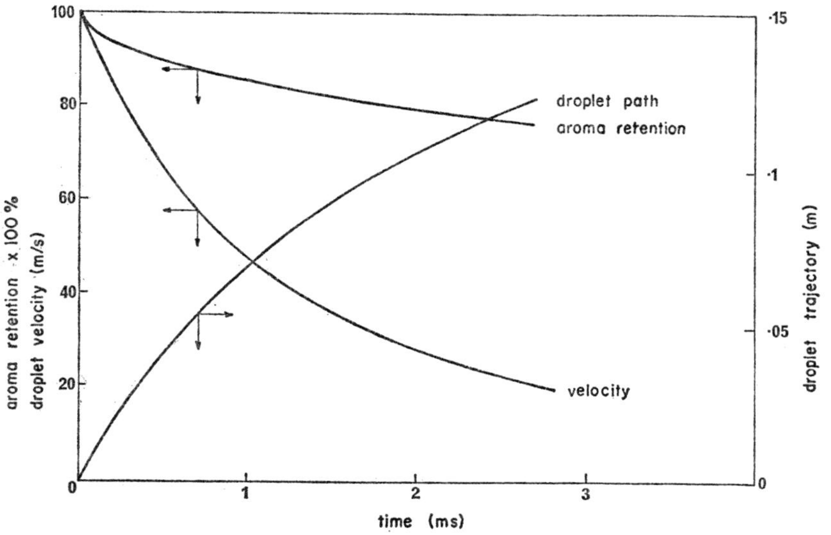

FIG. 6. Dependence of velocity, trajectory and aroma content on time for an internally circulating droplet moving vertically in still air. Droplet diameter 50 μm, difference in surface tension, $(\Delta\sigma)_{eff}$: 0.01 N/m.

TABLE II

RESULTS OF INTERNAL CIRCULATION CALCULATIONS. AIR TEMPERATURE 140°C, AIR HUMIDITY 0.01 kg/kg

$(\Delta\sigma)_{eff}$ N/m	d_0 μm	U_{cr} m/s	t_{cr} ms	S_{cr} m	$1 - E_a$	E_w
0.02	25	41.3	0.45	2.87	0.81	0.09
0.02	50	31.3	1.79	9.95	0.81	0.09
0.02	100	23.3	6.60	31.05	0.81	0.08
0.02	200	17.0	23.1	90.0	0.83	0.08
0.005	50	13.3	3.91	14.25	0.72	0.11
0.01	50	20.6	2.71	12.27	0.76	0.10
0.04	50	46.5	1.05	7.12	0.85	0.07
0.07	50	63.4	0.57	4.53	0.89	0.05

circulation may occur in spray drying, as well as an indication of the amounts of water and aroma lost during circulation. As one of the major factors in the model is the presence of surface active agents, experimental investigations into the effect of the presence of such agents on aroma retention would be of interest for a better understanding of the phenomena during the first short drying stage of the spray drying process.

THE DRYING OF RIGID DROPS: LOSS OF WATER AND AROMA COMPONENTS, TEMPERATURE HISTORY AND THERMAL DEGRADATION

For the calculation of the drying rate of a droplet the diffusion equations (70) and (84) were written in finite difference form and programmed in Algol-60 on a Philips Electrologica EL-X8 digital computer. Details of the numerical procedures will not be given here, but are given by van der Lijn (1974) and Schoeber (1973). The physical data used in the diffusion equations were:

Water diffusion coefficient D_{ww}

$$D_{ww} = \exp\left(-13.923 + 21.664X_s - \frac{4.5791 + 23.603X_s}{RT_K}\right) \tag{101}$$

as reported by van der Lijn for the maltose–water system. In this equation X_s is the mole fraction of dissolved solid, and R the gas constant given in kcal/mol K. T_K is the absolute temperature (K).

Aroma diffusion coefficient D_{aa}

For the aroma diffusion coefficient D_{aa} a model diffusion coefficient is used:

$$D_{aa} = \exp\left(-13.394 + 7.601X_s^2 + 24.314X_s - \frac{4.500 + 34.312X_s}{RT_K}\right) \tag{102}$$

Relation between $\underset{\cdot}{D}_{aw}$ *and* $\underset{\cdot}{D}_{as}$

It is assumed that the mobility of aroma with respect to water is higher than with respect to the dissolved solids; the relation between $\underset{\cdot}{D}_{aw}$ and $\underset{\cdot}{D}_{as}$ is approximated by

$$\underset{\cdot}{D}_{aw} = 10\underset{\cdot}{D}_{as} \tag{103}$$

Water vapour sorption isotherm

$$A_w = X_w \exp(-4.49X_s^2 + 5.53X_s^3) \tag{104}$$

in which X_w is the mole fraction of water. This relation was reported for maltose solutions (van der Lijn *et al.*, 1972).

Aroma Henry coefficient H_a

For the aroma Henry coefficient a model relation was used:

$$H_a = \frac{1}{d_w}\{20 + 100\exp(-7.5\rho_w/d_w)\} \tag{105}$$

after Kerkhof *et al.* (1972).

Thermal degradation reaction

$$k_1 = \exp(-34.9 - 1.39/\rho_w + 0.5183T) \tag{106}$$

which is a fit to results obtained by Verhey (1972) for the inactivation of phosphatase in skimmed milk.

The initial droplet temperature was taken to be 20°C, and uniform concentration profiles were assumed at $t = 0$. It was further assumed that the thermally unstable compound was immobile with respect to the dissolved solid. Further assumptions have been given in a previous section (p. 361). Calculations were performed for a number of process conditions. Several results were obtained:

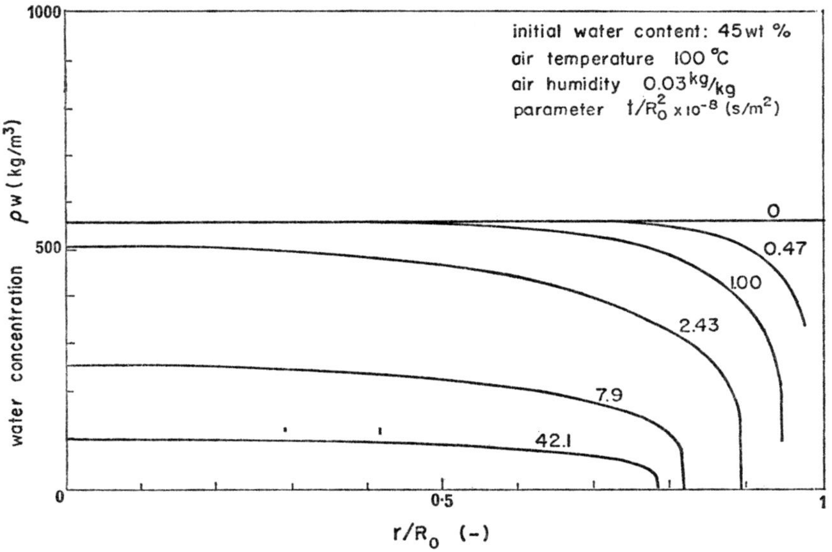

FIG. 7. Calculated water concentration profiles inside drying rigid droplets of aqueous maltose solution. Parameter is the reduced time t/R_0^2.

(1) concentration profiles;
(2) temperature, water content, aroma content and thermal degradation in relation to time;
(3) final aroma retention in relation to process variables.

Concentration profiles

In Figs. 7 and 8 the concentration profiles of water and of aroma are given in relation to reduced radius r/R_0 with the reduced time t/R_0^2 as parameter at 45 wt% initial water content. The interfacial water concentration decreases rapidly and at $t/R_0^2 \simeq 10^8$ a dry skin is formed.

The aroma concentration profiles at first resemble the penetration type profiles, but as soon as the dry skin is formed they become very steep. As no more aroma is lost, then the average aroma concentration starts to rise because of the shrinkage due to water loss.

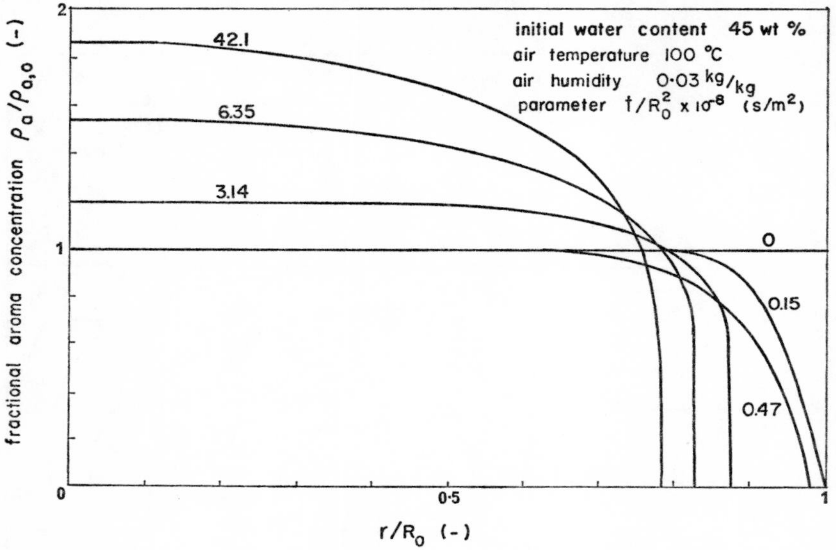

FIG. 8. Calculated aroma concentration profiles for same conditions as Fig. 7. Parameter is the reduced time t/R_0^2.

It is interesting to note that the surface water concentration has already dropped to a value of 100 kg/m^3 when the centre water concentration has not yet decreased appreciably. This means that the activity of water at the interface has decreased considerably, while the main part of the water is still in the droplet.

Temperature and aroma content, and thermal degradation in relation to time

In Fig. 9 the temperature, the fractional water content and the fractional aroma content are given in relation to the reduced time t/R_0^2 for the first part of the drying process under the same conditions as Figs. 7 and 8. After a warming-up period the droplet temperature remains constant for a short time and at $t/R_0^2 \simeq 0.5 \times 10^8$ the temperature begins to rise again. At that point the aroma retention starts to level off to a constant value, and the dry skin is formed.

In Fig. 10 the interfacial water flux, the water activity and the interfacial water concentration are given in relation to reduced time for the same conditions as Fig. 9.

The decrease in water activity at $t/R_0^2 = 0.5 \times 10^8$ occurs at 25 wt% water at the interface. This decrease is stronger than that in water flux, as the latter is partially compensated by the temperature rise.

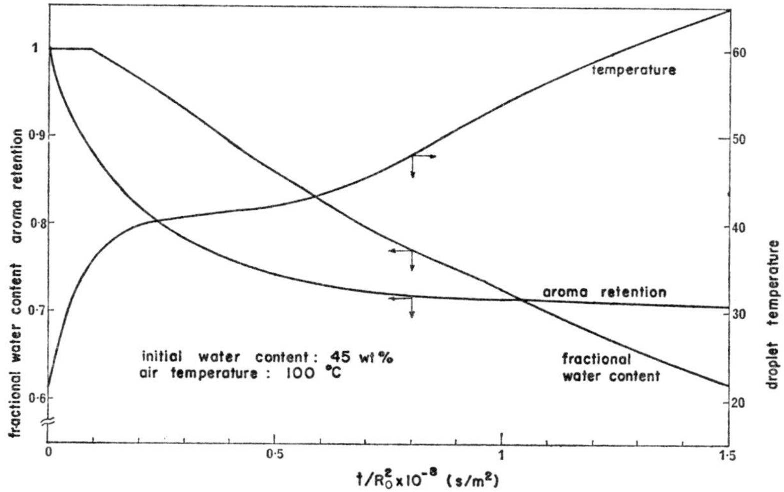

FIG. 9. Temperature, water content and aroma content in relation to reduced time t/R_0^2, for the first part of the drying process. Same conditions as Fig. 7.

Figure 11 illustrates the temperature and water content for several flow conditions of the drying air over a longer drying time. It can be seen that the fastest drying occurs in countercurrent flow, which also gives the highest temperature rise of the drying droplet. The drying in co-current flow is somewhat faster than in ideally mixed air flow.

Figure 12 gives temperature history and the thermal degradation versus reduced time. At 100°C air temperature the conversion is seen to run to completion in a very short time. At 80°C the reaction starts more slowly and the reaction rate in the last drying stage is decreased by the higher dissolved solids concentration. In both cases the reaction starts at a droplet temperature of approximately 70°C.

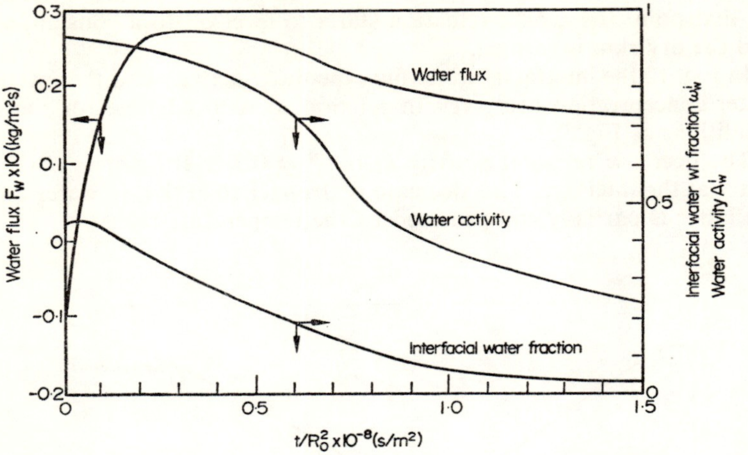

FIG. 10. Water flux, interfacial water activity and interfacial water concentration in relation to reduced time t/R_0^2, for the first part of the drying process. Same conditions as Fig. 7.

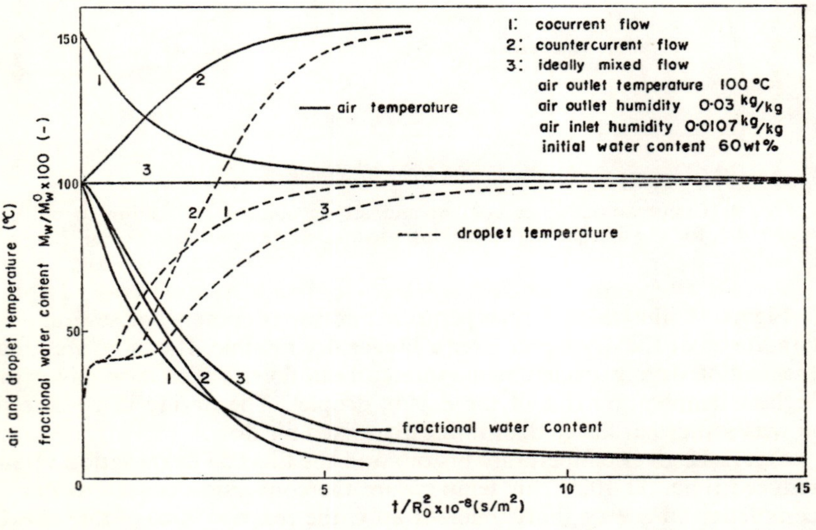

FIG. 11. Temperature of air and of droplets, and droplet water concentration in relation to reduced time t/R_0^2, for different air flow directions.

Effect of process variables on final aroma retention

In Tables III–VII inclusive is given the dependence of the aroma retention AR at the end of the drying process on several process variables.

From these tables the following conclusions can be drawn:

Aroma retention increases with
- increasing dissolved solids content in feed;
- increasing air outlet temperature;

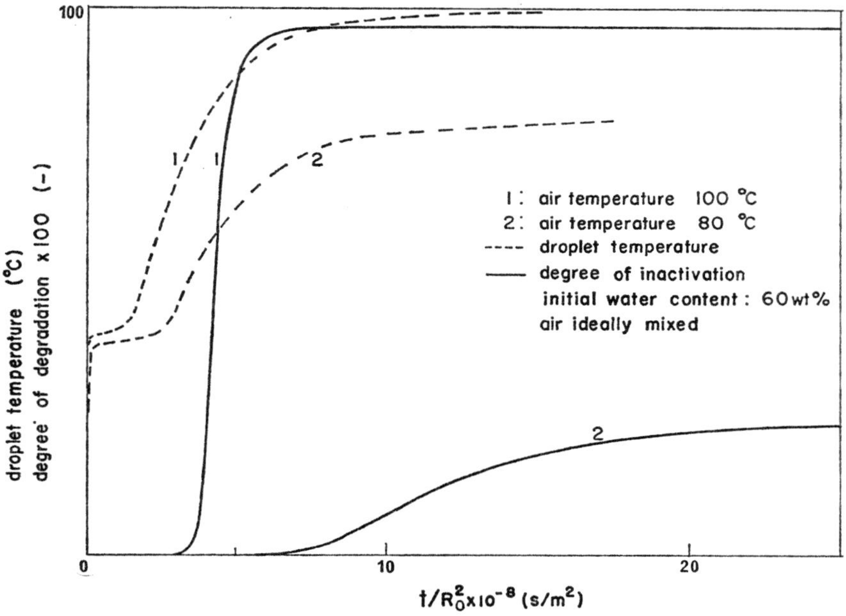

FIG. 12. Temperature and thermal degradation in relation to reduced time t/R_0^2, for two air temperatures.

- increasing feed temperature, especially at higher initial dissolved solids content;
- decreasing air humidity;
- increasing droplet radius at high initial velocity.

As seen from Figs. 9 and 10, the aroma loss stops when the interfacial water concentration has reached a value of about 25 wt%. A decreasing initial moisture content will shorten the time necessary to reach this critical interfacial moisture content, and hence the time available for aroma loss.

An increase in air temperature or feed temperature, or a decrease in air humidity, will result in higher initial drying rates, and hence also in a shorter time to reach the critical interfacial water concentration. The

TABLE III

EFFECT OF INITIAL DISSOLVED SOLIDS CONCENTRATION $\omega_{s,0}$ ON FINAL AROMA RETENTION AR

$\omega_{s,0}$	AR	
0.2	0.04	Air temperature: 100°C
0.3	0.16	Air humidity: 0.03 kg/kg
0.4	0.35	Ideally mixed air flow
0.5	0.58	Initial droplet temperature: 20°C
0.55	0.68	$Sh = Nu = 2$

TABLE IV

EFFECT OF AIR OUTLET TEMPERATURE $T_{air,out}$ ON FINAL AROMA RETENTION AR

$T_{air,out}$ °C	AR	
75	0.22	Initial dissolved solids content: 0.4
80	0.25	Air humidity: 0.03 kg/kg
100	0.35	Ideally mixed air flow
120	0.41	Initial droplet temperature: 20°C
125	0.43	$Sh = Nu = 2$
140	0.45	

TABLE V

EFFECT OF FEED TEMPERATURE T_{feed} ON FINAL AROMA RETENTION AR FOR DIFFERENT VALUES OF INITIAL DISSOLVED SOLIDS CONTENT

$\omega_{s,0}$	T_{feed}, °C	AR	
0.40	20	0.35	Air temperature: 100°C
0.40	80	0.42	Air humidity: 0.03 kg/kg
0.55	20	0.67	Ideally mixed air flow
0.55	80	0.79	$Sh = Nu = 2$

influence of the initial droplet radius is somewhat more complicated. By multiplying both sides of eqn (60) by $R_0{}^2$ we obtain

$$\frac{\partial \rho_w}{\partial t/R_0{}^2} = \frac{1}{(r/R_0)^2}\frac{\partial}{\partial(r/R_0)}\left(\frac{r^2}{R_0{}^2} D_{ww} \frac{\partial \rho_w}{\partial r/R_0}\right) \tag{107}$$

TABLE VI

FINAL AROMA RETENTION IN RELATION TO AIR OUTLET HUMIDITY H_{air}

H_{air} kg/kg	AR	
		Air temperature: 100°C
0	0.56	Initial dissolved solid content: $\omega_{s,0} = 0.4$
0.03	0.35	Ideally mixed air flow
0.06	0.23	Initial droplet temperature: 20°C
		$Sh = Nu = 2$

TABLE VII

FINAL AROMA RETENTION IN RELATION TO INITIAL DROPLET RADIUS R_0 FOR HIGH INITIAL DROPLET VELOCITY

R_0 μm	AR	
		Air temperature: 100°C
10	0.40	Air humidity: 0.03 kg/kg
25	0.43	Ideally mixed air flow
100	0.54	Initial droplet temperature: 20°C
		Initial droplet velocity: 75 m/s

The boundary conditions can be written as

$$t/R_0^2 = 0 \qquad 0 \leq r/R_0 \leq 1: \qquad \rho_w = \rho_{w0} \tag{108}$$

$$t/R_0^2 > 0 \qquad r/R_0 = 0: \qquad \frac{\partial \rho_w}{\partial r/R_0} = 0 \tag{109}$$

$$r/R_0 = R/R_0: \qquad \frac{-D_{ww}\dfrac{\partial \rho_w}{\delta r/R_0}}{1 - \rho_w/d_w} = k \cdot R_0(\rho_w'^{i} - \rho_w'^{\delta}) \tag{110}$$

As is visible from these equations, for a given set of independent conditions the concentration profiles of water on a dimensionless scale r/R_0 will be identical for equal reduced times t/R_0^2, provided that the value of $k \cdot R_0$ is constant, and independent of R_0. In the case of slowly moving droplets $Sh = 2$, so $k \cdot R_0$ is constant and independent of droplet diameter. Analogous equations follow for the transfer of heat and for aroma transport (Kerkhof, 1973), from which it can be concluded that the aroma retention is independent of the initial droplet diameter, provided that $Sh = 2$. At high relative velocity, however, the Sh number increases with increasing droplet diameter, which leads to sharper concentration profiles on a reduced distance scale for the same reduced time t/R_0^2, and hence to a shorter time t/R_0^2 necessary to reach the critical interfacial water concentration. As the aroma retention varies with t/R_0^2, it will increase with increasing droplet diameter if initial droplet velocity is high.

The qualitative influences found in these calculations agree well with the experimental results presented in the literature (Brooks, 1965; Reineccius and Coulter, 1969; Rulkens and Thijssen, 1972; Thijssen and Rulkens, 1968; Trommelen and Crosby, 1970). For further practical use, more data on transport properties of food liquids are necessary, as well as knowledge about the mixing and flow pattern in spray dryers.

CONCLUDING REMARKS

In the foregoing several aspects of spray drying have been treated.

The combined motion and evaporation of a pure water droplet was calculated. The equations governing the process of simultaneous evaporation and motion should preferably be solved without simplifying assumptions, as considerable errors may be introduced. The motion of evaporating water droplets in rotational air flow is a first step towards the calculation of drying phenomena in rotating air flow, such as the work of Stein (1972). However, no interaction between the flow of air and the droplets has been taken into account. This, and the quantitative influence of atomisation on air flow pattern, are fields of theoretical research which deserve more interest.

The model for the diffusional transport of water and aroma components in a drying rigid droplet of liquid foods can account for the effect of process variables on drying behaviour and aroma retention. Theoretically obtained results of drying behaviour agree well with the experimental results obtained for single droplets (Trommelen and Crosby, 1970). The important feature of this type of more sophisticated model is that it is easily extended to other influences. Van der Lijn (1974) includes also inflation of the droplets in his calculations. The influence of air flow pattern and degree of mixing, droplet size distribution, non-adiabatic drying and drying in several stages can easily be incorporated in the model without much increase in calculation time. This model can be very useful in the estimation of drying behaviour and aroma loss, but only if the physical data such as diffusion coefficients of water and of aroma are known. The determination of these data is very laborious. There is, however, good hope that from the theoretically calculated results correlations can be derived for the influence of process variables on typical drying data, such as length of the constant rate period, total drying time and aroma retention. Thus investigations leading to a more detailed description of the drying process, including particle–air interaction, droplet inflation and the determination of more physical data, on the one hand, and correlative methods for the derivation of short-cut methods from theoretically calculated data in order to decrease experimental work, on the other, will be of great importance in the future.

Contrary to the foregoing models, which are basic in nature and which can be extended to very realistic descriptions of motion and drying of droplets in spray dryers, the model for the internal circulation period is only qualitative in nature because of the many approximations made. Starting from very simple assumptions it was used to obtain a qualitative insight into the magnitude of the amounts of water and aroma evaporated during such an internal circulation period. As the results would predict a rather high aroma loss, it is interesting to obtain experimental data on the effect of internal circulation separately, and on the effect of surfactants on aroma retention.

As one of the unknown factors in the spray drying process is the evaporation and aroma loss at the atomising stage, this subject would also require more experimental work.

ACKNOWLEDGEMENTS

The authors are greatly indebted to Professor Dr. Ir. H. A. C. Thijssen for his guidance and stimulating interest and discussions. Many helpful contributions have been received from Dr. Ir. W. H. Rulkens and Ir. J. van der Lijn.

REFERENCES

Adler, D. and Lyn, W. T. (1971). *Int. J. Heat Mass Transfer*, **14**, 793.
Arni, V. R. S. (1959). Ph.D. Thesis, Univ. of Wisconsin.
Bailey, G. H. *et al.* (1970). *Brit. Chem. Engng*, **15**, 912.
Ball, C. O. and Olson, F. C. W. (1957). *Sterilization in Food Technology*, McGraw-Hill, New York.
Belcher, D. W. *et al.* (1963). *Chem. Eng.*, **70**, 1201.
Bentwich, M. *et al.* (1965). '8th Natl Heat Transfer Conf.', Los Angeles, ASME Paper 65-HT-38.
Bird, R. B. *et al.* (1960). *Transport Phenomena*, 3rd Edn, Wiley, New York.
Boussinesq, M. (1905). *J. Math. Pure Appl.*, **1**, 285.
Brauer, H. and Krüger, R. (1969). *Verfahrenstechnik*, **3**, 107.
Brooks, R. (1965). *Birmingham Univ. Chem. Eng.*, **16**, 11.
Calderbank, P. H. and Korchinski, I. J. O. (1956). *Chem. Engng Sci.*, **6**, 65.
Chaloud, J. H. *et al.* (1957). *Chem. Engng Prog.*, **53**, 593.
Chandrasekaran, S. K. and King, C. J. (1971). *Chem. Engng Prog.*, Symp. Ser., **67** (108), 122.
Chandrasekaran, S. K. and King, C. J. (1972a). *AIChEJ.*, **18**, 520.
Chandrasekaran, S. K. and King, C. J. (1972b). *AIChEJ.*, **18**, 513.
Coulson, J. M. and Richardson, J. F. (1968). *Chemical Engineering*, Vol. 2, 2nd Edn, Pergamon Press, Oxford, New York.
Danckwerts, P. V. (1951). *Trans. Farad. Soc.*, **47**, 1014.

Dlouhy, J. and Gauvin, W. H. (1960). *AIChEJ.*, **6**, 29.
Dombrowski, N. and Fraser, R. P. (1953). *Phil. Trans. R. Soc.*, **247A**, 101.
Dombrowski, N. *et al.* (1960). *Chem. Engng Sci.*, **12**, 35.
Dombrowski, N. and Munday, G. (1968). *Biochemical and Biological Engineering Science*, Vol. 2, Academic Press, New York, 207.
Edeling, C. (1950). *Beihefte Angew. Chemie*, **57**.
Elzinga, E. R. and Banchero, J. T. (1959). *Chem. Engng Prog.*, Symp. Ser., **55**, 149.
Elzinga, E. R. and Banchero, J. T. (1961). *AIChEJ.*, **7**, 394.
English, A. C. and Dole, M. (1950). *J. Am. Chem. Soc.*, **72**, 3261.
Fish, B. P. (1958). *Fundamental Aspects of the Dehydration of Foodstuffs*, Aberdeen, Society of Chemical Industry, London, 143.
Frossling, N. (1938). *Beitr. Geophys.*, **52**, 170.
Gladden, J. K. and Dole, M. (1953). *J. Am. Chem. Soc.*, **75**, 3900.
Gosting, L. J. and Morris, M. (1949). *J. Am. Chem. Soc.*, **71**, 1998.
Griffith, R. M. (1962). *Chem. Engng Sci.*, **17**, 1057.
Haas, U. *et al.* (1972). *Chem. Ing. Techn.*, **44**, 1060.
Hadamard, J. (1911). *Comp. Rend. Acad. Sci., Paris*, **152**, 1735.
Heiss, R. (1968). *Haltbarkeit und Sorptionsverfahren wasserarmer Lebensmittel*, Springer Verlag, Berlin.
Henrion, P. N. (1964). *Trans. Farad. Soc.*, **60**, 72.
Hortig, H. P. (1970). *Chem. Ing. Techn.*, **42**, 390.
Hughes, R. R. and Gilliland, E. R. (1952). *Chem. Engng Prog.*, **48**, 497.
Ihme, F. *et al.* (1972). *Chem. Ing. Techn.*, **44**, 306.
Jen, Y. *et al.* (1971). *J. Fd Sci.*, **36**, 692.
Kaskas, A. (1964). *Diplomarbeit*, Technische Universität, Berlin.
Kerkhof, P. J. A. M. (1973). Unpublished results.
Kerkhof, P. J. A. M. *et al.* (1972). 'Int. Symp. on Heat and Mass Transfer Problems in Food Engineering', Wageningen, Netherlands, Oct. 24–27, paper C2.
Kessler, H. G. (1967). *Chem. Ing. Techn.*, **39**, 601.
Klee, A. J. and Treybal, R. E. (1956). *AIChEJ.*, **2**, 444.
Kronig, R. and Brink, J. C. (1950). *Appl. Sci. Res.*, **A-2**, 142.
Krüger, R. (1973). *Ph.D. Thesis*, Technische Universität, Berlin.
Labuza, T. P. (1972). *CRC Critical Reviews in Food Technology*, September, 217.
Lapidus, L. (1962). *Digital Computation for Chemical Engineers*, McGraw-Hill, New York.
Lapple, C. E. and Shepherd, C. B. (1940). *Ind. Engng Chem.*, **32**, 605.
Levich, V. G. (1962). *Physicochemical Hydrodynamics*, Prentice-Hall, London.
Lightfoot, E. N. *et al.* (1962). *AIChEJ.*, **8**, 708.
Lijn, J. van der *et al.* (1972). 'Int. Symp. on Heat and Mass Transfer Problems in Food Engineering', Wageningen, Netherlands, Oct. 24–27, paper F3.
Lijn, J. van der (1972). Private communication.
Lijn, J. van der (1974). *Ph.D. Thesis*, Agr. University, Wageningen.
Lyne, C. W. (1971). *Brit. Chem. Engng*, **16**, 370.
Marshall, W. R. Jr. (1954). *Chem. Engng Prog. Mon. Ser.*, **50** (2).
Marshall, W. R. Jr. (1955). *Trans. Amer. Soc. Mech. Engng*, **77**, 1377.

Masters, K. (1968). *Ind. Engng Chem.*, **60**, 53.
Masters, K. (1972). *Spray-Drying*, Leonard Hill, London.
Menting, L. H. C. (1969). *Ph.D. Thesis*, Eindhoven University of Technology.
Menting, L. H. C. *et al.* (1970a). *J. Fd Technol.*, **5**, 111.
Menting, L. H. C. *et al.* (1970b). *J. Fd Technol.*, **5**, 127.
Miller, D. G. (1959). *J. Phys. Chem.*, **63**, 570.
Mühle, J. (1971). *Chem. Ing. Techn.*, **43**, 1158.
Mühle, J. (1972). *Chem. Ing. Techn.*, **44**, 889.
Nemitz, G. (1963). *Zeitschr. Leb. Unt. und Forschung*, **123**, 1.
Oelrich, L. *et al.* (1973). *Chem. Engng Sci.*, **28**, 711.
Peebles, F. N. and Garber, H. J. (1953). *Chem. Engng Prog.*, **49**, 88.
Perry, J. H. (1963). *Chemical Engineers Handbook*, 4th Edn, McGraw-Hill, New York.
Pita, E. G. (1969). *Ph.D. Thesis*, University of Maryland.
Ranz, W. E. and Marshall, W. R. Jr. (1952). *Chem. Engng Prog.*, **48**, 141, and **48**, 173.
Redfield, J. A. and Houghton, G. (1965). *Chem. Engng Sci.*, **20**, 131.
Reineccius, G. A. and Coulter, S. T. (1969). *J. Dairy Sci.*, **52**, 1219.
Richtmyer, R. O. and Morton, K. W. (1967). *Difference Methods for Initial-value Problems*, Interscience, Wiley, New York.
Rulkens, W. H. (1973). *Ph.D. Thesis*, Eindhoven University of Technology.
Rulkens, W. H. and Thijssen, H. A. C. (1969). *Trans. Inst. Chem. Engrs*, **47**, T292.
Rulkens, W. H. and Thijssen, H. A. C. (1972). *J. Fd Technol.*, **7**, 95.
Scatchard, G. *et al.* (1938). *J. Am. Chem. Soc.*, **60**, 3061.
Schoeber, W. J. A. H. (1973). *M.Sc. Thesis*, Eindhoven University of Technology (in Dutch).
Schoeber, W. J. A. H. and Kerkhof, P. J. A. M. (1973). To be published.
Seltzer, E. and Settlemeyer, J. T. (1949). *Advances in Food Research*, Vol. 2, 'Spray-drying of foods', Academic Press, New York, 399.
Sideman, S. (1966). *Adv. Chem. Engng*, **6**, 207.
Sideman, S. and Shabtai, H. (1964). *Can. J. Chem. Engng*, June, 107.
Sivetz, M. and Foote, M. E. (1963). *Coffee Processing Technology*, AVI Publ. Comp., Westport, Conn.
Sjenitzer, F. (1952). *Chem. Engng Sci.*, **1**, 101.
Stein, W. A. (1972). *Chem. Ing. Techn.*, **44**, 1241.
Tate, R. W. and Marshall, W. R. Jr. (1953). *Chem. Engng Prog.*, **49**, 161, and **49**, 266.
Thijssen, H. A. C. (1960). Private publication, P. de Gruyter & Sons NV, Netherlands.
Thijssen, H. A. C. (1965). Inaugural Address, Eindhoven University of Technology.
Thijssen, H. A. C. and Rulkens, W. H. (1968). *De Ingenieur*, **80**, Ch. 45.
Timmermans, J. (1960). *Physico-chemical Constants of Binary Systems*, Interscience, New York.
Torobin, L. B. and Gauvin, W. H. (1961) *Can. J. Chem. Engng*, **39**, 113; (1959) *Can. J. Chem. Engng*, **37**, 129, 167, 224; (1960) *Can. J. Chem. Engng*, **38**, 142, 189. As quoted by Mühle (1971).

Trommelen, A. M. and Crosby, E. J. (1970). *AIChEJ.*, **5**, 857.
Verhey, J. G. (1972). Private communication.
Weast, R. C. (1970). *Handbook of Chemistry and Physics*, 50th Edn, Chemical Rubber Company, Cleveland.

DISCUSSION

J. D. Mellor (C.S.I.R.O. Australia): In your kinetic energy term in the equation of motion I notice you write $\bar{w}$ as $|\bar{w}|$. Now by taking the modulus of your relative velocity are you taking this for the relative expansion velocity between particles and air, whereas the ordinary $\bar{w}$ term is for your relative stream velocity? If so, aren't you in fact wanting another term or vector—a weighting factor, which you might embody in your constant c_d? Anyway, what is the reason for writing it this way instead of just $\bar{w}^2$?
Kerkhof: If the term is split into one vector term and one scalar term, then the equation can be manipulated in any way you want. You can do it in Cartesian co-ordinates, but then only the $\bar{w}$ term which represents a vector quantity is split and the modulus term will remain the same. In the case when the droplet is moving in a vertical direction only the term becomes the square of the relative velocity.
Mellor: Exactly, but in the spray drying initially you have an expansion velocity and your particle velocity would be greater than the air velocity, in which case it would be $\bar{w}$ negative but you would have to reiterate the modulus.
Kerkhof: Well the effect of the sign of the terminal velocity is accounted for.
Mellor: Are you in this term differentiating two separate phenomena with regard to velocity or only one?
Kerkhof: I considered that the droplet has some direction of motion of which the velocity vector is given by $\bar{w}$. For convenience in order to be able to split it into several directions, the modulus term was taken. I watched the phenomena only as a droplet moving in the same direction with some relative velocity with respect to the air. I don't know if that answers your question.
Mellor: Partly.
C. J. King (University of California, U.S.A.): First of all I would like to compliment you on a most impressive piece of work. I think you have handled a difficult problem very well. I was particularly impressed by the photographs of the droplet formation period which scare me in that I can see all sorts of horrible things happening there for the aroma loss and I notice you mentioned that you compute that up to 15% of aroma loss might occur during that period.
Kerkhof: It was not computed; it was determined experimentally by catching the spray droplets at several distances from the nozzle in liquid nitrogen and then determining the amount of aroma present.
King: Well then you have answered my question.
A. L. Moeller (Ecal Nateco AB, Sweden): Do you always calculate water as H_2O, because much has been done in the last 10–20 years to show water contains aggregates of molecules? You showed a diagram where the diffusion was very different when you used a carbohydrate and coffee extract. Have you studied the complexity of water in these systems—first regarding different

substances which should be dried, secondly the grading of vapour when the concentration of the substances is increasing relative to the water?

Kerkhof: We have assumed that water itself was present as simple water and we could treat the diffusion as occurring equally in every direction and not depending on some place or angle at the sphere. We did not include the binding energy of water to the salt substance which in the last stage of drying may even be doubled. We did not study phenomena at the molecular level and we did not include, for instance, possible crystallisation phenomena which would change all the figures shown here.

Thijssen: You say that the effective diffusion coefficient is a quantity measured in a large volume and differences between different locations in the specimen would not be detected. It might be that if you start to study a 20% solution in very thin droplets that there can be an enormous difference between one droplet of 20% and another droplet of 20%. Is that what you mean?

Moeller: I meant that according to the substance dissolved there could be different behaviours of the water molecules at the beginning and the end of the process.

Whisky Powders and their Applications to Foods

T. YAMADA

Chuo University, Tokyo, Japan

Chairman: The preparation of powders containing alcohol has recently been developed by N. Sato (President of Sato Foods Company, Komaki). To whisky, brandy, gin, etc., dextrin is added, the solution is then spray dried and powders containing alcohol are obtained. Dr. Yamada will speak on this development.

T. Yamada: Ladies and Gentlemen, it is my privilege to introduce this technology to you, though the technical know-how belongs to the Company and I have not been involved in the experiments. Two methods can be employed, the first involves spray drying sodium glutamate and thus obtaining the anhydrous salt to which 99% pure alcohol is added under constant stirring at a ratio of 10:4. Continuing agitation will turn the slurry into a microcrystalline free flowing powder containing about 28% alcohol. An alternative to anhydrous sodium glutamate would be anhydrous lactose.

TABLE I

	Alcohol %	*Sugar* %	*Protein* %	*Amino acid* %	*Acid* %	*Water* %
Alcoholic powder	30.0	65.0	0.3			3.0
Sake powder	28.0	65.0	0.8	0.5	0.6	3.0
Wine powder	28.0	65.0	0.7	0.7	1.0	3.0

The second method involves encapsulation. Substances such as dextrin, soluble polysaccharides or gelatine are added to a 30–60% alcohol concentration. The constantly stirred slurry is spray dried under controlled low temperature conditions producing a powder of a particle size from 20 to 40 μm.

The composition of various powders is shown in Table I.

About 10% alcohol is lost during the spray drying. The encapsulation method results in an absolutely stable powder, whilst the crystallisation method gives an unstable material, which continues to lose alcohol unless packed in suitable wrappers made from aluminium foil and cellophane, PVC, polypropylene or polyethylene. The encapsulation method also preserves the flavour of the alcoholic beverage.

Whisky and brandy powders can thus be produced which should have a wide application in the food and confectionery industry.

Chairman: We have just learnt of a novel and rather exciting development to which we should devote a few minutes of our discussion time.

Karel: I don't see the need for this technology. To produce solid liquors, brandy and so on, I can merely add to these liquids a mixture of alpha and beta dextrose. In the transition the water is absorbed by converting the anhydrous dextrose into the monohydrate. The resultant product is a free flowing powder with about 30% alcohol and 70% sugar.

Chairman: Dr. Yamada says that protein is quite important for flavour preservation.

Karel: Proteins may be of benefit only if you are spray drying the powder.

Chairman: This reminds me of how dry flavours were produced many years ago by using dextrose as the encapsulating agent. To the anhydrate concentrated flavour was added and stirring continued until the dihydrate was formed. This was added to sugar and gelatine and a first-class flavoured jelly powder obtained.

N. Wookey: What is the alcohol concentration in the Japanese process, which appears to be quite a simple one?

Yamada: The alcohol content may vary from 30 to 60% according to concentration in the original solution although about 10% alcohol is always lost in the operation. I cannot unfortunately go into further details without divulging company secrets, which I am not allowed to do.

Spray Dry Coating in Fluidised Beds

T. YAMADA

Chuo University, Tokyo, Japan

Chairman: I would like you to hear the abstract of the second paper by Dr. Yamada. It concerns the spray injection of dissolved salts on to a heated bed of fluidised food particles, although in order to study the coating mechanics, glass beads were used in the reported experiments. The effects of temperature, particle concentration and particle size are reviewed and examples of food applications—such as the coating of sesame seeds with thin layers of salts—are discussed.

K. Hashimoto (speaking for Dr. Yamada): The application of the fluidised beds technique for granulation and coating has been investigated by many workers; they have mainly concentrated on the process of calcination. Our apparatus consists of a mixing zone and a spouted zone. The design of the mixing zone is a cylindrical column of 79 mm inner diameter and 140 mm high equipped with rotating multiple paddles to ensure the thorough mixing of the particles, and to avoid bubble formation in the bed. The spouted zone has an opening angle of 60°, a diameter of 228 mm and a height of 800 mm. Our experiment concerns the injection through a pneumatic atomising nozzle of an aqueous salt solution onto a bed of granular solids. The feed solution was pumped from a tank at 60°C and the flow rate measured by an in-line rotameter. Air from a blower was heated to 120°C before passing through the column, atomising air and fluidising air were measured by the orifice meters. Thermocouples were provided for measuring the temperature of the top and the bottom of the bed, and humidity of the air was measured by hygrometer. From the size distribution of the coated particles we see that the coating proceeds uniformly and plotting of of the mean particle diameter against operating time shows that the dependence of particle diameter on time is linear. The particle growth rate $\mathrm{d}x/\mathrm{d}t$ therefore equals a constant where x represents the mean radius and t the time. Assuming particle size is uniform and the sprayed solution falls on all particles uniformly then the following equation could be tabulated:

$$N 4\pi x^2 \, \mathrm{d}x = cq \, \mathrm{d}t$$

In this equation $\mathrm{d}x$ is the increment of radius after $\mathrm{d}t$, q is the flow rate of the injected solution, c is the concentration of salt in solution and N is the number of particles. Particle growth rate is then given by the following equation:

$$\frac{dx}{dt} = \frac{cq}{4\pi x^2}$$

The experimental results on calcination by Grimmet confirm the linear relationship between diameter and time, though Solovjeva assumed that the growth rate is proportional to x^n, where n is found experimentally to be 0.7. The reason for these differences may be explained by the fact that in this experiment the coated layer is very thin and the particle growth rate is therefore not affected by the curvature of the particles.

P. J. A. M. Kerkhof (Eindhoven University): I would like to know whether the same kind of procedure applies to fluidised bed crystallisation?

Yamada: I believe this is the case.

Chairman: The Bedell Institute has undertaken a similar kind of work and obtained a number of patents.

General Discussion

Chairman: In the time left to us, Professor Karel would like to take us back to another subject we discussed in detail in the morning session, namely the factors that favour relative aroma retention in foods during dehydration.

Karel: Thank you, Professor Seltzer, for allowing me to take issue with something you said this morning. You mentioned your work on the mass transfer potential and the application of adsorption isotherms to calculations. This is a worthwhile effort but not a very novel one. In my paper on Monday, I explained that under a limiting moisture content the way the partial pressure varies with moisture content has to be considered. The drop in moisture content is then proportional to a coefficient divided by the bulk density of the solids and the second derivative of the partial pressure in relation to distance.

Now if I may proceed to the second topic. I feel very honoured that I may speak for Dr. Thijssen's group as well as mine. Our two groups have been friendly competitors and collaborators and I will try to tell you the way we picture the retention of flavour compounds during freeze dehydration.

I will use the situation in freeze drying but somewhat similar conclusions could be reached in spray drying. During freeze drying there occurs a partition between an ice zone and a very concentrated solution. In that solution we have both volatile and non-volatile components and as the dehydration progresses the parts vacated by the ice are relatively porous and the vapour diffuses through the solid. Everybody now agrees that the entrapment of flavours occurs in the concentrated solution region, which contains sugars, proteins or other hydrophilic polymers. As the moisture content of these regions decreases the aroma compounds contained in them are no longer able to escape.

Professor Thijssen's group has looked at the ratio of the diffusion coefficient for the aroma compound to the diffusion coefficient for water in substances with a very low water content. In their work they show that this tends to go to a very low value, perhaps zero.

The *n*-propanol is entrapped in polyvinyl pyrolidone and when rehumidified to 32% it is close to the monolayer value. This then is the amount of water which we pick up. Note that we begin to lose the propanol only significantly when the water content comes close to that value.

In freeze drying, in contrast to spray drying, ice is present. It is very possible that, even under ideal conditions, one could not retain 100% of the flavour. One of the reasons for it is as follows: we have shown that when you just freeze and then expose to activated carbon a frozen solution of butanol and water with no other solutes present—conditions in which no water evaporates—only a small amount of butanol is absorbed by the carbon, most of it cannot escape from the ice. There is a possibility that some of the flavour compounds may be entrapped between the growing ice but they are not in the ice. Clearly, under normal conditions of freeze-concentration, Professor Thijssen obtains pure water though under these slow freezing conditions some flavour may be entrapped.

Professor Rey suggested some years ago that the reason for retention is the re-adsorption of the flavour as it passes through the sieve of the dry layer. There is the apparent contribution of the re-adsorption to the total retention. And the way that we have shown this is by freeze drying an aqueous solution of solids and *n*-propanol and covering this with a frozen layer of the solids only in solution. During drying the vapour and propanol of the bottom layer must diffuse through the upper layer, where only a small amount of re-absorption occurs.

Thijssen: May I remind you of a patent of Nestlé dating back to the early 1940s. They claimed improvement in flavour retention if 20% dextrose was added to a dilute coffee extract. They had to do this because with the percolators available at that time, they weren't able to get a stronger extract than about 20–25% at the most. Nestlé found that components like dextrose had the property of binding aromas to their molecules. And that is in fact the same explanation which Professor Rey gave initially to the aroma retention in spray drying. Now it has been proved that aroma retention has little to do with absorption binding at all: it is just a physical hindering of the mobility of the aromas in the systems. If you start with a too high dilute solution, you just lose more flavour than if you start with a higher concentration.

The second thing is that we saw some interesting powders from Dr. Yamada which will probably bring about some new work in our group at Eindhoven University: because it still worried us that with 30 parts of alcohol in the final product you still need 60–65% dextrin. We haven't proved it yet experimentally, but from our theoretical models we would expect that you wouldn't need more than about 10% dextrin in the finished product if you prepared it in the right way. So you can end up with about 90% of the material you want to encapsulate (alcohol and aromas) and only 10% of the carrier material. This is because, if you disperse your aromas finely enough, so the dispersed droplets are small compared with the droplets produced in spray drying, and you are doing it in such a way that your aromas (alcohol and others) are still present in the discontinuous phase, you can indeed go up to a very high rate of volume fractions in your final product.

Chairman: Then you think that 30% is credible?

Thijssen: We expect, at least theoretically, that much higher concentrations are feasible, yes.

F. Bramsnaes (Food Technology Laboratory, Denmark): Work performed by Professor Blakeling shows that volatiles in a model material exposed to a relative humidity of 61% are mainly released in the first 20 h and after about 40 h a constant level of volatiles in the material is achieved. This, in my opinion, is contradictory to Professor Thijssen's theory on diffusion coefficients. If you take this diffusion coefficient theory wou should expect the volatile concentration to fall continuously with time. I would just like to ask the Professor for his comments.

Thijssen: We thought we had to marry the theory of the group of Professor

Karel and our work. Probably you can see that a certain fraction of the aromas in the solution, at a certain temperature and water content, is totally immobilised within microregions, let us say a fraction A, and a fraction $1 - A$ is still mobile outside these structures. Our normal theory of selective diffusivity still applies to the aroma present outside these microregions—these caves. So you can say that in dehydrating a solution or hydrating a solid at a certain water concentration and temperature a certain fraction is not participating in the diffusion process, and only that fraction $1 - A$ can diffuse out of the system. What we are measuring, in fact, is that fraction outside the microregions.

Karel: I should like to add to that. We have many curves for a number of materials. In every case, there is very little volatiles lost below the monolayer—in which case, the diffusion coefficient in these regions is practically zero. Then, above that monolayer we lose some, but there is a certain fraction at this new relative humidity at which apparently the diffusion coefficient again goes to zero.

P. J. A. M. Kerkhof (Eindhoven University of Technology, The Netherlands): I would just like to make one remark. I think there is one possible theory which could explain the phenomena, and that is a selective diffusion theory. When you look at this picture you are looking at the re-humidification of freeze-dried cake which is a very complicated diffusion process, I think, with the simultaneous transport of water and flavour volatiles. It may be assumed that there is a certain distribution of the aroma components in the direction perpendicular to the ice pores. And I think it may be assumed that when you start in the freeze-drying process and in the freezing itself, accumulation of aroma near the ice pores takes place. In the process of freeze-drying some of these aromas are lost. But in the process of re-humidification you lose first from the region where there is much aroma present, so there is a rapid decline of aroma content; and then you come to the regions where there isn't so much aroma present.

This might be an explanation and a reason for further investigation into the field of dextrose.

Chairman: Perhaps we should just let Professor Thijssen terminate this particular discussion.

Thijssen: I should like to give just a small example which proves the validity of the selective diffusivity concept—it will only take two minutes.

A patent which worked out in practice but didn't work out commercially concerned a concentration apparatus for the absolute selective dehydration of liquid foods. It consisted in principle of two very fine stainless steel gauzes having openings of about 40–45 μm and very thin wires. Between these gauzes which are large surfaces, we maintained a certain sub-pressure of the liquid. The liquid film formed a concave miniscus. In this way, we get stagnant pools of liquid in the gauze and here the liquid was flowing in a laminar turbulent flow. Outside the gauze we had a stream of hot dry air, and due to that, we got a very steep concentration gradient in these pools of stagnant liquid. And in this way we got a surface concentration of the layer well below 10% water: whereas the water concentration of the main stream between the courses was somewhere between 80 and 40%. The dehydration rate of this apparatus was somewhere around 225 kg/m^2/h. We succeeded in keeping the surface concentration of the liquid layer low enough to obtain an absolute selective dehydration. And this could be kept in operation for weeks. We removed the water—let's say the initial water concentration was 20%—to a final concentration of 50–60% without losing very much at all of our most volatile components, including the C_2 and C_3 aromas.

I think this is a very good and practical illustration of what selective diffusivity can do.

Chairman: That is very good.

Saravacos: I just wanted to ask if you have really measured the diffusivity coefficient or not? ('Yes, of course') And in which patent?

Thijssen: We measured the diffusivity of both water and aromas over a water concentration range between 100% and 5%. And you have already heard this week that the diffusivity decreased over that range by about five or six orders of magnitude.

We believe that in the lower concentration range our errors haven't been bigger than about 100%, but that is still quite accurate over the enormous range of changes!

Chairman: Very good. I think it should be recognised here that the leaders in the study of flavour conservation, flavour fixation during dehydration are either represented or are in this room. It is quite a privilege to have this opportunity made available by Dr. Spicer and his colleagues to assemble here in this manner. I think we owe Dr. Spicer a vote of thanks—it's a little premature as it should possibly come tomorrow night—for giving us the occasion to argue and to reconcile our views as we did tonight. I think we did reconcile them!

(The chairman adjourned the meeting until Friday morning.)

Session V

FREEZE-DRYING AND OTHER NOVEL DEHYDRATION TECHNIQUES

Introduction

NILS BENGTSSON

Swedish Institute for Food Preservation Research, Göteborg, Sweden

The object of this introduction is to set up a general framework for discussion and present some thoughts around advances in dehydration techniques, raising questions rather than giving answers and, it is hoped, not forestalling either the papers to be presented or the following discussions.

The word dehydration is usually used synonymously with drying to signify removal of most of the moisture in a food material. In addition, water removal by evaporation or sublimation in air, gas or vacuum is usually implied, even if dehydration of solid food matererials may also be achieved by other means, such as by heating in edible oil, by azeotropic distillation or by osmosis. Besides being an important preservation technique, drying offers advantages of reduced weight or bulk and of producing unique properties or convenience, such as in 'instant' products. In relation to preservation by freezing or canning, market volume and product range is more limited and quality mostly lower, at least for products of quite comparable type, with certain exceptions for freeze-drying. Processing costs for the major dehydration methods (air, drum and spray drying) are about comparable to those for freezing, while 'minor' methods like foam mat, vacuum drying and freeze-drying are increasingly more expensive in that order. For liquids or slurries, spray drying is by far the dominant method, as is shown by the fact that a separate day of this symposium has been entirely devoted to it. The circumstances here enumerated are important as a general background when considering new directions and advances in dehydration and dehydration techniques.

A discussion of novel dehydration techniques is likely to centre on the engineering and equipment design aspects of improved and entirely new methods, and on the experience gained with them so far with regard to product quality and economy. However, advances in fundamental dehydration theory and its application, and new knowledge on quality determining factors as well as novel application areas and novel product concepts will be of great importance to advances in dehydration techniques and should not be excluded from discussion.

The criterion of a practical advancement in dehydration techniques should, in my opinion, be that a novel and/or improved product results, that costs are reduced or that both of these happy events occur. Figure 1 is an attempt to illustrate some of the interrelated factors of main importance in the general course of process advancement.

We will first concern ourselves with the directions open to process advances by influencing drying rate and time–temperature exposure as based on elementary dehydration theory. Whether drying is done on

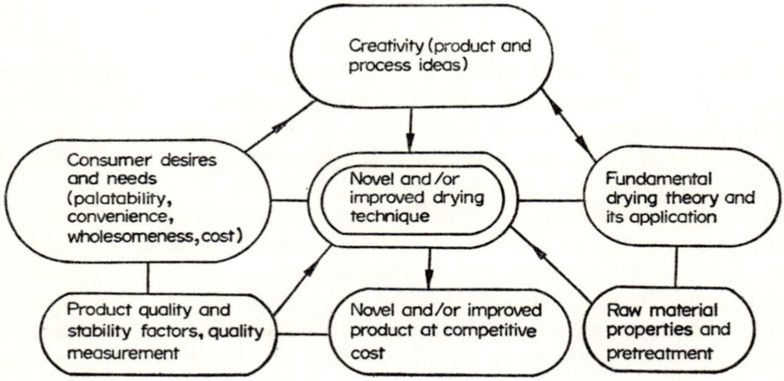

FIG. 1.

shelves, drums, bands or carrier bodies or by suspension or transport in air, heat must normally be transported from outside inwards and moisture from the interior to the surface. To increase drying rate, heat and mass transfer coefficients, conductivity and driving potentials as well as effective surface may be increased and product thickness decreased, all within the limitations imposed by permissible time–temperature and oxygen exposure, aroma retention, appearance and cost. Continuous removal of the drying surface offers another means of accelerating drying. When heat transfer is the limiting factor, use of electromagnetic energy or ultrasound may result in improvements, while atomisation, disintegration, puffing, foaming or pressure freezing may be used when mass transfer is rate limiting. No doubt all these factors will be discussed in the context of new dehydration techniques in the review papers.

In process development the advance and application of fundamental drying theory should be of great practical significance. The situation a decade ago was described by Van Arsdel in the following terms. 'Only the simpler practical drying methods are now accessible to theoretical analysis. In equipment design and process operation, empirical procedures have so far been much more useful than the predictions of basic physical theory.' However, he expected that the situation would be reversed in the long run, resulting in major process developments. Since that time our understanding of the fundamental drying mechanisms has increased considerably. It remains, however, to be demonstrated today to what extent this has as yet led to significant practical process advances, and to what extent process development and plant design is still mainly a question of empirical or semi-empirical procedures. What are in fact the chances that, in the future, we shall be able to optimise entire drying processes by computer simulation, based on a combination of drying theory, data-logging experience in the pilot plant and physical and chemical food data?

Considering briefly the importance of pre-treatment, considerable work has been done in freeze-drying to find optimal combinations of pre-concentration and dehydration, and to study the influence of special food additives on drying performance and quality. To what extent are these factors being generally considered in drying process development today?

The extensive work done in recent years on factors governing product quality and stability has greatly increased our knowledge and already led to important product and process improvements. Rapid and accurate methods of determining physical properties and quality on line may also warrant some discussion, being important tools in process development.

Ingenuity and creativity may still be the most important elements of all, provided the proper conditions and tools are present, as will be illustrated by some of the novel dehydration methods to be discussed.

Finally, to close this review of factors important to developments in dehydration techniques, improved methods of measuring and quantifying consumer attitudes and needs may greatly influence the direction of future product and process development.

New Directions in Freeze-drying

J. LORENTZEN

Manager R & D, A/S Atlas,
Ballerup, Denmark

General application of the freeze-drying technique started in the medical field in the years around 1935 and in the field of food preservation the industrial development started around 20 years later. Since that time Professor Louis Rey with collaborators has organised six international courses in freeze-drying. The sixth one took place at Burgenstock, Switzerland from the 17th to the 23rd of June, 1973. It was organised by Professor Louis Rey together with Professor S. A. Goldblith and dipl. engineer W. W. Rothmayr. On that occasion new research work and the developments in this field were reported. This paper will mainly review some of the main topics of that course.

THE FREEZE-DRYING PROCESS

To achieve freeze-dried products it is necessary to fulfill a series of production steps in the plant, from the receipt of the raw material, through several preparatory steps, of which an important one is the freezing operation, then the sublimation drying, the desorption drying, the conditioning of the product with packaging and storage. In the discussion here, the freeze-drying process is considered in the meaning of the freezing step, sublimation drying step and desorption drying step of this total operation.

The freezing step

We will first consider the freezing process, and as an illustration discuss it as it takes place with a product like orange juice. The composition of orange juice—as with other natural products—is somewhat variable but the figures of Table I can be taken as normal (Flink, 1973a).

As fruit juices are mainly sugar solutions, the freezing behaviour of them is quite similar to that of sugar solutions. Figure 1 (King, 1973) shows a simplified sucrose–water phase diagram.

To freeze a 10% sucrose solution the temperature is lowered gradually. At about −1°C the freezing starts, and pure ice crystals are formed, leaving a more concentrated solution between the ice crystals. When −2°C is reached about half of the water content is crystallised, and the intercrystalline liquid has reached double concentration. This process

TABLE I

Total solids		12%
Acid (as citric)		1%
Reducing sugars	4 %	
Sucrose	3.5%	
Total sugars		7.5%
Others		3.5%

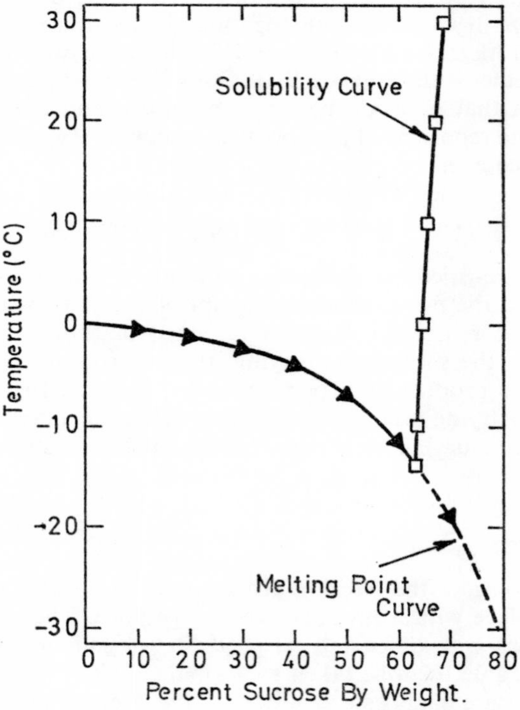

FIG. 1. Sucrose–water phase diagram with extension of ice-equilibrium curve (King, 1973).

goes on until at a temperature of $-13°C$ the intercrystalline solution has reached a concentration of approx. 60%.

In a normal solution the rest of the liquid should solidify as a eutectic, with further lowering of the temperature. But in the sucrose solution as in fruit juices, this is not the case. The concentrated liquid will not crystallise, but it is gradually subcooled as the viscosity increases. It forms a glasslike structure of increasing hardness between the ice crystals (vitrification).

To some extent this glasslike structure can be changed by thermal treatment; by further lowering the temperature of the product below a certain level and then gradually reheating it, but still keeping it at a very low temperature, a partial crystallisation—'devitrification'—may take place.

These phenomena may be studied calorimetrically by differential thermal analysis (DTA) (Rey, 1960) or differential scanning microcalorimetry (DSC) (Maltini, 1973) and also dynamometric measurements of the kind used in tenderometry have now been used.

When doing differential thermal analysis of juices of different concentrations, one discovers how the different fractions of water act in different ways in the product. It has been found for lemon juice that as much as 25% water (dry basis) is so firmly chemically bound that it is absolutely unfreezable. Another 20% (dry basis) may be frozen only by devitrification, and the rest, which is still as much as 70% of the total water content in a 40°Bx concentrated juice, will crystallise as ice during the lowering of the temperature in the normal freezing operation.

The freezing velocity influences the ice crystallisation generally in the way that a slow freezing produces large ice crystals, and a quick freezing produces many small crystals. Since the ice crystals in the frozen mass show up in the finished freeze-dried product as the void pores, the freezing velocity is decisive for the ultimate structure of the freeze-dried product.

When freezing living cells, such as yeast, bacteria and viruses it is found that a certain optimum freezing velocity gives the best result in preserving the viability of the cell culture (Simatos and Turc, 1973). The explanation is that there are different injurious factors, some of which increase with increasing freezing velocity, others increasing with decreasing freezing velocity, such as the drying effect on the cells by the concentrated interstitial fluids, have longer in which to act; whereas with high freezing velocity other factors may be detrimental, like intracellular crystallisation.

In many cases the best compromise for the freezing velocity is an extremely rapid one, but cryoprotecting agents, like glycerol, can shift this optimum to slower, more easily obtainable freezing rates.

Sublimation drying

This step of the freeze-drying production has been given much theoretical attention.

The heat and mass transfer conditions in the product during freeze-drying have often been a subject for study. Now new calculations of the progress of the sublimation process, taking into account the simultaneous mass flow and heat flow conditions, have been presented (Rothmayr, 1973), and some computer results are shown. The refinement over earlier calculation models is that the variation in transfer properties of the dried layer of product with time is included in the model.

Heat and mass transfer in freeze-drying of stirred granular particles have been investigated, calculated theoretically and tested experimentally (Kessler, 1973). The stirring brings an exchange of the particles in contact with the heater plate and makes higher heat transfer rates possible, particularly for small particle sizes.

When utilising this increase in heat transfer for an increase in drying rate, an increase in the water vapour flow out of the product will result. The mass flow velocity of the vapour is proportional to the specific drying capacity at the product surface. Such a vapour flow will carry away all particles of the product smaller than a certain size (Dauvois, 1971). This 'flying off size' is dependent on vapour mass flow velocity, on vapour pressure, on particle density, and on particle shape. With many types of products, when freeze-dried in a stirred bed at increased specific capacity, excessive product losses will occur. Separation of the fine product particles from the vapour stream is not easily achieved.

In the previous discussion was described how the freezing of a fluid food will produce a number of pure ice crystals in a matrix of concentrated glasslike material. In cellular material, the same freezing mechanism is working, but here also the cellular structure of the material will be at least partly preserved. When freeze-drying such a frozen sample of fluid food, the sublimation of the ice crystals starts at the product surface, and successively goes on into the interior, leaving void pores in the material, through which pores the water vapour reaches the product surface. Between the pores is left the matrix of concentrated material, and as the sublimation front recedes into the material, a successive dehumidification of this structure is also advancing. This latter dehumidification may be described as diffusion. The water vapour diffuses from the interior of the matrix material to the pore surface, and then escapes from the product through the pore. In this way the residual moisture content of the matrix gradually is brought down from the original of say 40% to the ultimate of say 2% simultaneously with the temperature rise of the product. Bearing in mind that this matrix is a solution of extremely high viscosity—a glass—and that this viscosity is a strong function of the concentration and of the temperature (King, 1973), a very illustrative theory is given of the collapse of the product structure, which may occur during freeze-drying in adverse conditions, and which will lead to partial melting and puffing and to a product of very poor quality.

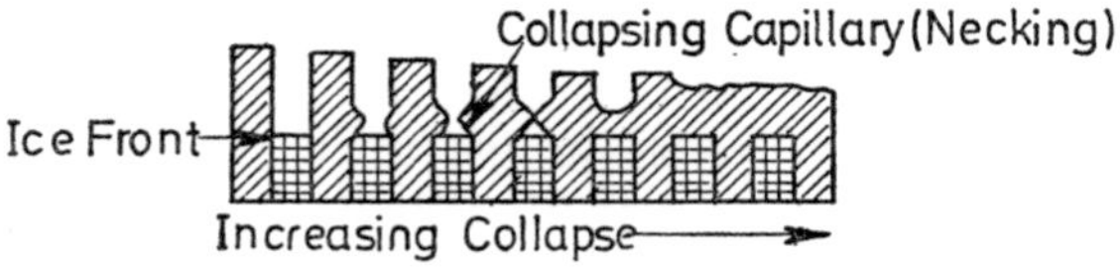

FIG. 2. Progressive stages of collapse (left to right) (King, 1973).

Figure 2 illustrates this theory, which says that collapse will occur when the matrix viscosity—because of too high temperature—gets so low that the structure loses its rigidity to a degree that the pores become closed before the main drying can be finished. This sagging of the structure results from the dominance of the surface tension forces over the gravity forces.

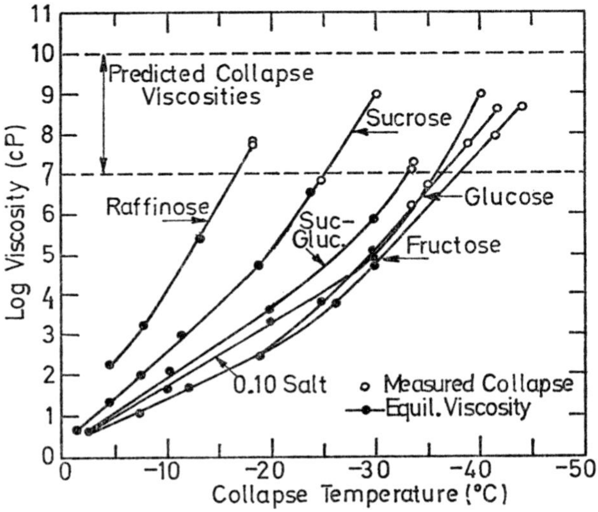

FIG. 3. Viscosity of interstitial concentrate *v.* temperature, predicted and observed collapse conditions (King, 1973).

Taking a surface tension of 70 dyn/cm, a pore radius of 20 μm and structure lifetime of 1–1000 s, a viscosity of 10^7–10^{10} cP can be predicted, which compares well with observations on collapse temperatures of solutions (Fig. 3). In food extracts often other structure-reinforcing constituents may help to retain the structures at higher temperatures than in the pure solutions.

Now we will turn to the question of aroma retention in freeze-dried products. In many food products the aroma is a very important quality factor. Food aromas have been shown to contain hundreds of different volatile components. Their volatility relative to water is usually higher than one, which means that an aqueous solution containing volatile aromas, when concentrated by straightforward evaporation, will lose relatively more of the aromatic compounds than of the water. Likewise in ordinary air drying and vacuum drying volatile aroma components will largely be lost.

From Table II it can be seen that the relative volatilities of aromatic compounds are often very large compared to one, and especially when

TABLE II

RELATIVE VOLATILITIES FOR SOME AROMA COMPONENTS AT INFINITE DILUTION (Chandrasekaran and King, 1971)

Aroma component	*Temp.* (°C)	*Relative volatility* 0% *sucrose*	*Relative volatility* 70% *sucrose*
Ethanol	25	7.79	15.5
Ethyl acetate	20	269	1 200
n-hexanal	25	390	2 200
n-butylacetate	20	689	4 070
Ethyl-2-methyl butyrate	20	1 160	9 170
n-hexyl acetate	20	1 580	23 900

present in carbohydrate solutions. In freeze-drying, however, an astonishing amount of the aroma components may be retained in the product although practically the whole water content is evaporated.

In the earlier days of freeze-drying this astonishing effect was ascribed to an adsorption of aromatic molecules on the large internal surfaces created in the product during the freeze-drying. Later investigations, however, have led to quite other explanations (Thijssen, 1973; Flink, 1973b).

Much of this work has been done on model systems containing just one simple volatile component in an aqueous solution of one simple carbohydrate. A typical example of such a model system is 18.8% maltose, 0.75% 2-propanol, 80.4% water, i.e. volatile component 4% of the carbohydrate component.

One experiment with this model material was conducted as follows: one layer of the model material was frozen first. Above this frozen layer was frozen another layer of the same maltose solution, but without the volatile component. This block was freeze-dried with evaporation from the top surface. After freeze-drying, the content of aroma component in each layer was measured using gas chromatography. The bottom layer was found to have retained 68% of the original content of 2-propanol. The rest, 32% had—together with the water vapour—passed through the upper layer of the already freeze-dried maltose. This layer when checked for volatile content contained only very little of the 2-propanol, about 1% of what was originally in the bottom layer. This shows clearly that the volatile is mainly kept in its original place in the product, and hardly caught on the pore surfaces when the water vapour is filtered out through the previously freeze-dried surface layers.

TABLE III

LOSS OF 1-BUTANOL DURING FREEZE-DRYING OF MALTOSE SOLUTIONS
(Flink, 1973b)

Time (h)	*Mean moisture content* (% *DS*)	*Mean volatile content* (% *DS*)	*Volatile retention* (%)
0	430	4	100
3	178	3.30	83
6	36	2.20	55
12	11	2.45	61
24	0.7	2.50	62

Another experiment compares the volatile retention after quick freezing with the retention after slow freezing, the result being that after freezing in 30–45 s a retention of 2-propanol in the maltose of 70% was achieved as compared to 87% when slow freezing in hours was used.

At which stages during the freeze-drying is the volatile loss important? This was investigated in a series of experiments, the result of which is illustrated in Table III.

This shows clearly that the main loss of volatiles takes place in the earlier stages of freeze-drying, and not in the later stage of desorption.

Other tests have confirmed that prolonged desorption time for days has no influence on the volatile retention, as long as the product temperature is not so high that it effects the structure of the freeze-dried carbohydrate.

Tests for volatile losses during storage at different moisture levels in the product, have shown that the volatiles are retained for a long time, as the moisture level is kept below a certain value.

Table IV shows for this model material that up to a moisture content of approximately the B.E.T. monolayer value, the volatile content is not significantly reduced. But above this level a reduction is observed, and the more pronounced it is the higher is the water content. This reduction mainly takes place within the first 20 h, and thereafter the volatile content seems to reach a new stable level.

Another characteristic observation in these tests was that any time there was a noticeable change of material structure, there was observed a marked loss of volatiles. So it was concluded that formation and maintenance of matrix structure are critical for retention of volatiles.

These matrixes have been observed in optical and electron microscopes. They may show up as networks of comparatively thin solid platelets, separated by larger cavities, left by the ice crystals. In some cases small circular droplets of sizes 2–6 μm have been discernible, and sufficient evidence found that those were droplets of liquid volatiles. The droplet volume found could be correlated to the partial or complete lack of solubility of the component in question in the carbohydrate solution, when concentrated during the growth of the ice crystals at the freezing stage.

TABLE IV

2-PROPANOL LOSS AFTER HUMIDIFICATION OF MALTOSE FOR 20 h AT 32°C (Flink, 1973b)

Relative humidity (%)	*Water content* (% *DS*)	*Volatile content* (% *DS*)
0	0.42	2.60
11	2.41	2.62
20	3.93	2.53
32	5.21	2.44
52	8.04	1.80
75	23.15	0.70

B.E.T. monolayer value is 5.0 g water/100 g solids.

The locking in of the volatile components in the matrix structure has been given two somewhat complementary interpretations.

One is the microregion concept (Flink, 1973b) assuming that the lowering of water content in the concentrated solution results in molecular bonding between carbohydrate molecules, forming a complex structure entrapping molecular dispersed volatiles as well as droplets. The other explains the entrapment as a selective diffusion mechanism (Thijssen, 1973). The diffusion coefficient of an aroma component in a carbohydrate–water solution is strongly dependent on the water concentration. So when water concentrations drop below a certain critical value, the diffusion coefficient for water is so much higher than the coefficient for the volatile, that the water will depart by diffusion, while the volatile will practically remain.

For acetone in malto-dextrin, the relation between the diffusion coefficient of acetone, D_a, to the diffusion coefficient of water, D_w, is shown as a function of the water content. As may be seen from Fig. 4, this relation is also dependent on the temperature. If a relation of 10^{-2} is regarded as close enough to 0, then the critical water content for different temperatures

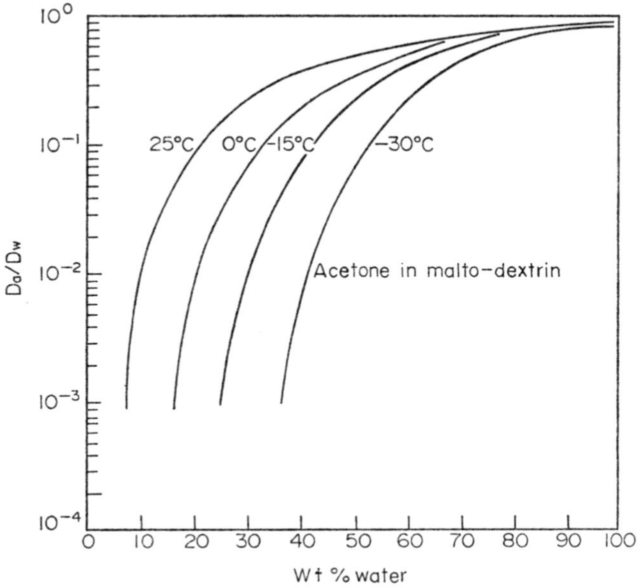

FIG. 4. Relation between diffusion coefficient of acetone D_a to diffusion coefficient of water D_w *v.* weight percent of water in malto-dextrin solution (Thijssen, 1973).

can be read as approximately 40% at −30°C, 30% at −15°C, 20% at 0°C and 10% at +25°C.

APPLICATIONS IN THE BIOLOGICAL AND MEDICAL FIELDS

When freezing or freeze-drying living cell cultures, the aim is to preserve the viability of the cells. Much new information about the different viruses has been gained in the latter years (Greiff, 1973). Application of electron microscope techniques and others have led to the characterisation and identification of many viruses. In extensive test programmes with different viruses the influence of the relevant factors of freezing and freeze-drying on infective power has been studied. Such important factors are suspension medium with cryoprotective agents, rate of cooling and terminal temperature of freezing, freeze-drying temperature, protective gas in storage and storage temperature.

The different kinds of viruses can be classified according to physical and chemical characteristics as follows.

(1) Nucleic acid type. They have a core of either ribonucleic acid, RNA or deoxyribonucleic acid, DNA.
(2) Solvent sensitivity. Solvents for lipids destroy the infectiousness of some viruses but have no effect on others.
(3) Limiting membrane. Some viruses have a covering envelope around the nucleocapcid, others are naked.
(4) pH-sensitivity. Some viruses are acid labile (pH = 3.0 for 30 min), others are stable.
(5) Heat-sensitivity. Some viruses are heat labile (50°C for 30 min), others are stable.

Classification according to these five characteristics, theoretically gives 32 possible combinations, out of which eight have been found in nature.

TABLE V

GROUPING OF VIRUSES (Greiff, 1973)

	Characteristics					
Group	*Nucleic acid*	*Solvent*	*Membrane*	*pH*	*Temp.*	*Cryo-sensitivity*
I	RNA	Sensit.	Envelope	Labile	Labile	Labile
II	RNA	Resist.	Naked	Labile	Labile	Stable
III	RNA	Resist.	Naked	Stable	Labile	Stable
IV	RNA	Resist.	Naked	Stable	Stable	
V	DNA	Sensit.	Envelope	Labile	Labile	Stable
VI	DNA	Sensit.	Envelope	Labile	Stable	Stable
VII	DNA	Resist.	Naked	Stable	Labile	
VIII	DNA	Resist.	Naked	Stable	Stable	Labile

The effect of the freezing factors on the different viruses follows the pattern of this classification closely, as indicated in Table V.

As for protective gases in the storage of freeze-dried samples, it was found that nitrogen, normally regarded as an inactive gas, had almost the same effect as air or oxygen, whereas hydrogen, helium and argon generally possessed protective properties.

Freeze-drying has been used to preserve human tissues for transplantation surgery (Sell *et al.*, 1973). Non-living tissues of skin and bone provide a substitute structure for the living part, into which the patient's own tissue can regrow for gradual replacement. Important characteristics of such grafts are strength and antigenicity or rejection behaviour.

In a test series with skin grafts on mice it was shown that grafts freeze-dried at temperatures below −30°C had better properties than grafts freeze-dried at higher temperature. This was ascribed to a better structural preservation at the lower temperature. More important seems the result that freeze-drying basically alters the antigenic properties of the grafts. These freeze-dried grafts are not rejected by antigenic action, as are fresh grafts or grafts preserved by ordinary freezing.

A topic of special interest in the medical and biological fields is the possibility of contamination of the freeze-dried products (Amoignon, 1973). The following possible sources of contamination should be considered.

(1) Diffusion of oil from the vacuum pumping system.
(2) Contamination of vials during the manufacturing process.
(3) Contamination of vials by exposure to atmosphere.
(4) Contamination by protecting gas source.
(5) Contamination from pipe system, carried by protecting gas.
(6) Contamination of stopper.
(7) Insufficient tightness of stopper.

Food products

Freeze-dried beef has not developed into a successful commercial product, although the freeze-drying cost is reasonable compared to product value and no quality products will result from conventional drying. The main reason may be that commercial products so far have not been up to quality standards obtainable (Bengtsson, 1973).

Later investigations have shown that it is possible to produce freeze-dried sliced raw or pre-cooked beef with good quality and shelf-life. The following recommendations may be given for sliced raw beef.

Raw material from young animals and from muscles with low fat content and well defined fibre direction should be chosen. Good ageing and careful trimming of fat are important. A freezing rate should be chosen between 0.1 cm/h and 2 cm/h and then slices of no more than 15 mm thickness should be cut across the fibres. The ice temperature during freeze-drying should be maintained below −20°C and the maximum temperature below +50°C. Final moisture content should be below 2%. Vacuum breaking and packaging under pure N_2 without intermediate exposure to air is necessary. Packaging material of really good protective characteristics is necessary, and even oxygen scavenging means are recommended, especially if air exposure after freeze-drying has been unavoidable. The storage temperature should be low, if long shelf-life is wanted. Rehydration in cold water is preferable and a good equilibration time should elapse before cooking. For pre-cooked beef similar conditions prevail, but somewhat more liberal freeze-drying temperatures may be utilised.

Equipment for freeze-drying of foods

Basically the economic conditions for an industrial freeze-drying plant are exactly the same for any manufacturing company. This may be illustrated

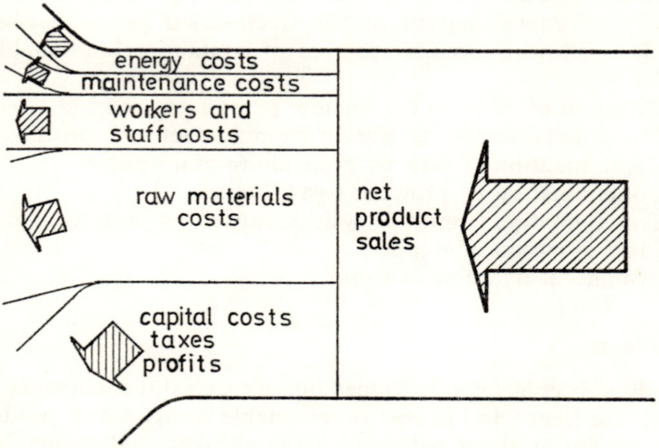

FIG. 5. Simplified cashflow diagram for normal freeze-drying plant.

with a simplified cashflow diagram, as in Fig. 5, for a typical freeze-drying company. From this figure is seen right away the dominating importance of product sales (Lorentzen, 1973).

This leads obviously to the following conclusions for the equipment of the plant.

(1) The plant should have good operation reliability.
(2) The process should be easily controlled, safeguarding the product quality.
(3) Product losses in the process should be avoided.

On the cost side of the figure, maintenance costs and workers' costs will likewise be favourably influenced by points No. 1 and No. 2 above. Also the energy costs may be largely influenced by the equipment design, so let us consider this point further.

In industrial freeze-drying the main part of the energy consumption is caused by the removal of the vast volumes of water vapour released by the product. So here is where any important savings on the power bill are possible.

In modern freeze-drying plants the water vapour from the product is condensed as ice on refrigerated surfaces in connection with the vacuum

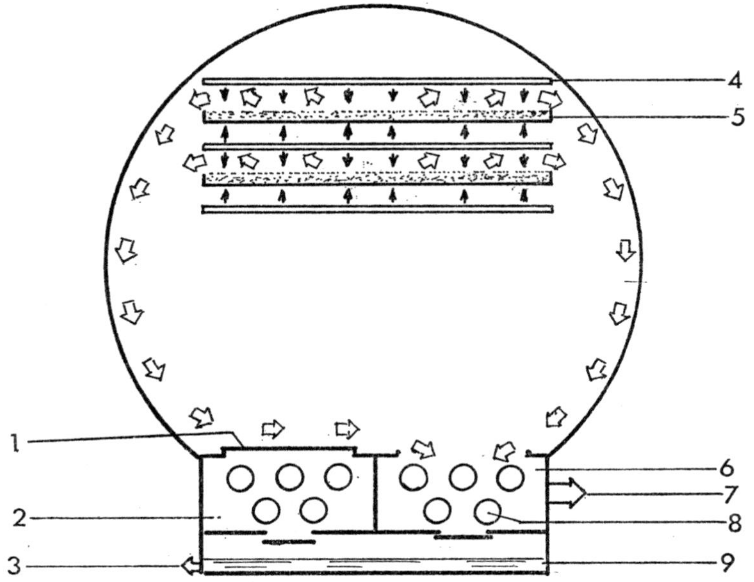

FIG. 6. Modern ice condensing system.

chamber. The main energy consumption is by the refrigeration plant for these ice condensers. Figure 6 illustrates the ice condensing system for a modern freeze-drier.

The product in each of the simple straight trays 5 receives heat by radiation from the heater plate below and from the one above 4. The water vapour escaping up from the product, flows out between the tray and the heater plate above and then down into the vapour trap chamber 6, where it is condensed as ice on the refrigerated tube 8. Non-condensable gases 7 are sucked off to the vacuum pumping plant. The other vapour trap chamber 2 is shut off from the freeze-drying cabinet by the sliding gate 1, but is connected, for deicing, to the bottom chamber, which contains water 9 at approximately 20°C.

From here water vapour flows up to the cold ice surfaces in chamber 2, where it will condense and give off heat for the melting of the ice. The water formed by condensation and by melting flows down into the bottom chamber, from where the surplus is drained off 3. During operation a frequent change between the sides 2 and 6 is effected, so that the ice on the refrigerated tubes is not allowed to get too thick.

The product quality demands that a certain vacuum, for example 0.5 torr, is maintained in the freeze-drying cabinet. At this pressure water vapour can be condensed on a cold surface at the saturation temperature; for the example above this temperature is −24.5°C. But the vapour has to flow to the cold surface to be condensed, and to achieve this flow a certain pressure drop is necessary, so around the cold surface the vapour pressure may be 0.4 torr, but at this pressure condensation will not take place unless the temperature is lowered to −26.7°C.

Furthermore, to keep this temperature on the ice surface when extracting the heat of condensation needs a temperature drop through the ice layer to the tube wall and another temperature drop from the tube wall to the refrigerant inside the tube to, say −35°C. An ice thickness of already 1 mm gives a considerable temperature drop, and any inefficiencies on the refrigerant side of the tube will also increase the temperature difference.

To preserve the required vacuum in the cabinet, a refrigerant temperature lower than the saturation temperature corresponding to this vacuum must be maintained. The difference between these two temperatures indicates the inefficiency of the vapour removal system. The efficiency of this system is the most important factor controlling the energy consumption of freeze-drying plants, for running the refrigeration plant at an evaporation temperature 10°C below optimum results in a 50% increase in energy consumption.

By far the largest number of industrial freeze-driers in operation are of the vacuum batch type which freeze-dry the product in trays. These plants have a number of features in common. The process takes place within a strongly built cabinet with a large door, through which charging and discharging is effected with the product trays transported in a carrier. The product trays are interleaved between horizontal heater plates, the temperature of which is automatically controlled. The cabinet can be quickly evacuated by a large capacity vacuum pumping plant, and it is equipped with a system for removal of the vast volumes of water vapour released by the products.

In Fig. 7 we see a trolley with product trays being charged into the cylindrical vacuum cabinet. The vacuum cabinet is served by the following main auxiliary systems: hot oil system for the heater plates, refrigeration system for the ice condensers, deicing system for the ice condensers, vacuum pumping plant, electrical equipment and control systems.

To achieve a simple and reliable operation of such complicated systems, automatic control of high standard is necessary. Modern plants of this kind are automated to a degree that all processes take place automatically, after the cabinet has been charged with the tray trolleys, the door has been closed, and the starter button has been operated on the control panel.

All important information about the operating conditions can be read on the instruments and signal lamps of the control panel. If anything abnormal should occur during the operation, it is possible at any time manually to make the necessary corrections in the control systems to get

FIG. 7. Batch type freeze-drier being charged with trolley load of product trays.

back to normal conditions. In this way the supervision of the process has been so simplified, that it may normally be left to a worker with other main tasks in the plant.

Figure 8 shows the control panel for such a plant. Recent years have shown a growing interest in freeze-drying plants operating with a continuous flow of material. Particularly in industries working with one standardised product and where the preparation of the product is by a continuous process, such plants are really profitable. They give continuity in processing throughout and constant operating conditions which are more easily controlled. They also require less manual operation and supervision.

CONRAD is a continuous freeze-drier for treating products in trays. Each tray when filled with pre-frozen material is carried through an entrance vacuum lock into the vacuum chamber, passing through several temperature zones and when the freeze-drying process is concluded, is brought out through the exit lock.

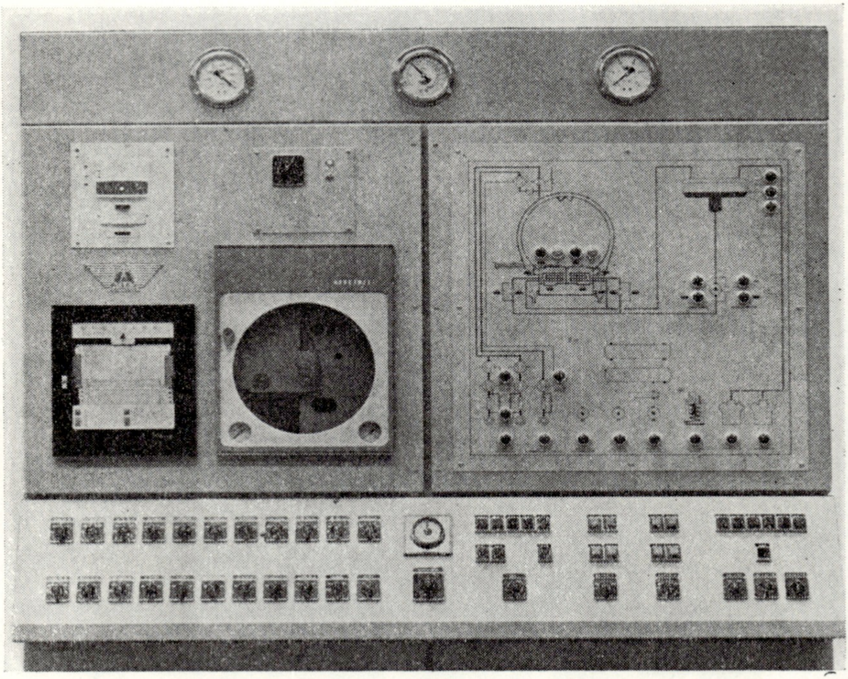

FIG. 8. Control panel for automatic control of batch type freeze-drier.

The simple, straight trays with a shallow layer of product give a rapid freeze-drying without any product loss. The heat transfer to the product and to the trays is by radiation, which in the easiest way provides a correct and an evenly distributed heat transfer to the material during the process. The radiant heat is produced by horizontal heater plates which are grouped in temperature zones. Each tray stays for a fixed period of time in each temperature zone, in a way so that the drying time is minimised.

The exterior transport system for the trays is synchronised with the operation of the vacuum locks. Electronic initiators control each movement so that it is precisely carried through, that pressures are correct, and that all trays are in position, reporting back to a central control unit, and here, by electronic logics, the next steps are initiated and controlled accordingly.

The system is built for an integrated control of the whole tray transport system, internally as well as externally. If some fault may arise, an alarm will immediately be activated, and the control lamps will indicate the source of trouble, so that immediate precautions may be taken.

This continuous freeze-drier has fulfilled the expectations of economic and reliable operation.

NEW FIELDS FOR FREEZE-DRYING

When applying the freeze-drying technique for the removal of solvents other than water, i.e. freeze-drying of non-aqueous solutions, many new possibilities arise (Rey, 1971). Only some examples of such applications will be mentioned.

For the preservation of biological specimens that are not soluble in water, an extract can be prepared with an organic solvent. The sample can be frozen, and the solvent sublimated away at low temperature. Examples of such solvents tested are dioxane, carbon tetrachloride, chloroform, benzol and diethylamine.

The operation temperatures are usually low, so liquid nitrogen is often used for freezing as well as for the cooling of solvent condensers. The heat of sublimation for most solvents, however, is small compared to that of ice, so the speed of the process is comparatively high. The pressure is normally in the range of 0.1–0.01 torr.

Complex material structures can be prepared by a series of freeze-drying operations, for example:

(1) A 5% solution of dextran in water is freeze-dried.
(2) The porous cake is soaked with a 2% solution of polystyrene in benzol. After freezing the benzol is sublimated off.
(3) The cake is now dipped in a solution of maltose in diethylamine. It is frozen again and the solvent sublimated off.

The result is a porous material with a main structure of dextran, containing substructures of polystyrene and maltose.

Also mineral solvents can be similarly used, and development work has started on the application of ammonia and carbon dioxide. CO_2 is interesting as a good solvent for many aromatic constituents and one compatible with foodstuffs. A technical complication is that the triple point pressure is as high as 5.1 atm, so liquid extraction has to take place in pressure vessels.

Interesting developments of new applications of this technique may be expected.

REFERENCES

Amoignon, J. (1973). 'Contamination physico-chimique des produits pharmaceutiques et biologiques lyophilisés', International course, Burgenstock.

Bengtsson, N. E. (1973). 'Freeze-drying of sliced beef', International course, Burgenstock.

Chandrasekaran, S. K. and King, C. J. (1971). *J. of Food Sci.*, **36**, 699.

Dauvois, Ph. (1971). 'Envols en lyophilisation de particules', *Revue générale du froid*, 827, Sept.

Flink, J. M. (1973a). 'Application of freeze-drying for preparation of dehydrated powders from liquid food extracts', International course, Burgenstock.

Flink, J. M. (1973b). 'The retention of volatile components during freeze-drying: a structurally based mechanism', International course, Burgenstock.

Greiff, D. (1973). 'Stabilities of viruses following freezing and storage at low temperatures', International course, Burgenstock.

Kessler, H. G. (1973). 'Heat and mass transfer in freeze-drying of mixed granular particles', International course, Burgenstock.

King, C. J. (1973). 'Application of freeze-drying to food products', International course, Burgenstock.

Lorentzen, J. (1973). 'Industrial freeze-drying plants for foods', International course, Burgenstock.

Maltini, E. (1973). 'Thermal phenomena and structural behaviour of fruit juices in the pre-freezing stage of the freeze-drying process', International course, Burgenstock.

Rey, L. R. (1960). *Ann. N.Y. Acad. Sci.*, **85**, 510.

Rey, L. R. (1971). Underwood Prescott Memorial Award lecture, M.I.T.

Rothmayr, W. W. (1973). 'Basic knowledge of freeze-drying. Heat and mass transfer', International course, Burgenstock.

Sell, K. W. *et al.* (1973). 'Evaluation of freeze-dried tissue allografts: strength and antigenicity', International course, Burgenstock.

Simatos, D. and Turc, J. M. (1973). 'Fundamental phenomena of freezing in biological systems', International course, Burgenstock.

Thijssen, H. A. C. (1973). 'Effect of process conditions in freeze-drying on retention of volatile components', International course, Burgenstock.

DISCUSSION

C. J. King (University of California, U.S.A.): I wonder if you could comment on the difficulties and problems associated with feedback control of freeze-drying processes and whether you see any prospect for a greater use of that in the future?

Lorentzen: I gather you are thinking of some sensor which is in the product, checking the product conditions during drying and thus adjusting the heat input. The difficulty really is to have a sensor which truthfully tells the conditions you have. If you put it in one place, you get one result: if you put it in another place, you get another: and if you then try to regulate the system according to such information, you usually run wild. When the freeze-drying process for a certain layer of a certain product has been defined, then we can run the operation with a large number of sensors. Some of them run wild right away but, with our knowledge of the process, we can disregard them.

Pressure in the cabinet is one parameter which is very important. System trouble of any of your auxiliaries ends up with you losing the vacuum in the cabinet, so that an alarm system coupled to the vacuum gives the best warning.

S. Barnett (General Foods Inc., U.S.A.): With respect to the Conrad system, are there any commercial-scale installations in operation? And what is your operating experience with a relatively complex tray movement system?

Lorentzen: The first version of this Conrad system has been in operation in Switzerland for more than a year, and the experiences completely fulfil our expectations. We have a freeze-drier with a higher degree of safety operation than the old batch type so there is considerably better utilisation of the plant throughout the year.

D. E. Blenford (RHM, U.K.): Basically there are two systems for transporting the material through the various dehydration stages in freeze-drying: either in a tray, or out of the tray, as it were. Out of the tray, the only system you described is the rotary disc method where an obvious degree of attrition takes place and could lead to a particle breakdown. The first question therefore is: Is there a system that you know of operating commercially which works out of the tray but without leading to particle attrition?

The second question refers to the implication you made that the rotary shelf system (if I may call it that) leads to a particle carry-over into the vapour stream, but you did not mention the potential loss of material taking place in tray emptying in conventional tray systems. Would you care to comment on the comparative product loss between the two systems, because in your costing breakdown the raw material is a very major factor?

Lorentzen: Other types of plants have been described with vibration trays instead of the circular plate trays, but I have no knowledge of any of those plants in operation.

As to product loss in operation, I would say that in emptying the trays we think that the product loss is quite negligible, compared with the percentage that you have to reckon on with products flying out of the beds. I think they have difficulty in keeping down to 1% on the non-tray system.

As to product breakdown, I would say that this has to be expected also with vibrating tray methods. During freeze-drying of particles, you always start by freeze-drying the outer layers of the particle; and you have a particle with a hard kernel and you lose considerable surface material due to particle–particle and particle–equipment impacts. With vibration it is quite easy to grind them down to a fine powder.

N. Meisel (Les Micro Ondes Industrielles, France): You were stressing the importance of as high a condenser temperature as possible in order to economise on energy. Could you give us the numbers concerning electricity, or steam, or other energy consumption per litre of water evaporated?

Lorentzen: I cannot rattle off the figures out of my head. But I would say, in answer to that question, you should not have more than 15°C between your ice temperature at the pressure in the cabinet and the refrigerant temperature. In a good system you should be able to operate around −35°C refrigerant temperature: but you see many systems where they operate around −50°C. If we say that we have 100% energy consumption at −35°C, the energy consumption at −50°C will be about 150%. That is the situation in a nutshell.

J. G. Brennan (National College of Food Technology, U.K.): The vacuum systems described all relate to condensers and steam pump systems. Where do the old ejector systems stand these days? Are they still worthy of consideration and, if so, how do they stand up in performance and costs against the refrigerated condensers?

Lorentzen: For various reasons the old ejector systems seem now to be out of favour. If you're operating at 1 torr it is O.K. and you get some capacity out of it. If you operate at 0.9, or 0.8, it is all right, but if you go below this it becomes much less efficient and you have to have more steps in your ejector system which has an adverse effect on the economy. Efficiency is dependent on the water temperature and with the old systems it has been shown that you can get into difficulties at ordinary freeze-drying pressures: they are good enough at higher pressures but not good enough in the lower ranges.

Chairman: Going back just a little, to a continuous system, I think that one advantage of this system over the tray system is that it would be easier to adapt it to in line process control.

We have Mr. Donovan who would like to give a brief description of a novel continuous system.

J. Donovan (Artisan Industries Inc., U.S.A.): The essential concept that Horace Smith has in this particular drying method is that he is freezing in one compartment, designed and operated for optimum freezing conditions. He has a belt conveying device which goes through to a drying chamber with optimum conditions for drying on a belt. His troubles—which have not been solved because this is a concept—would be involved solely in the design of the seals between the chambers and how much leakage there is through the seals from the relatively higher pressure of the freezing chamber to the relatively lower pressure of the drying chamber. There should be some considerable flow of vapour at limited velocity.

E. Seltzer (Rutgers University): The limitation I would expect would be the thickness of the film, because that determines the length of the drying and hence the length of the chamber. Therefore one has to co-ordinate and synchronise the film thickness with the length of the chamber.

Lorentzen: The freezing preparation of the product is a very important part of the operation in terms of the final product and this system is very bad for making any special freezing preparations such as granulation of the product. It should be operated in a vacuum around 30–34°C to prevent excess boiling off of too much water. Lack of access to the product after freezing is a bad weakness of such a system.

S. Barnett (General Foods Inc., U.S.A.): On the question of the difficulty of release from the freezing belt—this is no problem. If one lays down a very thin layer of water or water ice prior to putting on the extract, such as coffee extract, the release is very simple.

My second comment concerns the freeze-drier itself. This is a rather expensive piece of vacuum apparatus, and the economics depend on getting the maximum amount of square footage of drying area within the volume of the chamber: and certainly a belt within the vacuum chamber would not offer much drying area.

Novel Dehydration Techniques

C. JUDSON KING

Department of Chemical Engineering,
University of California, Berkeley, California, U.S.A.

INTRODUCTION

Novel dehydration techniques are, virtually by definition, those approaches which have been conceived but are not in common use. This may be because of some fundamental difficulty or failing of the process, or because of the need for some further critical technical concept or discovery. More optimistically, a process may be 'novel' because implementation is awaiting fuller process development or spreading of knowledge about it, because of a reluctance of processors to replace a proven process with something that is new and therefore risky, or because use of the process is held up until new processing capacity for the particular product is required.

Anyone familiar with the literature of food dehydration and concentration is well aware that a large number of new and different processing approaches have been suggested and brought to various stages of development. It is a major challenge to sift among these and decide which one or more would be the most attractive for further exploration in connection with one's own processing application. A primary purpose of this paper is to provide a logical framework for evaluating these novel processes and, hopefully, for conceiving new and improved processing approaches which have previously not been suggested for foods. Another main purpose is to provide a review and critical examination of some of the more prominent novel techniques.

Morphological approach

As the logical framework for exploring novel dehydration methods I have chosen to use a variant of *morphological analysis*. Morphology in its most general sense is the study of structure, shape and form—in this case the structure, shape and form of the problem of composing processes for the removal of water from foods. Morphological analysis as an approach for problem solving was initiated by Zwicky (1948). The approach is systematically presented by Norris (1963) and by Alger and Hays (1964). The basic idea is to identify the principal different *goals*, or criteria of success, which will be used to judge the worth of any proposed solution to the problem. Next, one identifies each of the major *functions* which must be incorporated in combination into any solution or process, and then conceives several different methods, or *alternatives*, for accomplishing each function. One then considers all different combinations of various alternatives for carrying out the various functions, and judges the resulting processes on the basis of the predetermined goals. Promising combinations of alternatives can then be turned into more specific sub-problems which are dissected by further morphological analyses. Clark (1968) has suggested that the alternatives can, in general, be identified with different physical and chemical phenomena (e.g. radiation, convection and conduction as means of accomplishing heat transfer), and has also suggested that it is profitable to look for particular combinations of alternatives satisfying the different functions in a synergistic way (e.g. one step which satisfies more than one function).

Alger and Hays (1964) present a morphological analysis of the problem of designing a home clothes dryer, where the functions are (1) the need for a heat source, (2) mechanisms for heat and mass transfer, and (3) the environment around the clothes in the dryer. Various alternatives for fulfilling these functions are considered in all combinations, with the aims being (1) to avoid wrinkling the clothes, (2) to prevent uneven drying, and (3) to achieve a pleasant smell of the clothes after drying. These aims are not unlike those of a food dehydration process.

The benefits of a morphological organisation to such a study are several. First, it provides a logical structure which can replace random thinking; second, it enables one to be more confident of considering all existing alternatives; third, it gives one a logical basis for comparisons in assessing reasons for the greater success of certain processes as opposed to others; finally, it can lead to the concept of entirely new processing approaches, such as those represented by combinations of alternatives which have not been previously considered.

Kröll (1964) carried out a systematic categorisation of existing dryers (air-, vacuum-, contact-, and freeze-dryers), based upon differences in operating temperatures and pressures, methods of energy supply, methods of conveyance of material within the dryer, types of auxiliary mechanical action, methods of steam or air movement, and means of preparation and feeding of material. Clark (1968) carried out a morphological analysis of design concepts for freeze-dryers, and Omran (1972) performed a similar analysis for freeze concentration processes.

ALTERNATIVES—DEHYDRATION AND CONCENTRATION AS SEPARATION PROCESSES

Dehydration and concentration are both separation problems, in that a separation must be made between water and the rest of the food. Since it is water that is usually the sole substance to be added for reconstitution, it is desirable to remove water and nothing else during the separation. In principle, one can picture removing the water from everything else (such as by selective evaporation), or one can picture removing everything else

TABLE I

DIFFERENT APPROACHES FOR THE SEPARATION OF WATER FROM FOOD MATERIALS

Initial phase of water	*Receiving phase of water*	*Means of lowering chemical potential of water in receiving phase*	*Example*
I. *Equilibration processes*			
Liquid	Vapour	Vacuum (pressure)	Flash evaporation
		Superheat of feed (temperature)	Drying with superheated steam
		Carrier (composition)	Air drying
	Immiscible liquid	Solvent	Extraction
	Solid	Freeze (temperature)	Freeze concentration
		Precipitate (composition)	Clathrate processes
		Adsorb	—
Solid	Vapour	Vacuum	Ordinary freeze-drying
		Carrier	Carrier gas freeze-drying
	Immiscible liquid	Solvent	—
II. *Rate-governed processes*			
Liquid	Vapour	Vacuum	Aroma-retentive vacuum drying
		Superheat	—
		Carrier	Pervaporation
	Liquid	Added solute	Osmotic dehydration
		Solvent	Perstraction
III. *Mechanical processes*		*Principle for separating phases*	
Liquid	—	Screening	Pulp removal
		Density difference	Centrifugation before evaporation of juices
Solid	—	Screening	Grinding of frozen solutions and pastes
		Density difference	Grinding of frozen solutions and pastes

from the water (such as by a process in which all solutes but water are adsorbed onto a solid adsorbent, or by a process in which a chemical is added to cause precipitation of all solutes from water). However, food substances are invariably exceedingly complex mixtures, and it is most unlikely that a process of the latter sort can be conceived which will, for example, remove all the solutes which are essential to consumer appeal from coffee and allow for their subsequent recovery as coffee powder or concentrate. Consequently the water must be removed selectively from all other species.

If the chief function to be fulfilled is that of separation, then one approach to the generation of novel processes is to review a list of the various different known separation processes. A list of 42 separation processes has been given by the author (King, 1971, Table 1–2). Of these, roughly half can be discarded as being inherently unsuited for this separation, for reasons such as requiring a gaseous feed or the fundamental principle not being capable of effecting a separation on foods. The remaining 20 processes could then be considered as alternatives for fulfilling the separation function.

Another approach to the generation and categorisation of alternatives is to consider that for water to be removed from a food there is a thermodynamic necessity that the fugacity, or chemical potential, of water in the environment surrounding the food must be made lower than that within the food. Table I outlines ways of accomplishing this, both for the case where the water is principally present in the liquid phase within the food and for the case where most of the water within the food has been converted to ice through preliminary freezing.

Separation processes for homogeneous mixtures may be divided into two fundamentally different categories—*equilibration* processes and *rate-governed* processes (King, 1971). Equilibration processes achieve separation by virtue of the equilibration of different phases of matter, while rate-governed processes achieve separation through differences in rates of permeation or diffusion of substances through some medium, usually a barrier.

In Table I the alternatives are divided with respect to the phase of matter into which the water is received, and are further sub-divided with respect to the means of lowering the chemical potential of water in that receiving phase. The chemical potential in the receiving phase may, in general, be altering by changing the pressure, the temperature or the composition. For equilibration processes involving equilibrating the food with a second vapour phase these three alternatives correspond to running the vapour phase under vacuum, with the feed superheated above the boiling point of water, and with a carrier gas diluent, respectively. For two liquid phases in contact, variation of temperature and pressure do not cause substantial changes in the ratio of chemical potentials of water between the two phases; hence altering the composition of the second phase is the one alternative left. For equilibration processes this corresponds to the use of a solvent which has a significant water capacity (extraction). For rate-governed processes the reduced chemical potential in the receiving phase can be achieved by adding substantial quantities of a solute such as sugar (osmotic dehydration) or by using a solvent. Equilibration processes are based upon the existence of immiscible phases; hence the water cannot be transferred into a liquid phase that is water- or food-miscible. Because of the semi-permeability of the barrier, rate-governed processes are not necessarily subject to this restriction.

Examples of embodiments of these different alternatives which have been studied or developed for food dehydration or concentration are also listed in Table I.

TABLE II

BARRIERS POTENTIALLY USEFUL FOR RATE-GOVERNED SEPARATION PROCESSES

Filter media with extremely fine micropores:
Ultrafiltration membranes
Millipore
Nuclepore
Membranes:
Synthetic polymer membranes
Natural cell walls
Surface layers:
Natural or dynamically formed membranes
Added surfactants

Some coarse, preliminary screening has also been performed on the alternatives presented in Table I. Rates of mass transfer in solid phases are infinitesimally slow; hence processes involving equilibration of a solid with a solid are not listed. Processes requiring a gaseous feed are omitted, since foods cannot be gasified without severe damage. Rate-governed processes involving a feed with solidified water are not included, because of mass-transport limitations within the solid and because of the difficulty of keeping intimate contact between a solid and an artificial membrane.

Mixtures which are heterogeneous rather than homogeneous are subject to separation by *mechanical* separation processes, which simply segregate phases of matter that are already present within the mixture. In fact, any of the equilibration separation processes considered for homogeneous mixtures must end with a mechanical separation process. Separations based upon density differences between phases (settling) and the finite sizes of solid phase particles (screening) are most often used for this purpose. Examples of cases where a mechanical separation may be used as at least a preliminary step in dehydration and concentration are given in Table I. These include centrifugation of citrus juices to separate pulp (and/or cloud) from serum before dehydration (Peleg and Mannheim, 1970), and grinding of frozen food liquids to a fine size commensurate with the average ice crystal size so as to take advantage of the fact that a frozen food has already undergone a separation on the microscale.

Table II gives a list of various barriers which can be considered for rate-governed separation processes with foods. These range from synthetic membranes or microporous materials to naturally formed surface layers. The latter can be cell walls of a structured food, accumulations of surface active species (Li, 1971), or surface layers of different composition from the bulk formed dynamically because of the mass transfer processes occurring during water removal.

GOALS OF DEHYDRATION AND CONCENTRATION PROCESSES

Dehydration and concentration of foods are used for the purposes of accomplishing preservation for extended periods of time under less severe conditions of packaging and storage than would otherwise be required, and for the reduction of weight and volume during shipping and storage. The goals of the process must be stated more specifically, however, and a more intensive list is given in Table III. There are two principal subdivisions of goals—those related to product quality and those related to process economy. In general product quality and process economy cannot be judged on a common basis. Depending upon the product, the location, and the circumstances a comparatively poor quality product may be acceptable if the costs are low enough, or a high quality product may be desired even though the costs of obtaining it are high.

MEANS OF SATISFYING GOALS

Minimal chemical and biochemical degradation reactions

The extent of degradation reactions during water removal processes is a function of the temperature, water activity and time. Indexes to the literature covering these phenomena in relation to dehydration are given by Loncin (1969, 1973) and by Karel (1973).

In general, low temperatures retard the kinetics of degradative reactions; however, it is not immediately to be concluded that dehydration at lower temperatures will give less degradation, since many drying processes also take a longer time when carried out at lower temperatures. It is the relative temperature coefficients of rate that are important, as shown in Fig. 1. This figure gives an Arrhenius plot of the rate constants of reactions of different activation energies. As shown by King (1968), the effective activation energy for most drying processes involving vaporisation of water is about 10 kcal/g mol (the latent heat of vaporisation of water). For most degradative reactions the activation energy is greater, e.g. Song *et al.* (1966, 1967) and Kluge and Heiss (1967) report an activation

TABLE III

GOALS OF A FOOD DEHYDRATION OR CONCENTRATION PROCESS

Product quality
Minimal chemical and biochemical degradation reactions
Selective removal of water —cf. other solutes
—cf. volatile flavour and aroma substances
Maintenance of product structure (if a structured food)
Control of density
Rapid and simple rehydration or redispersion
Storage stability—less deep refrigeration required
—simple packaging requirements
Desired colour
Lack of contamination or adulteration
Process economics
Minimal product loss
Rapid rate of water removal (high capacity per unit amount of equipment)
—heat transfer
—mass transfer
Inexpensive energy source (if phase change involved)
Inexpensive regeneration of mass separating agents
Minimal solids handling problems
Facility of continuous operation
Non-complex apparatus (reliability and minimal labour requirement)
Other
Minimal environmental impact

energy of about 30 kcal/g mol for the Maillard non-enzymatic browning reaction. The higher activation energy process has a rate which increases percentwise more rapidly with increasing temperature. Since the time required for dehydration is inversely proportional to the rate constant for water removal, the competing reactions will occur to a lesser extent during a given amount of water removal when the *ratio* of their kinetic rate constant to that for water removal is less. For the case where the degradation reaction has a higher activation energy than does the rate of water removal, this occurs at lower temperatures. Hence it is the relative values of the activation energies that make low temperatures generally preferable for evaporative drying processes.

The low water activity reached upon concentration or dehydration is one of the main reasons for using these processes for preservation. For many degradative reactions the rate tends to decrease monotonically as

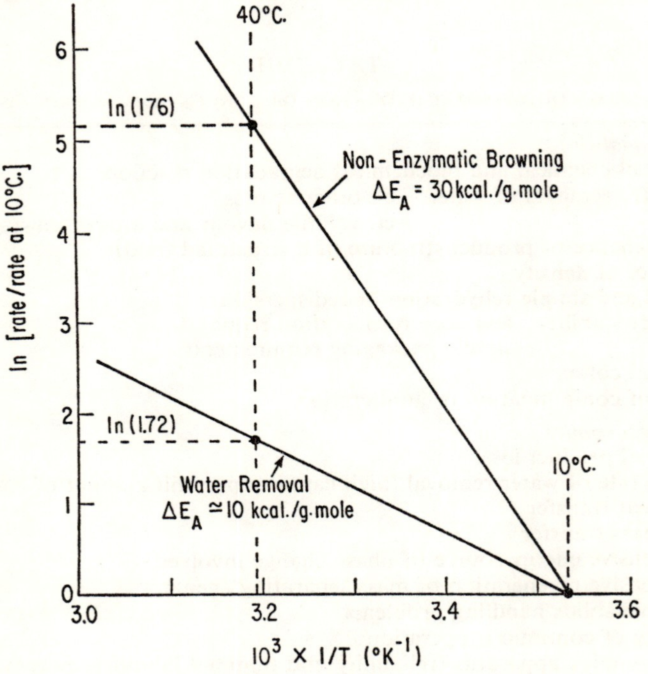

FIG. 1. Relative rates of processes with different energies.

the water content, and hence the water activity, is decreased (Karel, 1973; Loncin, 1973). Some very important exceptions to this trend exist, however. Lipid oxidation rates increase considerably at very low water activities; water appears to inhibit these reactions. For the Maillard reaction between proteins and sugars in solid foods water exerts both promoting and inhibiting effects, thereby giving a rate constant which reaches a maximum at an intermediate moisture content, typically at a water activity (equilibrium relative humidity) of 0.6–0.7.

Table IV summarises means of minimising chemical and biochemical degradation reactions.

TABLE IV

MINIMISING CHEMICAL AND BIOCHEMICAL DEGRADATION REACTIONS

Low temperature (as long as activation energy for water removal < activation energy for degradation reaction)
Low water activity
Protection against lipid oxidation
—low internal surface area
—exclusion of oxygen
Rapid transition through zone of water activity (0.5–0.8) most critical for non-enzymatic browning

Selective removal of water

In equilibration separation processes a principal factor governing the selectivity is the phase–equilibrium relationship between the two phases employed. This determines the *separation factor* (King, 1971), which is the equivalent of the relative volatility between components in distillation or evaporation processes. Ideally, one would like the water to be completely removed into the second phase, while other species are not removed at all. Freezing satisfies this criterion insofar as thermodynamic equilibrium is concerned. Ice formed from a complex food mixture incorporates only water molecules into the true crystal structure. This, and the concomitant low temperatures, are the principal reasons why so many of the processes providing a good product quality incorporate a freezing step. In evaporation processes, on the other hand, water is not the substance of highest volatility, and many of the important volatile flavour and aroma components enter the second phase with the water. True equilibrium between the product phases is rarely reached in equilibration processes, and hence rate factors governing the migration of the different components into the other phase can also be important in determining the selectivity.

In rate-governed processes, the mass transfer rates through the barrier and adjacent concentration polarisation layers determine the selectivity. It is difficult to identify a synthetic membrane which will allow water to pass while at the same time rejecting all other important components. The closest approach to a truly water-selective barrier that has been developed is the formation of a carbohydrate-rich layer immediately adjacent to the surface of an evaporating liquid whose surface is unagitated (see Table II), resulting from the build-up of a water-depleted region near the surface as a natural consequence of the mass transfer accompanying water removal (Thijssen and Rulkens, 1968; Menting *et al.*, 1970; Chandrasekaran and King, 1972). In effect, the development of such a layer turns an equilibration process into a rate-governed process. The formation of a semi-permeable layer can be aided by the presence at the surface of an immobilising structure (Thijssen and Paardekooper, 1968).

Gross loss of the feed phase into the water recipient phase is always detrimental to selectivity. Thus it is important to minimise entrainment into the water-receiving phase in an equilibration process, and leakage through larger holes across the barrier in a rate-governed process.

TABLE V

SELECTIVE REMOVAL OF WATER

Equilibration separation processes
Phase equilibrium (separation factor)—e.g. freezing
Rate limits
Avoidance of entrainment
Rate-governed separation processes
Rate limits
Avoidance of leaks through barrier
Compensating for volatiles loss
Essence—recovery
—distillation
—freezing
—absorption
—adsorption
Add-back of unconcentrated material
Addition of peel oil

Because of the difficulty of retaining volatiles during economical water-removal processes, there have been several approaches developed for returning volatile aromatic material to a concentrated or dehydrated product. These are also outlined in Table V, and may be considered as additional component parts of processes.

Maintenance of product structure

If a structured food is to retain its shape, texture and appearance, special precautions must be taken during drying. Freeze-drying retains structure because of the presence of rigid ice at the very point of water removal. Another approach for avoiding collapse and shrinkage is *puffing*, which is accomplished through vapour release within the substance. Puffing may be carried out by generation of water vapour through sudden vacuum, by release of bubbles of a gas initially dissolved in the substance being dried, or by a sudden acceleration of drying such as through the addition of microwave energy (Huxsoll and Morgan, 1968).

Collapse and shrinkage result from surface tension forces (Bellows and King, 1972, 1973). Alteration of surface tension is therefore another potential means of combating these phenomena.

Control of density

Density of structured foods is influenced by the same factors which govern the retention of structure. For food liquids density can be controlled through foaming the material to a greater or lesser extent before drying (Ettrup Petersen *et al.*, 1970), as well as through freeze-drying and puffing.

Rapid and simple rehydration or redispersion

Shrinkage, collapse of structure and density increase all tend to make the rehydration of a structured food or the redispersion of a liquid food more difficult. In addition, redispersion of a food liquid depends upon the wetting qualities of the surface of the dried material (Karel, 1973). A fine powder will tend to retain a layer of air between a particle and water or within a cluster of particles, thus causing the particles to float and/or clump together rather than redisperse readily in water. Approaches to overcoming that problem include agglomeration and chemical modification of the surface, e.g. with emulsifiers such as lecithin, or surfactants. High-temperature drying can also lessen rehydration tendencies, because of degradations such as denaturation of proteins.

For concentrates, the dispersability can be poor if the viscosity of the concentrate is very high. Everyone is familiar with the difficulties of dispersing a very thick sugar solution from a spoon into water by stirring. The high viscosity corresponds to a low diffusivity of water into the viscous mass. The mass also has a low surface area for water entry per unit volume if it is a large 'glob'. Thus the rate at which water can diffuse in and reduce the viscosity of the surface layers so as to allow mixing into solution is low. Concentrates containing 70–80% dissolved solids can be prepared from fruit juices in various ways, but because of their viscosity are difficult to reconstitute.

The factors affecting the related quality factors of structure retention, density and rehydration or dispersion are summaried in Table VI.

Storage stability

The storage stability has to do with deteriorative reactions occurring during storage and, to a lesser extent, with other factors such as friability. In general, storage stability is increased by lowering the temperature and/or lowering the water activity. This can be a factor favouring dehydration over concentration. Because of susceptibility to lipid oxidation at low water contents, fat-containing dried substances of high internal surface area (and hence of high structure retention), such as freeze-dried meats, require packaging that is impermeable to oxygen—for example, heavy foil or aluminium cans. Similar precautions must be taken to guard hygroscopic products, such as freeze-dried fruits, from water vapour.

TABLE VI

FACTORS INFLUENCING STRUCTURE RETENTION, SHRINKAGE, DENSITY, AND EASE OF REHYDRATION OR DISPERSION

Presence of ice structure where water is removed
Puffing—vacuum
—release of dissolved gas
—sudden, intense increase in drying rate
Foaming
Surface tension
Low-temperature processing
Agglomeration
Surface coating (redispersion)
Viscosity of product, in a concentrate

Desired colour

For many foods colour changes signal the occurrence of deteriorative reactions. In freeze-dried coffee extract, however, colour is controlled by other factors (Barnett, 1973; Ettrup Petersen, *et al.*, 1970). The darkness of the product is related to the number of reflecting surfaces left after drying; consequently a product with very many fine pores has a light colour that is usually judged undesirable. Slow freezing (so as to make large ice crystals) and partial melting at points during drying are ways of darkening freeze-dried coffee.

Lack of contamination or adulteration

Contamination can come from deteriorative reactions, such as micro-organism growth. Another potential form of contamination relates even more directly to the type of separation process used. A number of candidate processes (e.g. solvent extraction, adsorption, stripping) use a *mass separating agent*, which is a stream of material added to serve as the principal constituent of the second phase—in our case, the phase that receives the water. Any additional stream of material entering the process and coming into direct contact with the food is a potential contaminant of the food itself. For this reason, the number of potential mass separating agents which can be considered is limited. This factor discourages the use of mass separating agent processes.

Process economics

Since food products have a high unit value compared to most other processed materials, loss of product during the separation of water may contribute highly to the effective processing cost. Hence it is important for the final mechanical separation of product from the water-receiving phase to be as complete as possible.

Mass transfer of water is a potential rate-limiting factor in any water-removal process. If the rates of mass transfer provided by the process are not fast, the amount of product dehydrated per unit time will be low, the process inventory will be large, and a large amount of equipment will be required per unit capacity.

When the water undergoes a phase change upon removal, it is necessary to supply or withdraw the latent heat associated with the phase change in order to sustain the process. This heat transfer must proceed by radiation, convection or conduction, and can easily become the rate-limiting factor. Convection is generally the simplest and cheapest mechanism; that is one reason for the generally high costs of vacuum processes, where convection is either not possible or slow. When a phase change of water is involved, it is also essential to consider the way in which the necessary source or sink for energy is to be supplied in the process. Depending upon the source, the unit cost of energy can cover a large range, from the low cost associated with combustion to the high cost associated with microwaves and other forms of electrical energy.

Mass separating agents, if employed, introduce costs for continuous supply, and are often sufficiently expensive that regeneration is warranted. The process must then involve another homogeneous separation, and one should search for mass separating agents for which regeneration is relatively simple and inexpensive.

The remaining economic factors relate to the nature of the equipment. Obviously one would like to have simple equipment, from the standpoints of less investment, less labour requirement, less maintenance, and greater on-stream reliability. Continuous operation is usually preferable to batch operation in terms of labour and capacity; however, this can cease to be an advantage in a multi-product plant. Some separation processes are not conducive to continuous operation. Processes which require handling of one or more solid phases usually present additional difficulties, since solids are more difficult to transport continuously than are gases and liquids.

Minimum environmental impact

This factor relates, in part, to the question of product losses, since any lost product is likely to show up as BOD in a waste water stream. It is also associated with the question of the selectivity of water removal. If other substances are removed along with the water, then they too are likely to become BOD.

THE MORE ESTABLISHED PROCESSES

A brief review will be given of the ways in which the more established water removal processes meet and fail to meet the desired goals, to serve as a basis for consideration of the more novel processes.

Concentration processes

EVAPORATION

Evaporation is by far the most common concentration process for food liquids. It admirably fulfills the economic goals of rapid water removal (high capacity), relatively simple equipment, no solids handling, continuous operation and low product loss. The energy source is inexpensive, especially in view of the savings in energy consumption which are accomplished with multiple effects. Quality problems arise in connection with the removal of volatile flavour and aroma species along with the water, and thermal degradation reactions. Evaporation is frequently carried out in conjunction with one of the methods for compensating the loss of volatiles listed in Table V. The thermal degradation problem is often compensated for by operating under vacuum, which gives lower process temperatures, and by the use of high-temperature, short-time (HTST) or flash evaporators (Tressler and Joslyn, 1971).

Because of the high viscosities of concentrated food liquids and because of the tendency toward fouling which can considerably reduce heat transfer coefficients and increase residence time, a large number of different evaporator designs have been developed. These have been reviewed by Armerding (1966). Morgan (1967) presents novel evaporator systems designed to cope more effectively with the fouling problem.

FREEZE-CONCENTRATION

In freeze-concentration, part of the water present is frozen as ice, which is then removed by means of mechanical separation. In terms of the equilibrium separation factor, freeze-concentration is the process which fulfills the goal of selective water removal best. At equilibrium no water enters the ice phase, and no vaporisation need occur to cause loss of volatiles. The low temperature is also conducive to minimum degradation. If an indirect contact freezer is employed, there is no contamination or mass separating agent problem. Freeze-concentrators are also amenable to continuous operation. The energy sink is refrigeration, which is relatively expensive, but the latent heat of fusion of water is much less than the latent heat of vaporisation. Freeze-concentration does involve solids formation, and therefore incurs a penalty of equipment complexity to handle the solid phase effectively.

The principal disadvantage of freeze-concentration is the tendency for product loss through entrainment of viscous concentrate with the separated ice. In the light of this problem, it is interesting to note that the two most prominent present applications of freeze-concentration to foods—concentration of vinegar and preconcentration of coffee extract before freeze-drying—are both situations where the separated water can be recycled for re-use in creating the feed stream. Vorstman and Thijssen (1972) and van Pelt (1973) describe a pressurised wash column which they indicate will give a good separation of ice from concentrate without excessive use of wash water. Other novel design approaches have been suggested by Omran and King (1973) and Omran (1972).

In addition to this conference, the use of freeze-concentration processes for food liquids has been reviewed by Muller (1967) and by Thijssen (1969, 1973). Thijssen (1970) has also evaluated the economics of freeze-concention in comparison to evaporation with essence recovery, and reverse osmosis.

REVERSE OSMOSIS

In reverse osmosis the feed is taken to a sufficiently high pressure so that the chemical potential of water is raised by the Poynting effect (see, e.g. Prausnitz, 1969) to a value higher than that of pure water at low pressure. Water will then flow from the pressurised feed side, across a membrane semi-permeable to water, to the low-pressure, pure-water side. The elevation of the feed pressure is equivalent to overcoming its osmotic pressure. General reviews of reverse osmosis as a separation process are given by Lonsdale (1970), and, with emphasis on foods, by Merson and Ginnette (1972), in addition to this conference.

Reverse osmosis is potentially a very gentle process since it involves no phase change and occurs entirely at ambient temperature; thermal degradation is not a worry. There is no energy consumption other than that for pumping; hence heat transfer is not a concern. It is conducive to continuous operation and does not generate solids.

The limitations of reverse osmosis arise from two basic sources. First, because of the high viscosities and consequent low diffusivities of food concentrates, there tends to be severe concentration polarisation adjacent to the membrane surface. Secondly, present-day reverse osmosis membranes are not totally selective for water.

The concentration polarisation effect lowers the attainable water flux and thereby necessitates large membrane surface areas. The large membrane area, in turn, leads to complex devices and to more substantial pumping costs. The sharp change of viscosity with concentration in most food liquids can lead to a distorted velocity profile which gives most of the flow down the centre of the channel, leaving a stagnant zone near the wall and thereby accentuating the concentration polarisation problem. Monge *et al.* (1973), Kennedy *et al.* (1973) have recently demonstrated that higher-temperature operation can lead to higher membrane fluxes (20–50% flux increase for 10–20°C temperature increase).

Concentration polarisation also raises the amount by which the feed must be pressurised to overcome the osmotic pressure at the feed–membrane interface; 100 atm and more may be required. Pump-up and let-down to and from such pressures requires careful design for many food substances (e.g. egg white), so as to avoid degradation from severe shear action.

Although they reject sugars and higher-molecular-weight solutes, as well as many solutes of very low water solubility, most available membranes, e.g. cellulose acetate (Feberwee and Evers, 1970), give as much permeability to low-molecular-weight oxygenated organics as to water. Thus Merson and Morgan (1968), Bolin and Salunkhe (1971) and others have observed substantial loss of water-soluble volatile aroma from fruit juices subjected to reverse osmosis. Also, membrane non-selectivity can give colour loss for some substances.

Dehydration processes

EVAPORATIVE DRYING

Vaporisation of liquid water is by far the most commonly used method for full dehydration of food. Heat is supplied by air or other gases (air drying), by contact with heated surfaces (drum drying), or by radiation (sun drying). Evaporative drying processes give relatively inexpensive and rapid water removal, using low-cost energy sources and relatively simple, high-capacity apparatus. It is necessary, however, to supply an amount of energy equal to the latent heat of vaporisation of all the water that is to be removed. Evaporative drying gives selective removal of water in comparison to non-volatile substances, but is subject to substantial loss of volatiles. Most forms of evaporative drying are carried out at high temperatures and give some chemical and biochemical degradation and/or colour loss. Most important, evaporative drying leads to strong surface-tension forces, promoting shrinkage and collapse. As a result, rehydration of structured foods very often becomes quite difficult or impossible. In some products—notably prunes and raisins—the changes resulting from evaporative drying are desirable.

The various alternatives available for evaporative drying have been surveyed thoroughly by Van Arsdel *et al.* (1973), Krischer (1963), Krischer and Kröll (1959), Marshall (1963), Hertzendorf (1967), Loncin (1969), and by Holdsworth (1971). Research and development in Russia in this area is considerable (Fulford, 1969). Recent developments in evaporative drying are reviewed by MacDonald (1973).

Fluid-bed drying is a newer form of evaporative drying, in which advantage is taken of the very high wall heat-transfer coefficients attainable in fluidised beds. This approach is covered comprehensively by Vanecek *et al.* (1966) and by Holdsworth (1971). Farkas, Lazar, and co-workers (Farkas *et al.*, 1969; Lazar and Farkas, 1971; Brown *et al.*, 1972) have explored the use of a fluidised bed in a centrifugal field for drying foods, with the aim of attaining a more homogeneous fluidisation and higher allowable gas velocities. Pneumatic drying uses a high-flow gas stream to entrain particulate solids (Sutherland, 1972). Both fluid-bed and pneumatic dryer could cause attrition with fragile foods.

Spray drying is another implementation of evaporative drying. It removes water rapidly, at a relatively low cost and in relatively simple (although large) equipment. Thermal degradation may occur, and loss of volatiles is a major problem, unless the material has reached a high enough dissolved solids content before drying (Thijssen and Rulkens, 1968; Menting *et al.*, 1970). Some products, notably fruit juices, are difficult to spray dry because they become sticky and cannot stand even moderately high temperatures.

Vacuum drying is another variant of evaporative drying. The use of vacuum incurs higher operating costs and locking problems, but gives lower-temperature operation and less thermal degradation. A vacuum system gives improved mass transfer, but, as noted above, complicates heat supply since convection is no longer a convenient mode.

PUFF AND FOAM DRYING

Explosion puff drying (Eskew *et al.*, 1963) was recently reviewed by Holdsworth (1971). In it, pieces of structured food are partially dried, and then are heated to a high pressure in an enclosed chamber. The pressure is suddenly released, causing some of the remaining water in the food to flash into vapour. This flashing makes the structure more porous and allows more rapid finish drying than could otherwise occur, without as much structure collapse in the final product.

There are also benefits to drying liquids from a foamed state. The open, porous structure resulting from a foam allows a high rate of mass transfer which, in turn, permits shorter residence times. The structure remaining after drying is porous, and because of that foaming is sometimes used for control of product bulk density. The porous structure of the dried product from a foamed food liquid can also be of considerable help in promoting redispersion in water. Because of the increased internal surface area of the dried product, storage stability can become more of a problem. The porous nature of foams also makes them poor conductors of heat; however, evaporative drying processes tend to be rate-limited by mass transfer of water for the greater portion of a drying cycle.

Foam and puff drying processes for food liquids have been reviewed by Hertzendorf and Moshy (1970) and by Holdsworth (1971). Common ways of foaming are to draw a vacuum on the liquid, either with or without dissolved gas having been added beforehand (vacuum-puff drying), or to whip the food liquid with air before drying, a foam-supportive surfactant being added when needed (foam-mat drying). Drying is then carried out on trays, porous screens, or other extended-surface devices. Warman and Reichel (1970) report methods of foaming which can be used as a precursor to freeze-drying.

Another implementation of drying from a foamed or aerated state is the Groth drying process (Groth, 1967; Anon., 1968). In this approach, carried out batchwise, the feed liquid is supported upon a polymeric membrane containing 10–100 μm pores. Air or a more inert gas is then passed upward in the form of fine bubbles. The gas becomes saturated as the bubbles rise, and thereby serves to remove moisture continuously. Because of the limited water capacity of the gas, the process requires a long time (3–8 h) for dehydration. Water removal is carried out at ambient temperature, which serves to minimise thermal degradation. As in many other evaporative drying processes there is a tendency for almost complete loss of volatile flavour and aroma; Groth incorporates a silica-gel adsorbent bed for volatiles through which the gas passes after leaving the liquid. It is not apparent, however, how volatiles can be desorbed from this collector and reincorporated into or onto the dried product in a simple way.

There is no apparent heat source provided for this process other than convection and conduction from the surrounding room environment; yet water must consume latent heat of vaporisation upon evaporating into the gas bubbles. In a scaled-up form of the process the heat supply should become a more important problem. The product form is another question. Although it is stated that the product is a porous, easily re-dispersed material, it would seem difficult to achieve such a homogeneous dispersion in the latter stages of drying so that thick chunks of relatively non-porous material are avoided.

FREEZE-DRYING

Freeze-drying provides a very high quality product but at a cost that is greater than for other common dehydration processes. Because of the low temperatures of operation, thermal degradation reactions are avoided. Any particular local region within a particle undergoes a rapid transition from a fully hydrated and frozen state to a nearly completely dry state, because of the passage of a relatively sharp ice front inward during freeze-drying. Consequently degradation reactions which have a maximum rate at intermediate moisture contents, such as non-enzymatic browning, are minimised. Because of the presence of a rigid ice structure at the very point where the water is vaporised, shrinkage during drying occurs to a very small extent. Except for whatever damage occurs from ice crystal growth during preliminary freezing, the appearance and texture of a structured food are largely preserved. The high resultant porosity also gives very rapid rates of rehydration or redispersion, typically the most rapid known. There is a packaging stability problem, however, which comes from the same factor of the very porous structure. Oxygen exclusion through packaging in quite impermeable containers is necessary for lipid-containing foods. On the other hand, storage can be carried out at room temperature for relatively long periods of time.

Freeze-drying also gives very good retention of volatiles, which is particularly important for food liquids. This is a result of freeze-drying being, in fact, two separation processes in series. Freezing before drying serves to separate the substance on a microscale, giving a heterogeneous mixture of pure ice crystals and either residual interstitial concentrate (in most food liquids) or dehydrated cellular matter (structured foods). The diffusion coefficient of volatiles in the residual interstitial concentrate or the dehydrated cellular matter is much lower in proportion to the diffusion coefficient of water than in the substance before freezing. As a result the loss of volatile matter is highly restricted by diffusion, and volatile retentions are high. An exception occurs for volatile substances which exceed their solubility limit and are present as an additional emulsified phase. Emulsion droplets collect preferentially at the interface between ice crystals and the residual material during freezing, and are therefore lost to a degree that is proportionately much greater than for homogeneously dissolved volatiles (Massaldi and King, 1973).

The high cost of freeze-drying results from the simultaneous necessity of freezing, vacuum, batch operation (except in newer, continuous freeze-dryers), and long drying times. The long drying times are the result of heat and mass transfer factors. The available driving forces for heat and mass transfer are limited by the nature of freeze-drying, and the thermal conductivity of porous freeze-dried material is low. Increasing the pressure with inert gas so as to assist heat transfer gives some benefit, but a substantial mass-transfer limitation within the food is encountered at pressures above about 20 mm Hg, even if steps are taken to design the drying chamber configuration so as to avoid mass transfer resistance external to the food (Sandall *et al.*, 1967). This severe mass-transfer limit is encountered at much lower inert gas pressures (e.g. above 1 or 2 mm Hg) for standard batch, tray freeze-dryers, because of the thin vapour channels between the trays.

In addition to this Conference, freeze-drying has been reviewed by the author (King, 1970), with subsequent updatings by Holdsworth (1971) and Karel (1973).

NEEDS FOR NOVEL PROCESSES

In view of the foregoing discussion, the following needs appear.

(1) Development of processes giving product-structure retention equal to or approaching that from freeze-drying, but at a significantly lower cost.
(2) Development of processes utilising the ultraselective water separation which can be achieved with partial freezing, but somehow overcoming the product loss associated with most current freeze-concentration processes.
(3) Development of processes giving retention of volatiles equivalent to freeze-drying, but at a lower cost. This avenue is promising, in view of the considerable fundamental knowledge of volatiles loss mechanisms gained in recent years.
(4) Modification of freeze-drying leading to less stringent packaging requirements, while at the same time not losing the other good product-quality attributes.
(5) Development of improved means for retarding product degradation and structural collapse during evaporative drying processes.
(6) Development of rapid but gentle dehydration processes, operative at or near ambient temperature.
(7) Development of improved membranes which will be more selective for water as opposed to all other compounds.

NOVEL TECHNIQUES OF DEHYDRATION AND CONCENTRATION

In this section a number of novel and currently less-common process approaches for water removal from foods will be evaluated in the light of the previous discussions of process goals, process alternatives, and existing processes.

Membrane separation processes

Most efforts to overcome the effects of concentration polarisation in reverse osmosis involve alterations of the flow pattern. One approach that achieves some success is pulsing of the flow. Lowe and Durkee (1971) obtained improved fluxes in the dehydration of orange juice concentrate by reverse osmosis by pulsing the flow and including large spheres in the stream. Kennedy *et al.* (1973) have demonstrated experimentally that large-amplitude pulsation of the flow in tubular membrane systems gives an increased flux, at the expense of increased pumping power. They present a theoretical interpretation of this effect, based upon the observation that the average absolute flow velocity is increased if the pulsing amplitude is large enough so as to reverse the flow during part of the pulsing cycle.

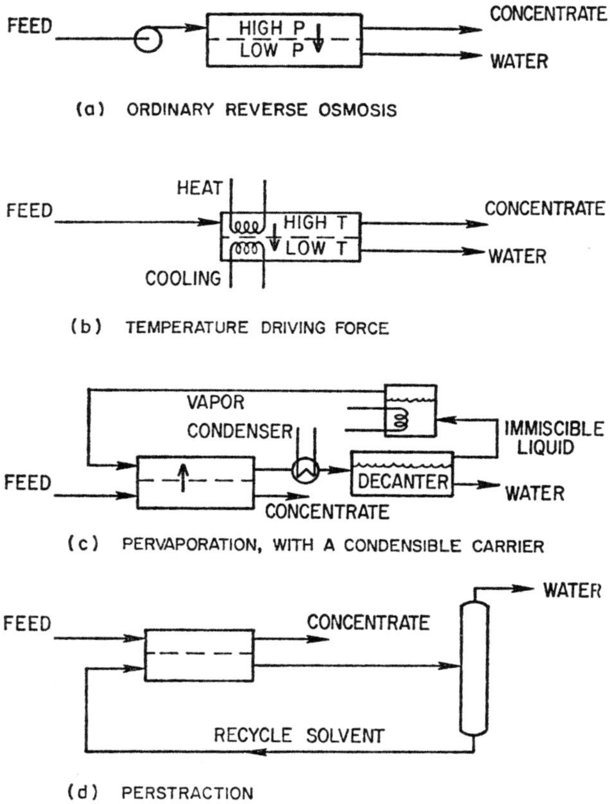

FIG. 2. Alternate driving forces for membrane concentration processes.

The high pressures required for reverse osmosis of most food liquids (up to 100 atm) present problems of high-pressure design and of damage to the feed during pressurisation and let-down. Figure 2 shows some other concepts for creating a positive chemical potential driving force for water from the feed side of the membrane to the permeate side. In addition to pressure, temperature and composition are other variables affecting the chemical potential of water. Figure 2b postulates the existence of a temperature gradient, with the chemical potential of the water in the feed solution being raised above that of the water on the permeate side by virtue of being at a higher temperature. This process is probably unworkable, since membrane thermal conductivities are large enough and membranes must be thin enough so that the rate of heat dissipation through the membrane would be much too large for a temperature driving force of appropriate size. It is the temperatures on either side of the rate-limiting section of the membrane which are important for this purpose; hence even anisotropic membranes would not be of help.

Another way of reducing the chemical potential of the water is to allow it to vaporise upon crossing the membrane. In *pervaporation* the partial pressure of water vapour on the permeate side is kept low enough so that it does not condense on the membrane and so that the chemical potential of the water is less than on the feed side. This is accomplished by dilution with a carrier gas, by vacuum and/or by the use of a cold condenser on the permeate side. Problems with pervaporation as originally suggested (Michaels *et al.*, 1962; Binning *et al.*, 1961) are the necessity of somehow supplying the latent heat of water to all points along the membrane surface in an extended-surface device, and handling large gas-phase flow rates which will have to be circulated and removed by pumping and/or a partial condenser. Furthermore, membrane materials tend to become less permeable when exposed to gas rather than liquid. These problems occur for any membrane process involving vaporisation. Yuan and Schwartzberg (1972) suggest using a heated air stream as both the permeate-side carrier and the heat source; however, this requires very large air circulation rates if thermal degradation problems are to be avoided, and such large volumes of air would require much pumping power and would operate against the desire for a compact, extended-area device.

Both the vapour-handling and heat-supply problems can be alleviated by using a condensable vapour as the sweep gas on the permeate side, as is shown in Fig. 2c (Michaels and Bixler, 1968). For the best results, this condensable vapour should form a liquid immiscible with water, and the membrane should not be permeable in the reverse direction to this substance. This vapour condenses on the permeate side in an amount necessary to furnish the latent heat of vaporisation of the water. If the condensed vapour is immiscible with water, it and the water can be separated principally by decantation, as shown in Fig. 2c, and the immiscible liquid can be revaporised and returned as sweep gas. If the liquids are miscible, distillation is necessary for the separation.

It is probably simpler, however, to eliminate the phase transition within the membrane by using a liquid solvent to take up the permeating water, as shown in Fig. 2d (Michaels and Bixler, 1968). Again, the water should be kept dilute enough in the solvent so that a positive chemical potential driving force exists. The solvent is then regenerated by distillation or simple flashing, and is recycled. Requirements for the solvent are that it is incapable of any appreciable permeation through the membrane in the reverse direction, that it has a volatility very different from that of water, and that it is capable of dissolving a substantial quantity of water. Strong solutions of sugars and higher glycols appear to be possibilities, as well as strong salt solutions with salt-impermeable cellulose acetate membranes.

Kennedy *et al.* (1973) report attempts to increase the flux of water at a given pressure differential in reverse osmosis by circulating a strong sucrose solution on the permeate side. No increase in water flux was observed, however, possibly because of extreme concentration polarisation within the membrane support on the permeate side, and/or because the lower activity of the water on the permeate side tended to dry out their cellulose acetate membrane, thereby lowering its permeability. If either of these is the cause of the lack of flux increase, it should be possible to overcome the problem by properly choosing other membranes, system designs and/or solvents.

Direct osmosis

Immersing a food substance in a hypertonic aqueous solution (one of higher osmotic pressure and hence lower water activity) creates a driving force for water removal. For lowering the water activity, the solution should contain a highly soluble, relatively low-molecular-weight solute. Concentrated sucrose solutions (50–75° Brix) have been most commonly used. There is only a thin conceptual dividing line between such a process and solvent extraction, which will be discussed subsequently. In dehydration by direct osmosis, there is also a driving force for the solute in the water-receiving solution (e.g. the sucrose) to penetrate into the food. Some sort of semi-permeable membrane must be present to prevent or retard that.

Ponting *et al.* (1966) and Farkas and Lazar (1969) demonstrated the use of direct osmosis with sugar solutions or powdered sugar for partial water removal from fruit pieces. In this case the semi-permeable membrane discouraging sugar entry into the fruit is the natural cell wall. This membrane is only partly selective, however, and does allow some sugar to enter. Typically, this may be about 25% sugar, on a water-free basis, for removal of enough water to give a 50% reduction in weight (Karel, 1973). This has led to the suggestion that the product be used as a confection or snack food, and in at least one instance commercial development is proceeding along that line (Hirschberg, 1973). The invasion of the sugar does raise the apparent stability of the product, principally by making it less hygroscopic (Ponting *et al.*, 1966). The osmotically-dehydrated product also regains less water upon rehydration.

The cell membranes have a finite permeability to other components, too. For example, acids are significantly leached out of the fruit, and this also contributes to increased sweetness of the product. One would also worry that the more water-soluble, low-molecular-weight flavour compounds would be leached out. Conclusive quantititative tests gauging the extent of retention of known important flavour compounds do not appear to have been carried out. Any components other than water leached from the fruit create a pollution problem as well. Water must somehow continuously be removed from any food dehydration plant, and the most apparent way to accomplish this for an osmotic-dehydration plant is to feed the water uptake solution to an evaporator. Concentrated solution can be recirculated to the process, and in that way can build up non-volatile extracted species so as to reduce the driving force for their removal. This can only be effective up to a point, however. One cannot give the water-uptake medium too long a residence time in the plant, and a slip-stream must be discarded and replaced by a make-up solution.

If the increased sweetness of the product from sucrose uptake is a drawback for a given food substance, solutions of lactose would seem to be another possibility as the water-uptake medium, since large quantities can be obtained from whey-processing plants.

Osmotic dehydration of piece-form foods can only give partial dehydration. There are obvious processing advantages of the use of a solution rather than a solid, e.g. powdered sugar, as the uptake medium for water. Thus the water removal can be no greater than corresponding to equilibrium with a saturated solution, and will be less to the extent that a finite amount of solution and a finite driving force for mass transfer are employed. It has been suggested (Ponting *et al.*, 1966; Farkas and Lazar, 1969) that the product of intermediate moisture content be used to manufacture a dehydrofrozen product, in addition to the possible use as a relatively stable intermediate-moisture snack food. Alternatively, the osmotically-dehydrated substance can be subjected to air drying to give a fully dehydrated product with less shrinkage and better texture than complete air drying would yield.

Advantages of dehydration by direct osmosis are that it is an ambient temperature process with no opportunity for cell-wall damage from freezing. Enzymatic browning does not occur during processing, but the product may require protection to prevent subsequent browning.

One disadvantage is the slow rate of the process, which appears to result from slow internal water transport. Typical time requirements for 50% weight reduction with apple slices and peaches are 4–6 h (Farkas and Lazar, 1969; Hirschberg, 1973). This can be lessened somewhat by operating at higher temperatures, but thermal damage has been observed above 30°C with fruits for these long processing times. The process is rate-limited by internal mass transfer for piece foods even though the surrounding medium is viscous. This is demonstrated by the absence of any large effect of agitation level in the surrounding fluid (Ponting *et al.*, 1966).

Because of the marginal selectivity of the natural cell-wall membranes, a promising step is to seek a way of imposing a more selective barrier on the mass transfer process. This was done by Camirand *et al.* (1968), who investigated osmotic dehydration of various structured foods, coated with a layer of calcium pectate or related substances and immersed in strong partially invert sugar solution. It was found that the water removal was both more complete and more selective with respect to sugar uptake. If food can be protected from solute entry by applying a coating membrane, the osmotic pressure within the food during dehydration will not rise to as great an extent per unit amount of water removed, and the product can therefore reach a lower final water content. Further thought and research can probably produce coating membranes of still better selectivity, but it must also be remembered that any added substance is a potential adulterant. Hence the membrane must either be edible and tasteless, or else it must be inexpensively removable. At the same time it should be close-fitting or adherent to the surface of the food.

It has been pointed out (Karel, 1973) that the times used by Camirand *et al.* (1968) for their osmotic dehydration (72–144 h) are undesirable for a medium where microbial growth can occur. On the other hand, Camirand *et al.* do not give rates of approach to this final state. Recalling that the process without a coating membrane tends to be rate-limited by internal mass transfer, it would seem likely that one can find coatings of sufficient thinness and/or permeability so that the added resistance to mass transfer will be relatively small.

Direct osmosis as a means of concentration has also been reported for food liquids. In this case use of an added membrane to separate the food from the water-uptake medium is necessary to avoid mixing of the two miscible liquids. Bolin and Salunkhe (1971) report experiments wherein fruit juices in cellulose acetate tubes were agitated in 71° Brix solution. Removal of about half the water was achieved, with relatively low retention of volatiles and poor taste panel scores in comparison to concentration by freeze-concentration, reverse osmosis and a pervaporation-like process. The reason for this loss and/or change of flavour in comparison to the other membrane processes is not immediately apparent, however, and it may result from some aspect of handling that is not related to the osmotic concentration process itself. Osmotic dehydration of food liquids across a synthetic membrane is essentially the same concept as perstraction (Fig. 2d), and from an engineering point of view is best carried out as a continuous-flow operation in a device similar to those used for reverse osmosis.

Clathrate formation

Clathrate compounds involving water solidify at a higher temperature than ice. For sea water conversion, their formation has been studied as an alternative to the formation of ice in freezing processes, primarily because the temperature of operation would be nearer to ambient. Clathrates, or hydrates, of C_3 and C_4 hydrocarbons and Freons have been used for sea water conversion tests, but processes involving them have been largely abandoned because the crystals formed tend to be even finer than those from an ordinary freezing process.

Huang *et al.* (1965, 1966) carried out experiments with CH_3Br and CCl_3F as hydrating agents for removing water from orange juice, apple juice and tomato juice by the formation of clathrates. Their results show considerable entrainment of the juice concentrate by the hydrate crystals and difficulty in washing the hydrate crystals free of this entrained concentrate. You may recall that loss of concentrate by entrainment is already a severe problem for freeze-concentration processes. Most probably, this difficulty reflects very small crystal sizes, which were also found in the sea water experiments. The small crystal sizes result from slow crystal growth rates, which in turn come from slow mass transfer of the clathrating agent through the solution to the growing crystal. The hydrating agents used were gaseous, with limited water solubility; consequently the rate of mass transfer of the gas dissolved in the liquid phase is slow because the concentration driving force (the solubility of the gas) is low. Slow mass transfer of the dissolved hydrating gas is also indicated by the observations reported by Huang *et al.* for the influences of stirring speed and of the rate at which the gas was made available. Werezak (1969) reports making much larger clathrate crystals, which were also more readily washable, from NaCl and sucrose solutions and from coffee extract. This was accomplished by using much more soluble hydrating agents, notably ethylene oxide.

The potential advantages of a clathrate process in comparison to freeze-concentration would lie in the need for less-cold refrigeration and in the presence of less viscous solutions because of the higher equilibrium solidification temperature corresponding to any given solution concentration. Conceivably, this could simplify the washing process. However, large enough crystals to be of interest are obtained only with the more soluble hydrating agents, and these appear to be largely incompatible with foods. In any event, the addition of a clathrating agent as a mass separating agent represents a source of adulteration of the food, and there is also the likelihood of loss of volatile flavour species with the dissolved clathrating agent when it is separated from the concentrate. In view of these drawbacks, it does not appear that clathrate processes for food concentration are as promising as other new approaches.

Slush evaporation

As has been pointed out, partial freezing of a food liquid creates a separation on a microscale. Freeze-concentration uses this separation directly by a simple mechanical separation of the phases. Freeze-drying takes advantage of it in another way, with sublimation of the ice crystals after more complete freezing to leave behind a rigid matrix in which the volatiles have a very low diffusivity and are therefore preserved.

Chandrasekaran and King (1971, 1973) have put forward another process utilising the separation accomplished by partial freezing in a food liquid. In this approach, sufficient freezing is carried out to solidify a large fraction (20–90%) of the water present into ice, but the temperature is kept high enough so that a semi-liquid slush is formed, rather than the deeper frozen state that would be used for freeze-drying. The slush is then subjected to conditions which give combined evaporation and sublimation of water, and which keep the slush partly frozen until a substantial fraction of the water is removed. With techniques such as foam or puff drying, it is likely that vaporisation of water from a slush can be used for full dehydration; the name 'slush drying' will be reserved for such a process. However, experimental work to date has focused on use of the process for concentration rather than full dehydration. The concentration process will be denoted as 'slush evaporation'.

A typical slush temperature for slush evaporation is −3 to −10°C, at which 40–80% of the water in most extracts and juices is present as ice (Riedel, 1949). These temperatures are considerably higher than those of −20 to −50°C required for satisfactory freeze-drying below the collapse temperature (Bellows and King, 1972, 1973). Since the vapour pressure of water determines the mass-transfer driving force, higher rates of drying are achievable at higher temperatures, and slush evaporation can therefore be carried out at a much faster rate than freeze-drying. Also the condenser temperature can be held much higher for slush evaporation than for freeze-drying. For these reasons slush evaporation should be a much cheaper water-removal process than freeze-drying, even though the equipment employed can be similar in nature.

For food liquids rich in carbohydrates it has been established that the pertinent diffusion coefficients vary in such a way that retention of volatile flavour and aroma species during evaporative water removal becomes much better as the initial dissolved solids content increases (Thijssen and Rulkens, 1968; Chandrasekaran and King, 1972). The separation of ice crystals by partial freezing in a slush serves to increase the dissolved solids content of the remaining liquid concentrate. Hence the dissolved solids content of the liquid regions in slush evaporation is higher than in the original liquid before freezing, and one would expect a better retention of volatiles for such evaporation than for vaporisation of water from the unfrozen state. Furthermore, any removal of water by direct sublimation from ice should occur without much volatile loss, since volatiles are excluded from the ice phase. Chandrasekaran and King (1971) tested this prediction by comparing retentions of natural volatiles, monitored by vapour headspace chromatography, during vaporisation of water from apple juice of various concentrations held in an open dish in a vacuum chamber. A radiant heater was used as necessary. Retentions observed for slush evaporation were considerably better than those occurring for evaporation from the unfrozen state, and were comparable to the retentions observed during transient freeze-drying during the period of time before collapse of the freeze-dried product was observed.

Recent work in our laboratory (Lowe and King, 1974) involved systematic measurements of the retention of natural volatile components

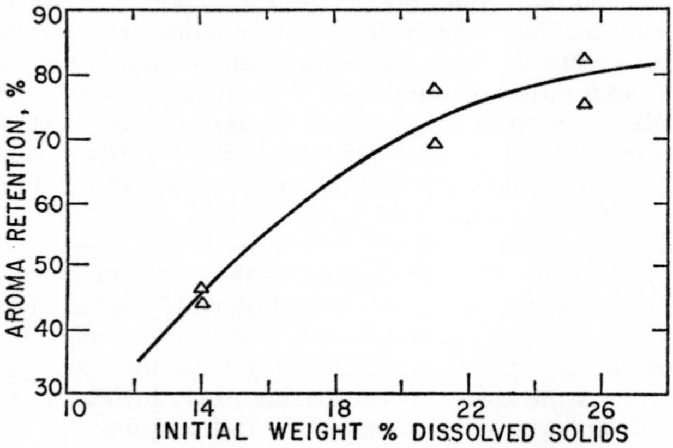

FIG. 3. Retention of volatiles during slush evaporation, as influenced by initial dissolved solids content.

during slush evaporation of apple juice. Again the slush was contained in an open dish in a controlled, laboratory freeze-drying chamber. Retention of volatiles was measured as the average of the three different percentage retentions computed for the ethanol, hexanal and ethyl-2-methyl butyrate peaks, after reconsitution of the concentrated product. Figure 3 shows retentions obtained as a function of the initial strength of the apple juice, for the removal of 10 g of water from a 50 g sample at a low evaporation rate of 0.195 lb/h ft^2, under temperature conditions (different for each run) such that 68% of the water was frozen as ice. The marked increase in retention at higher initial dissolved solids contents probably reflects the beneficial effect of increasing the dissolved solids content within the concentrate portion of the slush.

Figure 4 shows retentions for slush evaporation of apple juice as a function of the slush temperature, for 14% initial-dissolved solids natural juice and 10 g water removal from a 50 g sample at the same evaporation rate.

Lower temperatures give a greater percentage of the ice frozen and hence give a concentrate of higher dissolved solids content within the slush, again affording an explanation for the improved retentions at lower temperatures.

Figure 5 shows the influence of the amount of water removed upon the retention, for slush evaporation of intially 14% apple juice at a constant temperature of $-6.5°C$ and the same evaporation rate. The levelling-off

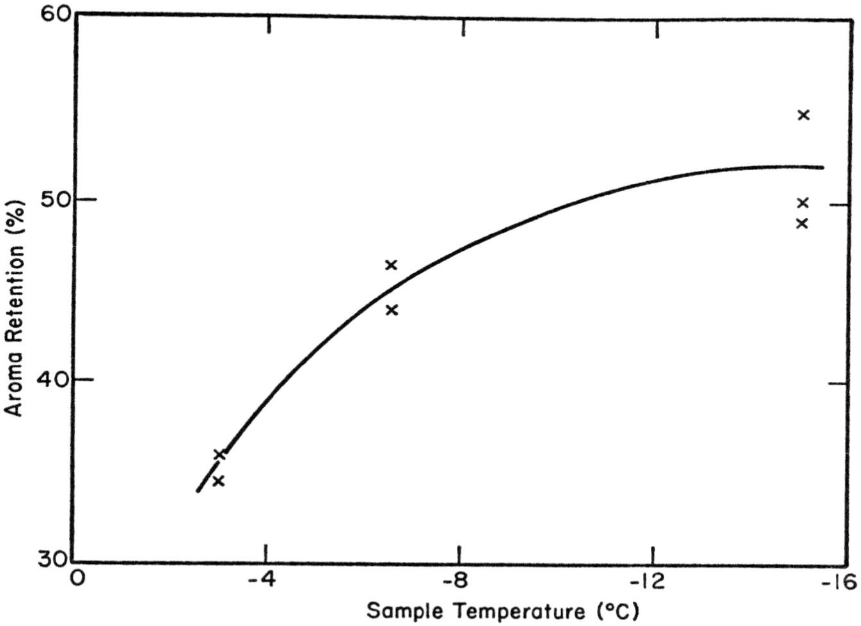

FIG. 4. Retention of volatiles during slush evaporation, as influenced by operating temperature.

of the retention at a nearly constant value of 40% up to very high product concentrations is striking, although the scatter in the retentions for the three individual components is sufficient so that one cannot say for sure that no loss at all is occurring beyond 25% dissolved solids. This behaviour appears to be a result of an inhomogeneity developing within the slush, such that the surface layer becomes more concentrated than the bulk. Water would then migrate internally to this surface region from the bulk slush before evaporation, and the retention would be characteristic of the higher dissolved solids content of the surface region. The rapid early loss of volatiles may result from start-up conditions in the laboratory chamber.

The retention was not much affected by changes in the evaporation rate (up to 0.6 lb/h ft^2), slush thickness, or by heating by conduction from below as opposed to radiation from above.

In another series of experiments in our laboratory (Lowe and King, 1974) we have used differential interferometry analyses to determine the fractional loss of the concentrate by entrainment and splattering during slush evaporation of solutions of sugars from an open dish in the same vacuum-drying chamber. Figure 6 shows the effect of the rate of evaporation upon the percentage of the concentrate lost, for 10 g removal from a

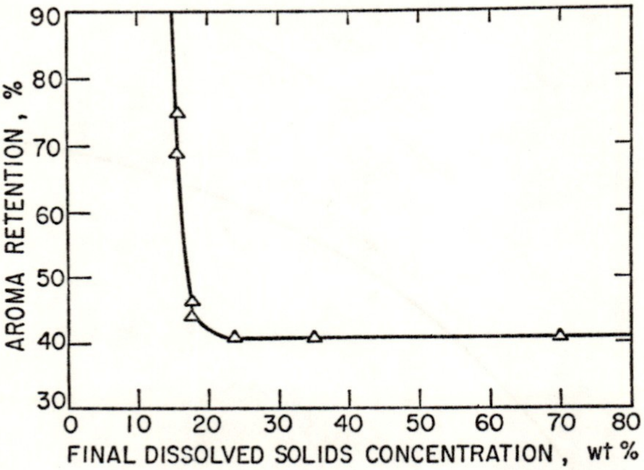

FIG. 5. Retention of volatiles during slush evaporation, as influenced by the fractional removal of water.

50 g specimen initially containing 14% dissolved solids and held at −5°C. The higher losses at higher rates are accompanied by some visible bubbling and splattering, but the losses are still not great, even for a rate of 3.3 lb/h ft^2. Varying the amount of water removed showed that the loss of dissolved solids occurs at a relatively even rate as more water is evaporated from the slush. For a drying rate of 0.32 lb/h ft^2, 0.8% of the concentrate was lost when the slush was taken from an initial 14% to a final 70% dissolved solids content at −5°C. In this case all of the ice disappears before the final concentration is reached. Figure 7 shows the influence of slush temperature on the loss of concentrate for an evaporation rate of 0.32 lb/h ft^2 and 10 g water removal from a 50 g sample initially containing 14% dissolved solids. Other tests showed (1) that the loss was much higher if the heat was supplied by conduction from below (boiling) rather than by radiation from above as in the other experiments, (2) that higher initial dissolved solids contents gave higher losses for a given weight reduction, and (3) that the presence of pectin increased losses somewhat.

Slush evaporation holds promise as a concentration method which gives retentions of volatile flavour and aroma that are superior to those achievable by other evaporative concentration methods, while requiring a cost substantially less than that for freeze-drying. The relatively low solids losses found experimentally offer economic promise. At temperatures just

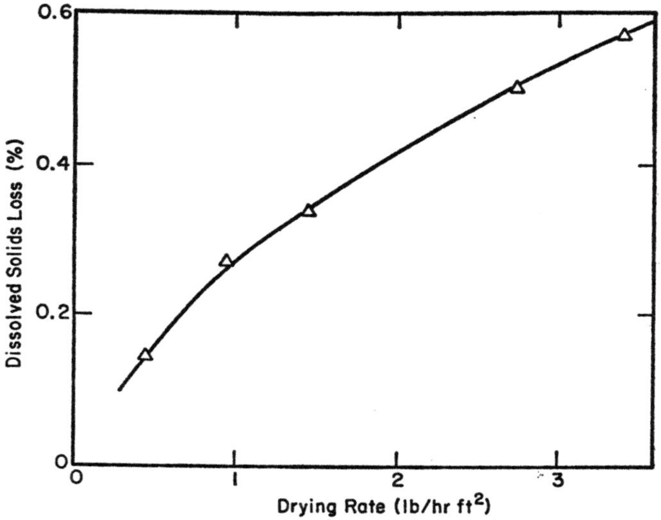

FIG. 6. Loss of concentrate during slush evaporation.

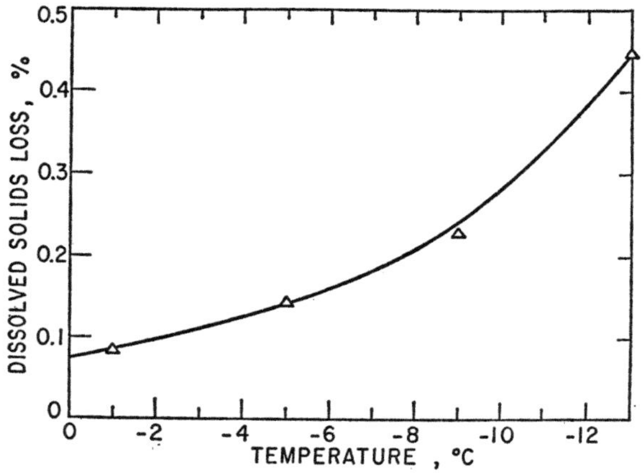

FIG. 7. Loss of concentrate during slush evaporation, as influenced by operating temperature.

above the collapse temperature for freeze-drying very high losses of dissolved solids are encountered. The reduction of solids losses with increasing temperature shown in Fig. 7 indicates a maximum solids loss somewhere between the lower temperatures studied by Lowe and King and the freeze-drying collapse temperature. This indictes that slush evaporation (or slush drying) and freeze-drying both offer better economy in avoiding solids losses than would be the case for drying in the intermediate temperature region. Slush evaporation may be looked upon as a way of carrying out freeze-concentration, replacing the difficult step of mechanically separating ice and concentrate by a separation based on selective evaporation and sublimation.

Fine grinding

Still another approach for making use of the microscale separation of water from concentrate which occurs upon freezing has been suggested by Spiess and co-workers (Spiess *et al.*, 1973; Spiess and Wien, 1973). Grinding to a scale of particle sizes comparable to the dimensions of the ice crystals in a product which is made fully rigid by freezing should give a heterogeneous mixture of particles, some being mostly or completely ice and others being mostly or completely rigid concentrate.

Spiess *et al.* (1973) used a variety of pulverising techniques to convert frozen coffee extract, starch paste, gelatine solution, apple puree, and curds at $-50°C$ into powders with average particle sizes in the range of 100–700 μm. In order to evaluate the potential benefit of obtaining a preconcentration in this way, they carried out a separation of the product particles by size, using standard screens. For gelatine and starch they found an appreciable separation, with the large particles having a smaller water content. For a feed containing 20 wt% gelatine the particles above 400 μm contained 60 wt% gelatine, and the particles below 200 μm contained about 2 wt% gelatine, with a transition range containing about 5% of the mixture. For coffee extract, apple puree and curds the separation was much poorer, however. The better separation for gelatine and starch was attributed to the presence of strong cross-linking bonds in those substances, which made the concentrate much more resistant to fracture than the ice crystals.

In considering a preliminary separation of this sort, one should keep in mind the usual economic situation for food liquids whereby even small product losses can add markedly to the cost of the process. Thus mechanical separation of a frozen mixture after grinding might better be considered as a means of dividing the feed into two portions of different compositions, both of which will be processed into the final product. This is advantageous if the division gains something in comparison to processing of the unseparated mixture. For example, prior separation by grinding could give a higher solids-content fraction which gives better volatiles retention upon drying; the lesser solids-content portion could then be dried in a harsher way that gives no retention of volatiles, but the mixed product would still have much aroma.

Ice crystals and interstitial concentrate regions will have a range of sizes, and even with slow freezing rates a substantial number will probably still be below 100 μm. Thus the separation attainable by grinding and sieving could perhaps be improved by grinding to a finer size. Also, a better approach may be a separation based upon density, since the concentrate will have a specific gravity appreciably higher than ice (greater than 1.0, compared to 0.915 for ice). One can picture flotation in an immiscible liquid of intermediate density for this purpose, although that liquid would then be subject to the usual constraints for a mass separating agent. Another approach could be to allow the particles to settle to their stable height in an upflowing, divergent gas stream, giving a separation based upon both size and density.

Spiess and Wien (1973) have considered grinding to a fine size, without necessarily using mechanical separation, as a precursor to freeze-drying. The reduction of particle size would lessen resistances to heat and mass transfer within the particles, and could even bring about the situation where much of the ice is directly exposed at the surface without any dry layer impeding heat and mass transfer. After computing that the residence time required for freeze-drying in such a case should be of the order of one second, they have endeavoured to utilise a dryer wherein the particles rain downward through a tower. Because of the need for vacuum to assist mass transfer external to the particles, convective heat transfer as used in spray drying was not felt to be practical, and radiation from hot walls was used instead. The particles occupy a very small fraction of the tower, and so very high wall temperatures are needed to deliver enough heat to accomplish the drying in a reasonable height. This then brings about a problem of thermal damage to particles striking the wall.

Extraction and distillative processes

A number of processes involving contact with solvents or boiling liquids have been suggested. Usually, these are mass separating agent processes, where the purpose of the mass separating agent is to lower the chemical potential of water, by dilution, in the water-receiving phase. These processes may be categorised in a number of different ways for purposes of evaluation, including:

Processing sequence: Extraction *v.* distillative drying
State of water in food during water removal: Frozen *v.* non-frozen

Class of solvent: Penetrating *v.* non-penetrating (or water-miscible *v.* water-immiscible, polar *v.* non-polar, volatile *v.* non-volatile, etc.)

Figure 8 illustrates the distinction between water removal, by extraction on the one hand, and distillative drying on the other. In extraction, water

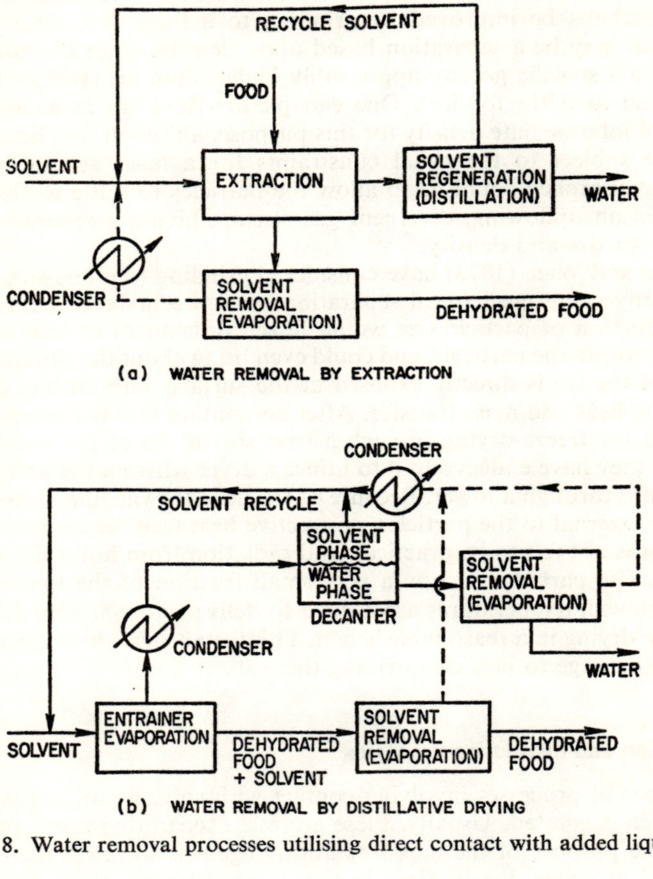

FIG. 8. Water removal processes utilising direct contact with added liquids.

removal from the food into the solvent takes place in an extractor with no vaporisation, and the solvent is then regenerated in a separate piece of equipment, usually by removing water from it through distillation. In distillative drying, boiling of the solvent and water removal from the food occur together, with the water leaving in the vapour phase. If the solvent is immiscible with water, the solvent can be recovered primarily through decantation, with distillation to remove the rest of the solvent from the effluent water; otherwise all the solvent must be recovered by distillation.

Table VII illustrates the alternatives regarding the state of water in the food and the solvent characteristics for the two basic schemes, each subdivided according to whether or not the solvent employed is one that penetrates into the food. Potential advantages for each alternative are also noted. A potential advantage common to all approaches is the removal of water at low temperatures—ambient in the case of extraction, and whatever lower boiling temperature results from the volatility of the solvent in the case of distillative drying. For distillative drying with a non-penetrating liquid, the liquid serves as a heat-transfer medium and, if it is volatile, as a carrier for evolved water vapour. The pores either collapse or are filled with vapour after the water is removed. The combination of a non-penetrating extraction solvent and a high solvent capacity for water would seem to be unlikely, but Kerkhof and Thijssen (1973) indicate that polyethylene glycol has these characteristics. A penetrating solvent replaces the water in the food by a liquid diffusion (extraction) process, with the solvent then being removed by vaporisation either simultaneously (distillative drying) or subsequently (extraction). A penetrating solvent with a surface tension lower than water can reduce the surface forces tending to cause collapse and shrinkage. The categories of penetrating and non-penetrating solvent can overlap for a solvent having properties mid-way between the extremes. Removing water from the frozen state could give the structure retention characteristic of freeze-drying.

Levin and Finn (1955) describe a distillative drying system for unfrozen substances, using non-penetrating solvents. Freeman *et al.* (1957) report both extraction with acetone and distillative drying with butyl acetate, both for unfrozen substances. Wistreich and Blake (1967) carried out distillative drying with toluene, removing water from the frozen state. Bohrer (1967) describes distillative drying using a variety of solvents for both frozen and unfrozen substances, although true freeze-drying for the conditions indicated is questionable (King, 1970). Dorsey *et al.* (1968) used a non-volatile liquid as a heat-transfer medium for evaporative drying. Hieu and Schwartzberg (1973) examined distillative drying of unfrozen shrimp, using ethanol, ethyl acetate and benzene as solvents, and found it possible to model the drying process successfully through a mathematical simulation. Kerkhof and Thijssen (1973) have explored the use of polyethylene glycol for the extraction of water from food liquids and aroma concentrates. Flink (1973) has indexed other work on extraction and distillative drying.

Extraction with highly volatile solvents has been used for some years by biologists as a first step in preparing specimens for the electron microscope in a way that retains fine details of structure (Anderson, 1951; Koller and

TABLE VII

ALTERNATIVES AND POTENTIAL ADVANTAGES OF EXTRACTION AND DISTILLATIVE DRYING

	Extraction		*Distillative drying*	
	penetrating solvent	*non-penetrating solvent*	*penetrating solvent*	*non-penetrating liquid*
Solvent volatility	High	Much greater than or much less than that of water	Similar to that of water	Similar to or less than that of water
State of water in food	Frozen or non-frozen	Non-frozen	Frozen or non-frozen	Frozen or non-frozen
Restrictions	—	Solvent must still have high capacity for water	—	—
Potential advantages	Replace water with solvent which can be removed more rapidly and/or at lower temperature, or which has lower surface tension	Little or no need for solvent removal from product; could be aroma-retentive	Liquid is heat-transfer medium; accomplish some extraction along with vaporisation of water; lower surface tension within food	Liquid is heat-transfer medium; lower temperature for water vaporisation (if liquid is volatile); little or no need for solvent removal from product; could be aroma-retentive; probably can recover solvent largely through decantation

Bernard, 1964; Cohen *et al.*, 1968). Solvents used are compounds under high pressure which are liquids, but are very near the critical point at ambient temperature. Liquid solvents commonly used are shown in Table VIII.

Because of insufficient miscibility with water, liquid carbon dioxide extraction is typically preceded by extractions with ethanol and amyl acetate, and Freon 13 extractions are preceded by a series of graded extractions with ethanol and Freon 113. After substitution with the volatile solvent, the system is heated just above the critical temperature. This frees

TABLE VIII

SOLVENTS FOR CRITICAL-POINT DRYING

	Critical temperature (°C)	*Critical pressure* (psi)
Carbon dioxide	31.1	1 070
Freon 13	28.9	561
Nitrous oxide	36.5	1 050

the specimen of any optically evident liquid. Surface tensions of liquids become extremely low near the critical point; hence the surface-tension forces tending to cause structure distortion are greatly reduced. The success of this method in retaining ultra-fine structure is striking. One can picture a process for food employing one of these solvents, with the super-critical solvent being purged by a non-condensable gas after having been raised above the critical point. The temperature would be close to ambient, which is good; however, operation at quite high pressure would be required, and the volatile replacement solvent would probably also turn out to be a good solvent for volatile flavour and aroma substances. Liquid CO_2 is already known to have that property.

It is instructive to evaluate the extent to which extraction and distillative drying processes meet the goals for dehydration processes, set forth in Table III:

(1) Degradative reactions should be minimised by the low temperatures involved, although this may be compensated for by the relatively long times which can be expected for extraction processes. Distillative processes can be run under vacuum to lower the temperature.
(2) Selective water removal is one of the most difficult goals to meet. Other components of the food (volatile and non-volatile) will probably dissolve in the solvent; in fact the distillative process described by Levin and Finn (1955) also serves as a fat-extraction process.

(3) The lower surface tension of a penetrating solvent can assist in the preservation of product structure and porosity, but some collapse can still be expected, except near the critical point. A non-penetrating solvent should be less successful in preserving structure, as is borne out by the reported results of Hieu and Schwartzberg (1973) on the use of benzene in distillative drying. Carrying out the process with water frozen in the food should help structure preservation considerably at the expense of slower rates.

(4) If fats are not essential to the organoleptic properties of a food, their removal by simultaneous extraction could aid storage stability of the product.

(5) Colour can be harmed by solvent action. Hieu and Schwartzberg report that the red colour of shrimp is transformed into a homogeneous pink by solvents.

(6) Contamination and adulteration of the food are major problems with most solvent processes. It has been found to be extremely difficult to remove the last vestiges of solvent. Apparently the only good solution is for the solvent to be undetectable and harmless. These restrictions considerably limit solvent possibilities.

(7) If non-aqueous portions of the food are extracted into the solvent, some product loss will occur.

(8) These processes require water migration to the food surface by liquid or vapour diffusion. These processes are both slow. Hieu and Schwartzberg (1973) found that at least 6 h were required for water removal from shrimp by either a penetrating or a non-penetrating solvent. Kerkhof and Thijssen (1973) allow 30 min for extraction of water from droplets of size 1 mm or less, and find that air drying is still necessary to remove the last traces of water. The long times required for an extraction process can be estimated from solutions to the transient diffusion equation. Hieu and Schwartzberg inferred diffusivities within shrimps from their drying rate data with penetrating solvents and found values of the same order of magnitude as those in free liquids.

(9) Energy is required for solvent regeneration and circulation. For a distillative drying process the solvent should not be so volatile as to require circulation rates much greater than the water removal rate. Solvents should be picked so that regeneration is relatively simple.

(10) Solvent processes do lend themselves to continuous operation in at least some cases, but require a rather complex array of equipment, as may be seen from Fig. 8.

(11) Solvent loss or extracted food components can pose pollution problems.

The problems of non-selective water removal and of contamination from the solvent are the two factors which most discourage the use of solvent processes, and can be reduced for food liquids by using a selective membrane between the food substance and the solvent. With this modification, extraction and distillative drying become perstraction and pervaporation, respectively (Fig. 2). Kerkhof and Thijssen (1973) have found conditions which lead to the formation of a water-depleted selective surface layer during extraction of water from aroma-laden dextrin and maltodextrin solutions, using polyethylene glycol as a non-penetrating solvent. For solutions initially containing 60% or more dissolved carbohydrate, retentions of model aromas (alcohols) as high as 95% could be achieved. The formation of the selective surface layer is favoured by a high distribution coefficient for water into the solvent, rapid mass transfer in the continuous (solvent) phase achieved by agitation, small droplets as the feed phase, and a high initial dissolved solids concentration. For solutions containing 50% dissolved solids or less, however, the retention of the alcohols was much poorer (5–50%). Kerkhof and Thijssen attribute this to internal circulation and droplet distortion when the viscosity of the feed becomes too low. The approach of extraction into polyethylene glycol through a dynamically-formed surface layer is attractive, but has yet to be shown to give high water selectivities for real foods at concentrations of interest for dehydration. The amount of penetration of polyethylene glycol into the food phase is also a significant question.

Adsorption

Sorbents are known which selectively take up smaller molecules, notably molecular sieves and Sephadex gel. Conceivably water removal into either of these substances could be used as the basis for a concentration process. The use of molecular sieves is prevented by the extremely high heat of adsorption which is released upon water take-up; this would unavoidably cause thermal damage to food products. Use of any adsorbent or solid absorbent is hampered severely by the fact that a large fraction of a food liquid will be water, thereby requiring very frequent regeneration, and by the difficulty of achieving so sharp a separation that dissolved solids losses with the removed water are not an imposing economic factor.

Superheated steam

As was noted in Table I, superheated stream may be used as a drying medium so as to gain a positive chemical potential driving force for vaporisation by adjusting the temperature. Heat will be transferred from the superheated steam to the substance being dried until the surface of the substance just exceeds the equilibrium vaporisation temperature. The approach is not new (Hausbrand, 1908; Walker *et al.*, 1937) but it has not seen significant application to food materials.

The potential advantages of using superheated steam, as opposed to air, for drying (Trommelen and Crosby, 1970) are:

(1) An increased thermal efficiency.
(2) A non-oxidising drying environment.
(3) Simplified dust collection, since a total condenser can be used.
(4) Less volume of exhaust than in air drying, again because the steam can be condensed.

At temperatures just above 100°C air drying will clearly give more rapid rates of heat transfer and hence more rapid rates of water removal than will drying with superheated steam, because of the temperature driving force coming from the wet-bulb depression. Whether or not drying with superheated steam will give higher rates than air drying if both are carried out at some higher temperature has been the subject of some dispute (see, e.g. Yoshida and Hyodo, 1970). The answer is one of definition: if the two processes are compared at the same *mass* velocity (lb/h ft^2), then above a certain inversion temperature superheated steam will give the faster drying rate in the constant-rate period, because the steam will give a higher heat-transfer coefficient. If the comparison is made at the same *volumetric* flow of steam and air, the heat-transfer coefficients will be about the same, and air drying will be faster in the constant-rate period because of the increased driving force coming from the wet-bulb depression.

Disadvantages of drying with superheated steam include the need for guarding against condensation at unwanted places, the need for leaving some moisture in the product (although this can be quite small), and the need for drying at a temperature at least as high as the condensation temperature of steam (100°C at atmospheric pressure). This temperature limitation can be helped for thermally sensitive substances by operating with superheated steam under partial vacuum, which is a form of vacuum drying.

Lazar (1972) carried out experiments on partial dehydration of potato and carrot cubes with superheated steam while at the same time steam blanching. Steam temperatures up to 143°C could be used without visible product damage.

Temperatures of 120–260°C are commonly used for the inlet air in spray dryers; consequently it is also logical to consider superheated steam as a replacement for air in spray dryers. Trommelen and Crosby (1970) did extensive experiments on drying of single drops of water, sucrose solutions, tomato juice, coffee concentrate and skim milk, as well as a number of other substances, using both air and superheated steam. They monitored temperature, weight loss and appearance, finding a complex interaction of a number of effects. At vapour-phase temperatures of 150°C drying rates in air were substantially faster than those in steam at the same volumetric velocity except for milk. At vapour-phase temperatures of 250°C the rates were close to one another, with the rate in steam generally being somewhat faster than in air at the same volumetric velocity. Thermal degradation effects, judged visually, were similar for the two heating media. The more rapid rates with steam for milk, and with steam for most substances at 250°C, were attributed to the formation of a skin at the surface of the drop which is more pliable with steam than with air. This more pliable skin allowed more inflation, bursting, and rupture of droplets in the case of steam-heating. These effects, in turn, serve to increase the interfacial area for heat transfer and serve to delay the development of strong moisture gradients within the drop, thereby prolonging the constant-rate period and accelerating drying.

One would expect the greater tendencies toward drop inflation and rupture with steam-heating to promote dispersability of the product in water. At the same time, however, the development of a less-rigid and wetter surface layer on the drops with steam should lose much or all of the aroma-selective action of the low-moisture surface layers which can develop in rapid spray drying of concentrated solutions with air (Menting *et al.*, 1970). It appears that conditions giving high retention of volatiles for spray drying with air should give much lower retention with steam, although this is not confirmed.

Pressurised freezing

Haas and Prescott (1972), Haas *et al.* (1972) and Haas (1971) have recently reported that pressurised freezing before drying is an effective means of avoiding collapse, shrinkage and slow rates during air dying, and can also improve freeze-drying characteristics in some cases. The beneficial action is specific to the use of certain gases, and requires that the sequence of steps be pressurisation (typically to 1000 psig), then freezing, then depressurisation, and then thawing before or during drying. The expansion effect occurs even after storage for a few days in the thawed state before air drying, but it does not occur to a significant extent if the freezing step is omitted between pressurisation and depressurisation, or if the pressurisation step is omitted.

Haas and co-workers report that microscope sections of pressure-frozen tissues and foods are opaque, suggesting the presence of very fine gas bubbles. Other symptoms, such as sponginess and slightly expanded size, also suggest the presence of gas. Thawed, previously pressure-frozen specimens show the presence of large gas bubbles with diameters comparable to cell dimensions. Opacity and bubbles after thawing are not observed in pressure-frozen samples for which the shrinkage-retarding effect does not occur, or if freezing is omitted. After observing the degree of effectiveness of different gases, Haas and co-workers conclude that gases which do not give the effect are too soluble and/or too diffusive. Carbon dioxide pressurisation before freezing gave an adverse effect of actually increasing shrivelling during air drying. Microscopic examination of thawing for vegetable products revealed that gas bubble growth appeared to displace liquid from cells (Haas and Prescott, 1972).

The following mechanism extends somewhat the qualitative mechanism put forward by Haas and Prescott (1972) and places it on a quantitative footing. The basis for the mechanism and comparison of it with observed results are discussed more fully elsewhere (King, 1974).

(1) Upon pressurisation, the substance becomes saturated with the gas at the high pressure. To a good approximation, the amount of gas absorbed can be judged from the solubility of the gas in water. This is given by the value of the Bunsen coefficient at the high pressure, β_H (volume of gas, measured at 0°C and 1 atm, dissolved in one volume of water).

(2) Upon freezing, the presence of ice crystals within cells furnishes nucleation sites (probably at crystal-surface irregularities) for the formation of gas bubbles. Since freezing selectively removes water, the liquid phase becomes supersaturated, and therefore the gas bubbles can form and grow. The amount of gas so liberated into bubbles will be in direct proportion to the amount of gas present, β_H.

(3) The ice crystals in a sufficiently frozen substance provide a rigid, impermeable structure so that there is no path for gas to diffuse out to or through the cell walls. This continues to be true upon depressurisation, when the rigid ice structure also keeps the bubbles from expanding.

(4) Thawing removes the ice structure and allows the bubbles to increase in size, in accordance with Boyle's law. If the initial pressurisation was to 1000 psig, the bubbles will grow by about a factor of 70 as thawing occurs at atmospheric pressure.

(5) The growth of the bubbles creates a positive pressure within the cell, which causes a substantial amount of cell water to be displaced outward through the cell wall by reverse osmosis. Most likely, only a fraction of the gas bubbles accomplish this effect; other gas will leak directly out of the substance.

(6) Considering the above steps, β_H should have a value that is comparable to unity in order for the desired amount of cell water displacement to occur. This follows since β_H is the amount of gas dissolved at high pressure, measured as volume when released to atmospheric pressure. On the basis of measurements reported by Haas *et al.* (1972) and physical reasoning, β_H should have a value of approximately 1.5 or better if the full effect is to be observed. Thus for adequate gas solubility at the high pressure,

we can say that a value of β_H of 1.5 or greater should have the potential of giving the full expansion effect, while the potential degree of expansion from β_H less than 1.5 should be $\beta_H/1.5$.

(7) No expansion occurs if the freezing step is omitted, because there is no nucleating agent for gas bubbles within the cells. The cells become supersaturated with gas upon depressurisation, but the gas remains in solution. Most of the gas then diffuses out of the cells.

(8) Once thawing has been completed and displacement of liquid from the cells by gas bubble growth has ceased, the bubbles within the cells are filled with the gas used for pressurising. There is now an opportunity for diffuse gas exchange with the atmosphere through the cell wall membranes. The rate of loss of the gas and the rate of counterdiffusion of air inwards to the bubble within the cell are characterised by the permeability of the cell wall and the internal solution to the gas and to air, respectively. The permeability in such a situation is given by the product of the solubility of the gas and the diffusivity of the gas when dissolved in the liquid. These, in turn, may to a good approximation be taken as the solubility and diffusivity in water. The pertinent solubility is that at atmospheric pressure, which may be expressed through the atmospheric-pressure Bunsen coefficient, β.

If there is no appreciable pressure differential across the cell membrane, the partial-pressure-difference driving forces for the gas exiting and for air entering will be the same. In such a situation it is easily shown that the ratio of the inward molar flux of air to the outward molar flux of the gas will be given by the ratio of the permeability for air to the permeability for the gas. If the permeability for air is *greater* than that for the gas, a pressure difference will develop which will preclude the bubbles from growing much in size. On the other hand, if the permeability for air is *less* than that for the gas, the bubbles will shrink and the expansion effect will be reduced in proportion to the ratio of permeabilities. Thus the full expansion effect can potentially be obtained if βD for the gas is less than βD for air, but the potential expansion effect is reduced by a factor $(\beta D)_{air}/(\beta D)_{gas}$ if βD for the gas is greater than that for air. Thus the amount of expansion effect, in comparison to the maximum possible effect, should be given by the product of (1) $\beta_H/1.5$, or 1 if $\beta_H \geq 1.5$, and (2) $\beta D_{air}/\beta D_{gas}$, or 1 if $\beta D_{gas} \leq \beta D_{air}$.

Table IX shows the various gases considered by Haas and Prescott (1972), along with the pressure P_H to which each was raised before freezing. Also shown are values of β, β_H and D, taken from various sources (King, 1974). β_H is at 0°C, and β and D are at 25°C. The expansion criterion, obtained by the rule just developed, for each gas is shown in the right-hand column. Haas and Prescott divided the gases into three categories—effective, less-effective and ineffective—based upon their success as a promoter of expansion in pressure freezing-air drying. Notice that, with the

TABLE IX

BUNSEN COEFFICIENTS AND DIFFUSIVITIES IN WATER OF VARIOUS GASES

Gas	*Bunsen coefficient at* 1 atm and 25°C (β)	*Pressure employed* psig P_H	*Bunsen coefficient at* P_H β_H	10^5 × *diffusivity* (cm^2/s) D	$D\beta$	*Expansion criterion* (*see text*)
Best						
Methane	0.030	1 000	3.8	1.8	0.054	0.76
Nitrogen	0.014	1 000	1.6	2.3	0.033	1.00
Carbon monoxide	0.021	1 000	2.4	2.2	0.045	0.91
Air	0.017	1 000	2.0	2.4	0.041	1.00
Freon-13	0.019	450	0.69	1.4	0.026	0.46
Ethane	0.041	540	2.4	1.44	0.059	0.69
Less effective						
Neon	0.010	1 000	0.83	3.0	0.030	0.55
Argon	0.032	1 000	3.9	2.1	0.067	0.61
Ethylene	0.108	840	9.9	1.9	0.20	0.20
Propane	0.030	110	0.63	1.2	0.035	0.42
Freon-12	0.059	70	0.40	1.2	0.071	0.15
Freon-115	0.013	102	0.082	1.1	0.014	0.06
Ineffective						
Carbon dioxide	0.759	25–800	4.6–50	2.0	1.48	0.03
Nitrous oxide	0.54	745	66	1.7	0.91	0.04
Helium	0.008 7	1 000	0.54	6.3	0.055	0.43
n-Butane	0.028	16	0.095	1.0	0.028	0.06
iso-Butane	0.017	30	0.068	1.1	0.019	0.05
Freon-318	Very low	25	Very low	0.9	Low	Very low
Ethylene oxide	195	7	Very high	1.7	Very high	Very low

exceptions of helium and Freon-13, the gases in the effective category have expansion criteria of 0.69–1.0, those in the less effective category have expansion criteria of 0.06–0.61, and those in the ineffective category have expansion criteria of 0.00–0.06. This offers good confirmation of the proposed quantitative mechanism.

Helium showed poorer expansion capabilities than predicted by the mechanism. Possible explanations include the small enough molecular size of helium allowing it to escape from the frozen matrix or giving a membrane diffusivity which bears an even greater ratio to that for air than is predicted for water solution. Also the solubility data for Freon-13 are not well established.

Finally, the adverse effect of CO_2 on product shrivelling may correspond to β_H for CO_2 being so much greater than unity. The growing gas bubbles upon thawing could cause substantial rupturing of cell walls.

Use of penetrating radiation

Radiation in the radio-frequency and microwave wavelength regions, and infra-red radiation to a lesser extent, have the property of releasing heat directly within a material that is being dried. For all other forms of heat delivery, it is the piece surface that is heated, and heat then travels to the interior by means of a conduction process. Dehydration with radio-frequency and microwave energy has been thoroughly reviewed recently (Decareau, 1970; Goldblith, 1973; Goldblith and Decareau, 1973).

Microwave or radio-frequency energy is expensive; Huxsoll and Morgan (1968) estimate that the cost of delivered microwave energy for food dehydration is at least 20 times that from combustion sources, per calorie. Consequently, the prospective applications of microwave or radio-frequency dehydration are for situations where a product quality can be achieved that is not attainable with less expensive forms of drying, or where rates of drying can be very greatly accelerated. Applications which can achieve goals of this sort while using microwave energy for the removal of only a small portion of the total water are attractive.

Benefits which can result from using penetrating radiation for evaporative drying include a puffing effect from the sudden vaporisation which results from intensive release of energy within the food substance, and more uniform heating which avoids such effects as case-hardening. The application of microwave energy for finish-drying of potato chips has been well established (Decareau, 1970), and results from the desirability for consumer appeal of avoiding non-enzymatic browning. This is overcome by drying with microwaves at maximum temperatures of only 210°F, as opposed to those of 325–350°F which are required in a potato-chip fryer. Since microwaves are used only for finish-drying, the moisture removal is small and the energy costs are thereby kept low. Huxsoll and Morgan (1968) have demonstrated the utility of a short but intense treatment with microwave energy mid-way along in air drying. The benefit is one of puffing, which accelerates subsequent air drying and improves product rehydration and appearance.

Research on microwave freeze-drying is reviewed by Decareau (1970) and King (1970). Direct, internal heat generation would be most desirable for freeze-drying, but several complications arise, among them being the relatively low absorptive power of ice for microwave energy.

Ginzburg (1969) has presented a large amount of information on the use of infra-red radiation for food dehydration. This form of energy is cheaper than microwave or radio-frequency energy, but has less penetrating power. Penetration depths for various foods range from 1 to 12 mm. Thus infra-red heating can be an effective form of uniform heat delivery for relatively thin substances. Microwave energy, on the other hand, has quite large penetration depths and thus probably finds its greatest incentive, in comparison to other forms of heating, for materials that are very thick.

Use of sonic and ultrasonic energy

A number of studies have been made of the use of sonic and ultrasonic energy to accelerate rates of air-drying and freeze-drying. The results of these studies are conflicting; this appears to come in part from the difficulty of achieving close control of experimental conditions.

It is clear that sonic energy can accelerate drying in the constant-rate period considerably (Boucher, 1961; Bartolome *et al.* 1969). Under these conditions the rate-limiting resistances to heat and/or mass transfer lie outside the substance being dried. If convective mass transfer or convective heat transfer controls, then sonic energy can accelerate the rate by the introduction of acoustic streaming or turbulence into the boundary layer adjacent to the substance (Fogler, 1971). External resistances prevail more and the constant-rate period contributes a greater fraction of the total drying time for substances of very high surface-to-volume ratio, e.g. fibrous or powdered materials such as textiles and pulverised coal. Substantial increases in drying rate with sonic energy have been found for such substances (Boucher, 1961; Purdy *et al.*, 1971; Wilson *et al.*, 1971).

Food materials are generally much more dominated by internal mass or heat transfer resistances, and hence exhibit drying times largely or completely controlled by the falling-rate period. Boucher (1961) has pointed out that it is much more difficult to achieve an effect of sonic energy upon internal mass transfer because of the impedance mismatch in transmitting sonic energy across a gas–solid interface. Nonetheless, mechanisms have been suggested, as reviewed by Fogler (1971), for the effects of sonic energy in accelerating internal mass transfer. These include possible increased diffusivities, increased mobility of liquid within capillaries, and formation of bubbles which could expel liquid. To these should be added any retarding effect that sonic energy has upon the collapse of pores.

An apparent direct conflict occurs between the results of Bartolome *et al.* (1969) and those of Carlson *et al.* (1972) for the effect of sonic energy from non-pneumatic generators upon air-drying rates for pieces of potato and apple. Bartolome *et al.* found significant increases in the rate of drying in the falling-rate period of as much as a factor of two and extending down to low residual moisture contents. Operating under seemingly identical conditions, Carlson *et al.* found no effect at all for drying taking place entirely in the falling-rate period. Conceivable reasons for this dichotomy are the need for very fine tuning of the sonic source and sample, different chamber geometry, or unrecognised secondary effects from the energy of the sound field in the experiments where an increase was observed.

Moy and DiMarco (1970) report increases of 10–100% in rates of freeze-drying of extracts and of ice in a circulating, atmospheric-pressure air stream when sonic energy was used. For ice this is a system with an external mass transfer limit for which increases would be expected, and for the extracts the authors report some flaking off of the dry product, which may explain the higher apparent drying rates. Moy and DiMarco (1972) also examined the influence of ultrasonic energy on vacuum freeze-drying of beef, with the generator directly coupled to the specimen. Increases in drying rate were found to be negligible or very small.

It appears that sonic and ultrasonic energy is of less use in food drying than it is for the dehydration of substances which are less limited by internal resistances to mass and heat transfer. Further, any consideration of the use of sonic energy should take into account the operating and amortised cost of energy so applied, and compare that with rate increases achievable by expending more energy in other ways.

Compressed foods

Compression of dried foods has been investigated for a number of years by the U.S. Army Natick Laboratories with the aim of reducing the storage and shipping volume of freeze-dried foods in particular (see, e.g. Rahman *et al.*, 1971). Satisfactory preservation of structure, texture and rehydration ability during compression and storage has been found for a number of freeze-dried foods when compression is carried out from a uniformly distributed residual water content corresponding to equilibrium with 10–50% relative humidity.

In addition to the benefit of reduced volume, compression serves to obstruct access to and eliminate the large internal surface area of freeze-dried foods. This should give improved resistance to lipid oxidation for foods such as meats, and should permit less severe packaging requirements. Further, a compressed, freeze-dried food is considerably less friable than a non-compressed, fully dry product. Finally, there are indications (e.g. Rockland, 1969) that a finite residual moisture content is preferable to complete dryness for storage stability in many cases.

Processing approaches for dehydration leading to compressed foods are currently being investigated in both our own laboratory and the Western Regional Laboratory of the U.S.D.A. At least one commercial processor (Hirschberg, 1973) is also actively engaged in this area.

REFERENCES

Alger, J. R. M. and Hays, C. V. (1964). *Creative Synthesis in Design*, Prentice-Hall, Englewood Cliffs, N.J., pp. 39–42.
Anderson, T. F. (1951). *Trans. New York Acad. Sci.*, **13**, 130.
Anon. (1968). *Chem. Eng.*, 104–106, May 6.
Armerding, G. F. (1966). *Adv. Food Research*, **15**, 305.
Barnett, S. (1973). *AIChE Symp. Ser.*, **69**(132), 26.
Bartolome, L. G., Hoff, J. E. and Purdy, K. R. (1969). *Food Technol.*, **23**, 321.
Bellows, R. J. and King, C. J. (1972). *Cryobiology*, **9**, 559.
Bellows, R. J. and King C. J. (1973). *AIChE Symp. Ser.*, **69**(132), 33.
Binning, R. C., Lee, R. J., Jennings, J. F. and Martin, E. C. (1961). *Ind. Eng. Chem.*, **53**, 45.
Bohrer, B. (1967). U.S. Patent No. 3,298,109.
Bolin, H. R. and Salunkhe, D. K. (1971). *J. Food Sci.*, **36**, 665.
Boucher, R. M. G. (1961). *Chem. Eng.*, 97, October 2.
Brown, G. E., Farkas, D. F. and DeMarchena, E. S. (1972). *Food Technol.*, **26**(12), 23.
Camirand, W. H., Forrey, R. R., Popper, K., Boyle, F. P. and Stanley, W. L. (1968). *J. Sci. Food Agric.*, **19**, 472.
Carlson, R. A., Farkas, D. F. and Curtis, R. M. (1972). *J. Food Sci.*, **37**, 793.
Chandrasekaran, S. K. and King, C. J. (1971). *Chem. Eng. Prog. Symp. Ser.*, **67**(108), 122.
Chandrasekaran, S. K. and King, C. J. (1972). *AIChE J.*, **18**, 520.
Chandrasekaran, S. K. and King, C. J. (1973). U.S. Patent No. 3,716,382. Assigned to U.S. Secretary of Agriculture.
Clark, J. P., Ph.D. (1968). Dissertation in Chemical Engineering, University of California, Berkeley.
Cohen, A. L., Marlow, D. P. and Garner, G. E. (1968). *J. Microscopie*, **7**, 331.
Decareau, R. V. (1970). *CRC Critical Rev. Food Technol.*, **1**, 199.
Dorsey, W. R., Roberts, R. L. and Strashun, S. I. (1968). Canadian Patent No. 777–749.
Eskew, R. K., Cording, J. Jr. and Sullivan, J. F. (1963). *Food Eng.*, **35**(4), 91.
Ettrup Petersen, E., Lorentzen, J. and Føsbol, P. (1970). A/S Atlas Symposium on Freeze-Drying, Copenhagen, August.
Farkas, D. F. and Lazar, M. E. (1969). *Food Technol.*, **23**, 688.

Farkas, D. F., Lazar, M. E. and Butterworth, T. A. (1969). *Food Technol.*, **23**, 1457.
Feberwee, A. and Evers, G. H. (1970). *Lebensm.-Wiss. Technol.*, **3**(2), 41.
Flink, J. M. (1973). *AIChE Symp. Ser.*, **69**(132), 63.
Fogler, H. S. (1971). *Chem. Eng. Prog. Symp. Ser.*, **67**(109), 1.
Freeman, R. R., Auro, M. A., Dashiell, T. R., Murphy, J. E., Oshrine, I. and Smith, R. F. (1957). *Chem. Eng. Prog.*, **53**, 590.
Fulford, G. D. (1969). *Canad. J. Chem. Eng.*, **47**, 378.
Ginzburg, A. S. (1969). *Applications of Infra-Red Radiation in Food Processing* (Engl. trans.), Chemical Rubber Press, Cleveland.
Goldblith, S. A. (1973). Pres. at 6th Intl Course on Freeze-Drying and Adv. Food Technol., Int. Inst. Refrig., Comm. Cl, Bürgenstock, Switzerland, June.
Goldblith, S. A. and Decareau, R. V. (1973). *An Annotated Bibliography on Microwaves, their Properties, Production and Applications to Food Processing*, M.I.T. Press, Cambridge, Mass.
Groth, W. (1967), *Chem.-Ing. Techn.*, **39**, 53.
Haas, G. J. (1971). *Food Eng.*, 58, November.
Haas, G. J. and Prescott, H. E., Jr. (1972). *Cryobiology*, **9**, 101.
Haas, G. J., Prescott, H. E., Jr. and D'Intino, J. (1972). *J. Food Sci.*, **37**, 430.
Hausbrand, E. (1908). *Das Trocknen mit Luft und Dampf*, Springer Verlag, Berlin.
Hertzendorf, M. S. (1967). *ASHRAE J.*, **7**(2), 65.
Hertzendorf, M. S. and Moshy, R. J. (1970). *CRC Critical Rev, Food Technol.*, **1**, 25.
Hieu, T. C. and Schwartzberg, H. G. (1973). *AIChE Symp. Ser.*, **69**(132), 70.
Hirschberg, E. (1973). Personal communication, South San Francisco, California.
Holdsworth, S. D. (1971). *J. Food Technol.*, **6**, 331.
Huang, C. P., Fennema, O. and Powrie, W. D. (1965). *Cryobiology*, **2**, 109 *ibid.*, **2**, 240 (1966).
Huxsoll, C. C. and Morgan, A. I., Jr. (1968). *Food Technol.*, **22**, 705.
Karel, M. (1973). *CRC Critical Rev. Food Technol.*, **4**, 329.
Kennedy, T. J., Monge, L. E., McCoy, B. J. and Merson, R. L. (1973). *AIChE Symp. Ser.*, **69**(132), 81.
Kerkhof, P. J. A. M. and Thijssen, H. A. C. (1973). Personal communication and unpublished manuscript, Eindhoven University of Technology, Eindhoven, Netherlands.
King, C. J. (1968). *Food Technol.*, **22**, 165.
King, C. J. (1970). *CRC Critical Rev. Food Technol.*, **1**, 379 (1970); also republished as 'Freeze-Drying of Foods', Chemical Rubber Co., Cleveland, 1971.
King, C. J. (1971). *Separation Processes*, McGraw-Hill, New York.
King, C. J. (1974). *Cryobiology*, **11**, 121.
Kluge, G. and Heiss, R. (1967). *Verfahrenstechnik*, **1**, 251.
Koller, T. and Bernard, W. (1964). *J. Microscopie*, **3**, 589.
Krischer, O. (1963). *Trocknungstechnik. I. Die Wissenschaftlichen Grundlagen der Trocknungstechnik*, 2nd ed., Springer Verlag, Berlin.

Krischer, O. and K. Kröll. (1959). *Trocknungstechnik. II. Trockner und Trocknungsverfahren*, Springer Verlag, Berlin.
Kröll, K. (1964). *Aufbereitungstechnik*, **4**, 287.
Lazar, M. E. (1972). *J. Food Sci.*, **37**, 163.
Lazar, M. E. and Farkas, D. F. (1971). *J. Food Sci.*, **36**, 315.
Levin, E. and Finn, R. K. (1955). *Chem. Eng. Prog.*, **51**, 223.
Li, N. N. (1971). *Ind. Eng. Chem. Proc. Des. Devel.*, **10**, 215.
Loncin, M. (1969). *Die Grundlagen der Verfahrenstechnik in der Lebensmittelindustrie*, Verlag Sauerlander, Aarau, Switzerland, 1969.
Loncin, M. (1973). Pres. at 6th Intl Course on Freeze-Drying and Adv. Food Technol., Int. Inst. Refrig., Comm. Cl, Bürgenstock, Switzerland, June.
Loncin, M., Bimbenet, J. J. and Lenges, J. (1968). *J. Food Technol.* **3**, 131.
Lonsdale, H. K. (1970). 'Separation and purification by reverse osmosis', in *Progress in Separation and Purification*, vol. 3 (Eds. E. S. Perry and C. J. van Oss), Interscience, New York.
Lowe, C. M. and King, C. J. (1974). *J. Food Sci.*, **39**, 248.
Lowe, E. and Durkee, E. L. (1971). *J. Food Sci.*, **36**, 31.
MacDonald, J. O. S. (1973). *Process Technol. Intl.*, **18**, 203.
Marshall, W. R., Jr. (1963). 'Drying', in *Kirk-Othmer Encyclopedia of Chemical Technology*, 2nd ed., vol. 7 (Ed. A. Standen), Interscience, New York, 1963, pp. 326–378.
Massaldi, H. A. and King, C. J. (1973). Pres. at 6th Intl Course on Freeze-Drying and Adv. Food Technol., Int. Inst. Refrig., Comm. Cl, Bürgenstock, Switzerland, June.
Menting, L. C., Hoogstad, B. and Thijssen, H. A. C. (1970). *J. Food Technol.*, **5**, 127.
Merson, R. L. and Ginnette, L. F. (1972). 'Reverse osmosis in the Food Industry', in *Industrial Processing with Membranes* (Ed. R. E. Lacey and S. Loeb), Wiley, New York.
Merson, R. L. and Morgan, A. I., Jr. (1968). *Food Technol.*, **22**, 631.
Michaels, A. S., Baddour, R. F., Bixler, H. J. and Choo, C. Y. (1962). *Ind. Eng. Chem. Process Des. Devel.*, **1**, 14 (1962).
Michaels, A. S. and Bixler, H. J. (1968). 'Membrane permeation', in *Progress in Separation and Purification*, vol. 1 (Ed. E. S. Perry), Interscience, New York.
Monge, L. E., McCoy, B. J. and Merson, R. L. (1973). *J. Food Sci.*, **38**, 633.
Morgan, A. I., Jr. (1967). *Food Technol.*, **21**, 1353.
Moy, J. H. and DiMarco, G. R. (1970). *J. Food Sci.*, **35**, 811.
Moy, J. H. and DiMarco, G. R. (1972). *Trans. ASAE*, **15**, 373.
Muller, J. G. (1967). *Food Technol.*, **21**, 49.
Norris, K. W. (1963). 'The morphological approach to engineering design', in *Conference on Design Methods* (Ed. J. C. Jones and D. G. Thornley), Macmillan, New York.
Omran, A. M. (1972). Ph.D. Dissertation in Chemical Engineering, University of California, Berkeley.
Omran, A. M. and King, C. J. (1973). Pres. at AIChE Natl Mtg, New Orleans, March 1973; *AIChE J.*, in press.
Peleg, M. and Mannheim, C. H. (1970). *J. Food Sci.*, **35**, 649.

Ponting, J. D., Watters, G. G., Forrey, R. R., Jackson, R. and Stanley, W. L. (1966). *Food Technol.*, **20**, 1365.

Prausnitz, J. M. (1969). *Molecular Thermodynamics of Fluid-Phase Equilibria*, Prentice-Hall, Englewood Cliffs, N.J.

Purdy, K. R., Simmons, G. W., Hribar, A. E. and Griggs, E. I. (1971). *Chem. Eng. Prog. Symp. Ser.*, **67**(109), 55.

Rahman, A. R., Henning, W. L. and Westcott, D. E. (1971). *J. Food Sci.*, **36**, 500.

Riedel, L. (1949). *Z. Lebensm. Untersuch. Forsch.*, **89**, 298.

Rockland, L. B. (1969). *Food Technol.*, **23**, 1241.

Sandall, O. C., King, C. J. and Wilke, C. R. (1967). *AIChE J.*, **13**, 428.

Song, P. S., Chichester, C. O. and Stadtman, F. H. (1966). *J. Food Sci.*, **31**, 906; Song, P. S. and Chichester, C. O. (1966) *ibid.* **31**, 914; (1967) **32**, 98, 107

Spiess, W. E. L. and Wien, H. J. (1973). Pres. at 6th Intl Course on Freeze-Drying and Adv. Food Technol., Intl Inst. Refrig., Comm. Cl., Bürgenstock, Switzerland, June.

Spiess, W. E. L., Wolf, W., Buttmi, W. and Jung, C. (1973). *Chem.-Ing. Techn.*, **45**, 498.

Sutherland, K. S. (1972). *Brit. Chem. Eng. Proc. Technol. Intl*, **17**, 55 (1972).

Thijssen, H. A. C. (1969). *Dechema Monographien*, **63**, 153.

Thijssen, H. A. C. (1970). *J. Food Technol.*, **5**, 211.

Thijssen, H. A. C. (1973). Pres. at 6th Intl Course on Freeze-Drying and Adv. Food Technol., Intl Inst. Refrig., Comm. Cl., Bürgenstock, Switzerland, June.

Thijssen, H. A. C. and Paardekooper, E. J. C. (1968). U.S. Patent No. 3,367,787.

Thijssen, H. A. C. and Rulkens, W. H. (1968). *De Ingenieur*, **80**(47), CH45.

Tressler, D. K. and Joslyn, M. A. (1971). *Fruit and Vegetable Juice Processing Technology*, 2nd edn, AVI Publ. Co., Westport, Conn.

Trommelen, A. M. and Crosby, E. J. (1970). *AIChE J.*, **16**, 857.

Van Arsdel, W. B., Copley, M. J. and Morgan, A. I., Jr. (1973). *Food Dehydration*, 2nd edn, AVI Publ. Co., Westport, Conn.

van Pelt, W. H. J. M. (1973). Pres. at 6th Intl Course on Freeze-Drying and Adv. Food Technol., Intl Inst. Refrig., Comm. Cl., Bürgenstock, Switzerland, June.

Vanecek, V., Markvart, M. and Drobohlav, R. (1966). *Fluidized Bed Drying*, Engl. Trans., Leonard Hill, London.

Vorstman, M. A. G. and Thijssen, H. A. C. (1972). *De Ingenieur*, **84**, (45), CH65.

Walker, W. H., Lewis, W. K., McAdams, W. H. and Gilliland, E. R. (1937). *Principles of Chemical Engineering*, 4th edn, McGraw-Hill, New York, p. 635.

Warman, K. G. and Reichel, A. J. (1970). *The Chemical Engineer*, CE134, May.

Werezak, G. N. (1969). *Chem. Eng. Prog. Symp. Ser.*, **65**(91), 6.

Wilson, J. S., Moore, A. S. and Bowie, W. S. (1971). *Chem. Eng. Prog. Symp. Ser.*, **67**(109), 68.

Wistreich, H. E. and Blake, J. A. (1967). *Science* **138**, 138.
Yoshida, T. and Hyodo, T. (1970). *Ind. Eng. Chem. Proc. Des. Devel.*, **9**, 207.
Yuan, S. and Schwartzberg, H. G. (1972). *AIChE Symp. Ser.*, **68**(120), 41.
Zwicky, F. (1948). In *Courant*, Anniversary Volume, Interscience, New York.

DISCUSSION

D. Simatos (Dijon University, France): I would like to report our work on the effect of dry matter content on the aroma retention in freeze-drying.

We measured the head space volatiles content after freeze-drying orange and raspberry juices with different dry matter contents. For some volatiles we observed higher retention in the more concentrated juice; but for others the result was the opposite.

Recently we tested the effect of dry matter content on flavour by freeze-drying orange juice with different dry matter contents (from about 12 to 48%) after grinding in the frozen state. Freeze-drying was performed under what we call optimal conditions: that is, temperature and pressure were adjusted in order to give the shortest drying time without apparent collapse. We compared, through sensory analysis, the frozen and the freeze-dried juice and looked for differences in flavour. Our test panel noticed no difference between frozen and freeze-dried juice with the different initial dry matter contents.

We can give some idea of the sensitivity of the sensory testing technique by considering the effect of other processing parameters. For example, we have a highly significant difference between juices which have been dried by short or long processes. The flavour score is much higher for the juice which has been dried in the shorter time—without apparent collapse.

These experiments are not yet quite complete, but it seems that initial dry matter content is not of prime importance in flavour retention in freeze-drying fruit juices. So we can choose the optimal dry matter content determined by other criteria.

A. S. Michaels (Alza Research, U.S.A.): First of all, perstraction really is a special case of dialysis—the difference being that the liquid phases on the two sides of the membrane are different, but in dialysis they're the same. The principle is a very intriguing one, particularly in the context of the use of relatively high molecular weight extracting liquids, e.g. polyethylene glycols, which cannot penetrate either artificial membranes or cell membranes.

I was surprised when you described the use of supported membranes for this particular purpose, when it is obvious there is no hydrostatic pressure difference across a membrane in perstraction: so there's no reason why you can't use a pretty flimsy system. I am prompted to wonder what would happen if we took conventional cellulose dialysis tubing—flattened it out, wrapped it around some sort of a drum and pumped fluid through it immersed in a polyglycol bath.

Another aspect of the extracting fluid which intrigues me is that polyethers have inverted temperature coefficients of water solubility: that is to say, water is very soluble in them at low temperatures, in the neighbourhood of 0–10 °C. But, if you heat them up to 40–50°, the water splits up as a separate liquid phase. So here is an interesting case where you could effect a phase separation by simple decantation without having the necessity of vaporisation as a means of getting the water out of the system.

J. D. Mellor (C.S.I.R.O. Australia): The question I want to ask is: Is your slush dehydration process a refinement of the standard rotary laboratory preconcentrator—when you are able to draw a vacuum on it as the glass evaporator turns round slowly and, if you adjust the vacuum, you are able to freeze-evaporate a portion of the liquor, and then finally, if you draw a better vacuum you can do some freeze-drying?

King: I think probably that it is. But there are a couple of differences here. One is, you want to stop at a degree of frozenness whereby you're definitely not at freeze-drying—you're at a higher temperature. Secondly, for foods, you would want to carry out a preliminary freezing in a way that was more protective of aromas than is evaporative freezing.

A. Spicer (Scientific Adviser to RHM, U.K.): I should like clarification of the statement you made right at the beginning of your paper. You mentioned that one of the reasons for going into freeze-drying was the attention to low temperature processing. Could we accept, as a result of the discussion we have had during these last few days, that there is a temperature time coefficient—that high temperature, low residence time, could be equivalent to low temperature, long residence time—and that an emphasis on low temperatures alone could be wrong? Could I have a clarification on this point?

King: I think the answer to your question is one of relative activation energies and processes. The emphasis *should* be on low temperatures completely—if the deteriorative reactions you are talking about have higher activation energies than does the water removal. In a plot of the logarithms for rate coefficient for some kinetic process versus reciprocal temperature, a high energy reaction will have a high slope on such a plot, and a low energy reaction will have a low slope. What is really of interest is the relative rates of the two processes. If water removal is a low activation energy process and deterioration is a high activation energy process, then you want to go to low temperatures where a ratio of deteriorative reaction to water removal rate will be the lowest.

M. Karel (M.I.T., U.S.A.): I agree entirely on the plot; but I would like to add some additional data on that map.

If we are talking about dehydration of proteins, then the 'activation energy' for it can be as high as 100 kcal/mol.

Next, let's say we go down to browning reactions. If they are primarily sugar-based, rather than involving lipid-originating carbonyls—we are talking now about 20–50 kcal/mol with a strong dependence of the activation energy on moisture content. Water evaporation is at 10 kcal/mol and if we talk of deterioration due to surface retention effects that can be very low indeed.

A. L. Moeller (Ecal Nateco, AB, Sweden): My question concerns the pressure in dry freezing and air drying. Is this a process which has been studied in reality with different materials, or not? Freezing and thawing are used for opening up micro-organisms to release their contents, e.g. enzymes: and with over-pressure just a small rupture would let the whole contents out and no drying could take place.

King: I have never carried out this process, but find it an interesting one, and I think Mr. Barnett would like to comment as he is a representative of the company that has produced this.

S. Barnett (General Foods Inc., U.S.A.): First of all, as co-author with Dr. Prescott on work on the latest research on pressure air freeze-drying, I'd like to thank Dr. King for his analytical study of our rather empirical work. But to answer Dr. Moeller's question, we have run a wide variety of foodstuffs through this: ones that come to mind are celery, carrots and strawberries. In all cases, the dried material closely resembled freeze-dried—there was no rupture. I should go on to say that I recently studied the economics of the process; and we can take the range of drying costs from air drying at 1 c/lb of water removed to freeze drying at 10 c/lb. Then a gross projection (from bench scale to commercial scale of the process) would be in the order of 4–5 c/lb of water removed. So it is certainly not a cheap process.

E. Seltzer (Rutgers University): Many years ago, Mr. Sherwood of M.I.T. tried to help the leather people by initiating a dehydration method that would not shrink the surface and thereby make the leather harder to dry. What he did was to spread kieselguhr on the leather and then build it in layers, which kept the surface of the leather just as though it were an interior piece; therefore the capillaries did not shrink, and it gave good uniform drying. I wonder, is that still prevalent?

Karel (M.I.T., U.S.A.): I would like to add to your osmotic drying work. We did osmotic drying on some large pieces of water melon prior to freeze-drying to try to improve the aroma retention and palatability. In these conditions we had a fairly short osmotic drying period to prevent sugar entering because the diffusion could effect the rate limit rather than the equilibrium limit. We had quite an improvement in aroma after the complete cycle for osmotic drying and freeze-drying. So not too much, apparently, is lost—at least not from this—into the osmotic process.

F. Bramsnaes (Food Technology Laboratory, Denmark): Have you examined the influence of the freezing rate in the slush freezing method?

King: Flink and others from the Western Regional Lab. had some success in preserving cell viability by extremely slow freezing rather than the very rapid freezing usually employed. As far as I can see for the slush evaporation, the freezing rate is going to be of importance only as it determines the size of the ice crystals which you get. We have not noticed over a reasonable range of ice crystal sizes any particular effect. I can think of ways in which it might have an effect: it might have to do with the extent to which the surface dries out preferentially to the bulk and thereby exerts this aroma loss limiting action.

Cyclic-pressure Freeze-drying in Practice

J. D. MELLOR

Division of Food Research, C.S.I.R.O.,
North Ryde, N.S.W., Australia

INTRODUCTION

Freeze-drying is a process in which ice is sublimed as water vapour from frozen materials, usually under vacuum. Sublimation occurs by heating the frozen products, and the water vapour released diffuses out through the drying layer formed around the frozen core, to the vacuum space, where it is condensed at a temperature of about −40°C.

In freeze-drying a constant vacuum pressure is maintained below 1 torr, at which pressure ice may be made to sublime at −20°C when heated by radiation. A disadvantage of freeze-drying in this way is that shortly after the beginning, the sublimation rate falls appreciably to a steady value. At this stage the sublimation rate could be increased by applying more heat, but the danger of over-heating the surface and thereby denaturing the product sets a limit to this procedure. Also, if the vacuum pressure is further reduced to facilitate vapour flow from the product, the thermal conductivity of many freeze-dried products then becomes too low (e.g. 0.044 W/m K at 0.5 torr for beef) for this to be effective.

To overcome these difficulties, a new method of freeze-drying has been developed (Mellor, 1967), offering substantial improvements in overall efficiency and in drying rate by controlling the vacuum pressure in such a way that the initial sublimation rate tends to be sustained. Briefly, the product is subjected to repeated cyclic changes in vacuum pressure during the course of freeze-drying. Mellor and Middlehurst (1971) have shown that the optimum pressures and their periods of duration depend upon the nature of the product and its permeability, the gas composition and thermal conductivity, and they can be determined either by experiment or by calculation. When the heat transfer in a pressure cycle is considered in terms of an equivalent normal freeze-drying process, they showed that the ratio of the thermal conductivities: effective (pressure cycle)/normal (constant pressure) equals 0.9 times the ratio of the drying times for the two processes.

A high sublimation rate occurs in each pressure cycle as a flash of vapour; the system behaves as if the initial sublimation rate is sustained whilst the pressure is being lowered. The mass of ice sublimed in this part of the cycle is proportional to time, whereas it is proportional to the square root of time in conventional freeze-drying (Mellor, 1966).

INDUSTRIAL PLANT

Description

A drying chamber, 1.4 m in diameter by 3.5 m in length, is reinforced by several steel rings attached to the outside shell. Two of the rings at the ends of the chamber are flanged to seal a domed cover, which is bolted in position, and a door at the front which slides on an overhead rail. A guide on the bottom of the door ensures that an 'O'-ring seal can be seated when the door is in the closing position and prevents it from pinching in the sliding position. Four lever-action catches hold the door tightly to the chamber ready to be evacuated.

Two sets of radiant heating platens located side by side are made from roll-bonded aluminium and anodised to a black matt finish. They are arranged horizontally to interleave between trays of frozen food situated on a trolley in the loading position, and are in two sets of 17 with a narrow gap between them to permit the central vertical frame of the trolley to enter the chamber. Glycol is continuously circulated in the platens for heating. It is pumped through a steam-heated shell-and-tube exchanger and by-pass *via* a three-way valve to control the glycol temperature. The valve is a modulating type fitted with an electric drive which responds to a thermal resistance element attached to one of the platens.

The condensation load in the cyclic-pressure process is extremely variable, being negligible during high pressures and very high during low pressures. Thus, the important considerations in the design of the condenser with a critical spacing between rows of coils and baffles (Mellor and Greenfield, 1972) are to ensure that the latent heat can be absorbed quickly by the refrigeration system and to allow a high moisture vapour flow rate from the product down the length of the coil system. In practice, the design has resulted in a working condenser on which ice builds up uniformly and the high condensation load can be absorbed.

The condenser is housed in the other half of the chamber, and is a flooded-type refrigerant evaporator equipped with a surge drum above the drying chamber. It is defrosted by water cascading down the coils, and is cooled by a compound refrigeration plant comprising a six-cylinder ammonia compressor, evaporative condenser, and a flash-type interstage desuperheater. The compressor is driven by a two-speed electric motor (22 or 36 kW) and is fitted with a mechanical unloading mechanism. The higher speed is generally required for the conventional freeze-drying in order to maintain lower condenser temperatures.

The vacuum pumps, connected to the back of the chamber, include a Roots pump and an oil-seal rotary piston pump, each driven from separate 5.4 kW electric motors. When the chamber is first closed after loading, it takes five to six minutes to reduce the pressure to 20 torr (2.7 kPa), after which the Roots pump reduces the pressure down to 0.2 torr (27 Pa) in a few minutes. The pumps are kept running whilst atmospheric air is alternately admitted through a pair of bleed and solenoid valves which are adjusted for fixed flow rates to give a pressure cycle having a maximum of 20 torr (2.7 kPa) and a minimum of 0.2 torr (27 Pa). Lower pressures are used with helium and this gas is recycled via a reservoir and special oil-free pumps.

Operation

Except for periods of loading and unloading, the plant can be left operating unattended because of a programmed control system which, once it has been preset, will automatically control and reduce the heating and refrigeration requirements during freeze-drying.

The prepared food is spread on trays which are stacked on a trolley for freezing, then quickly transferred into the chamber. The trays, 0.8 m square, are made from aluminium; the bottoms being perforated to allow the water vapour to escape. Granulated products are frozen on plain trays, but liquid products are first frozen in plain trays lined with plastic sheets and afterwards the frozen slabs are transferred to perforated trays for freeze-drying. In the loading position the trays on the trolley are located evenly between the horizontal heating platens at 64-mm centres. The trolley nominally holds 300 kg of food in 32 trays and can be loaded from both sides.

Performance

The results of a typical cyclic process on yoghurt (Fig. 1) show the following advantages:

(1) Reductions in drying time amount to 20 and 36% for cyclic air and cyclic helium pressures respectively.
(2) The spread of sublimation temperatures in a stack of trays is considerably reduced by the pressure cycles—a 3-hour spread for normal drying can be reduced to 1 hour using helium. Usually the temperature fluctuations are about double those in normal drying, yet below the prescribed operating temperature.

(3) Maintaining a residual level of heating for longer in normal drying represents about 9% of the total heating for cyclic-pressure freeze-drying with helium. This additional amount of energy is just about equivalent to the power needed for cycling the pressure.
(4) During the approach to the end point, the drying rate is about double that obtained under conventional freeze-drying.
(5) Lower final moisture contents are possible, especially with helium.
(6) Helium losses due to recycling have been estimated to cost very little. Although the pumping rate decreases by about 30% because of air leakage during the course of drying, the pressure range 0.2–20 torr (27 Pa–2.7 kPa) is not affected.
(7) Apart from savings in fixed costs because of the reduction in drying time, significant savings in running costs are also possible owing to higher condenser temperatures. Estimates, Table I, show a 57% decrease in energy for refrigeration.

END-POINT MEASUREMENT

A long secondary drying stage is characteristic of all freeze-drying. Unless the end-point is determined accurately and reliably, freeze-dried products may retain too much moisture for successful storage at ordinary temperatures, or they may be overdried.

Of all the simple methods for determining the end-point the best seems to be the one using the trapped pressure ratio (TPR). The gas space above the product is sampled on several occasions during the final stages of secondary drying and the partial pressure of water vapour is estimated in terms of a ratio, as an indication of the moisture content of the product. The measurements actually made are the total pressure in the chamber and the partial pressure of air when the water vapour is condensed out; the ratio of this partial pressure to the total pressure is called the TPR.

The product is unloaded when the TPR approaches a constant value in the range 0.6–0.8. The actual value in this range depends on the type of food being freeze-dried, and is defined on the basis of prior determinations of moisture contents in the product at several points in the range.

The apparatus consists of a copper U-tube; one arm is bent at right angles and attached horizontally to a manual vacuum valve which in turn is connected by an elbow to the side of the drying chamber. A Pirani vacuum gauge head (range $10–10^{-3}$ torr (1.3 kPa–0.13 Pa)) is connected to the top of the other arm. The whole assembly is mounted on the outside wall of the drying chamber so that the tube enters at a point in the middle of the drying end and has a short tube on the inside which can be directed close to the product. A Dewar flask, containing a cold refrigerant, e.g. dry ice (CO_2) chips and alcohol, is used for condensing out the water vapour in the gas sample. After closing the valve, the Dewar flask is

TABLE I

ESTIMATED ENERGY REQUIREMENTS FOR AVERAGE LOADS IN FREEZE-DRYING PROCESSES—VACUUM PUMPING AND COOLING

		Drying time		*Vacuum pumps*		*Refrigeration compressor*		
Type of operation		*total* (h)	*decrease* (%)	*energy* (kWh)	*increase* (%)	*condenser temp.* (°C)	*energy* (kWh)	*decrease* (%)
Constant pressure		10.5	—	113	—	−45	454	—
Cyclic pressure	air	8.5	20	145	28	−36	193	57
	helium	6.7	36					

slowly moved up from below the U-tube and clamped in position when the tube is fully immersed.

To operate the apparatus, the absolute pressure (water and air) is first read on the Pirani gauge after closing the valve. The U-tube is then cooled by immersion in the Dewar flask and the gauge is again read when steady. The reading may be corrected for the dry ice temperature if accurate results are required. The TPR is then the ratio of the second to the first reading.

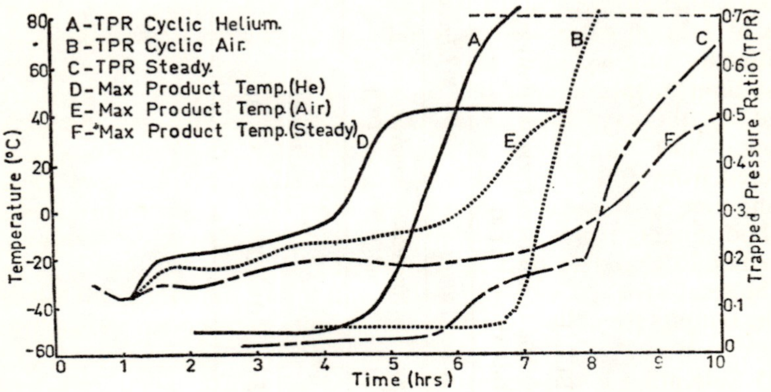

FIG. 1.

After the readings have been taken the manual valve is opened again to receive the next sample and the Dewar flask removed to allow the tube to warm up. Normally about 30 min is allowed to elapse between readings, so that the gas sample in the tube may again come to equilibrium with the gas in the chamber. Typical response curves of TPR against drying time are shown in Fig. 1. It is seen from these curves that the approach to the end-point under cyclic pressures is about twice as fast as under constant pressures.

PRODUCT QUALITY

There is a range of materials which can be added to food to reduce irreversible changes during preparation, freezing, storage, and rehydration. These materials are also of value during freeze-drying for improving the quality of the product (Mellor and Irving, 1969).

Protective agents or excipients for improving the quality of food products freeze-dried by the cyclic process include those used in the preparation of fruit for freeze-drying, in the storage of freeze-dried vegetables, and the incorporation of fruit flavours in milk products.

Sulphite treatment for peas

Contrary to original expectations about the storage of vegetables without sulphite treatment (Gooding, 1962), recent storage trials with peas freeze-dried after a dip in a solution of sodium metabisulphite showed that quality had deteriorated substantially, while an initial puncturing of the skin into the cotyledons to ensure uptake of sulphite resulted in a product of good quality.

Two lots of peas were blanched in water at 96°C for 80 s, cooled in water for 60 s, then drained, and frozen at −20°C. One lot was punctured with pins of 1.2-mm diameter before blanching and freezing. When sulphite treatments were included, blanched peas were dipped in a 0.3% solution of sodium metabisulphite for 30–60 s. All the frozen peas were freeze-dried by the cyclic-pressure process with a sublimation temperature below −14°C, and with a heating platen temperature progressively lowered from 100°C to 40°C. The tray loading was 4.1 kg/m^2 and the drying time 5.5 h, the end-point being determined by the trapped-pressure method. Finally the peas were packed in cans, sealed under vacuum, and stored at 20°C and at 37°C for 7 months.

The results for tasting tests (Mellor and Irving, 1969), carried out on all of the samples for colour, texture, and flavour, are summarised as follows:

(1) Sulphiting is important for storage at high temperature providing the peas are punctured, but low moisture contents (below 1%) could also be important.
(2) The difference in scores for punctured and sulphite treated peas stored at 20°C and 37°C were not significantly different.
(3) Grading seems to influence only colour scores, but puncturing is just as important, since during freeze-drying the cotyledons do not shrink if the skin of the pea is punctured to allow the water vapour to escape during drying. Reconstitution is better also and tasters tend to give weight to the shrunken appearance more often than to the green colour under such circumstances.

A comparison with grades of commercial frozen peas and freeze-dried punctured, sulphited, and graded peas stored at 20°C shows that these freeze-dried peas are about as good as second grade frozen peas.

Flavour retention and enhancement

Some of the more general observations of Saravacos and Moyer (1968), Flink and Karel (1970), and Thijssen and Rulkens (1969) on retention of aroma constituents in freeze-dried products have been tested using the cyclic-pressure process.

One example has been a product consisting of 10% sucrose as the excipient added to pineapple pulp before freeze-drying. This was found to reduce the loss of the volatile esters making up pineapple flavour. Such a product was freeze dried in under 12 h at a sublimation temperature below −25°C and heater temperature from 79° to 40°C. Final moisture content for the normal product was 1.8% and for the sweetened 2.8%. Tasting tests have shown that the sweetened product was good to excellent over 2 years storage at 20°C.

Another example concerns natural excipients incorporated during the drying and storage of yoghurt. Here trials have indicated that it is better to cyclic-pressure freeze-dry a mixture of yoghurt and fruit together rather than to freeze-dry them separately, and then mix. Adsorption of the volatile esters from the fruit to the highly porous dried yoghurt plays an important role during freeze-drying and storage to enhance the quality of the product. Yoghurts flavoured with pineapple, apricot, passionfruit and strawberry have been tested, and all have potentially good prospects for freeze-drying, since the storage life of yoghurt can be extended over many years.

REFERENCES

Flink, J. M. and Karel, M. (1970). *J. agric. Fd. Chem.*, **18**, 295.

Gooding, E. G. B. (1962). in *Recent Advances in Food Science* (Ed. J. Hawthorn and J. Muil Leitch), Butterworths, London, pp. 27–38.

Mellor, J. D. (1966). in *Lyophilisation—Recherches et Applications Nouvelles* (Ed. L. Rey), Hermann Press, Paris, pp. 75–88.

Mellor, J. D. (1967). British Patent No. 1,083,244.

Mellor, J. D. and Irving, A. R. (1969). Proc. Internat. Instit. Refrig. Comm X, Paris, Vol. II, pp. 225–232.

Mellor, J. D. and Middlehurst, J. (1971). Proc. XIII Internat. Congress Refrig., Washington, D.C., Vol. III, pp. 717–723.

Mellor, J. D. and Greenfield, P. F. (1972). Proc. Internat. Symp. Heat and Mass Transfer Problems in Food Engng, Wageningen, Vol. II, pp. E8/1–23.

Saravacos, G. D. and Moyer, J. C. (1968). *Chem. Engng. Prog. Symp. Ser.*, **64**, 37–42.

Thijssen, H. A. C. and Rulkens, W. H. (1969). Bull. Int. Inst. Refrig., Annexe 69–4, 99–113.

DISCUSSION

E. Seltzer (Rutgers University, U.S.A.): I notice that where you showed the helium and the air curve diverging in favour of the helium, you were in a high pressure region—up around 3 and 4 torr, which is pretty risky for most foodstuffs.

Mellor: Actually, with helium, because of the extent of the effect you are prevented from taking advantage of the full change in thermal conductivity with pressure. Consequently, in practice, with helium you only use a very small change in pressure compared with the air.

E. Seltzer: We're currently doing freeze-drying of certain kinds of meat with some government experience which specifies that we should operate at 130°F. With temperatures up to 135°F and 140°F a discriminating examination of a product indicates that there is less porosity, and more browning reaction damage in the initial quality. We submit this to panels of 20 or higher, and we very carefully try to determine what the effect has been. I suspect that, if you took the liberties you did with some of the materials I happen to be working with, there would be an appreciable decrease in quality.

Mellor: Most of our taste test evaluations show the difference between processing normally and with cyclic pressure is not really significant.

The Ecal Process—A Novel Dehydration Process

A. L. MOELLER

Technical Director, Ecal
Nateko AB,
Munka, Ljungby, Sweden

I would like to present a completely new drying method called the Ecal method of the 'Slow Motion Gravity Bed Low Temperature Air Drier'. In fact, we call the drier itself the Ecal Drier for Food; but for sludge or air pollution and industrial waste problems we call it the Nateko Drier.

The Ecal method is a very new method compared with the spray drying process. It was innovated by Mr. Eric Carlsson in Sweden in 1966–67. Thirty-one plants have now been ordered since the industrial start in 1971. Patents were granted in 40 countries and still much patent work is going on.

Several plants up to 2000 kg H_2O/h have been built for food; we are now up to 6000 kg H_2O/h for sludge.

Important factors for an air drier are low temperature but good heat economy; good dryness without dust; easily regulated drying times and processing of solutions of very different dry solid content and viscosity. It is also advantageous if solid sliced matter could be dried in the unit. Both drying and concentration should be able to be carried out. It should be a compact unit working automatically with an easily regulated final dryness, in an open or a closed system with inert gas.

The principles of the Ecal drier are as follows (see Fig. 1) where we have tried to live up to the demands outlined above. It dries with air or gas at low temperature down to +50°C of the inlet air. Spherical bodies are used to enlarge the liquid surfaces; but if solid sliced matter is dried these spherical bodies hold the matter. These spheres are designated MCBs (Material Carrying Bodies). By continuous circulation of the MCB, a slow motion gravity bed without turbulent flow back but with multi-layer equilibrium is obtained. A separate air stream (see Fig. 2) upwards from the main drying chest permits the use of the entire heat content of the air down to the dew point, i.e. the relative humidity reaches 100%. A second co-current air stream (about 10% of the total inlet) is provided for final drying. Drying times for the same unit are between 50 s and 5000 s: that explains why we can dry thin solutions and onion in the same unit.

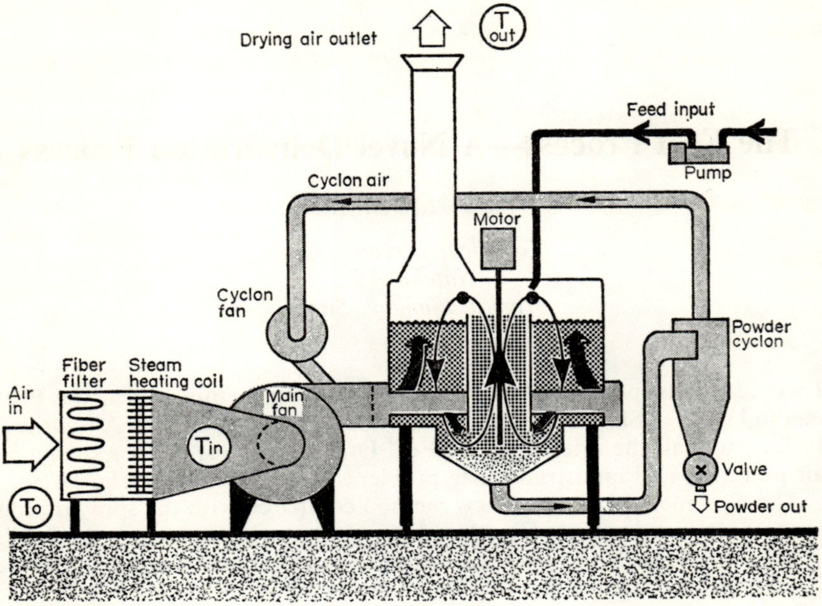

FIG. 1. Ecal-Nateko food drier.

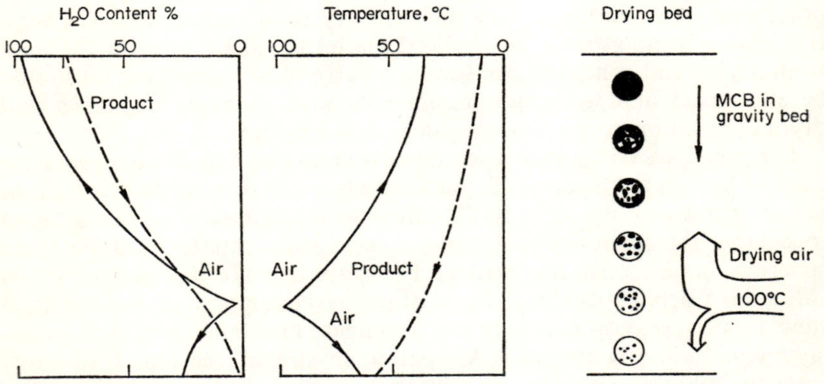

FIG. 2. Temperature and humidity of air and product in the drying bed of an Ecal-drier.

Dust absorption is achieved because all the outgoing air travels through the wet top layer of the gravity bed.

The two-way air system provides simple final dryness and temperature regulation because of the secondary air flow. The main drying air flows in countercurrent to the MCBs with liquid on their surface but goes co-current with the dry materials. The air with the powder entrained goes to the cyclone where separation occurs before the air goes back into the process again so utilising all available heat.

The drying unit which is cylindrical has radial air inlet ducts. In the centre there is a screw operating at 10–30 rev/min which screws up the clean MCBs so that they can present fresh surfaces to the feed before they very slowly descend through the drying zone which is 70–80 cm high.

At the bottom there is a screen to separate the MCB and the dry product which is taken off to a cyclone where the powder can be removed from the air stream. The MCBs are recycled to the top of the bed.

The feed is distributed to the MCBs generally by rolling them in the liquid on a collar before they are distributed into the upper part of the bed. The wet upper part will in fact act as a scrubber, preventing dust from going out with the counter-current air. On the other hand, we don't dry the material to a powder before it reaches the co-current air stream. The powder should be entrained in the downward flowing air.

Air inlet temperatures vary from 150°C down to 65°C and outlet temperatures from 50°C to 25°C. The capacity depends upon the air temperature drop being used (see Fig. 3).

The drier's action is to enlarge the liquid surface on the MCB. The normal surface area is 200 m^2/m^3 of MCB when using balls with a diameter of 18 mm. The diameter of MCBs can be up to 40 mm. The thickness of solution on the MCB is from 0.1 mm up to 1 mm.

The product temperature can be kept low. We are drying some dairy materials as for instance, buttermilk; where the product comes out at a temperature of 50°C or even lower. We have dried living yeast cells when the temperature of material was never above 25°C; we then introduced extra cold dry air at the bottom part to bring the cyclone air temperature down. Economy is dependent on the difference between inlet and outlet air temperatures and the ambient air temperature of +25°C; we can have an 80% use of the supplied heat even if we introduce air at only 60°C.

With the Ecal method, heat is supplied by the air, but also by the MCBs—depending upon what they are made of. If they are of stainless steel plastics, or glass they have a different heat capacity which can be useful when drying sliced matter. We can also dry sticky materials which are very viscous. These MCBs are of great interest, and there are many important demands to be met. We must have good adhesion of wet material, and a good release of the dry; a high mechanical strength; high abrasion resistance; high chemical resistance; and good thermal resistance. If we are drying food with a 75°C inlet, it is no thermal problem; but to dry at 300°C will cause trouble with polymers, and one must use aluminium or stainless steel. We have manufactured about 10–15 million MCBs; they must be easy to manufacture in different sizes in low cost raw materials. The most used materials today are polymers, e.g. polycarbonate and polypropylene.

With the low viscosity feed, where the MCBs act as a matrix, the end product is a powder or crystals. With a high viscosity feed we will end up either with a thicker product or flakes. With bad adhesion at the surface,

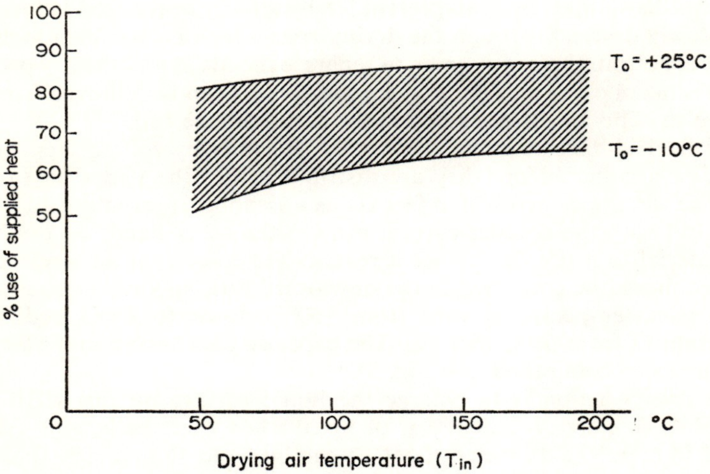

FIG. 3. Thermal efficiency of an Ecal drier. Use of supplied heat at different drying air and ambient air temperatures.

we will have small flakes or powder. Solid materials such as parsley, leeks, onions and apples have been processed. The interesting thing is that all these are different feeds dried in the same drier. Whole egg, yeast, whey and protein from whey are examples of liquid food. Parsley is very interesting because it retains its structure: we cut the parsley and it comes out dry and looking just like fresh parsley.

To summarise: we have developed a completely new drying process. We have completed the design and construction and built 25 driers up to now, we have demonstrated the product. We are marketing pioneers and have received enormous interest in the method. Blood, buttermilk, whey and yeast besides sludge and wastes are in routine production. In our company we work mostly with design, but we must also follow the market, economy and patents, investigate possible MCB materials and do all things necessary when developing a great, new product and bringing it to a world use.

We are willing to develop modifications together with the people who are buying our driers, because they know much more about the products than we do; they also know far better what the market requires.

DISCUSSION

E. Seltzer (Rutgers University, U.S.A.): What is a typical cycle time? What is the time distribution for a typical residence? It may be in there all day!

Moeller: Our normal cycle time is roughly 4–6 min depending on the coating thickness. There is a holding time in the drying bed of some 100 s. The time distribution was followed by balls of different colours in different places in the bed, and we found it to be roughly $\pm 10\%$ of the expected time.

I won't say that the balls are completely clean when they come up the screw, but they are very clean; it is sometimes a question of some electrostatically held fines that could adhere some of the time, but it is only a very small fraction of the whole; it's not going round and round.

A. S. Michaels (Alza Research, U.S.A.): What about the attrition loss of the balls? How long do they last? Do they change weight or diameter and so cause contamination of the product by the plastic or the metal of the balls?

Moeller: With most processes on the food side, the material lubricates the balls, but if they are drying material with silica, it is a little harder for the balls. We haven't used them long enough but we expect 2–5 years lifetime for silica and 5–10 for foods and normal products.

Mellor (C.S.I.R.O., Australia): I suppose you accept the fact that you're getting the breaking up of some particular products creating a dust problem? Of course, with liquid products drying as a powder, it doesn't matter. Why isn't the dust blown out with the airstream in the tall vertical pipeline?

Moeller: The upper layers of MCB are wet layers and they absorb about 92–95% of atmospheric dust. All the powder goes out the bottom way, through the cyclone. It is a process failure if something is going upwards in the countercurrent air stream. The material should be wet enough in the countercurrent stream not to lose solids.

J. G. Brennan (National College of Food Technology, U.K.): When you have to resort to using metals for the MCBs, what is the noise level like? It sounds as though it could be high.

Moeller: Yes, you're right, but it is very easy to avoid noise by coating the whole construction with a heavy material as for sludge driers. When we are drying yeast, which dries like flakes, the material itself acts like a damper but some other products could be rather noisy. Perhaps 85 dB.

H. A. C. Thijssen: How do you release the dried layer from the spheres?

Moeller: Normally, it is easy to release it, but it depends on the product. For instance, if we take the sliced materials they are not attached to the MCBs and can be released easily. If we take whole egg powder or yeast or protein, when it is dry enough, up to 94–98% dryness, then it normally loosens very easily. At the bottom the air speed is accelerated and there can be impacts on the balls, also the surface of the MCBs is important. We ran into some trouble with trying to dry orange juice because it was so sticky, so we had to cool it. We introduced cold air into the bottom which broke it up, but it is a question of economy.

D. Simatos (Dijon University, France): May I ask what the Ecal drier cannot do! For instance, can you dry a thermoplastic?

Moeller: Yes—we have been drying several carbohydrate products which are very sticky—like tomato paste, for instance. But it breaks up in the co-current stream which is much better. In fact, we have now dried hundreds of different foods, and our trouble is now to live up to expectations. We have sold most of these units in Sweden because of the necessity to be rather close to the units. But there are two for blood in Ireland; one is in France and there are two in Denmark.

The same unit is used for the drying of many different structural types of food: it is used for the drying of chemicals, for sludge and waste from industries and for gas cleaning and for humidifying air. If you create something that has a large contact surface between gas and liquid, gas and solid, solid and liquid, then you can do very many things with it.

Brennan: Can it pregelatinise and dry starch as on a drum drier?

Moeller: We have dried potato mash to make a powder to be dissolved in water, and we achieved good results. It was a good powder, but on the other hand, using a higher temperature and longer time, we make the pregelatinisation first, then produce the powder.

King: Is the moisture content of the material determined by what will fall off the MCB, or is there appreciable drying after it comes off?

Moeller: Not exactly—we cannot peel it off with 30% water left because it is too sticky, but between 80 and 99% we can select the value. We can also remove crystal water for instance from dextrose which was advantageous, because we can use air for the final drying in the bottom part, which has practically no humidity.

Microwave Heating in Vacuum Drying

N. MEISEL

Les Micro Ondes Industrielles,
Epône, France

I shall be talking to you today about vacuum dehydration with microwaves. We have achieved good results in our continuous system for microwave vacuum dehydration. The energy dissipation in $W/cm^3/s$ is proportional to the frequency, to the dielectric constant and to the square of the field value.

It is a known fact that one of the main difficulties of practically all dehydration methods using heat is heat transfer. It was therefore obvious that dielectric heating would be the perfect answer to that problem. Unfortunately, the first tests were carried out with low frequencies in freeze-drying—a situation which has two drawbacks—poor field distribution and rapid glow discharge or ionisation. Furthermore, the most favourable pressure for freeze-drying, in order to achieve good sublimation of the product, is between 0.5 and 1 torr, which pressure—by a most unfavourable coincidence—also corresponds to the minimum microwave field value before glow discharge occurs. This is the reason why researchers who have tried to use microwaves as a means of accelerating heat transfer in processes have reached negative conclusions.

The more important work done in the United States was done at 915 MHz, in spite of the fact that existing data showed that 2450 MHz would have been more suitable for two reasons: it is easier to obtain better field distribution and we can have a much higher critical value for the microwave field before glow discharge. We can use practically double the energy by just changing the frequency from 915 to 2450 Mc/s.

For these reasons one would hope to get acceptable results without having to freeze the product, and it has prompted us to undertake some very primitive exploratory tests in a transformed static microwave oven.

We first tried with coffee, sugar, milk concentrate and orange juice concentrate, and we found that, in spite of some burned areas, we were getting some nice foam with good flavour and colour retention and excellent solubility, even in cold water. Thus encouraged, we decided to build the first static evaporator especially to further our experiments.

The idea here was to have just a vacuum drum divided in two by a platform, which would separate the microwave cavity and the cavity underneath, where we had a metric balance. In order to avoid high field values, we had a pre-resonating microwave chamber with a very big porthole which acted as a microwave window to the vacuum drum.

A looking glass mirror was fitted to enable us to read the value on the metric balance through the porthole in the front.

The next stage involved fitting an electronic balance in place of the metric one so we were able to register the loss in weight with time. This new equipment has enabled us to carry out numerous tests on coffee extract, sugared milk, strawberry and raspberry purées, orange juice concentrate, passionfruit concentrate, enzymes, protein hydrolysates, etc. Each time, we got a very quick foaming—as soon as the vacuum reached about 5 torr: and we got dehydration to 1.5–3% within 30 min. The product was introduced at ambient temperature; it fell to 5–10°C during the linear evaporation period, and reached about 40–50°C during the last few minutes. The dried foam was about 8–10 cm high, with an apparent density of 0.05. The initial colour and taste were not altered, and the fact that the initial amount of dry matter could be as high as 80% allows, in most cases, an excellent aroma retention. This was particularly true of purées of fresh fruit to which sucrose was added.

Why does ionisation happen? Well, in a static oven when you start with a mixture where there is a lot of water, the power can be easily dissipated into the water and cause evaporation of the water. As the product gets drier, less energy can be absorbed, for if all other parameters are constant, the smaller amount of water which absorbs the power causes the dissipation to decrease and then the field value will increase by a power of 2. And because of this rapid increase in field power, we have to turn the microwave power down. You probably know that each microwave applicator has a most favourable load, into which we get the best coupling efficiency. Therefore, the best way of using microwave evaporators is in constant optimum load conditions. These optimum load conditions, as you have seen, cannot be obtained in a static oven, and that is why we decided to build a continuous microwave drier.

It consists of a metal cylinder with a conveyor belt in it and there is a tube with many holes in it to feed the material on to the belt.

The belt slowly advances in front of the microwave windows carrying the material as a foam which is produced on entering the vacuum chamber. Therefore, when the feed comes in at a temperature of 30°C, it will go down to 5–10°C—depending on the pressure within the chamber—and will stay there until it has passed the two windows and until approximately 70–80% of the water is evaporated and then the temperature will start to climb slightly until it reaches 30–40°C before it reaches a rotary scraper. The powder falls into a hopper and through into a container. When the container is full, it can be isolated from the chamber and exchanged for another one, although the drying goes on with the product discharging into the hopper. The process is continuous, the unloading is discontinuous.

Here we have the ideal microwave conditions, because we always have the same load inside the tunnel. We can determine the equivalent amount of water for each product we want to dry, and therefore we can work at optimum conditions, having a generally low field value and so avoid glow

FIG. 1.

discharge, but having excellent efficiency. The temperature of the product can be measured through infra-red germanium windows, and we can measure the surface temperature of the foam very easily, and even have it recorded. This enables us to make the whole process completely automatic; because by measuring the temperature of a known product, we know exactly if it is going to be dry enough and what kind of dryness we can achieve. If it is too moist or too dry, we can vary the microwave power instantaneously; we can have the belt move faster; or we can reduce the amount of heat into the chamber.

There is also another excellent means of measurement of the amount of energy which is absorbed by the product, and that is that we can measure the input energy and the reflected energy and so conclude the amount of energy taken up by the product. We can see what happens through the portholes (see Fig. 1).

Products with much higher dry matter content than for freeze-drying can be used, freezing being superfluous; drying times vary between 25 and 40 min—hence, a big economy; the operation is simple because we can have direct optical inspection; we can have 100% automation. When anything happens there is no spoilage due to stoppage: if something went wrong and we had to release the vacuum to repair something we could restart and not lose one pound of product.

What is more, it is very easy to clean by flushing it with water because everything is tight in there. Again we use an industrial vacuum: 5–20 torr, and can therefore use steam ejectors if necessary.

The products that have been dried are extremely varied but, for the time being, are only in paste form. Whatever we have put through we have succeeded very quickly in getting ultra-satisfactory products because of the ease with which all these parameters can be changed. We have also dried protein hydrolysates, enzymes and other materials.

The microwave power is 5 kW: we take approximately 12 kW from the mains in order to evaporate about 5 litres of water per hour.

DISCUSSION

Chairman: This is not the first plant or type of equipment for microwave heating of foods that Mr. Meisel has developed. He has also developed equipment for industrial thawing of foods—primarily meat—and there are two, possibly more, plants operating that process, on that principle.

S. Barnett (General Foods Inc., U.S.A.): What is the thickness of the foam layer?

Meisel: We start with a layer of about 2–3 mm, and it increases to about 12–15 cm before we crush it a little in removing it from the belt. The apparent density is about 0.05; and this is one of the difficulties of the process because we have to grind it back to a density of 0.5.

J. D. Mellor (C.S.I.R.O., Australia): Is the density of the foam controlled by the vacuum pressure initially?

Meisel: No, not really. It depends on the viscosity and on the microwave power used.

C. Varsel (Coca Cola Company, U.S.A.): What is the normal starting solids content?

Meisel: There is no minimum, but the process will be much too expensive if we start with very low dry matter content. We normally run between 50 and 60% dry matter.

Varsel: What is the belt material? Is it stainless steel?

Meisel: No, the belt material is teflon—PTFE with woven glass fibre reinforcement.

A. S. Michaels (Alza Research, U.S.A.): During what part of the travel of the belt through the tube is it receiving microwave radiation? I am a little puzzled that the temperature continues to climb during the latter part of the process.

Meisel: It is exposed to the field all the time but in order to prevent glow discharge, reflectors have been installed where we have a lot of exits and passages.

Varsel: Do you have the cost of evaporation of a pound of water?

Meisel: The cost is very difficult to estimate, if you don't run it 20 h a day. The capital investment is the most important part of the costs.

We are now working on developing a 30 litre/h evaporator on this same principle, and the projected price is somewhere around $120 000 for the whole equipment. If we work it for about 20 h a day, then the evaporation cost would be around 15–16 c/litre of water.

C. H. Mannheim (Technion, Haifa): How do you measure temperature?

Seltzer (Rutgers University, U.S.A.): What type of sensors do you have?

Meisel: Infra-red. Infra-red sensors are mounted over the main windows and we take the surface temperature of the foam: we have verified that the same temperature is also inside the foam due to the vacuum—as there is very little loss of heat from the surface to the surroundings.

Seltzer: What pressure do you maintain?

Meisel: It depends on the product, but between 5 and 20 torr.

P. Van Twisk (National Food Research Institute, South Africa): Have you tried drying vegetables with this type of equipment?

Meisel: Yes, we dried asparagus tips. This is an expensive way of drying, so I don't think it would be suitable for the less expensive vegetables. But we did asparagus tips, whole strawberries and mushrooms, and we got quite good results.

General Discussion

J. G. Brennan (National College of Food Technology, U.K.): I should like to ask a general question. We haven't heard anything about the combined processes, such as dehydro-freezing and dehydro-canning, and I wondered whether we could ask our American friends what state these processes are in: if they are still in existence, and to what extent they are used; and to what extent some of these novel methods of drying might be applied to them, for example, osmotic drying followed by freezing or canning.
C. J. King (University of California, U.S.A.): I think combination processes are certainly viable. Coffee is processed by a combination of freeze-concentration and freeze-drying; and this makes sense—to get the first part of the water out in an easy way and the last part in a more expensive way.
S. Barnett (General Foods Inc., U.S.A.): One combination process that we use commercially is a combined air-freeze-drying for certain vegetables, such as peas, in which we remove a proportion of the water in the early part of the drying cycle by air drying when there is little degradative result happening, and then we finish by freeze-drying. This gives a somewhat denser product than by straight freeze-drying, but it rehydrates very readily.
D. E. Blenford (RHM, U.K.): We have talked about aroma retention very extensively, but very little has been said about either the retention of structure *per se*, or its combination with aroma retention.
J. D. Mellor (C.S.I.R.O., Australia): I might perhaps amplify my talk a little on this. Regarding those peas which were sulphite treated and compared with the unsulphited ones, it was very important that the peas be punctured before blanching. With a single puncture with a millimetre pin in every pea (these were done in a special machine) the blanching was much more effective if they were punctured than if they were not. Also, the uptake of sodium metabilsulphite dip was much more, it penetrated into the cotyledons through this single hole in the skin, was retained in the inside and assisted the storage life markedly.

Turning to the structure and to the texture problems, with peas I found that the peas that were punctured could be reconstituted much more readily in the hot water in a pan, and after a final cooking, there was no noticeable shrivelling at all. Peas that hadn't been punctured usually ended up as a little wet mass inside a light green shell.
E. Seltzer: How do they compare with 'Surprise' peas which are air dried?
Mellor: Well, 'Surprise' peas are also punctured, and our measurements show that these cyclic pressure freeze-dried peas were superior to the 'Surprise' peas. 'Surprise' peas have a hay-like taste compared with the freeze-dried ones. Nevertheless, the structure of both was comparable—only the taste was a little different.

Chairman: I think there have been a number of articles in the Journal of Texture Research on dehydrated foods. I remember one by Szczesniak on the texture of freeze-dried meat.

I should perhaps add with regard to microwave heating for drying purposes, that there is a relatively new and evidently successful method used in Switzerland for the finish drying of pasta noodles. And at a lower frequency there is a very big unit operating in Sweden, and has been operating for ten years, for finish-drying for sugar cubes. The power level of that unit is 1000 kW.

J. Donovan (Artisan Industries Inc., U.S.A.): I wish to describe three pieces of equipment.

(1) A stage-wise continuous evaporator with rising film and finishing, if you wish, with a falling film. The feed goes through a single pass rising film, it flashes into a chamber; it is then picked up and goes through a second, and third—or more, for that matter—and then the more concentrated liquid comes down through a falling film evaporator; and you have a very short time evaporative concentration. It works very satisfactorily on a large number of materials.

(2) A material may be concentrated in this agitated thin film horizontal unit, starting with anything you want—1% solids and taking 99% overhead; or start with 90% solids and take 1–10% overhead. You could take any range that will handle without absolutely gumming up and sticking to everything. We have either countercurrent or concurrent liquid feed. With this unit we have fed solids and brought out liquids: we've fed liquids and brought out solids: so one can readily concentrate to anything which will flow; or it will go to the dry stage without boiling up and gumming too badly with anything which will come out dry as powder. And this will bring it in one stage from 1% feed up to a dry powder coming out. It isn't used normally over that wide range, of course.

The unit works on a very simple basis. Ahead of the blade is what amounts to a bow wave in a boat which scrubs, washes and replenishes the heat transfer surface, and one has a system which is always wet and which will take a material from virtually pure water to solid. It has a very high turndown ratio also. Most equipment that is designed for, say 10 000 lb/h will run at 10 000 lb/h, but to go down to 1000 lb/h is a little hard for it. This unit will take 10 000 lb/h or 1000 lb/h with the same equipment—one naturally changes the ΔT; but it will handle a wide range of turndown.

(3) It was mentioned here that sometimes you wish to separate a material from what you sometimes call a serum and a more concentrated solid. This is a new continuous filter that will concentrate a large number of materials from a relatively low or medium concentration. So one puts in a mixture in one end and, by going through the filter up and down with an agitator which tends to keep materials nicely suspended as they get more concentrated, as long as they will flow, they will come out at the far end through a pressure reducing valve—a solid material running from 30 to 70% concentration. And insofar as you have occasion to wish to pre-separate into a liquid and a thickened solid, this continuous filter might be of interest.

Index